ELECTRICAL SCIENCE

SECOND EDITION

TARLOK SINGH

ELECTRICAL SCIENCE

For
Engineering Diploma Students

By
Tarlok Singh
B.E. (Elec.)
Senior Lecturer Govt. Polytechnic for Women
Ambala City (Haryana)

S.K. KATARIA & SONS®
Publisher of Engineering Books
4885/109, Prakash Mahal, Dr. Subhash Bhargav Lane,
Daryaganj, New Delhi-110002 (India)
Phone: +91-11-43551243; Mobile: +91-9871775858, 7678425209
e-mail: katariabook@gmail.com, katariabooks@yahoo.com
Website : www.skkatariaandsons.com
(e-books, subscriptions and bulk buying also available on our website)

Published by :
S.K. Kataria & Sons®
4885/109, Prakash Mahal, Dr. Subhash Bhargav Lane,
Daryaganj, New Delhi-110002 (India)
Phone: +91-11-43551243; Mobile: +91-9871775858, 7678425209
e-mail: katariabook@gmail.com, katariabooks@yahoo.com

First Edition : September 1998, Second Edition : July 1999, Reprint: 2024

Price: ₹ 395/-

Printed at : N.K. Printers

PREFACE TO THE SECOND EDITION

The favourable reception given to the book by the students and the teachers have been a matter of great and pleasure to the author and the publishers. It has been a constant desire of the author that the book may prove and more useful to the readers. This edition has been thoroughly revised and updated.

Special efforts have been made to check the accuracy of answers of both solved and unsolved numericals included in the book. Most of the worked out examples, exercises and problems have been removed and replaced by new ones with thoroughly checked answers.

Accordingly, this revised edition includes new topic such as **Polyphase System.**

The author is highly thankful to the teachers and readers for recommendations, made by them in favour of this book.

July, 1999 **TARLOK SINGH**

PREFACE

This book **Electrical Science** has been written to meet the requirements of Third Semester students of Electrical Engineering of State Board of Technical Education Punjab in particular and Polytechnics of Haryana, Himachal Pradesh, Delhi and other States.

The book has been organised in a systematic manner such that each chapter covers a specific topics.

At the end of each chapter its Highlights, Short questions, Multiple choice questions, Numerical problems are given. This will help the students to revise the subject matter quickly.

Frequently use of figures have been made for clearity. Complete practicals according to the syllabus has been written at the end of book.

Author is also thankful to M/s S.K. Kataria & Sons, for bringing out the book in such a short time in nice way and pricing it so moderately inspite of heavy cost of paper and printing.

Suggestions regarding improvements, corrections and misprints of the book along with healthy comments from the teachers as well as students shall be greatfully received and acknowledged.

TARLOK SINGH

SYLLABUS
ELECTRICAL SCIENCE

1. D.C. CIRCUITS

1.1. Concept of electricity. Basic terms—Voltage, current, Potential Difference, Power, Energy and their units.
1.2. Ohm's law.
1.3. Effect of temperature on resistance, temperature coefficient of resistance.
1.4. Resistances in series and parallel.
1.5. Kirchhoff's current law and Kirchhoff's voltage law and their applications to simple circuits. Conversion of circuits from Star to Delta and Delta to Star.
1.6. Thevenin theorem, Superposition theorem, Norten theorem and their use to solve simple problems.

2. BATTERY

2.1. Basic idea about primary and secondary cells.
2.2. Construction working and applications of lead acid battery and Nickel Cadmium cells, Silver Oxide cells.
2.3. Charging methods used for lead acid accumulator.
2.4. Care and maintenance of lead acid battery.
2.5. Series and parallel connections of batteries.

3. MAGNETISM AND ELECTROMAGNETISM

3.1. Introduction to electrimagnetism, Magnetic field, around a straight current carrying condutor and a solenoid and methods to find its direction, force between two parallel current carrying conductors.
3.2. Force on a conductor placed in the magnetic field.
3.3. Series and parallel magnetic circuit, simple problems.
3.4. Importance of Air gap.
3.5. Concept of hysteresis and hysteresis loop.

4. ELECTROMAGNETIC INDUCTION

4.1. Faraday's laws of electromagnetic induction.
4.2. Lenz's law and Fleming's Right and Left hand rules.
4.3. Principle of self and mutual induction, self and mutually incuded e.m.f., simple numerical problems.
4.4. Energy stored in a magnetic field.
4.5. Eddy currents and their effect.
4.6. Rise and decay of current in RL and RC circuits.

5. **AC THEORY**

5.1. Concept of alternating voltage and current.

5.2. Difference between AC and DC.

5.3. Concept of cycle, frequency, period, amplitude, instantaneous value, average value, r.m.s. value, maximum value, form factor and peak factor.

5.4. Representation of sinusoidal quantities by phasors.

5.5. Derivation of equation of sinusoidal wave form.

5.6. Physical concept of alternating voltage applied to a pure resistance, pure inductance and pure capacitance.

5.7. Inductive reactance, capacitive reactance.

5.8. Alternating voltage applied to resistance and inductance in series.

5.9. Alternating voltage applied to resistance and capacitance in series.

5.10. Impedence triangle and phase angle.

5.11. Complex Numbers—Various forms; addition, subtraction, multiplication and division; additIon of sinusoidal terms.

5.12. Solutions and phasor diagrams for simple practical RLC circuits (series and parallel).

5.13. Introduction to series and parallel resonance and its conditions.

5.14. Power in pure resistance, inductance and capacitance, power in combined RLC circuits, Power factor, active and reactive currents and their significance, practical importance of power factor, solution of simple circuits using J-notation. Use of condutance, succeptance and admittance.

6. **POLYPHASE SYSTEM**

6.1. Concept of polyphase in power system.

6.2. Advantages of 3-phase system over single phase-system.

6.3. Generation of 3-phase voltages, relation between phase voltage and line voltage, phase current, line current and power in a star and delta connection system. Simple problem on 3-phase balanced loads and concept of unbalanced loads.

PRACTICALS

1. Determination of voltage-current relationship in a d.c. circuit under specific physical conditions and to draw conclusions.
2. To verfiy that $R_T = R_1 + R_2$ when resistances are in series, and $R_T = [(R_1 + R_2)/ (R_1 + R_2)]$ when resistances are in parallel.
3. Verification of Kirchhoff's current law and Kirchhoff's voltage law in a dc circuit.
4. To measure the value of unknown resistance with the help of a wheat stone bridge.

5. To observe the effect of temperature on resistance of a fan and hence calculate the value of temperature coefficient of the winding material.
6. To find the ratio of inductance values of coil having air-core and iron-core respectively and to observe the effect of introduction of a magnetic core on coil indutance.
7. To find the voltage current relationship in a single phase R-L and R-C Series circuits, draw their impedence triangles and determine the power factor in each case.
8. To find the voltage-current relationship in a single phase R.L.C. series circuit, draw its impedence triangle and determine the power factor.
9. To test a lead acid storage battery for charged and discharged conditions and to recharge it.
10. To find out resonance frequency in a devices R.L.C. circuit.
11. Measurement of power and power factor in single phase R.L.C. circuit and to calculate active and reactive power.
12. Plotting hystereis loop of a given magnetic material with the help of a CRO.

CONTENTS

1

D.C. Circuits

1.1. CONCEPT OF ELECTRICITY

Electricity is a gift of the modern science to mankind. It is well known that electricity play a very important part in our daily life. It can be used for lighting, cooking, heating, entertainment, welding and transportation purposes. The electricity is also used for traction, electroplating and battery charging. It is free from smoke, fumes and ashes.

Electricity is a form of matter which has been set in motion under certain conditions.

Matter is composed of tiny particles called atoms. The atom is further made of protons, neutrons and electrons.

The protons carry a very small positive electrical charge and in magnitude, just balance the negative charge of the electron.

Neutrons are neutral and carry no electrical charge whatsoever.

Electrons carry a unit negative charge.

In the normal state, the number of electrons is equal to the number of protons in an atom. Therefore, an atom is neutral as a whole. But there is difference in the placement of the two in the atom itself. The protons are at the centre, nucleus, the electrons. These orbits are of a fixed number for a particular atom.

The number of orbits in an for atom will depend upon the Atomic number of the material. The first orbit can not have more than to electrons, the second has 8 electrons maximum and third eighteen and the last one cannot have more than eight electrons.

As a rule, a material is stable if it has same number of protons and electrons. Under such conditions the outermost orbit has complete number of electrons. Thus if an atom has only two electrons, *i.e.*, its atomic number is 2, in a case of Helium gas they are only placed in that orbit which is complete. So the gas is stable.

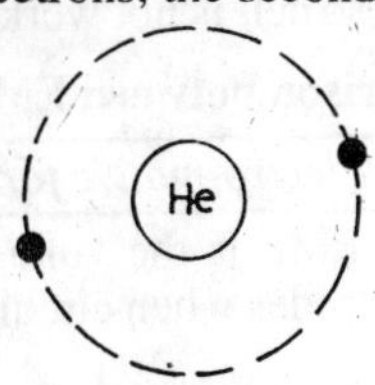

Fig. 1.1.
Helium At. No. 2
Number of protons = 2
Number of electrons = 2

If from a neutral material some electrons are removed, there is deficit of electrons in the material. Then the material attains a positive charge.

On the other hand, if a material is supplied with electrtons then there is excess of electrons. Consequently, material attains a negative charge.

The electric current is constituted by the electrons set in motion by applying a suitable force.

Flow of Current

Since the electrons are negatively charged, they are attracted by positive. So, the current flows from –ve to +ve (contrary to the popular notion that it flows from +ve to –ve).

1.2. BASIC TERMS

Electric Potential : The capacity of a charged body to do work is called Electric Potential.

Every charged body has the capacity to work by moving another charge. This ability of charged body to do work is its electric potential. Greater the charge on a body, large is the electric potential to the body.

$$\text{Electric potential} = \frac{\text{Work done}}{\text{Charge}}$$

or $$V = \frac{W}{Q} \text{ J/C}$$

Unit of Electric Potential : The practical unit of electri potential is volt.

Hence a body is said to have an electric potential of 1 volt if one joule of work is done to charge the body to one coulomb.

Potential Difference : When a body is charged to different electric potential from the other charged body, then the difference in the electric potential of two charged bodies is called potential difference.

Unit : The unit of potential difference is volt.

EMF : EMF of a cell is equal to PD between its terminals when the circuit is open, so cell is not working.

Comparison between EMF and Potential difference

Electro-motive force (EMF)	*Potential differenced (PD)*
1. EMF is the voltage between two electrodes when circuit is open.	1. Potential difference is the voltage between two points when circuit is closed.
2. It is measured at source of supply.	2. It is measured in between any part of the circuit.
3. It does not depends upon resistance.	3. It is directly proportional to resistance.

4. It is always greater than the potential difference between any two points of conductor.	4. It is less than EMF applied.
5. It is exists even if the circuit is not closed.	5. It exists only when circuit is closed.
6. It is measured in volts.	6. It is also measured in volts.

1.2.1. Electric Current

The flow of free electrons is called *electric current.* The flow of electric current can be explained by referring to Fig. 1.2. The copper strip has a large number of free electrons. When electric pressure of voltage is applied, then free electrons, being negatively charged, will start moving towards the positive terminal round the circuit as shown in Fig. 1.2. This directed flow of electrons is called electric current.

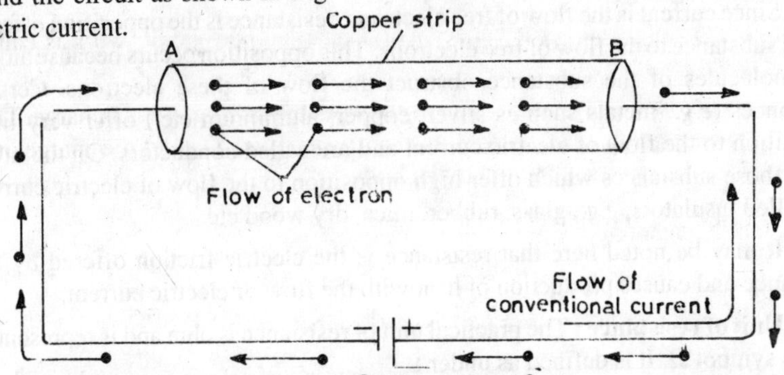

Fig. 1.2.

Conventional direction of flow of current

(i) Current is flow of electrons and electrons are the constituents of matter. Therefore, electric current is matter (*i.e.*, free electrons) in motion.

(ii) The actual direction of current (*i.e.*, flow of electrons) is from negative terminal to the positive terminal through that part of the circuit external to the cell. However, prior to Electron theory, it was assumed that current flowed from positive terminal to the negative terminal of the cell *via* the circuit. This convention is so firmly established that it is still in use. This assumed direction is now called conventional current.

(iii) Those substances which have large number of free electrons will permit current flow easily. Such substances are called *conducors, e.g.*, copper, zinc, silver, aluminium, etc. On the other hand, atoms of some substances have valence electrons that are tightly held to their nuclei, *i.e.*, they have a very few free electrons. Such substances will not permit the flow of electric current and are called *bad conductors* or *insulators, e.g.*, glass, mica, porcelain etc.

Unit of current : The strength of electric current I is the rate of flow of electrons, *i.e.*, charge flowing per second.

$\therefore$ Current, $I = \dfrac{Q}{t}$

The charge Q is measured in coulombs and time t in seconds. Therefore, the unit of electric current will be *coulombs/sec* or *ampere*. If $Q = 1$ coulomb, $t = 1$ sec, then $I = 1/1 = 1$ ampere.

One amphere of current is said of flow thorugh a wire if at any section one coulomb of charge flows in one second.

Thus, if 5 ampere current is flowing through a wire, it means that 5 coulombs per second flow past any point on the wire.

Resistance : The opposition offered by a substance to the flow of electric current is called *resistance.*

Since current is the flow of free electrons, resistance is the opposition offered by the substance to the flow of free electrons. This opposition occurs because atoms and molecules of the substance obstruct the flow of these electrons. Certain substances (*e.g.*, metals such as silver, copper, aluminium etc.) offer very little opposition to the flow of electric current and are called conductors. On the other hand, those substances which offer high opposition to the flow of electric current are called insulators, *e.g.*, glass, rubber, mica, dry wood etc.

It may be noted here that resistance is the electric friction offered by the substance and causes production of heat with the flow of electric current.

Unit of resistance : The practical unit of resistance is ohm and is represented by the symbol Ω. It is defined as under :

A wire is said to have a resistance of **1 ohm** if a p.d. of 1 volt across its ends causes 1 ampere to flow through it (see Fig. 1.3).

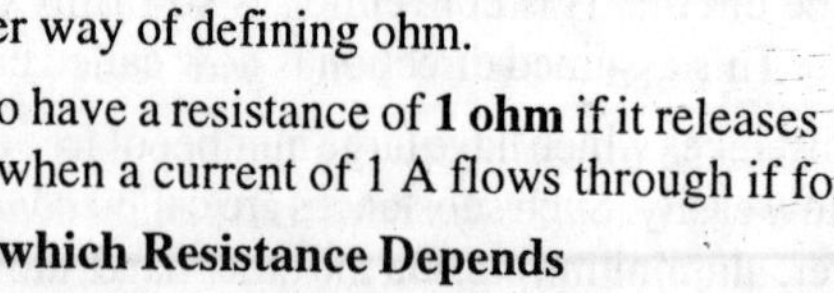

Fig. 1.3.

There is another way of defining ohm.

A wire is said to have a resistance of **1 ohm** if it releases 1 joule (or develops 0.24 calorie of heat) when a current of 1 A flows through if for 1 second.

1.2.2. Factors upon which Resistance Depends

The resistance R of a conductor

(*i*) is directly proportional to its length, *i.e.*, $R \propto l$

(*ii*) is inversely proportional to its area of X-section, *i.e.*,

$$R \propto \frac{l}{a}$$

(iii) depends upon the nature of material.

(iv) depends upon temperature.

From the first three points, we have

$$R \propto \frac{l}{a}$$

or

$$\boxed{R = \rho \frac{l}{a}}$$

where ρ (Greek letter 'Rho') is a constant and is known as *resistivity or specific resistance* of the material. Its value depends upon the nature of material.

1.3. RESISTIVITY

The resistance of a wire is given by the relation, $R = \rho \frac{l}{a}$

If $l = 1$ m ; $a = 1$ m^2 [see Fig. 1.4 *(a)*]; then $R = \rho$

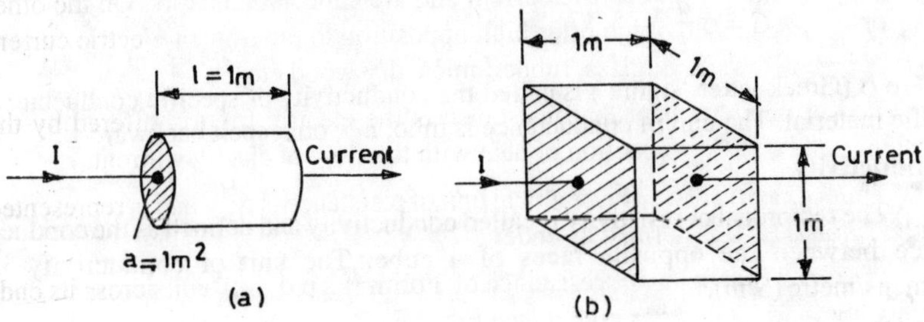

Fig. 1.4.

Hence, the resistance offered by one metre length of wire (of given material) having an area of cross-section of one square metre is called the **resistivity** of the material.

In place of wire, if a cube of one metre side of a given material is taken as shown in Fig. 1.4 *(b)*. Considering opposite two faces of the cube.

$$l = 1 \text{ m} ; a = 1 \times 1 = 1 \text{ m}^2 \text{ and } R = \rho$$

Hence, the resistance offered between the opposite two faces of one metre cube of the given material is called the **resistivity** of the material.

Unit : We know that $R = \rho \frac{l}{a}$ or $\rho = \frac{Ra}{l}$

Substituting the units of various quantities as per SI units, we get

$$\rho = \frac{\text{ohm} \times \text{m}^2}{\text{m}} = \text{ohm} - \text{m}$$

Hence, the unit of resistivity is **ohm–metre** in SI units.

1.4. SPECIFIC RESISTANCE

The resistivity of a material is also known as *specific resistance* of that material. It is, because it represents the resistance of a material having specific dimensions, *i.e.*, 1 metre length and 1 square metre area of cross section.

Hence, **specific resistance** of a material is defined as the resistance of the material having specific dimensions, *i.e.*, one metre length and one square metre as area of cross section.

1.4.1. Conductance

Conductance (G) is the reciprocal of resistance. While resistance is the opposition to current flow, conductance is the inducement to the flow of current.

We know $R = \rho \dfrac{l}{a}$

$$\therefore \quad G = \frac{1}{R} = \frac{1}{\rho}\frac{a}{l}$$

or
$$G = \sigma \frac{a}{l}$$

where σ (Greek letter 'sigma') is called the conductivity or specific conductance of the material. The unit of conductance is mho, *i.e.*, ohm spelt backward.

Conductivity

The reciprocal of resistivity is called conductivity and defined as the conductance between the opposite faces of a cube. The unit of conductivity is siemens/metre (S/m).

Example 1.1. Calculate the resistance of copper conductor having a length of 2 km and a cross section of 22 mm². Assume the resistivity is $18 \times 10^{-9}\ \Omega - m$.

Solution : Length of conductor, $l = 2$ km = 2000 m

Area of X-section, $a = 22\ \text{mm}^2 = 22 \times 10^{-6}\ \text{m}^2$

Resistivity of copper, $\rho = 18 \times 10^{-9}\ \Omega - \text{m}$

Resistance of copper conductor is given by :

$$R = \rho \frac{l}{a} = 18 \times 10^{-9} \times \frac{2000}{22 \times 10^{-6}} = \mathbf{1.64\ \Omega}$$

Example 1.2. Find the resistance of 1000 metres of a copper wire 25 sq. mm in cross-section. The resistance of copper is 1/58 ohm per metre length and 1 sq-mm cross-section. What will be the resistance of another wire of the same material, three times as long and one-half the cross-sectional area ?

Solution : For the first case, $R_1 = ?$; $a_1 = 25\ \text{mm}^2$; $l_1 = 1000$ m

For the second case, $R_2 = 1/58\ \Omega$; $a_2 = 1\ \text{mm}^2$; $l_2 = 1$ m

$$R_1 = \rho\,(l_1/a_1)$$

$$R_2 = \rho\,(l_2/a_2)$$

$$\therefore \qquad \frac{R_1}{R_1} = \frac{l_1}{l_2} \times \frac{a_2}{a_1} = \left(\frac{1000}{1}\right) \times \left(\frac{1}{25}\right) = 40$$

or $$R_1 = 40R_2 = 40 \times \frac{1}{58} = \frac{20}{29}\ \Omega \quad \textbf{Ans.}$$

For the third case, $R_3 = ?\ ;\ a_3 = a_1/2\ ;\ l_3 = 3l_1$

$$\therefore \qquad \frac{R_3}{R_1} = \left(\frac{l_3}{l_1}\right) \times \left(\frac{a_1}{a_3}\right) = (3) \times (2) = 6$$

or $$R_3 = 6R_1 = 6 \times \frac{20}{29} = \frac{120}{29}\ \Omega \quad \textbf{Ans.}$$

Example 1.3. A length of wire has a resistance of 4.5 ohms. Find the resistance of another wire of the same material three times as long and twice the cross-sectional area.

Solution : For the first case, $R_1 = 4.5\ \Omega\ ;\ l_1 = l\ ;\ a_1 = a$

For the second case, $R_2 = ?\ ;\ l_2 = 3l\ ;\ a_2 = 2a$

Now $$R_1 = \rho\frac{l_1}{a_1}\ ;\ R_2 = \rho\frac{l_2}{a_2}$$

$$\therefore \qquad \frac{R_2}{R_1} = \left(\frac{l_2}{l_1}\right)\left(\frac{a_1}{a_2}\right) = \left(\frac{3l}{l}\right) \times \left(\frac{a}{2a}\right) = 1.5$$

or $$R_2 = 1.5R_1 = 1.5 \times 4.5 = 6.75\ \Omega \quad \textbf{Ans.}$$

Example 1.4. There are two wires *A* and *B* of the same material. *A* is 20 times longer than *B* and has one fifth of the cross-section as that of *B*. If the resistance of *A* is one ohm, what is the resistance of *B* ?

Solution : Length of wire *A*, $l_A = 20\ l_B$

Area of cross-section of wire *A*, $a_A = (1/5)\ a_B$

Now $$R_A = \rho\frac{l_A}{a_A}\ ;\ R_B = \rho\frac{l_B}{a_B} \quad \therefore \quad \frac{R_B}{R_A} = \frac{l_B}{a_B} \times \frac{a_A}{l_A}$$

or $$R_B = \frac{l_B}{a_B} \times \frac{l_{aB}}{5 \times 20l_B} \times 1 = \frac{1}{100}\ \Omega = 10 \text{ milli-ohm} \quad \textbf{Ans.}$$

Example 1.5. A coil consists of 2,000 turns of copper wire having a cross-sectional area of 0.8 mm^2. The mean length per turn is 80 cm, and the resistivity of copper is 0.02 μ Ω – m. Find the resistance of the coil.

Solution : Length of the coil, l = Number of turns × mean length per turn

$$= 2000 \times 0.8 = 1{,}600 \text{ m}$$

Cross-sectional area of wire, $a = 0.8 \text{ mm}^2 = 0.8 \times 10^{-6} \text{ m}^2$

Resistivity of copper, $\rho = 0.02\ \mu\ \Omega - \text{m} = 2 \times 10^{-2} \times 10^{-6}\ \Omega - \text{m}$

$$= 2 \times 10^{-8}\ \Omega - \text{m}$$

Resistance of the coil, $R = \rho \dfrac{l}{a} = 2 \times 10^{-8} \times \dfrac{1600}{0.8 \times 10^{-6}} = 40\ \Omega$ **Ans.**

Example 1.6. A heater element is made of nichrome wire having resistivity equal to 100×10^{-8} Ω-m. The diameter of the wire is 0.4 mm. Calculate the length of the wire required to get a resistance of 40 Ω.

Solution : Resistance of nichrome wire, $R = 40\ \Omega$

Resistivity of nichrome wire, $\rho = 100 \times 10^{-8}\ \Omega - \text{m}$

Diameter of nichrome wire, $d = 0.4\ \text{mm} = 4 \times 10^{-4}\ \text{m}$

X-sectional area of nichrome wire, $a = \dfrac{\pi}{4} d^2 = \dfrac{\pi}{4} (4 \times 10^{-4})^2 = 4\ \pi \times 10^{-8}\ \text{m}^2$

Length of nichrome wire required,

$$l = \frac{R \times a}{\rho} = \frac{40 \times 4\ \pi \times 10^{-8}}{100 \times 10^{-8}} = 5.03\ \text{m}\ \textbf{Ans.}$$

Example 1.7. The resistance of a conductor 1 mm² in cross- section and 20 m long is 0.346 Ω. Determine the specific resistance of the conductor material.

Solution : Resistance of conductor, $R = 0.346\ \Omega$

Length of conductor, $l = 20$ m

Cross-sectional area of conductor, $a = 1\ \text{mm}^2 = 1 \times 10^{-6}\ \text{m}^2$

Specific resistance of conductor material,

$$\rho = \frac{R \times a}{l} = \frac{0.346 \times 1 \times 10^{-6}}{20} = 1.73 \times 10^{-8}\ \Omega - \text{m}\ \textbf{Ans.}$$

Example 1.8. A piece of silver has a resistance of 2 ohms. What will be the resistance of a manganin wire one-third the length and one-third the diameter, if the resistivity of manganin is 30 times that of silver ?

Solution : Resistivity of manganin, ρ_m = 3o times the resistivity of silver, *i.e.*, $30\ \rho_s$

Length of manganin wire, l_m = one-third the length of silver wire, *i.e.*, $\dfrac{1}{3}\ l_s$

Diameter of manganin wire, d_m = one-third the diameter of silver wire, *i.e.*, $\dfrac{1}{3}\ d_s$

Cross-sectional area of manganin wire,

$$a_m = \frac{\pi}{4}(d_m)^2 = \frac{\pi}{4}\left(\frac{d_s}{3}\right)^2 = \frac{\pi}{36} d_s^2 = \frac{1}{9} a_s$$

Resistance of silver wire, $R_s = \frac{\rho_s \, l_s}{a_s} = 2\ \Omega$

Resistance of manganin wire,

$$R_m = \rho_m \frac{l_m}{a_m} = 30\ \rho_s \frac{l_s/3}{a_s/9} = 90 \frac{\rho_s \, l_s}{a_s} = 90 \times 2 = 180\ \Omega \ \textbf{Ans.}$$

Example 1.9. Determine the resistance of the material in terms of volume *v*, length *l* and resistivity ρ. Hence determine the resistivity of the material of a conductor of volume 0.05 m³, length 300 metres and resistance 0.0306 Ω.

Solution : Cross-sectional area of material of volume *v* units and length *l* units.

$$a = \frac{v}{l} \qquad ...(i)$$

Resistance of metarial $R = \rho \frac{l}{a} = \rho \frac{l}{v/l} = \rho \frac{l^2}{v}$ **Ans.**

$\because$ from expression *(i)* $a = \frac{v}{l}$

Substituting, $v = 0.05\ \text{m}^3$, $l = 300$ m and $R = 0.0306\ \Omega$ in above expression we get,

$$0.0306 = \rho \frac{(300)^2}{0.05}$$

or

$$\rho = \frac{0.0306 \times 0.05}{(300)^2} = 1.7 \times 10^{-8}\ \Omega - \text{m} \ \textbf{Ans.}$$

Example 1.10. Determine *(i)* length and *(ii)* diameter of copper cylinder of volume *v*, resistance *R* and specific resistance ρ in terms of *v*, *R* and ρ.

Solution : Volume of cylinder, $v = a \times l$ ...*(i)*

Resistance of cylinder, $R = \rho \frac{l}{a}$ ...*(ii)*

Multiplying expressions *(i)* and *(ii)*, we get

$$vR = al \times \rho \frac{l}{a} = \rho\, l^2$$

or

$$l^2 = vR/\rho$$

or

$$l = \sqrt{\frac{vR}{\rho}} \ \textbf{Ans.}$$

Dividing expression *(i)* by expression *(ii)*, we get

$$\frac{v}{R} = \frac{al}{\rho\, l/a} = \frac{a^2}{\rho}$$

or $$a = \sqrt{\frac{v\,\rho}{R}}$$

or $$\frac{\pi}{4} d^2 = \sqrt{\frac{v\,\rho}{R}}$$

or diameter of cylinder, $d = \dfrac{16\, v\, \rho}{\pi^2 R}$ **Ans.**

Example 1.11. Two wires of the same metal are equal in length and their resistances are 2500 and 3600 ohms respectively. Find the ratio of their diameters. If the wires had the same diameters, what would be their relative lengths ?

Solution : We know that, $R \propto \dfrac{l}{a}$

or $$R = \frac{\rho\, l}{a}$$

or $$Ra = \rho\, l$$

or $$l = \frac{Ra}{\rho} = \frac{R\,\pi\, d^2}{4\,\rho} \qquad \left[\because\ a = \frac{\pi\, d^2}{4}\right]$$

$$\therefore \quad \frac{l_1}{l_2} = \frac{R_1}{R_2}$$

$$\therefore \quad \frac{l_1}{l_2} = \frac{1}{1.44}$$

$\therefore$ $l_1 : l_2 = 1 : 1.44$ **Ans.**

We know that, $l = \dfrac{R\,\pi\, d^2}{4\,\rho}$

or $$R\,\pi\, d^2 = 4\, l\, \rho$$

or $$d^2 = \frac{4\, l\, \rho}{R\,\pi} \quad \text{or } d = \sqrt{4\, l\, \rho / R\,\pi}$$

or $$\frac{d_1}{d_2} = \frac{\sqrt{4\, l\, \rho / R\,\pi}}{\sqrt{4\, l\, \rho / R\,\pi}} = \frac{\sqrt{1/2500}}{\sqrt{1/3600}}$$

$$= \sqrt{\frac{3600}{2500}} = \frac{6}{5} = 1.2 : 1 \textbf{ Ans.}$$

Example 1.12. There are two wires *A* and *B* of the same metal. *A* is 20 times as long as *B* and is one-third the cross-section of *B*. If the resistance of *A* is 1 ohm, what will be the resistance of *B* ?

Solution : Length of wire $A = 20l_B$

Length of wire $B = l$

Area of cross-section of A, $a_A = \frac{1}{3} a_B$

Resistivity is same

$$\frac{R_B}{R_A} = \rho \frac{l_B}{a_B}$$

$$\rho \frac{l_A}{a_A} = \frac{l_B}{a_B} \times \frac{a_A}{l_A}$$

Therefore, resistance of wire B, $R_B = \frac{l_B}{a_B} \times \frac{\frac{1}{3} a_B \times 1}{20 l_B} \times 1 = \frac{1}{60} \Omega$ **Ans.**

Example 1.13. Determine the resistance of a copper tube having external diameter 8 cm; thickness 5 mm and length 5 meters. It is given that specific resistance of copper is 1.70 μ Ω-cm.

Solution : Resistivity of copper, $\rho = 1.7 \times 10^{-8}\ \Omega - \text{m}$

Length of copper tube, $l = 5$ metres

External diameter of tube, $D = 8$ cm $= 0.08$ metre

Internal diameter of tube,

$$d = D - 2 \times \text{thickness of tube} = 0.08 - 2 \times 0.005 = 0.07 \text{ m}$$

Cross-sectional area of tube,

$$a = \frac{\pi}{4}(D^2 - d^2) = \frac{\pi}{4}(0.08^2 - 0.07^2) = 0.001178 \text{ m}^2$$

Resistance of copper tube,

$$R = \rho \frac{l}{a} = \frac{1.7 \times 10^{-8} \times 5}{0.001178} = 72.12\ \mu\Omega \quad \textbf{Ans.}$$

Example 1.14. Determine the conductivity and conductance of 1,000 metres of a cable. The *x*-sectional area of the cable core is 0.125 cm² and resistivity of copper is 1.7 × 10⁻⁸ Ω-m.

Solution : Conductivity, $\sigma = \frac{l}{\rho} = \frac{1}{1.7 \times 10^{-8}} = 58.8 \times 10^6$ S/m **Ans.**

Conductance, $G = \frac{\sigma a}{l} = \frac{58.8 \times 10^6 \times 0.125 \times 10^{-4}}{1{,}000} = 0.735$ S **Ans.**

Example 1.15. Calculate the resistance of a copper tube 0.005 m thick and 2 m long. The external diameter is 0.1 m. Resistivity of copper is 1.7×10^{-8} Ω metre.

Solution : Thickness = 0.005 m

Length, $l = 2$ m

External diameter, $D = 1$ m

$$d = .10 - 2 \times \text{thickness} = 0.1 - 2 \times 0.005 = 0.09 \text{ m}$$

Resistivity $\rho = 1.7 \times 10^{-8}$ Ω m

$$\therefore \text{Resistivity of tube, } R = \frac{4\rho l}{\pi (D^2 - d^2)} = \frac{4 \times 1.7 \times 10^{-8} \times 2}{3.14 \left[(0.1)^2 - (0.09)^2 \right]}$$

$$= 22.7 \times 10^{-6}\ \Omega \text{ or } 22.7\ \mu\Omega \quad \textbf{Ans.}$$

1.4.2. Electric Power

The rate of which work is done in an electric circuit is called **electric power,** *i.e.,*

$$\text{Electric power} = \frac{\text{Work done in electric circuit}}{\text{Time}}$$

When voltage is applied to a circuit, it causes current (*i.e.*, electrons) to flow through it. Clearly, work is being done in moving the electrons in the circuit. This work done in moving the electrons in a unit time is called the electric power. Thus referring to the part *AB* of the circuit (See Fig. 1.5).

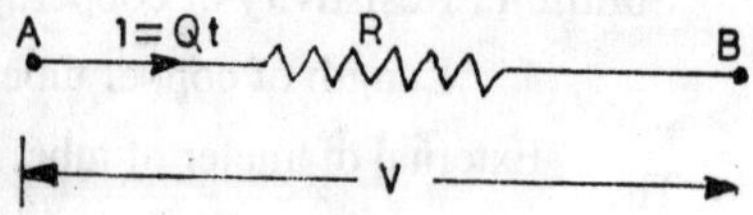

Fig. 1.5.

V = P.D. across *AB* in volts

I = Current in amperes

R = Resistance of *AB* in Ω

t = Time is sec. for which current flows.

The total charge that flows in t seconds is $Q = I \times t$ coulombs and by definition

$$V = \frac{\text{Work}}{Q}$$

or $\quad \text{Work} = VQ = VIt \qquad (\because Q = It)$

$$\therefore \text{Electric power, } P = \frac{\text{Work}}{t} = \frac{VIt}{t} = VI \text{ joules/sec or watts}$$

$$\therefore \quad P = VI \qquad ...(i)$$

$$= I^2 R \qquad ...(ii) \qquad (\because V = IR)$$

$$= \frac{V^2}{R} \qquad ...(iii) \qquad \left(\because I = \frac{V}{R}\right)$$

The above three formulae are equally valid for calculation of electric power in d.c. circuit. Which one is to be used depends simply on which quantities are known or most easily determined.

Unit of electric power : The basic unit of electric power is Joules/sec or watt. The power consumed in a circuit is 1 watt if a p.d. of 1 V causes 1 A current of flow through the circuit.

Power in watts = Voltage in volts × current in amperes. The bigger units of electric power are kilowatts (kW) and megawatts (MW).

1 kW = 1000 watts

1 MW = 10^6 watts or 10^3 kW

1.4.3. Electrical Energy

The total work done in an electric circuit is called **electrical energy**, *i.e.*,

Electrical energy = Electrical power × time

$$= VIt \qquad ...(i)$$

$$= I^2 Rt \qquad ...(ii)$$

$$= \frac{V^2}{R} t \qquad ...(iii)$$

The reader may note that formulae for electrical energy can be readily derived by multiplying the electric power by *t*, the time for which the current flows. The unit of electrical energy will depend upon the units of electric power and time.

(i) If power is taken in watts and time in seconds, then the unit of electrical energy will be watt-sec, *i.e.*,

Energy in watt-sec = Power in watts × Time in sec.

(ii) If power is expressed in watts and time in hours, then unit of electrical energy will be watt-hour, *i.e.*,

Energy in watt-hours = Power in watt × Time in hours

(iii) If power is expressed in kilowatts and time in hours, then unit of electrical energy will be kilowatt-hour (kWh), *i.e.*,

Energy in kWh = Power in kW × Time in hours.

It may be pointed out here that in practice, electrical energy is measured in kilowatt-hours (kWh). Therefore, it is profitable to define it.

One kilowatt-hour (kWh) of electrical energy is expeneded in a circuit of 1 kW (1000 watts) of power is supplied for 1 hour.

The electricity bills are made on the basis of total electrical energy consumed by the consumer. The unit for charge of electricity is 1 kWh. One kWh is also called Board of Trade (B.O.T.) unit or simply unit. Thus when we say that a consumer has consumed 100 units of electricity, it means that electrical energy consumption is 100 kWh.

Example 1.16. A 100 V lamp has a hot resistance of 250 Ω. Find the current taken by the lamp and its power rating in watts. Calculate also the energy it will consume in 24 hours.

Solution : Current taken by lamp, $I = V/R = 100/250 = 0.4$ A

Power rating of lamp, $P = VI = 100 \times 0.4 = 40$ W

Energy consumtion in 24 hours = Power × Time

$= 40 \times 24 = 960$ watt-hours **Ans.**

Example 1.17. A heating element supplies 300 kilojoules in 50 minutes. Find the p.d. across the element when current is 2 amperes.

Solution : Total charge, $Q = I \times t = 2 \times 50 \times 60 = 6000$ C

$$\text{P.D.,} \quad V = \frac{\text{Work}}{\text{Charge}} = \frac{300 \times 10^3}{6000} = 50 \text{ V} \quad \textbf{Ans.}$$

Example 1.18. A 10-watt resistor has a value of 120 Ω. What is the rated current through the resistor ?

Solution : Rated power, $P = I^2 R$

$$\therefore \quad \text{Rated current}, I = \sqrt{\frac{P}{R}} = \sqrt{\frac{10}{120}} = 0.2887 \text{ A} \quad \textbf{Ans.}$$

If current through the resistor exceeds this value, the resistor will be burnt due to excessive **heat**

Note : Every electrical equipment has power and current ratings marked on its body. While the equipment is in operation, care should be taken that neither of these limits is exceeded, otherwise the equipment may be damaged/burnt due to excessive heat.

Example 1.19. The following are the details of load on a circuit connected thorugh a supply meter :

(i) **Six lamps of 40 watts each working for 4 hours per day**

(ii) **Two flourescent tubes 125 watts each working for 2 hours per day.**

(iii) **One 1000 watt heater working for 3 hours per day.**

If each unit of energy costs 70 P, what will be the electricity bill for the month of june ?

Solution : Total wattage of lamps = 40 × 6/= 240 watts

Total wattage of tubes = 125 × 2 = 250 watts

Wattage of heater = 1000 watts

Energy consumed by the appliances per day

= (240 × 4) + (250 × 2) + (1000 × 3) = 4460 watt–hours = 4.46 kWh

Total energy consumed in the month of June (*i.e.*, in 30 days)

= 4.46 × 30 = 133.8 kWh

Bill for the month of June = Rs. 0.7 × 133.8 = **Rs. 93.66**

Example 1.20. A house is fitted with 5 lamps of 100 watts each, one electric press of 220 V drawing 2A of current, 4 fans of 110 watts each and a heater of 1120 watts. If all the appliances works for 2 hours a day, find the electricity bill for the month of September. Electrical energy is supplied at 50 paise per kWh.

Solution : Total wattage of lamps = 5 × 100 = 500 watts

Wattage of electric press = 220 × 2 = 440 watts

Total wattage of fans = 4 × 110 = 440 watts

Wattage of heater = 1120 watts

Total wattage of all the appliances = 500 + 440 + 440 + 1120 = 2500 watts

Energy consumed by all the appliances per day

= 2500 × 2 = 5000 watts-hours = 5 kWh

Energy consumed by all the appliances in the month of Sept.

= 5 × 30 = 150 kWh

Bill of the month of September = Rs. 0.50 × 150 = **Rs. 75**

Example 1.21. The name plate of an electric iron reads 200 volts, 350 watts. What current does it take and what is its hot resistance ?

Solution : $V = 200$ V

Wattage = 350 watt ; I = ?

$$I = \frac{V}{R} = \frac{200}{R}$$

Power $P = 350$ watt

$$R = \frac{V^2}{P} = \frac{(200)^2}{350} = \frac{800}{7} = 114.28\ \Omega \ \textbf{Ans.}$$

$$I = \frac{V}{R}$$

$$I = \frac{200}{114.28} = 1.75 \text{ A} \ \textbf{Ans.}$$

Example 1.22. The heating element of heater draws 10 A from 240 volts supply mains. How much power does the plate use and what is the resistance of the element ?

Solution : $I = 10$ A ; $V = 240$ V. $P = ?$; $R = ?$

$$R = \frac{V}{I} = \frac{240}{10} = 24\ \Omega$$

$$P = \frac{(240)^2}{24} = \frac{57600}{24} = 2400\ \text{W} \quad \textbf{Ans.}$$

Example 1.23. What will be the current drawn by a lamp rated at 250 V, 25 watt connected to a 230 V supply ?

Solution : $P = 25$ W ; $V = 250$ V.

$$R = \frac{V^2}{P} = \frac{(250)^2}{25} = 2500\ \Omega$$

$$I = \frac{230}{R} = \frac{230}{2500} = 0.092\ \text{A} \quad \textbf{Ans.}$$

Example 1.24. An electric iron takes the 3 amperes at 230 volts. What does it cost to operate this iron if the cost of energy is 30 paise per unit ?

Solution : Wattage of electric iron = 230 × 3 = 690 watts

Energy consumed in 2 hours = Wattage × Time

= 690 × 2 = 1380 watt-hr

= 1.38 kWh or units

Cost of one unit = 30 paise

Total cost = 1.38 × 30 = 41.4 paise **Ans.**

Example 1.25. The name plate of an electric iron reads 200 volts, 350 watts. What current does it take and what is its hot resistance ?

Solution : Voltage in iron, $V = 200$ volts

Power of iron, $P = 350$ watts

$I = ?$; $R = ?$

$P = VI$

$VI = 350$

$$I = \frac{350}{V} = \frac{350}{200} = 1.75\ \text{amp} \quad \textbf{Ans.}$$

$$R = \frac{V}{I} = \frac{200}{1.75} = 114.3\ \Omega \quad \textbf{Ans.}$$

Example 1.26. A house is fitted with 5 lamps of 100 watts, one electric press of 220 volts drawing 2 amp of current, 4 fans of 110 watts and a heater of 1120 watts. If all the application work for 2 hours a day, find the electric bill for Sept., 1987 at a rate of 25 paise per unit.

Solution : Total energy consumed by 5 lamps,

$$= 5\ VIt = 5 \times 100 \times 2 = 1000 \text{ wH}$$

Total energy consumed by electric press

$$= VIt = 220 \times 2 \times 2 = 880 \text{ wH}$$

Total energy consumed by 4 fans = $4\ VIt = 4 \times 110 \times 2 = 880$ wH

Total energy consumed by heater = $VIt = 1120 \times 2 = 2240$ wH

Total energy consumed in a day = $1000 + 880 + 880 + 2240 = 5000$ wH

$$= \frac{5000}{1000} = 5 \text{ kWh}$$

Total energy consumed in September = $30 \times 5 = 150$ kWh

Rate of electricity = 25 paise/kwH

Total cost of electricity = $\frac{150 \times 25}{100}$ = Rs. 37.50 **Ans.**

Example 1.27. Twenty lamps of 60 watts are run for 5 hrs from 230 volt supply mains. Calculate :

(i) **The units consumed.**

(ii) **The current through the mains.**

(iii) **The resistance of one lamp.**

Solution : *(i)* Voltage, $V = 230$ volts

Total wattage of 20 lamps = $60 \times 20 = 1200$ watts

Total wattage of 20 lamps for 5 hours = $1200 \times 5 = 6000$ wh or 6 kWh

(ii) We know that, $P = V \times I$

or $$I = \frac{P}{V} = \frac{1200}{230} = 5.22 \text{ amp} \quad \textbf{Ans.}$$

(iii) Power of 20 lamps = 1200 W

$$\text{Power of 1 lamp} = \frac{1200}{20} = 60 \text{ watts}$$

We know that, $R = \frac{V^2}{P}$ $\left[\because \frac{V^2}{P} = \frac{I^2 R^2}{I^2 R} = R\right]$

$$= \frac{230 \times 230}{60} = 881.7\ \Omega \quad \textbf{Ans.}$$

Example 1.28. A 2 H.P. motor, a 660 watts heater, an electric iron using 5 A are all used at the same time on 220 V line. Find the cost of operating for the three for 4 hours, if the energy cost 24 paise per kWh.

Solution : 1 H.P. = 746 watts

2 H.P. = 746 × 2 = 1492 watts

Total power of a motor in 4 horus = 1492 × 4 = 5968 wh

Total wattage of a heater in 4 hours = 660 × 4 = 2640 wh

Total wattage of an electric iron in 4 hours = 220 × 5 × 4 = 4400 wh

Total energy consumed in 4 hours

= 5968 + 2640 + 4400 = 13008 wh or 13.008 kWh

The cost of 1 kWh is = 0.24 paise

The cost of 13.008 kWh = .24 × 13.008 = 3.12 **Ans.**

1.5. OHM'S LAW

Ohm's law states that the current flowing through a conductor is directly proportional to the potential difference applied between the two ends of the conductor, provided the temperature remains constant, *i.e.*, $I \propto V$, where I is the current through the conductor and V the potential difference across it.

or $$I = V \times \text{constant}$$

This constant of proportionality is equal to $1/R$ where R is the resistance of the conductor. Hence as per the Ohm's law, equation relating the current flowing through a conductor and the voltage applied across it, is given by

$$I = \frac{V}{R} \qquad ...(i)$$

In eqn. *(i)*, the unit of current I is ampere, unit of voltage V is volts and that of resistance R is ohms. This relationship is quite useful for solving simple dc circuits, it can be applied to a complete circuit or any part of the circuit. This relationship can be represented in three different forms for the purpose of calculations :

(i) $I = \frac{V}{R}$ or current = $\frac{\text{voltage}}{\text{resistance}}$

(ii) $V = IR$ or voltage = current × resistance

(iii) $R = \frac{V}{I}$ or resistance = $\frac{\text{voltage}}{\text{current}}$

1.5.1. Ohm's Law — Verification

The circuit diagram for verification of ohm's law is shown in Fig. 1.6. A conductor having resistance R is connected in series with an ammeter A, Battery B, rheostat R_h and key K. The voltmeter is connected across the conductor is to measure the potential difference.

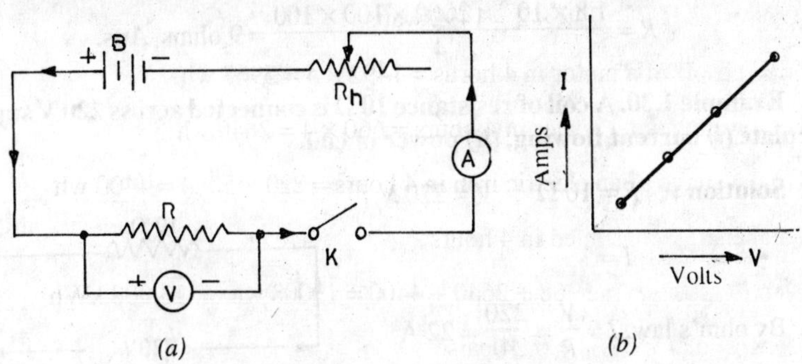

Fig. 1.6. Verification of Ohm's law

The ammeter measures the current through the resistance R. The purpose of rheostat is to change the circuit resistance and hence the current key K is to make or break the circuit.

To start the experiments, close the key K and set the rheostat at some value. Take readings of voltmeter and ammeter. Let it be V_1 and I_1 respectively. Find the ratio $\frac{V_1}{I_1}$. Now, change the setting of rheostat so that current in the circuit changes, Note down the readings of voltmeter and ammeter. Let it be V_2 and I_2 respectively. Find the ratio $\frac{V_2}{I_2}$ it will be found that :

$$\frac{V_1}{I_1} = \frac{V_2}{I_2}$$

Take a number of readings by changing the setting of the rheostat. Find the ratio of potential difference to current in each case. It will be observed that :

$$\frac{V_1}{I_1} = \frac{V_2}{I_2} = \frac{V_3}{I_3} = \text{constant}$$

Thus ohm's law is verified.

Example 1.29. Calculate the resistance of a coil of copper wire 2 km long and 4 mm² cross-sectional area. The resistivity of copper is 1.8×10^{-6} Ω cm.

Solution : $R = \rho \frac{l}{a}$ ohms

$$\rho = 1.8 \times 10^{-6}$$

$$l = 2 \text{ km} = 2000 \text{ m} = 2000 \times 10^3 \text{ cm}$$

$$a = 4 \text{ mm}^2 = \frac{4}{100} \text{ cm}^2$$

$$R = \frac{1.8 \times 10^{-6} \times 2000 \times 100 \times 100}{4} = 9 \text{ ohms } \textbf{Ans.}$$

Example 1.30. A coil of resistance 10 Ω is connected across 220 V supply. Calculate *(i)* current flowing, *(ii)* power of coil.

Solution : $R = 10\ \Omega$; $V = 220$ V ;

$I = ?$

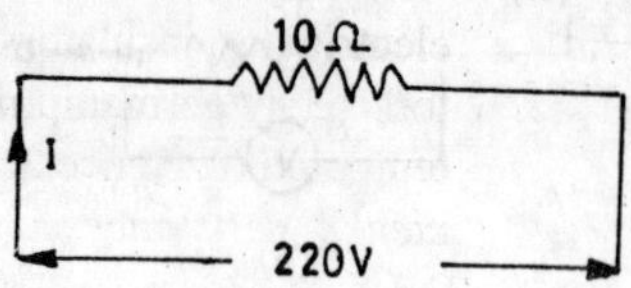

Fig. 1.7.

By ohm's law, $I = \frac{V}{R} = \frac{220}{10} = 22$ A

Power of coil $= I^2 R = 22 \times 22 \times 10$

$= 4840$ W $= 4.84$ kW

PROBLEMS

1. Calculate the resistance of 915 metres length of a wire having a uniform cross-sectional area of 0.77 cm^2 if the wire is made of copper having a resistivity of 1.7×10^{-6} Ω cm. **[Ans.** 0.08 Ω]
2. A wire of length 1 m has a resistance of 2 ohms. What is the resistance of second wire, whose specific resistance is double the first, if the length of wire is 3 metres and the diameter is double of the first ? **[Ans.** 3 Ω]
3. A rectangular copper strip is 20 cm long, 0.1 cm wide and 0.4 cm thick. Determine the resistance between *(i)* opposite ends *(ii)* opposite sides. The resistivity of copper is 1.7×10^{-6} Ω cm. **[Ans.** 0.85×10^{-4} Ω ; 0.212×10^{-6} Ω]
4. A cube of a material of side 1 cm has a resistance of 0.001 Ω between its opposite faces. If the same material has a length of 9 cm and a uniform cross-sectional area 1 cm^2, what will be the resistance of this length ? **[Ans.** 0.009 Ω]
5. An aluminium wire 10 metres long and 2 mm in diameter is connected in parallel with a copper wire 6 metres long. A total current of 2A is passed through the combination and it is found that the current through the aluminium wire is 1.25 A. Calculate the diameter of copper wire. Specific resistance of copper is 1.6×10^{-6} Ω cm and that of aluminium is 2.6×10^{-6} Ω m. **[Ans.** 0.94 mm]

1.6. EFFECT OF TEMPERATURE ON RESISTANCE

In general, the resistance of a material changes with the change in temperature. The effect of temperature upon resistance varies according to the type of material as discussed below :

(i) **Pure Metals :** The resistance of pure metals (*e.g.*, copper, aluminimum) increases with the increase of temperature. The change in resistance is fairly regular for normal range of temperatures so that temperature/resistance graph is a straight line for copper as shown in Fig. 1.7). Since the resistance of metals increases with the rise in temperature, they have *positive temperature co-efficient of resistance.*

(ii) **Electrolytes, Insulators and Semiconductor :** The resistance of electrolytes, insulators (*e.g.*, glass, mica, rubber etc.) and semiconductors (*e.g.*, germanium, silicon etc.) decreases with the increase in temperature. Hence these materials have *negative temperature co-efficient of resistance.*

(iii) **Alloys :** The resistance of alloys increases with the rise in temperature but this increase is very small and irregular. For some high resistance alloys (*e.g.*, Eureka, manganin, constant etc.) the change in resistance is practically negligible over a wide range of temperatures.

Fig. 1.7 shows temperature resistance graph for copper which is a straight line. If this line is extended backward, it would cut the temperature axis at –234.5°C. It means that theoretically, the resistance of copper wire is zero at –234.5°C. However, in actual practice, the curve departs (point A) from the straight line path at very low temperatures.

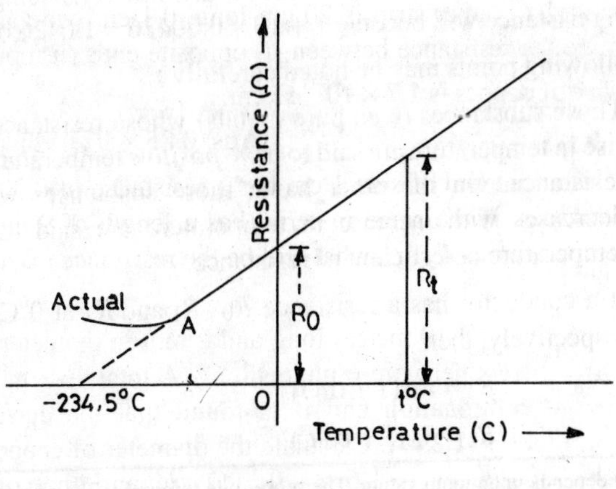

Fig. 1.7.

1.6.1. Temperature Co-efficient of Resistance

Consider a conductor having resistance R_0 at 0°C and R_t at t°C. It has been found that in the normal range of temperatures, the increase in resistance (*i.e.*, $R_t - R_0$)

(*i*) is directly proportional to the initial resistance, *i.e.*, $R_t - R_0 \propto R_0$

(*ii*) is directly proportional to the rise in temperature, *i.e.*, $R_t - R_0 \propto t$

(*iii*) depends upon the nature of material.

Combining the first two, we get

$$R_t - R_0 \propto R_0\, t$$

or $$R_t - R_0 = {}^*\alpha_0 R_0\, t \qquad ...(i)$$

where α_0 is a constant and is called temperature co- efficient of resistance at 0°C. Its value depends upon the nature of material and temperature.

Rearranging eq. (*i*), we get,

$$R_t = R_0 (1 + \alpha_0 t) \qquad ...(ii)$$

Definition of α_0. From eq. (*i*) we get,

$$\alpha_0 = \frac{R_t - R_0}{R_0 \times t}$$

= increase in resistance/ohm original resistance/°C rise in temperature.

Hence **temperature co-efficient of resistance** of a material is the increase in resistance per ohm original resistance per °C rise in temperature.

The unit of α is ohms/ohm/°C, *i.e.*, /° C. Thus, copper has a temperature co-efficient of resistance of 0.00426/°C. It means that if a copper wire has a resistance of 1 Ω at 0°C, then it will increase by 0.00426 Ω for 1°C rise in temperature, *i.e.*, it will become 1.00426 Ω at 1°C. Similarly, if temperature is raised to 10°C, then resistance will become 1 + 10 × 0.00426 = 1.0426 ohms.

The following points may be noted carefully :

(*i*) Those substances (*e.g.*, pure metals) whose resistance increases with rise in temperature are said to have *positive* temperature co-efficient of resistance. On the other hand, those susbtances whose resistance decreases with increase in temperature are said to have *negative* temperature co-efficient of resistance.

(*ii*) If a conductor has a resistance R_0, R_1 and R_2 at 0°C, t_1°C and t_2°C respectively, then

$$R_1 = R_0 (1 + \alpha_0 t_1)$$

$$R_2 = R_0 (1 + \alpha_0 t_2)$$

*The value of α depends upon temperature. Therefore, it is referred to the original temperature, i.e., 0°C in this case. Hence the symbol α_0.

$$\therefore \quad \frac{R_2}{R_1} = \frac{1 + \alpha_0 t_2}{1 + \alpha_0 t_1} \quad ...(iii)$$

This relation is often utilised in determining the rise of temperature of the winding of an electrical machine. The resistance of the winding is measured both before and after the test run. Let R_1 and t_1 be the resistance and temperature before the commencement of the test. After the operation of the machine for a given period, let these values be R_2 and t_2. Since R_1 and R_2 can be measured and t_1 (ambient temperature) and α_0 are known, the value of t_2 can be calculated from eq. *(iii)*. The average rise in temperature of the winding will be $(t_2 - t_1)$°C.

1.6.2. Summary of Temperature Co-efficient Relations

(i) If R_0 and α_0 are the resistance and temperature co-efficient of a conductor at 0°C, then its temperature R_t at t°C is given by; $R_t = R_0 (1 + \alpha_0 t)$

(ii) If α_0, α_1 and α_2 are the temperature co-efficient at 0°C, t_1°C and t_2°C respectively, then

$$\alpha_1 = \frac{\alpha_0}{1 + \alpha_0 t_1}$$

$$\alpha_2 = \frac{\alpha_0}{1 + \alpha_0 t_2}$$

$$\alpha_2 = \frac{1}{\frac{1}{\alpha_1} + (t_2 - t_1)}$$

(iii) Suppose R_1 and R_2 are the resistances of a conductor at t_1°C and t_2°C respectively. If α_1 is the temperature co-efficient at t_1°C, then

$$R_2 = R_1 [1 + \alpha_1 (t_2 - t_1)]$$

Example 1.31. A coil made of copper wire has a resistance of 25 Ω when immersed in ice water (0°C). What would its resistance be in boiling water (100°C) ? Assume the temperature co-efficient is 0.004/°C at 0°C.

Solution : $R_{100} = R_0 (1 + \alpha_0 t)$

$= 25 (1 + 0.004 \times 100) = 35\ \Omega$ **Ans.**

Example 1.32. A tungsten lamp has a resistance of 150 Ω at 2850°C. What is its resistance at room temperature (20°C) ? Temperature co-efficient of tungsten is 0.0049/°C at 0°C.

Solution :
$$\frac{R_{20}}{R_{2850}} = \frac{R_0 (1 + \alpha_0 \times 20)}{R_0 (1 + \alpha_0 \times 2850)}$$

or
$$\frac{R_{20}}{150} = \frac{1 + 0.0049 \times 20}{1 + 0.0049 \times 2850} = \frac{1.098}{14.965}$$

$$\therefore \quad R_{20} = \frac{1.098}{14.965} \times 150 = 11\ \Omega \ \textbf{Ans.}$$

Example 1.33. A coil has a resistance of 10 Ω at 20°C. What is its resistance at 60°C ? Resistance-temperature co- efficient at 20°C is 0.0039/°C.

Solution : $R_{60} = R_{20}\,[1 + \alpha_{20}\,(t_2 - t_1)]$

$= 10\,[1 + 0.0039\,(60 - 20)] = 11.56\ \Omega$ **Ans.**

Example 1.34. The resistance of the field coils of a dynamo is 173 Ω at 16°C. After working for 6 hours on full-load, the resistance of the coils increases to 212 Ω. Calculate *(i)* the temperature of the coils *(ii)* mean rise of temperature of the coils. Assume temperature co-efficient of resistance of copper is 0.00426/°C at 0°C.

Solution : *(i)* Let t°C be the final temperature.

$$\frac{R_{16}}{R_t} = \frac{R_0\,(1 + \alpha_0 \times 16)}{R_0\,(1 + \alpha_0 \times t)}$$

or
$$\frac{173}{212} = \frac{1 + 0.00426 \times 16}{1 + 0.00426 \times t}$$

or
$$0.816 = \frac{1.068}{1 + 0.0046\,t}$$

$$\therefore \quad t = 72.5^\circ\text{C} \ \textbf{Ans.}$$

(ii) Rise in temperature $= t - 16 = 72.5 - 16 = 56.5$°C **Ans.**

Example 1.35. The resistance of a transformer winding is 460 Ω at room temperature of 25°C. When the transformer is running and the final temperature is reached, the resistance of the winding increases to 520 Ω. Find the average temperature rise of winding, assuming that α_{20} = 1/250 per °C.

Solution :
$$\alpha_{25} = \frac{1}{1/\alpha_{20} + (25 - 20)} = \frac{1}{250 + 5} = \frac{1}{255}/^\circ\text{C}$$

Let t °C be the final temperature of the winding. Then, the rise in temperature is $t - 25$.

Now, $R_{25} = 460\ \Omega$; $R_t = 520\ \Omega$

$$R_t = R_{25}\,[1 + \alpha_{25}\,(t - 25)]$$

or
$$t - 25 = \frac{1}{\alpha_{25}}\left(\frac{R_t}{R_{25}} - 1\right) = 255\left(\frac{520}{460} - 1\right) = 33.26^\circ\ \text{C}$$

∴ Temperature rise $= t - 25 = 33.26 - 25 = 8.26$° C **Ans.**

Example 1.36. The filament of a 60-watt, 230 V lamp has a normal working temperature of 2000°C. Find the current flowing in the filament at the instant of switching, when the lamp is cold. Assume the temperature of cold lamp to be 15°C and α_{15} = 0.005/°C.

Solution : Resistance of lamp at 2000°C

$$R_{2000} = \frac{V^2}{W} = \frac{(230)^2}{60} = 881.67\ \Omega$$

$$R_{2000} = R_{15}\,[1 + \alpha_{15}\,(2000 - 15)]$$

$$\therefore \qquad R_{15} = \frac{R_{2000}}{1 + 0.005\,(1985)} = \frac{881.67}{10.925} = 80.7\ \Omega$$

$\therefore$ Current taken by cold lamp (*i.e.*, at the time of switching)

$$I = \frac{V}{R_{15}} = \frac{230}{80.7} = 2.85\text{ A} \quad \textbf{Ans.}$$

Example 1.37. Show that if α_0 be the resistance temperature coefficient of a conductor at 0°C then resistance temperature coefficient at t_1 ° C is given as

$$\alpha_1 = \frac{1}{\dfrac{1}{\alpha_0} + t_1}$$

Solution : Let R_0 , R_1 and R_2 be the resistances of a conductor at 0°C, t_1 °C and t_2 °C temperatures respectively, then,

Resistance at t_1 °C, $R_1 = R_0\,(1 + \alpha_0\,t_1)$ ...(*i*)

Resistance at t_2 °C, $R_2 = R_0\,(1 + \alpha_0\,t_2)$ ...(*ii*)

Also $\quad R_2 = R_1\,[1 + \alpha_1\,(t_2 - t_1)]$

or $\quad \dfrac{R_2}{R_1} = 1 + \alpha_1\,(t_2 - t_1)$ ...(*iii*)

Dividing eqn. (*ii*) by eqn. (*i*), we get

$$\frac{R_2}{R_1} = \frac{1 + \alpha_0\,t_2}{1 + \alpha_0\,t_1} = \frac{1 + \alpha_0\,t_2 + \alpha_0\,t_1 - \alpha_0\,t_1}{1 + \alpha_0\,t_1}$$

$$= \frac{1 + \alpha_0\,t_1}{1 + \alpha_0\,t_1} + \frac{\alpha_0\,t_2 - \alpha_0\,t_1}{1 + \alpha_0\,t_1}$$

$$= 1 + \frac{\alpha_0\,(t_2 - t_1)}{1 + \alpha_0\,t_1} = 1 + \frac{\alpha_0}{1 + \alpha_0\,t_1}\,(t_2 - t_1) \qquad ...(iv)$$

Comparing expressions (*iii*) and (*iv*), we get

$$1 + \alpha_1\,(t_2 - t_1) = 1 + \frac{\alpha_0}{1 + \alpha_0\,t_1}\,(t_2 - t_1)$$

or $\quad \alpha_1 = \dfrac{\alpha_0}{1 + \alpha_0\,t_1} = \dfrac{1}{\dfrac{1}{\alpha_0} + t_1}$ **Proved**

Example 1.38. A coil has resistance of 18 Ω when its mean temperature is 20°C and of 20 Ω when its mean temperature is 50°C. Find its mean temperature rise when its resistance is 21 Ω and the surrounding temperature is 15°C.

Solution : Let R_0 be the resistance of the coil at 0°C and α_0 be its temperature coefficient is resistance at 0°C.

Resistance at 20°C, $18 = R_0 (1 + 20\,\alpha_0)$ ...*(i)*

Resistance at 50°C, $20 = (1 + 50\,\alpha_0)$ ...*(ii)*

Dividing expression *(ii)* by expression *(i)*, we have

$$\frac{20}{18} = \frac{1 + 50\,\alpha_0}{1 + 20\,\alpha_0}$$

or $$\alpha_0 = \frac{1}{250} = 0.004$$

It t °C is the temperature of coil when its resistance is 21 Ω, then

$$21 = R_0 (1 + 0.004\,t) \qquad ...(iii)$$

Dividing expression *(iii)* by expression *(ii)*, we have

$$\frac{21}{20} = \frac{1 + 0.004\,t}{1 + 50 \times 0.004}$$

or $$t = 65°\text{ C}$$

Temperature rise = t – surrounding temperature = 65 – 15 = 50°C **Ans.**

EXERCISE

1. What do you understand by electric current ? Give its unit. Explain why the flow of electron in an electric circuit is opposite to the direction of conventional current ?
2. What are the factors affecting the resistance of a conductor ? How they affect its value.
3. Write short note on :
   (i) Resistance and conductance
   (ii) Conductor and insulator.
4. Define the following and mention their units :
   (i) Specific resistance
   (ii) Conductivity.
5. State and explain Ohm's law.
6. State Ohm's law. Give the three forms in which they may be expressed.
7. Explain briefly the limitations of Ohm's law.

8. What do you understand by temperature co-efficient of resistance ? Specify its unit. Name five materials whose resistance decreases with rise in temperature.

9. Define the resistance temperature co-efficient of a conductor. If α_1 is the resistance temperature co-efficient of a conductor at t_1 °C, show that the co-efficient at t_2 °C is given by :

$$\alpha_2 = \frac{1}{\frac{1}{\alpha_1} + (t_2 - t_1)}$$

10. What do you understand by electrical energy and electrical power ? Give their practical units.

PROBLEMS

1. The resistance of 360 m of a certain wire is 90 Ω. What length of the same wire will have a resistance of 125 Ω ? **[Ans.** 500 m]

2. A wire has a resistance of 6 Ω. What will be the resistance of another wire of the same material, three times as long and one half the cross-sectional area? **[Ans.** 36 Ω]

3. The resistance of the field coil of a dc machine at 15°C is 120 Ω. During its full load run, the resistance increases to 135 Ω. Find the average temperature of the field coil. Take the resistance-temperature coefficient to be 0.00401 per °C at 15°C. **[Ans.** 46.2°C]

4. The field winding of a shunt motor takes 5 A at a constant supply voltage of 460 V, when started from cold, the temperature then being 16°C. After a run of 5 hours, the field current reduces to 4.55 A. Calculate the temperature of the winding at the end of the run. Assume the value of resistance-temperature coefficient as 0.00428 per °C at 0°C. **[Ans.** 65.23°C]

5. The resistance of shunt field winding of a dc generator is 45 Ω at 20°C. At the end of 6 hours run, its resistance increases to 48.5 Ω. Find the average temperature of the winding at the end of the run. Take resistance-temperature coefficient 0.004 per °C at 0°C. **[Ans.** 41°C]

6. A resistor of 50 Ω has a p.d. of 100 volts d.c. across it for 1 hour. Calculate *(i)* power and *(ii)* energy. **[Ans.** *(i)* 200 watts, *(ii)* 7.2×10^5 J]

7. An incandescent lamp draws a current of 0.5 A at 250 V. What is the resistance of lamp ? **[Ans.** 500 Ω]

8. 100 W lamp is switched on to 230 V supply. What is the resistance of the lamp. **[Ans.** 529 Ω]

9. The potential difference of 10 volts is applied across a 2.5 Ω. Calculate the current, the power dissipated and energy consumed in 5 minutes. **[Ans.** 12 kWh]

10. The following are the details of load on a circuit connected through a supply meter.

(i) Six lamps of 40 W each working for 4 hours per day.

(ii) Two flourscent tubes 125 watts each working for 2 hours per day.

(iii) One 1000 watts heater working for 3 hours per day. If each unit of energy cost Rs. 1.25 what will be the electricity bill for the month of June ? **[Ans. Rs. 167.25]**

1.7. RESISTANCE IN SERIES AND PARALLEL

1.7.1. Series Circuits

When a number of resistors are connected end to end, so that they form only one path for the flow of current, then resistors are said to be connected in series and such circuits are knwon as *series circuits*.

Let resistors R_1, R_2 and R_3 be connected in series as shown in Fig. 1.8, and the potential difference of V volts be applied between extreme ends A and D to cause flow of current of I amperes through all the resistors R_1, R_2 and R_3.

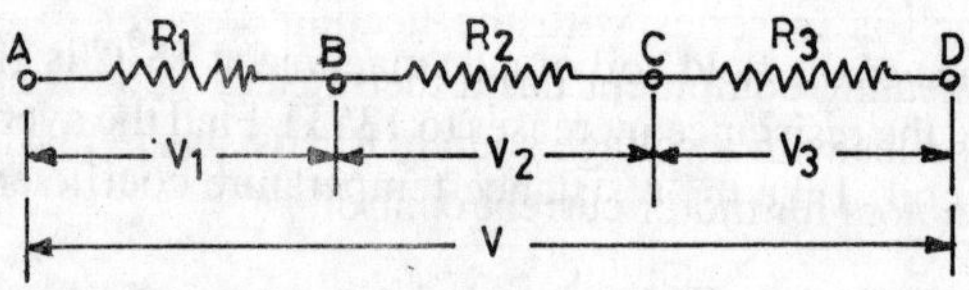

Fig. 1.8. Series circuit

Now according to Ohm's law

Voltage drop across resistor R_1, $V_1 = IR_1$

Voltage drop across resistor R_2, $V_2 = IR_2$

Voltage drop across resistor R_3, $V_3 = IR_3$

Voltage drop across whole circuit,

V = Voltage drop across resistor R_1 + voltage drop across resistor R_2 + voltage drop across resistor R_3

i.e., $V = IR_1 + IR_2 + IR_3 = I(R_1 + R_2 + R_3)$

or $\dfrac{V}{I} = R_1 + R_2 + R_3$ ...(i)

And according to Ohm's law $\dfrac{V}{I}$ gives the whole circuit resistance, say R

$\therefore$ Effective resistance of the series current, $R = R_1 + R_2 + R_3$ ..(ii)

Thus when a number of resistors are connected in series, the equivalent resistance is given by the arithmetic sum of their individual resistances.

i.e., $$R = R_1 + R_2 + R_3 + ... R_n$$

From the above discussions for a series circuit we conclude that

(i) same current flows through all parts of the circuit.

(ii) applied voltage is equal to the sum of voltage drops across the different parts of the circuit.

(iii) different resistors have their individual voltage drops.

(iv) voltage drop across individual resistor is directly proportional to its resistance, current being the same in each resistor.

(v) voltage drops are additive.

(vi) resistances are additive.

Series circuit are common in electrical equipment. The tube filaments in small radios are usually in series. Current controlling devices are wired in series with the controlled equipment. Fuses are in series with the equipment they protect. A thermostat switch is in series with the heating element in an electric iron. Automatic house-heating equipment has a thermostat, electro-magnet coils, and safety cut-outs in series with a voltage source. Rheostats are placed in series with the coils in large motors for motor current control.

Example 1.39. Four coils having resistance of 3, 5, 10 and 12 ohms are connected in series across 120 V. Determine *(i)* equivalent resistance of the circuit *(ii)* current flowing through the circuit and *(iii)* voltage drop across individual coils.

Solution : The circuit diagram is shown in Fig. 1.9.

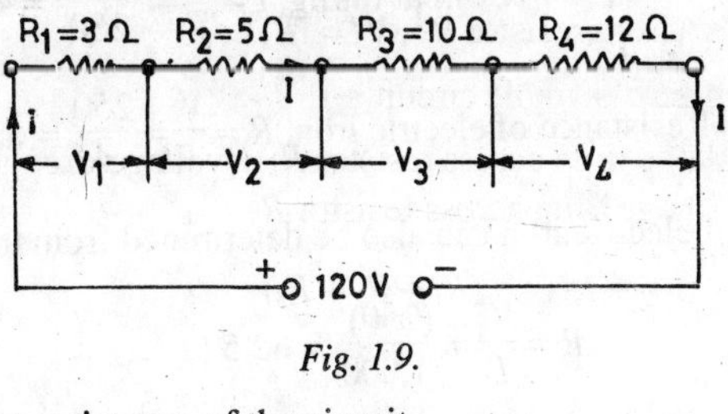

Fig. 1.9.

(i) Equivalent resistance of the circuit,

$$R = R_1 + R_2 + R_3 + R_4 = 3 + 5 + 10 + 12 = 30\ \Omega \quad \textbf{Ans.}$$

PD applied across the circuit, $V = 120$ volts

(ii) Current flowing through the circuit, $I = \frac{v}{R} = \frac{120}{30} = 4$ A **Ans.**

(iii) Voltage drop in 3 Ω resistance coil, $V_1 = IR_1 = 4 \times 3 = 12$ V **Ans.**

Voltage drop in 5 Ω resistance coil, $V_2 = IR_2 = 4 \times 5 = 20$ V **Ans.**

Voltage drop in 10 Ω resistance coil, $V_3 = IR_3 = 4 \times 10 = 40$ V **Ans.**

Voltage drop in 12 Ω resistance coil, $V_4 = IR_4 = 4 \times 12 = 48$ V **Ans.**

Example 1.40. A cell of emf 2 V and negligible internal resistance is connected in series with the resistance of 3, 5 and 2 ohms. Determine the current flowing through the circuit.

Solution : Equivalent resistance of the circuit,

$$R = R_1 + R_2 + R_3 = 3 + 5 + 2 = 10\ \Omega$$

Potential difference applied across the circuit, V = EMF of the cell = 2 V

Current flowing through the circuit,

$$I = \frac{V}{R} = \frac{2}{10} = 0.2 \text{ A } \textbf{Ans.}$$

Example 1.41. An electric iorn is rated as 1 Kw, 250 V. Calculate the current taken by it if it is connected to 220 V supply.

Solution : Power rating of electric iron, P = 1 Kw = 1,000 watts

Voltage rating, V = 250 V

$$\text{Current rating, } I = \frac{P}{V} = \frac{1000}{250} = 4 \text{ A}$$

$$\text{Resistance of electric iron, } R = \frac{V}{I} = \frac{250}{4} = 62.5\ \Omega$$

Resistance of electric iron can also be determined from the relation

$$R = \frac{V^2}{P} = \frac{(250)^2}{1,000} = 62.5\ \Omega$$

Current drawn by electric iron when connected across 220 V supply

$$I = \frac{V}{R} = \frac{220}{62.5} = 3.52 \text{ A } \textbf{Ans.}$$

Example 1.42. A 100 V, 60 watt bulb is to be operated from a 220 V supply. What is the resistance to be connected in series with the bulb to glow normally ?

Solution : Rated power of lamp, $P = 60$ W

Rated voltage of lamp, $V = 100$ V

Current drawn by the lamp, when operated on rated voltage, *i.e.*,

$$\text{Rated current } I = \frac{P}{V} = \frac{60}{100} = 0.6 \text{ A}$$

Lamp will operate normally on 220 V also if the current flowing through the lamp remains the rated current, *i.e.*, 0.6 A

Let the resistance connected in series with the lamp to make it glow normally to 220 V be of R ohms as shown in Fig. 1.10.

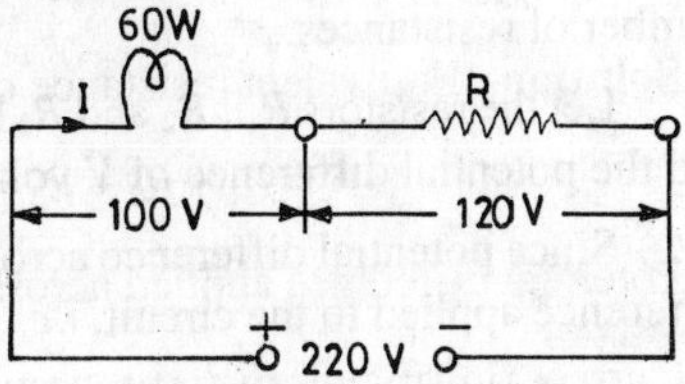

Fig. 1.10.

Now since the resistance R is in series with the lamp, therefore, the same current will flow through the resistance R, as in the lamp, *i.e.*, 0.6 A and voltage drop across series resistance R will be equal to supply voltage less voltage drop across the lamp (*i.e.*, rated voltage of the lamp).

or Voltage drop across the series resistace

$$IR = \text{Supply voltage} - \text{rated voltage of the lamp}$$

$$= 220 - 100 = 120 \text{ V}$$

or $$R = \frac{120}{I} = \frac{120}{0.6} = 200\ \Omega \quad \textbf{Ans.}$$

Example 1.43. Four lamps each rated for 230 V are connected in series and the combination is further connected across a 230 V supply. What will be the effect of connecting lamps in this manner on their illumination ?

Solution : Rated voltage of each lamp, $V = 230$ V

Assume that *(i)* all the four lamps are of same wattage rating (*i.e.* of same resistance) and *(ii)* lumens given out by the lamps vary as the fourth power of the voltage.

When the four lamps are connected in series across 230 V supply, the voltage across each lamp will be one-fourth of supply voltage, *i.e.*, $\frac{230}{4} = 57.5$ V

Now since lumens given out by the lamps varies as the fourth power of the voltage.

$\therefore$ The illumination due to each lamp will becomes $\left(\frac{57.5}{230}\right)^4 = \frac{1}{256}$ times of rated illumination. **Ans.**

1.7.2. Parallel Circuits

When a number of resistors are connected in such a way that one end of each of them is joined to a common point and the other ends being joined to another common point, as shown in Fig. 1.11 then resistors are said to be connected in parallel and such circuits are knwon as *parallel circuits*. In these circuits current is divided into as many paths as the number of resistances.

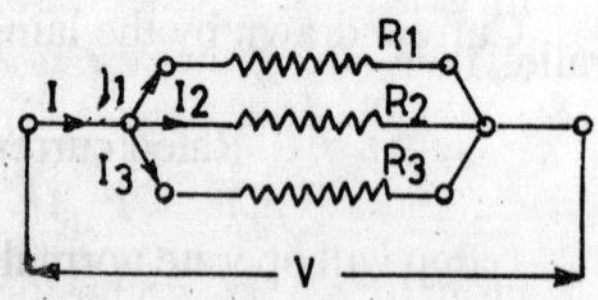

Fig. 1.11. Parallel circuits

Let the resistors R_1, R_2 and R_3 be connected in parallel as shown in Fig. 1.11, and the potential difference of V volts be applied across the circuit.

Since potential difference across each resistor is same and equal to potential difference applied to the circuit, *i.e.*, V

$\therefore$ According to Ohm's law

$$\text{Current in resistor } R_1,\ I_1 = \frac{V}{R_1} \qquad ...(i)$$

$$\text{Current in resistor } R_2,\ I_2 = \frac{V}{R_2} \qquad ...(ii)$$

$$\text{Current in resistor } R_3,\ I_3 = \frac{V}{R_3} \qquad ...(iii)$$

Adding expressions *(i)*, *(ii)* and *(iii)*, we have

$$I_1 + I_2 + I_3 = \frac{V}{R_1} + \frac{V}{R_2} + \frac{V}{R_3}$$

or $$I_1 + I_2 + I_3 = V\left(\frac{1}{R_1} + \frac{1}{R_2} + \frac{1}{R_3}\right) \qquad ...(iv)$$

and since $I_1 + I_2 + I_3 = I$, the total current flowing through the circuit

So $$I = I_1 + I_2 + I_3 = V\left(\frac{1}{R_1} + \frac{1}{R_2} + \frac{1}{R_3}\right)$$

or $$\frac{1}{V} = \frac{1}{R_1} + \frac{1}{R_2} + \frac{1}{R_3}$$

and since $\frac{I}{V} = \frac{I}{R}$ where R is the equivalent resistance of the whole circuit

$$\therefore \quad \frac{1}{R} = \frac{1}{R_1} + \frac{1}{R_2} + \frac{1}{R_3} \qquad ...(v)$$

Thus when a number of resistors are connected in parallel, the reciprocal of the equivalent resistance is given by the arithmetic sum of the reciprocals of their individual resistances.

In general if n resistors of resistances R_1, R_2, R_3 ... R_n are connected in parallel, then equivalent resistance R of the circuit is given by the expression

$$\frac{1}{R} = \frac{1}{R_1} + \frac{1}{R_2} + \frac{1}{R_3} + ... + \frac{1}{R_n} \qquad ...(vi)$$

From the above discussions for a parallel circuit we conclude that

(i) same voltage acts across all parts of the circuit,

(ii) different resistors (or branches) have their individual currents,

(iii) total circuit current is equal to the sum of individual currents through the various resistors (or branches),

(iv) branch currents are additive,

(v) the reciprocal of the equivalent or combined resistance is equal to the sum of the reciprocals of the resistances of the individual branches.

Parallel circuits are very common in use. Various lamps and appliances in a house are connected in parallel, so that each one can be operated independently. A series circuit is an "all or none" circuit, in which either every thing operates or nothing operates. For individual control, devices are wired in parallel.

Example 1.44. *(i)* Four coils having resistances of 2 , 5, 10 and 12 ohms respectively are connected in series across 120 V. Find the voltage drop in each coil and also the total current. *(ii)* If these coils are now connected in parallel across 120 volts, determine the current in each coil and the total current.

Solution : For part *(i)* refer example 1.40.

Voltage across each resistance = Voltage across the circuit = 120 V

Current flowing through 3 Ω resistance,

$$I_1 = \frac{V}{R_1} = \frac{120}{3} = 40 \text{ A} \quad \textbf{Ans.}$$

Current flowing through 5 Ω resistance,

$$I_2 = \frac{V}{R_2} = \frac{120}{5} = 24 \text{ A} \quad \textbf{Ans.}$$

Current flowing through 10 Ω resistance,

$$I_3 = \frac{V}{R_3} = \frac{120}{10} = 12 \text{ A} \quad \textbf{Ans.}$$

Current flowing through 12 Ω resistance,

$$I_4 = \frac{V}{R_4} = \frac{120}{12} = 10 \text{ A} \ \textbf{Ans.}$$

Total current, $I = I_1 + I_2 + I_3 + I_4 = 40 + 24 + 12 + 10 = 86$ A **Ans.**

Example 1.45. The equivalent resistance of four resistors joined in parallel is 20 Ω. The currents flowing through them are 0.6, 0.3, 0.2 and 0.1 ampere. Find the value of each resistor.

Solution : Equivalent resistance of the whole circuit $R = 20\ \Omega$

Total current flowing through the circuit,

$$I = I_1 + I_2 + I_3 + I_4 = 0.6 + 0.3 + 0.2 + 0.1 = 1.2 \text{ A}$$

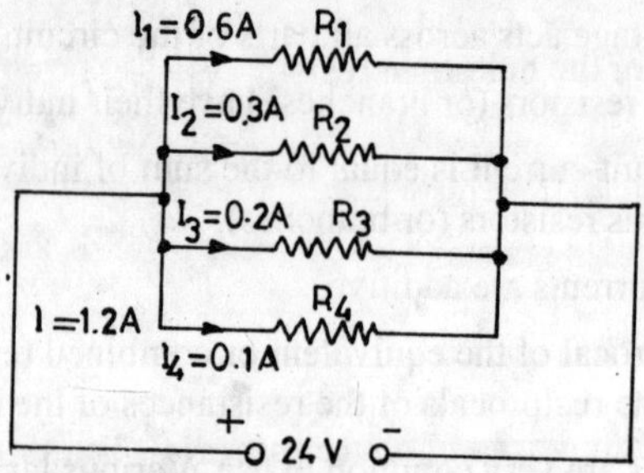

Fig. 1.12.

PD across the circuit, $V = IR = 1.2 \times 20 = 24$ V

$$\text{Resistance } R_1 = \frac{V}{I_1} = \frac{24}{0.6} = 40\ \Omega \ \textbf{Ans.}$$

$$\text{Resistance } R_2 = \frac{V}{I_2} = \frac{24}{0.3} = 80\ \Omega \ \textbf{Ans.}$$

$$\text{Resistance } R_3 = \frac{V}{I_3} = \frac{24}{0.2} = 120\ \Omega \ \textbf{Ans.}$$

$$\text{Resistance } R_4 = \frac{V}{I_4} = \frac{24}{0.1} = 240\ \Omega \ \textbf{Ans.}$$

Example 1.46. Two bulbs of 250 W, 230 V each, are connected in parallel across a 200 V supply. Calculate the total power drawn from the supply.

Solution : Wattage rating of bulbs, $P = 250$ watts

Voltage rating of bulbs, $V = 230$ V

$$\text{Resistance of each bulb, } R = \frac{V^2}{P} = \frac{(230)^2}{250} = 211.6\ \Omega$$

$$\text{Total resistance of parallel combination, } R' = \frac{R \times R}{R + R} = \frac{R}{2} = \frac{211.6}{2} = 105.8\ \Omega$$

Total power drawn from the supply $= \frac{(V')^2}{R'} = \frac{(200)^2}{105.8} = 387.07$ W **Ans.**

Example 1.47. Two resistances, one of 30 Ω and another of unknown value are connected in parallel, the total power dissipated in the circuit is 450 watts when the applied voltage is 90 volts. Find the value of unknown resistance.

Solution : Power dissipated in the circuit, $P = 450$ watts

Voltage applied across the circuit, $V = 90$ volts

Power dissipated in 30 Ω resistance

$$P_1 = \frac{V^2}{R_1} = \frac{(90)^2}{30} = 270 \text{ watts}$$

Power dissipated in the unknown resistance,

$$P_2 = P - P_1 = 450 - 270 = 180 \text{ watts}$$

The value of unknown resistance,

$$R_2 = \frac{V^2}{P_2} = \frac{(90)^2}{180} = 45\ \Omega \text{ Ans.}$$

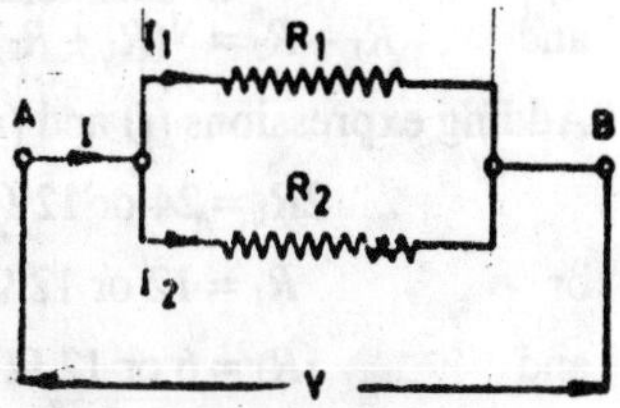

Fig. 1.13.

Example 1.48. Four wires of same material, the same cross- sectional area and the same length when placed in parallel give a total resistance of 0.25 Ω. If the same four wires are connected in series what will be the effective resistance ?

Solution : Since the length, area of cross-section and material of each wire are the same, therefore, resistance of each wire will be same, let it be R ohms.

When four wires, each of resistance R ohms are connected in parallel, the equivalent resistance will be $\frac{1}{1/R + 1/R + 1/R + 1/R} = \frac{R}{4}\ \Omega$

Since the total resistance of the circuit consisting of four similar wires is given as 0.25.

So $\frac{R}{4} = 0.25$

or $R = 4 \times 0.25 = 1\ \Omega$

When these wires are connected in seris, their equivalent resistance will become equal to $R + R + R + R$ *i.e.*, $4R$ or $4 \times 1 = 4\ \Omega$ **Ans.**

Example 1.49. Two coils connected in series have a resistance of 18 Ω and when connected in parallel have a resistance of 4 Ω. Find the resistance of each coil.

Solution : Let the resistance of the coils be R_1 and R_2 ohms.

Equivalent resistance when connected in series

$$R_1 + R_2 = 18\,\Omega \qquad \text{... (i)}$$

Equivalent resistance when connected in parallel

$$\frac{1}{\dfrac{1}{R_1} + \dfrac{1}{R_2}} = \frac{R_1 \times R_2}{R_1 + R_2} = 4\,\Omega \qquad \text{...(ii)}$$

Multiplying expressions *(i)* and *(ii)*, we get

$$R_1 R_2 = 72 \qquad \text{...(iii)}$$

and $$R_1 - R_2 = \sqrt{(R_1 + R_2)^2 - 4\,R_1 R_2} = \sqrt{18^2 - 4 \times 72} = \pm 6\,\Omega \qquad \text{...(iv)}$$

Adding expressions *(i)* and *(iv)* we get

$$2R_1 = 24 \text{ or } 12\,\Omega$$

or $R_1 = 12$ or $12\,\Omega$ **Ans.**

and $R_2 = 6$ or $12\,\Omega$ **Ans.**

1.7.3. Series–Parallel Circuits

So far, only simple series and simple parallel circuits have been considered. Practical electric circuits very often consists of combinations of series and parallel resistances. Such circuits may be solved by the proper application of Ohm's law and the rules for series and parallel circuits to the various parts of the complex circuit. There is no definite procedure to be followed in solving complex circuits, the solution depending on the known facts concerning the circuit and the quantities which one desires to find. One simple rule may usually be followed, however-reduce the parallel branches to an equivalent series branch and then solve the circuit as a simple series circuit.

For example, consider a series parallel circuit shown in Fig. 1.14 for solution.

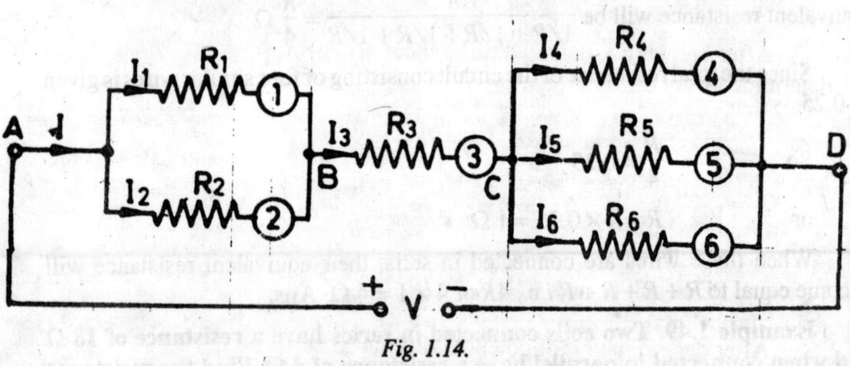

Fig. 1.14.

First of all equivalent resistances of all parallel branches are determined separately *e.g.*, of branches *AB* and *CD* by the laws of parallel circuits discussed in art 1.7.2.

Equivalent resistance of parallel branch *AB*,

$$R_{AB} = \frac{1}{\frac{1}{R_1} + \frac{1}{R_2}} = \frac{R_1 R_2}{R_1 + R_2}$$

and equivalent resistance of parallel branch *CD*,

$$R_{CD} = \frac{1}{\frac{1}{R_4} + \frac{1}{R_5} + \frac{1}{R_6}} = \frac{R_4 R_5 R_6}{R_5 R_6 + R_4 R_6 + R_4 R_5}$$

Now the circuit shown in Fig. 1.14 gets reduced to a simple series circuit shown in Fig. 1.15 consisting of three resistances,

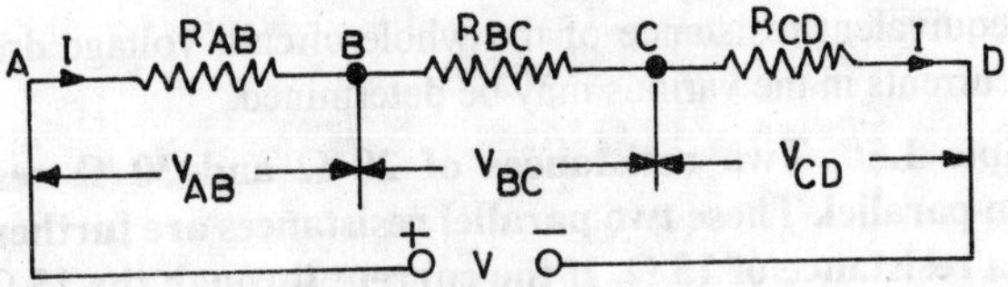

Fig. 1.15.

$$R_{AB} = \frac{R_1 R_2}{R_1 + R_2}, \quad R_{BC} = R_3$$

and $$R_{CD} = \frac{R_4 R_5 R_6}{R_5 R_6 + R_4 R_6 + R_4 R_5}$$

Total resistance of circuit $R_1 = R_{AB} + R_{BC} + R_{CD}$

Now circuit current may be determined from the relation

$$I = \frac{V}{R_T}$$

After knowing *I*, potential differences across branches *AB, BC* and *CD* are determined from the relations

PD across branch *AB*, $V_{AB} = IR_{AB} = \frac{V}{R_T} \cdot R_{AB}$

PD across branch *BC*, $V_{BC} = IR_{BC} = \frac{V}{R_T} \cdot R_{BC}$

and PD across branch *CD*, $V_{CD} = IR_{CD} = \frac{V}{R_T} \cdot R_{CD}$

After determination of potential difference across each parallel branch, the currents in the various resistances are determined from the relations

$$\text{Current in resistance } R_1 = I_1 = \frac{V_{AB}}{R_1}$$

$$\text{Current in resistance } R_2 = I_2 = \frac{V_{AB}}{R_2}$$

$$\text{Current in resistance } R_3 = I_3 = I$$

$$\text{Current in resistance } R_4 = I_4 = \frac{V_{CD}}{R_4}$$

$$\text{Current in resistance } R_5 = I_5 = \frac{V_{CD}}{R_5}$$

and $$\text{Current in resistance } R_6 = I_6 = \frac{V_{CD}}{R_6}$$

Thus equivalent resistance of the whole circuit, voltage drop across each branch and currents in the various may be determined.

Example 1.50. Two resistances of 20 Ω and 30 Ω respectively are connected in parallel. These two parallel resistances are further connected in series with a resistance of 15 Ω. If the current through the 15 Ω resistance is 3 A, find *(a)* the currents through the 20 Ω and 30 Ω resistances respectively, *(b)* the voltage across the whole circuit, *(c)* the total power consumed.

Solution : Equivalent resistance of branch AB, $R_{AB} = \dfrac{1}{\dfrac{1}{20}+\dfrac{1}{30}} = 12\ \Omega$

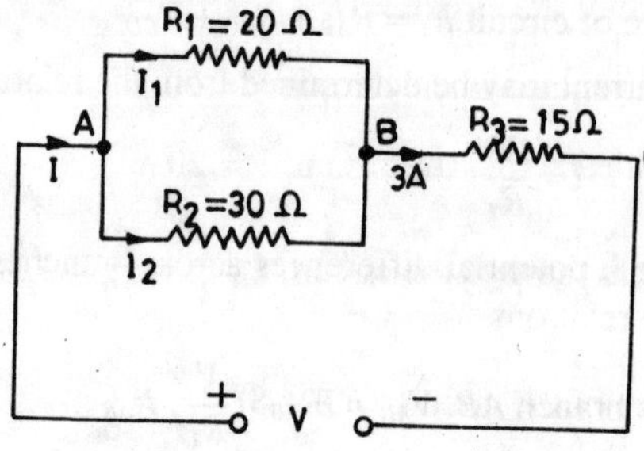

Fig. 1.16.

Effective resistance of the circuit, $R_{eff} = R_{AB} + R_{BC} = 12 + 15 = 27\ \Omega$

Circuit current I = Current through 15 Ω resistance = 3 A

(b) Voltage across the whole circuit, $V = IR_{eff} = 3 \times 27 = 81$ V **Ans.**

(c) Total power consumed, $P = VI = 81 \times 3 = 243$ watts **Ans.**

(a) Voltage drop across branch AB, $V_{AB} = IR_{AB} = 3 \times 12 = 36$ V

Current through 20 Ω resistance, $I_1 = \frac{V_{AB}}{R_1} = \frac{36}{20} = 1.8$ A **Ans.**

Current through 30 Ω resistance, $I_2 = \frac{V_{AB}}{R_1} = \frac{36}{30} = 1.2$ A **Ans.**

Example 1.51. A resistance *R* is connected in series with a parallel combination of two resistances 12 Ω and 8 Ω. Calculate *R* if the power dissipated in the circuit is 70 W when the applied voltage is 20 V across the cuicuit.

Solution : Equivalent resistance of the parallel circuit *AB*,

$$R_{AB} = \frac{12 \times 8}{12 + 8} = 4.8\ \Omega$$

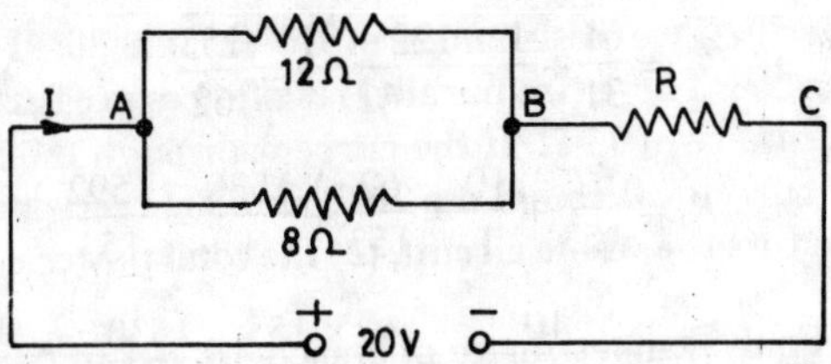

Fig. 1.17.

Total resistance of the circuit = (*R* + 4.8) ohms

Total power dissipated in the circuit, *P* = 70 W

Voltage applied across the circuit, *V* = 20 V

Current through the circuit, $I = \frac{P}{V} = \frac{70}{20} = 3.5$ A

Total resistance of the circuit, $(R + 4.8) = \frac{V}{I} = \frac{20}{3.5} = 5.714\ \Omega$

or $R = 5.714 - 4.8 = 0.914\ \Omega$ **Ans.**

Example 1.52. Calculate the current supplied by the battery in the following circuit.

Solution : $\frac{1}{3} + \frac{1}{5} + \frac{1}{2}$

$$\frac{1}{R_{AB}} = \frac{10 + 6 + 15}{30} = \frac{31}{30}\ \Omega$$

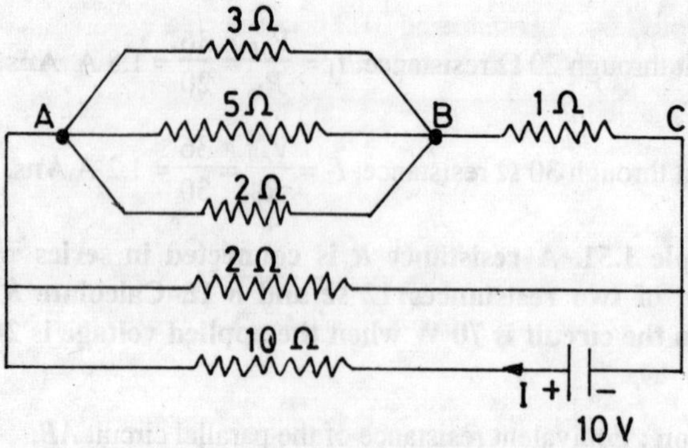

Fig. 1.18.

or $$R_{AB} = \frac{30}{31} + 1 = \frac{30 + 31}{31} = \frac{61}{31}\ \Omega$$

$$= \frac{61}{31} + \frac{1}{2} = \frac{122 + 31}{62} = \frac{153}{62}$$

or $$R_T = \frac{62}{153} + \frac{10}{1} = \frac{62}{153} + \frac{1530}{1} = \frac{1592}{153}$$

∴ $$I = \frac{10}{1592/153} = \frac{10 \times 153}{1592} = \frac{1530}{1592} = 0.906 \text{ A} \quad \textbf{Ans.}$$

Example 1.53. What is the value of an unknown resistor connected in series withy a resistor of 24 Ω, if the potential drop across 24 ohms resistor is 72 volts and that across the former is 48 volts ?

Solution :

$$\text{Current through } 24\ \Omega = I = \frac{V}{R} = \frac{72}{24} = 3 \text{ amps.}$$

$$\text{Resistor} = R = \frac{V}{I} = \frac{48}{3} = 16\ \Omega \quad \textbf{Ans.}$$

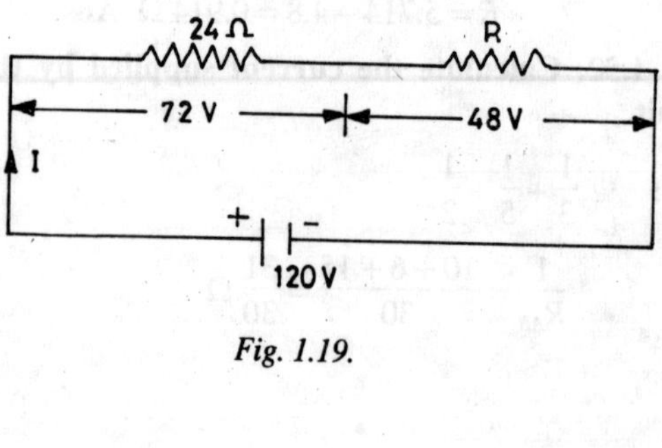

Fig. 1.19.

Example 1.54. A resistor of 10 Ω is connected in series with an unknown resistor *R*. Determine the value of unknown resistance if a current of 8 A flows through the circuit when connected across 200 volts supply mains.

Solution :

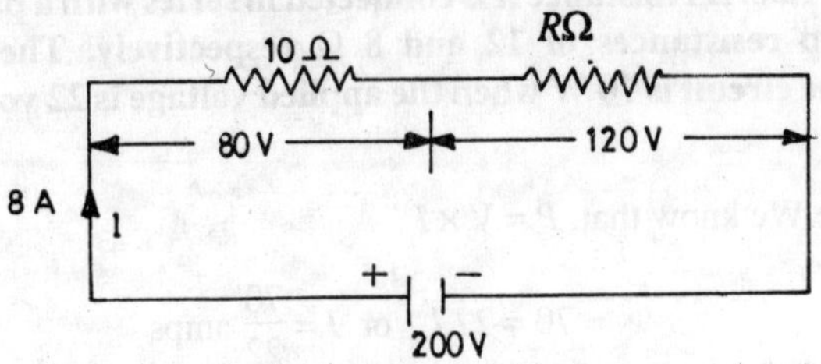

Fig. 1.20.

Voltage across 10 Ω, $V = IR = 10 \times 8 = 80$ volts

Voltage across $R = 200 - 80 = 120$ volts

$$\therefore \qquad R = \frac{V}{I} = \frac{120}{8} = 15\ \Omega \quad \textbf{Ans.}$$

Example 1.55. Two coils are connected in parallel across 100 V supply mains take 10 A from the line. The power dissipated in one coil is 600 W. What is the resistance of the other coil ?

Solution :

We know that, $P = V \times I_1$

$$600 = 100\ I_1$$

or $$I_1 = \frac{600}{100} = 6 \text{ amps.}$$

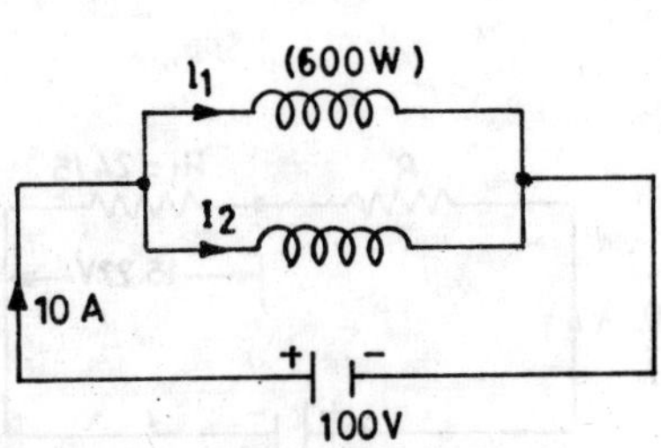

Fig. 1.21.

$$\therefore \quad I_2 = I - I_1 = 10 - 6 = 4 \text{ amps.}$$

$$\therefore \text{ Resistor,} \quad R_2 = \frac{V}{I} = \frac{100}{4} = 25\ \Omega \textbf{ Ans.}$$

Example 1.56. A resistance R is connected in series with a parallel circuit comprising two resistances of 12 and 8 Ω respectively. The total power dissipated in the circuit is 70 W when the applied voltage is 22 volts. Calculate R.

Solution : We know that, $P = V \times I$

$$70 = 22\,I \quad \text{or} \quad I = \frac{70}{22} \text{ amps}$$

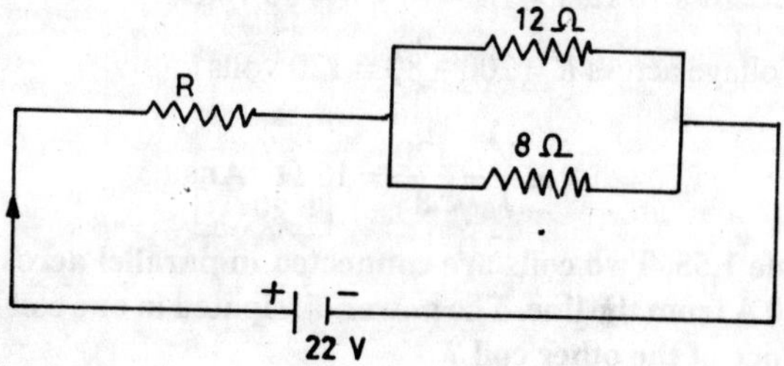

Fig. 1.22.

We know that, $\frac{1}{R} = \frac{1}{R_1} + \frac{1}{R_2}$

$$\frac{1}{R} = \frac{1}{12} + \frac{1}{8} = \frac{2+3}{24} = \frac{5}{24}$$

or $$R_1 = \frac{24}{5}\ \Omega$$

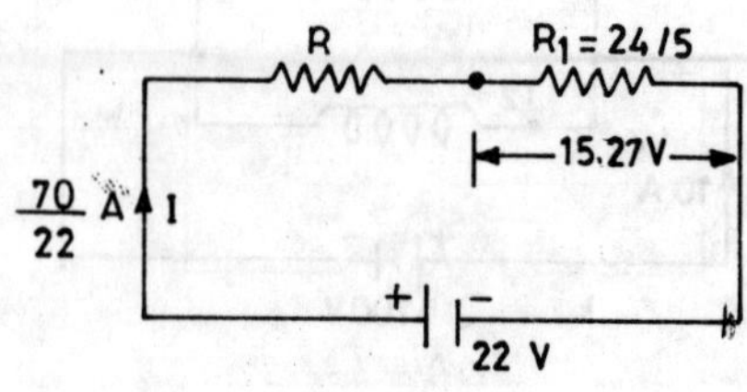

Fig. 1.23.

Now, voltage drop across R_1,

$$V = IR_1 = \frac{70}{22} \times \frac{24}{5} = 15.27 \text{ volts}$$

Now, voltage across, $R = 22 - 15.27 = 6.73$ volts

$$\therefore \qquad R = \frac{V}{I} = \frac{6.73 \times 22}{70} = 2.11\ \Omega \textbf{ Ans.}$$

Example 1.57. Find the resistance between the points *P* and *Q* in the series-parallel network shown in the Fig. 1.24.

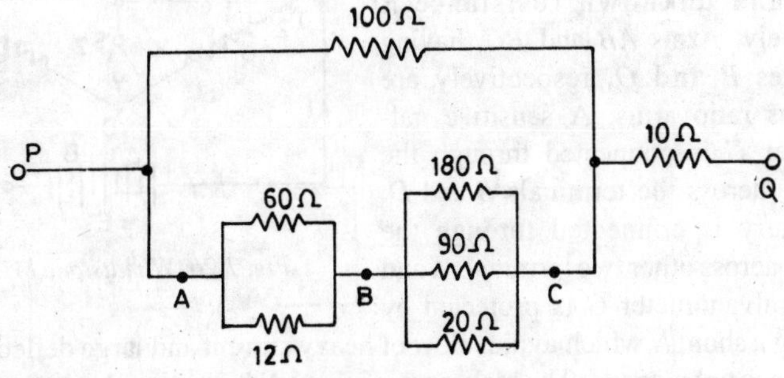

Fig. 1.24.

Solution : $R_{AB} = \frac{R_1 R_2}{R_1 + R_2} = \frac{60 \times 12}{60 + 12} = \frac{720}{72} = 10\ \Omega$

$$\frac{1}{R_{BC}} = \frac{1}{180} + \frac{1}{90} + \frac{1}{20} = \frac{12}{180}$$

$$R_{BC} = \frac{180}{12} = 15\ \Omega$$

$$R_{AC} = 10 + 15 = 25\ \Omega$$

$$R = \frac{100 \times 25}{100 + 25} = \frac{2500}{125} = 20\ \Omega$$

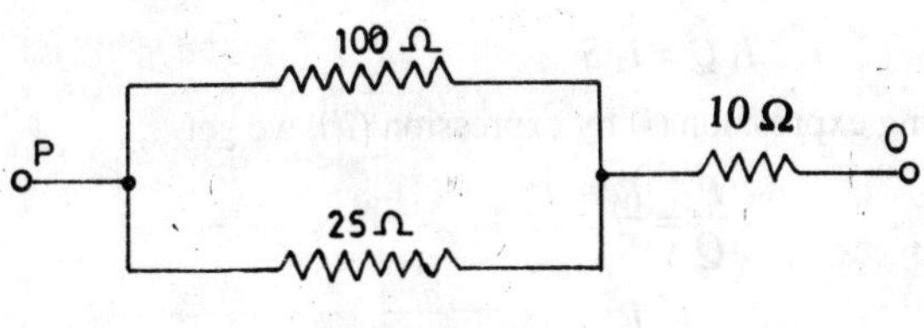

Fig. 1.25.

Fig. 1.25 (a)

Hence $R_{PQ} = 20 + 10 = 30\ \Omega$ **Ans.**

1.8. WHEATSTONE BRIDGE

The general arrangement of Wheatstone bridge is shwon in Fig. 1.26. It consists of four arms *AB, BC, CD* and *DA* consisting or known and fixed resistances *P* and *Q* known and variable resistanc *S* and unknown resistance *R* respectively. Arms *AB* and *BC*, having resistances *P* and *Q*, respectively are known as ratio arms. A sensitive galvanometer *G* is connected through the swtich S_2 across the terminals *B* and *D*. The battery is connected through the switch S_1 across other two terminals *A* and *C*. The galvanometer *G* is protected by providing a shunt *N* which avoids flow of heavy current and large deflection when the bridge is unbalanced. The resistance of shunt *N* is increased step by step as the bridge reaches the balance point so that more and more current may flow through the galvanometer and increases its sensitivity.

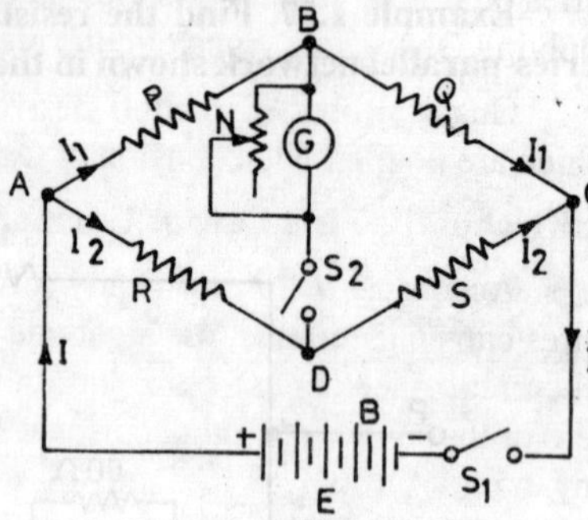

Fig. 1.26. Wheatstone bridge

While balancing the bridge, the switch S_1 is closed, the values of resistances *P, Q* and *S* are so adjusted that the deflection in galvanometer is nil when the switch S_2 is closed.

With the bridge balanced the current flowing through resistance *P* also flows through resistance *Q* (say I_1), the current flowing through resistance *R* also flows through resistance *S* (say, I_2) and the points *B* and *D* are at the same potential, *i.e.*, voltage drop from point *A* to point *B* must be equal to voltage drop from point *A* to point *D* and voltage drop from point *B* to point *C* must be equal to voltage drop from point *D* to point *C*.

Hence $$I_1 P = I_2 R \qquad ...(i)$$

$$I_1 Q = I_2 S \qquad ...(ii)$$

Dividing expression *(i)* by expression *(ii)*, we get

$$\frac{P}{Q} = \frac{R}{S}$$

or $$R = \frac{P}{Q} \times S \qquad ...(iii)$$

The resistances P and Q of the bridge are made of definite ratio as 1 : 1, 10 : 1 or 100 : 1 known as ratio arms and S, the rheostat arm is made continuously variable from 1 to 1,000 Ω or from 1 to 10,000 Ω.

1.9. LAWS GOVERNING DISTRIBUTION OF ELECTRIC CURRENT IN VARIOUS BRANCHES OF A NETWORK

Sometimes, complicated circuit cannot be reduced or simplified into a simple series or simple parallel or series-parallel circuit and cannot be solved simply by applying Ohm's law. Such circuits may be solved by applying *Kirchhoff's laws.*

Gustov Robert Kirchhoff derived the laws, the konw as *Kirchhoff's laws,* which are nothing but extension of Ohm's law.

1. Kirchhoff's First Law or Current Law (KCL) or Point Law

According to this law in any network of wires carrying currents, the algebraic sum of all currents meeting at a point (or junction) is zero or the sum of incoming currents towards any point is equal to the sum of outgoing currents away from that point. If I_1, I_2, I_3, I_4, I_5 and I_6 are the currents meeting at juntion O, flowing in the directions of arrow heads marked on them (Fig. 1.27), taking incoming currents as positive and outgoing currents as negative, according to Kirchhoff's first law (KCL).

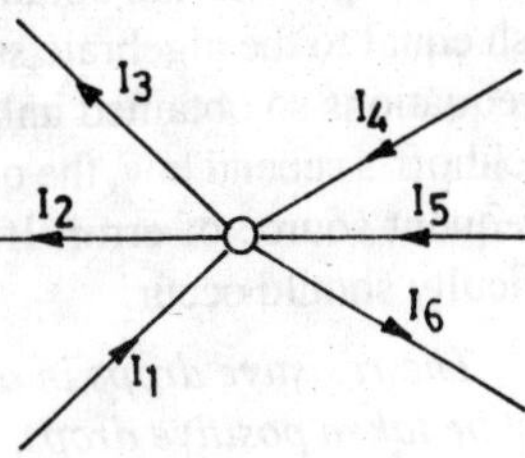

Fig. 1.27.

$$I_1 - I_2 - I_3 + I_4 + I_5 - I_6 = 0$$

or
$$I_1 + I_4 + I_5 = I_2 + I_3 + I_6$$

2. Kirchhoff's Second Law or Voltage Law (KVL) or Mesh Law

According to this law in any closed circuit or mesh the algebraic sum of emfs acting in that circuit or mesh is equal to the algebraic sum of the products of the currents and resistances of each part of the circuit.

If the circuit shown in Fig. 1.28 is considered, then according to Kirchoff's second law (or KVL).

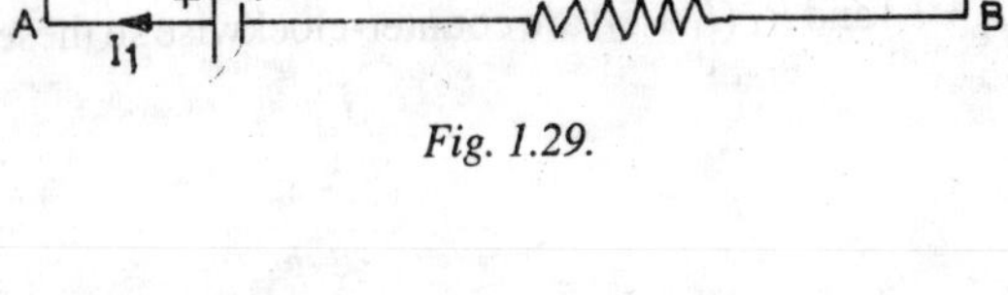

Fig. 1.29.

In mesh $AFCBA$, $E_1 = R_1 (I_1 + I_2) + R_2 (I_1 + I_2) + R_5 I_1$

or $E_1 = (R_1 + R_2) (I_1 + I_2) + R_5 I_1$

In mesh $FEDCF$, $-E_2 = -R_3 I_2 - R_4 I_2 - R_2 (I_1 + I_2) - R_1 (I_1 + I_2)$

or $E_2 = R_3 I_2 + R_4 I_2 + (R_1 + R_2) (I_1 + I_2)$

and in mesh $AFEDCBA$, $E_1 - E_2 = -R_3 I_2 - R_4 I_2 + R_5 I_1$

or $E_1 - E_2 = R_5 I_1 - (R_3 + R_4) I_2$

1.10. APPLICATION OF KIRCHHOFF'S LAW TO CIRCUIT SOLUTIONS

First of all the current distribution in various branches of the circuit is made with directions of their flow complying with first law of Kirchhoff. Then Kirchhoff's second law is applied to each mesh (one by one) separately and algebraic equations are obtained by equating the algebraic sum of emfs acting in a mesh equal to the algebraic sum of respective drops in the same mesh. By solving the equations so obtained unknown quantities can be determined. While applying Kirchhoff's seocnd law, the question of algebraic signs may be troublesome and is a frequent source of error. If, however, the following rules are kept in mind, no difficulty should occur.

The resistive drops in a mesh due to current flowing in clockwise direction must be taken positive drops.

The resistive drops in a mesh due to current flowing in counter-clockwise direciton must be taken as negative drops.

Similarly the battery emf causing current to flow in clockwise direction in a mesh must be taken as positive emf and the battery emf causing current to flow in counter-clockwise direction in a mesh must be taken as negative emf.

For example for the circuit shown in Fig. 1.28 let the current distribution be made as shown, which satisfy Kirchhoff's first law fully.

Taking first, mesh $AFCBA$ for the application of Kirchhoff's second law, wc see that there is only one emf acting in the mesh (E_1) and since it tries to send current in clockwise direction so E_1 be taken as positive, similarly all the resistive drops, *i.e.*, $R_1 (I_1 + I_2)$, $R_2 (I_1 + I_2)$ and $R_5 I_1$ are clockwise, so these must be taken as positive.

∴ According to Kirchhoff's second law in mesh $AFCBA$

$$E_1 = R_1 (I_1 + I_2) + R_2 (I_1 + I_2) + R_5 I_1$$

In mesh $FEDCF$, there is only one emf acting in the mesh (E_2) and since it tries to send current in counter- clockwise direction through the mesh under consideration so it may be taken as negative. Since all of the resistive drops $R_3 I_2$, $R_4 I_2$, $R_2 (I_1 + I_2)$ and $R_1 (I_1 + I_2)$ are counter-clockwise so these may be taken as negative.

Hence according to Kirchhoff's second law in mesh *FEDCF* we get

$$-E_2 = -R_3 I_2 - R_4 I_2 - R_2 (I_1 + I_2) - R_1 (I_1 + I_2)$$

or $$E_2 = (R_3 + R_4) I_2 + (R_1 + R_2)(I_1 + I_2)$$

In mesh *AFEDCBA*, emf E_1 tries to cause current in clockwise direction so be taken as positive and emf E_2 tries to cause current in counter-clockwise direction so be taken as negative. Similarly resistive drop $R_5 I_1$ being clockwise be taken as positive and resistive drops $R_3 I_2$ and $R_4 I_2$ being counter-clockwise be taken as negative.

Hence $$E_1 - E_2 = -R_3 I_2 - R_4 I_2 + R_5 I_1 = R_5 I_1 - (R_3 + R_4) I_2$$

Application of Kirchhoff's laws will be clear from the following workedout examples.

Example 1.58. State Kirchhoff's laws and find the current in the following circuit in each branch.

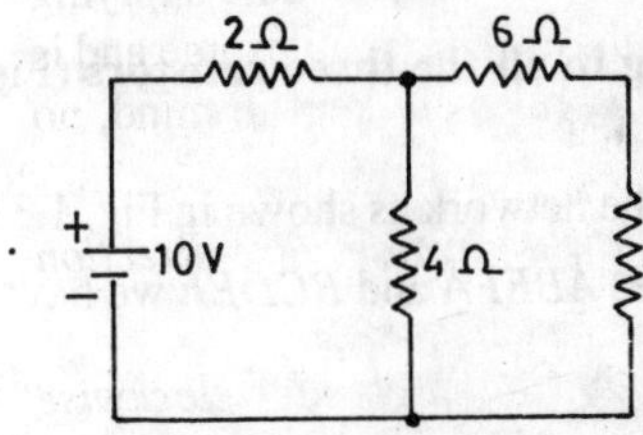

Fig. 1.29.

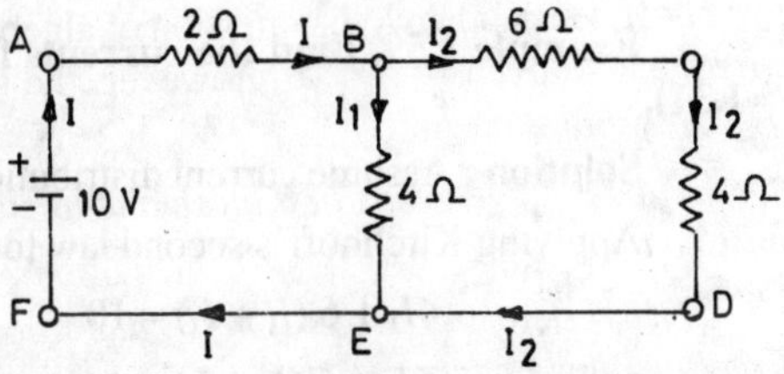

Fig. 1.30.

Solution : Assume current distribution in the network as shown in Fig. 1.30.

Applying Kirchhoff's first law at junction *B*

$$I = I_1 + I_2 \qquad ...(i)$$

Applying Kirchhoff's second law to meshes *ABEFA* and *BCDEB* we get

$$2I + 4I_1 = 10 \qquad ...(ii)$$

and $$6I_2 + 4I_2 - 4I_1 = 0$$

or $$10I_2 - 4I_1 = 0 \qquad ...(iii)$$

Substituting $I = I_1 + I_2$ from expression *(i)* in expression *(ii)* we have

$$2I_1 + 2I_2 + 4I_1 = 10$$

or $$6I_1 + 2I_2 = 10 \qquad ...(iv)$$

Multiplying expression *(iv)* by 5 we have

$$30I_1 + 10I_2 = 50 \qquad ...(v)$$

Subtracting expression *(iii)* from expression *(v)* we have

$$34I_1 = 50$$

or $$I_1 = \frac{50}{34} = \frac{25}{17} \text{ A}$$

Substituting $I_1 = \frac{25}{17}$ in expression *(iii)* we have

$$10I_2 - 4I_1 = 4 \times \frac{25}{17} = \frac{100}{17} \text{ A}$$

or $$I_2 = \frac{10}{17} \text{ A}$$

So current in branch *AB* (2 Ω-resistor), $I = I_1 + I_2 = \frac{25}{17} + \frac{10}{17} = \frac{35}{17}$ A **Ans.**

Current in branch *BE* (4 Ω-resistor), $I_1 = \frac{25}{17}$ A **Ans.**

Current in branch *BC* and *CD* (6 Ω and 4 Ω resistors), $I_2 = \frac{10}{17}$ A **Ans.**

Example 1.59. Find the currents flowing in all the three resistors (Fig. 1.31).

Solution : Assume current distribution in the network as shown in Fig. 1.31. Applying Kirchhoff's second law to meshes *ABEFA* and *BCDEF* we have

$$4I_1 + 6\,(I_1 + I_2) = 10 \quad \text{...(i)}$$

and $$5I_2 + 6\,(I_1 + I_2) = 14 \quad \text{...(ii)}$$

or $$10I_1 + 6I_2 = 10 \quad \text{...(iii)}$$

and $$6I_1 + 11I_2 = 14 \quad \text{...(iv)}$$

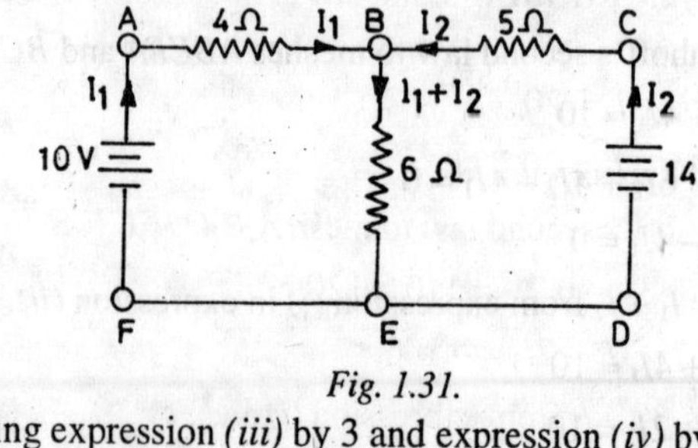

Fig. 1.31.

Multiplying expression *(iii)* by 3 and expression *(iv)* by 5 we have,

$$30I_1 + 18I_2 = 30 \quad \text{...(v)}$$

$$30I_1 + 55I_2 = 70 \quad \text{...(vi)}$$

Subtracting expression *(v)* from expression *(vi)* we have

$$37I_2 = 40$$

or $$I_2 = \frac{40}{37} \text{ A}$$

Substituting $I_2 = \frac{40}{37}$ in expression *(iii)* we have

$$I_1 = \frac{10 - 6I_2}{10} = 1 - \frac{6 \times 40}{10 \times 37} = \frac{13}{37} \text{ A}$$

So current in 4 Ω resistor $= I_1 = \frac{13}{37}$ A **Ans.**

Current in 5 Ω resistor $= I_2 = \frac{40}{37}$ A **Ans.**

Current in 6 Ω resistor $= I_1 + I_2 = \frac{13}{37} + \frac{40}{37} = 1\frac{16}{37}$ A **Ans.**

Example 1.60. Determine the value of R_1, R_2 and I_1, I_2 and I_3 for the following circuit (Fig. 1.32).

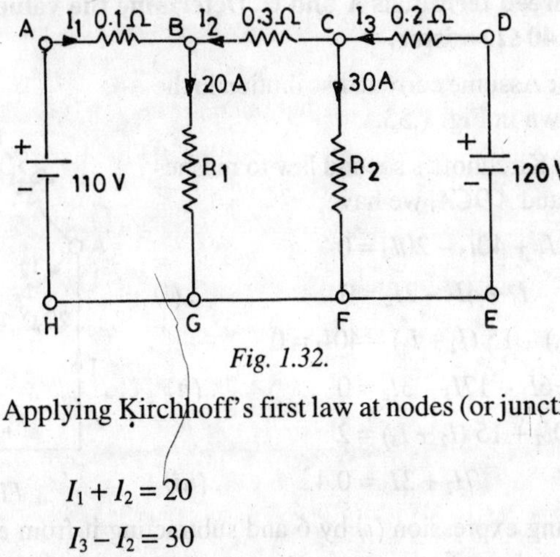

Fig. 1.32.

Solution : Applying Kirchhoff's first law at nodes (or junctions *B* and *C*) we have

$$I_1 + I_2 = 20 \quad ...(i)$$

and $$I_3 - I_2 = 30 \quad ...(ii)$$

Applying Kirchhoff's second law to mesh *A B C D E F G H A*, we have

$$0.1I_1 - 0.3I_2 - 0.2I_3 = 110 - 120$$

or $$I_1 - 3I_2 - 2I_3 = -100$$

Substituting, $I_1 = 20 - I_2$ from expression *(i)* and $I_3 = 30 + I_2$ from expression *(ii)* we have

$$20 - I_2 - 3I - 2(30 + I_2) = 100$$

or $$6I_2 = 60$$

or $$I_2 = 10 \text{ A } \textbf{Ans.}$$

From expressions *(i)* and *(ii)*, we have

$$I_1 = 20 - I_2 = 20 - 10 = 10 \text{ A} \;\; \textbf{Ans.}$$

and $$I_3 = 30 + I_2 = 30 + 10 = 40 \text{ A} \;\; \textbf{Ans.}$$

Applying Kirchhoff's second law to meshes *ABGHA*and*CEEFC,wehave*

$$0.1I_1 + 20R_1 = 110$$

and $$0.2I_3 + 30R_2 = 120$$

or $$R_1 = \frac{110 - 01I_1}{20} = \frac{110 - 0.1 \times 10}{20} = 5.45\ \Omega \;\; \textbf{Ans.}$$

and $$R_2 = \frac{120 - 0.2I_3}{30} = \frac{120 - 0.2 \times 40}{30} = 3.733\ \Omega \;\; \textbf{Ans.}$$

Example 1.61. A bridge network *ABCD* is arranged as follows :

Resistances between terminals *A-B, B-C, C-D, D-A* and *B- D* are 10, 30, 15, 20 and 40 Ω respectively. A 2-V battery negligible internal resistance is connected between terminals *A* and *C*. Determine the value and direction of the current in 40 Ω resistor.

Solution : Assume current distribution in the network as shown in Fig. 1.33.

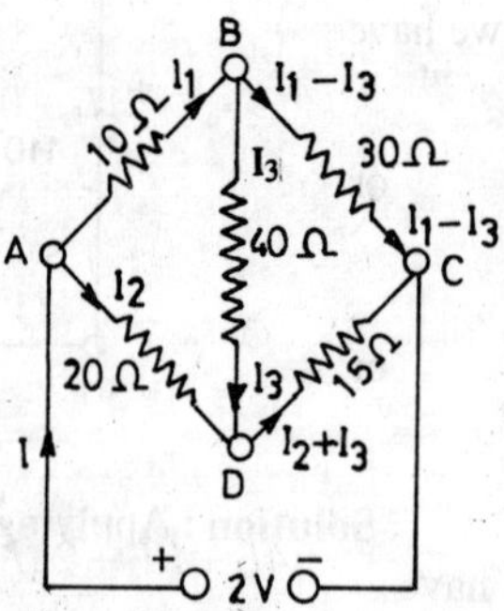

Fig. 1.33.

Applying Kirchhoff's second law to meshes *ABDA, BCDB* and *ADCA*, we have

$$10I_1 + 40I_3 - 20I_2 = 0$$

or $$I_1 + 4I_3 - 2I_2 = 0 \qquad ...(i)$$

$$30\,(I_1 - I_3) - 15\,(I_2 + I_3) - 40I_3 = 0$$

or $$6I_1 - 17I_3 - 3I_2 = 0 \qquad ...(ii)$$

and $$20I_2 + 15\,(I_2 + I_3) = 2$$

or $$7I_2 + 3I_3 = 0.4 \qquad ...(iii)$$

Multiplying expression *(i)* by 6 and subtracting it from expression *(ii)*, we have

$$41I_3 - 9I_2 = 0$$

or $$I_2 = \frac{41}{9} I_3$$

Substituting, $I_2 = \frac{41}{9} I_3$ in expression *(iii)* we have

$$7 \times \frac{41}{9} I_3 + 3I_3 = 0.4$$

or $$I_3 = \frac{0.4 \times 9}{314} \text{ A} = 11.465 \text{ mA}$$

i.e., Current flowing through a resistor of 40-Ω resistor is 11.465 m *A* from point *B* to point *D*. **Ans.**

Example 1.62. Calculate the value of unknown resistance *R* and the current flowing through it when the current in the branch *OC* is zero (see Fig. 1.34).

Solution : Let the current through the branches *AC* and *AO* be I_1 and I_2 amperes respectively. Applying Kirchhoff's first law to junctions *A*, *C* and *O* we have

Current supplied by battery or current in branch $BA = I_1 + I_2$

Current through branch *CB*
= Current through branch $AC = I_1$

∵ current through branch *OC* is zero.

Current through branch *OB*
= Current through branch $AO = I_2$

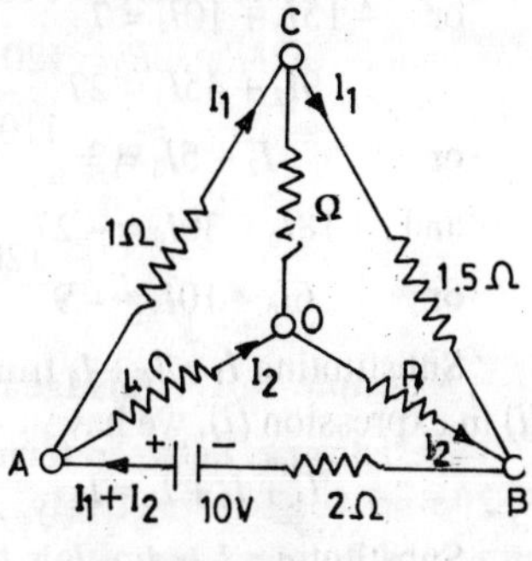

Fig. 1.34.

Now applying Kirchhoff's second law to meshes *ACOA*, *CBOC* and *AOBA*, we have

$$1I_1 + 0 - 4I_2 = 0$$

or $$I_1 = 4I_2 \quad ...(i)$$

$$1.5I_1 - RI_2 = 0 \quad ...(ii)$$

or $$R = \frac{1.5I_1}{I_2} = \frac{1.5 \times 4I_2}{I_2} = 6\ \Omega \ \textbf{Ans.}$$

∵ from expression *(i)* $I_1 = 4I_2$

$$4I_2 + RI_2 + 2\,(I_1 + I_2) = 10$$

Substituting $I_1 = 4I_2$ from expression *(i)* and $R = 6\ \Omega$ determined above we have

$$4I_2 + 6I_2 + 2\,(4I_2 + I_2) = 10$$

or $$20I_2 = 10$$

or $$I_2 = 0.5 \text{ A} \ \textbf{Ans.}$$

Example 1.63. Apply Kirchhoff's laws and find the currents in the branches of network of Fig. 1.35.

Solution : Applying Kirchhoff's first law to junctions *B*, *G*, *F* and *D* we have

$$I_1 + I_3 = I_2 \quad ...(i)$$

$$I_3 + I_4 = I_5 \quad ...(ii)$$

$$I_4 + I_6 = I_1 \quad ...(iii)$$

$$I_5 + I_6 = I_2 \quad ...(iv)$$

Applying Kirchhoff's second law to meshes *ABGFHA, BCDGB* and *GDEFG* we have

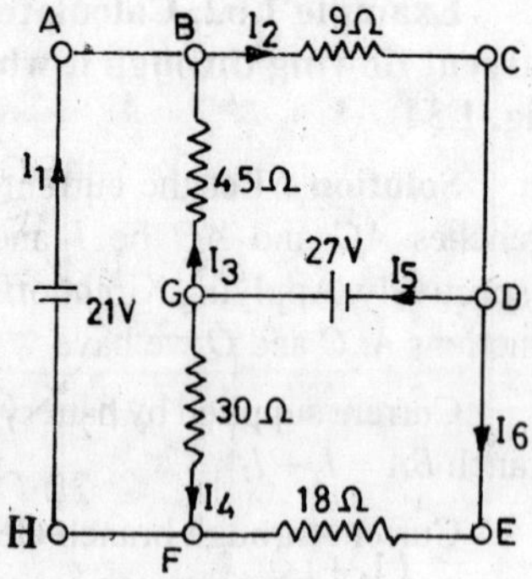

Fig. 1.35.

$$-45I_3 + 30I_4 = 21$$

or $$-15I_3 + 10I_4 = 7 \quad ...(v)$$

$$9I_2 + 45I_3 = 27$$

or $$I_2 + 5I_3 = 3 \quad ...(vi)$$

and $$18I_6 - 30I_4 = -27$$

or $$6I_6 - 10I_4 = -9 \quad ...(vii)$$

Substituting $I_1 = I_4 + I_6$ from expression *(iii)* in expression *(i)*, we have

$$I_4 + I_6 + I_3 = I_2 \quad ...(viii)$$

Substituting, $I_2 = I_4 + I_6 + I_3$ from expression *(viii)* in expression *(vi)* we have

$$6I_3 + I_4 + I_6 = 3 \quad ...(ix)$$

Substituting $I_3 = \dfrac{2I_4 - 1.4}{3}$ from expression (*v*) and $I_6 = \dfrac{-9 + 10I_4}{6}$ from expression *(vii)* in expression *(ix)* we have

$$\left(\frac{2I_4 - 1.4}{3}\right) \times 6 + I_4 + \frac{-9 + 10I_4}{6} = 3$$

or $$I_4 = \frac{43.8}{40} = 1.095 \text{ A} \quad \textbf{Ans.}$$

Substituting $I_4 = 1.095$ A in expressions (*v*) and (*vii*) we have

$$I_3 = \frac{2I_4 - 1.4}{3} = \frac{2 \times 1.095 - 1.4}{3} = 0.26333 \text{ A} \quad \textbf{Ans.}$$

$$I_6 = \frac{-9 + 10I_4}{6} = \frac{-9 + 10 \times 1.095}{6} = 0.325 \text{ A} \quad \textbf{Ans.}$$

From expression *(ii)*, $I_5 = I_3 + I_4 = 0.26333 + 1.095 = 1.35833$ A **Ans.**

From expression *(iii)*, $I_1 = I_4 = I_6 = 1.095 + 0.325 = 1.42$ A **Ans.**

From expression *(i)*, $I_2 = I_1 + I_3 = 1.42 + 0.26333 = 1.68333$ A **Ans.**

Example 1.64. The four arms of a wheatstone bridge are as under : arm *AB* a resistance of 15 Ω; arm *BC*, a resistance of 25 Ω; arm *CD*, a resistance of 30 Ω; and arm *DA*, a resistance of 20 Ω. A battery with an e.m.f. of 2V and internal resistance of 0.5 Ω is connected across *A* and *C* and a galvanometer having a resistance of 30 Ω is connected across *B* and *D*. Find the current flowing in the galvanometer.

Solution :

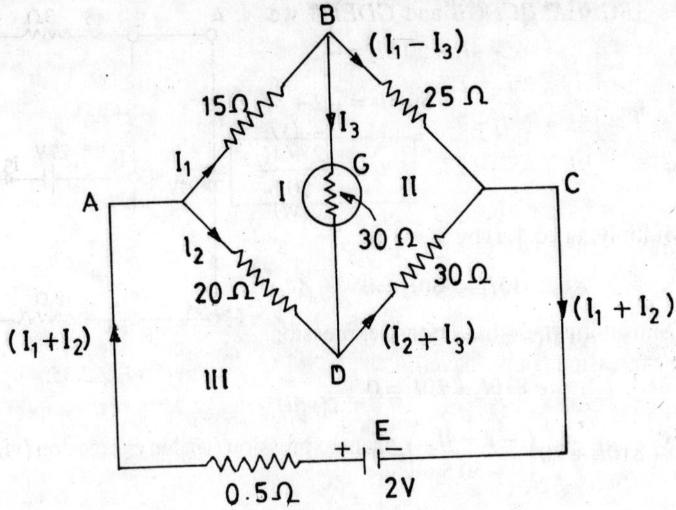

Fig. 1.36.

Ist mesh $ABDA$, $\quad -15I_1 - 30I_3 + 20I_2 = 0$ ...(i)

IInd mesh $BCDB$, $\quad -25(I_1 - I_3) + 30(I_2 + I_3) + 30I_3 = 0$

$$-25I_1 + 25I_3 + 30I_2 + 30I_3 + 30I_3 = 0$$

$$-25I_1 + 30I_2 + 85I_3 = 0 \qquad ...(ii)$$

IIIrd mesh $ADCEA$, $\quad -20I_2 - 30(I_2 + I_3) + 2 - 0.5(I_1 + I_2) = 0$

$$-20I_2 - 30I_2 - 30I_3 + 2 - 0.5I_1 - 0.5I_2 = 0$$

$$0.5I_1 - 50.5I_2 - 30I_3 = -2 \qquad ...(iii)$$

Multiplying eq. *(i)* by 3 and eq. *(ii)* by 2, we get

$$-45I_1 - 90I_3 + 60I_2 = 0 \qquad .(iv)$$

$$-50I_1 + 170I_3 + 60I_2 = 0 \qquad ...(v)$$

$$5I_1 - 260I_3 = 0 \qquad ...(vi)$$

$\therefore$ $$5I_1 = 260I_3$$

$$\boxed{I_1 = 52I_3}$$

Put this value in eq. *(iii)*, we get

$$0.5\,(52I_3) - 50.5I_2 - 30I_3 = -2$$

$$26I_3 - 50.5I_2 - 30I_3 = -2$$

$$-4I_3 - 50.5I_2 = -2$$

$$-50.5I_2 = -2 + 4I_3$$

or
$$\boxed{I_2 = \frac{-2 + 4I_3}{-50.5}}$$

Multiplying eq. *(vi)* by 3, we get

$$15I_1 - 780I_3 = 0 \qquad \text{...(vii)}$$

$$-15I_1 - 30I_3 + 20I_2 = 0$$

$$-810I_3 + 20I_2 = 0$$

$$-810I_3 + 20\left(\frac{-2 + 4I_3}{-50.5}\right) = 0$$

$$20\left(\frac{4I_3 - 2}{-50.5}\right) = 810I_3$$

or
$$\frac{80I_3 - 40}{-50.5} = 810I_3$$

or
$$80I_3 - 40 = -40905I_3$$

$$40985I_3 = 40$$

$\therefore$
$$I_3 = \frac{40}{40985} = \frac{8}{8197}\text{ A Ans.}$$

Example 1.65. The four arms of Wheatstone bridge have the following resistances :

***AB* = 2 Ω, *BC* = 6 Ω, *CD* = 8 Ω and *DA* = 3 Ω. There is a battery of 10 V and 4 Ω resistance in series with it connected across *B* and *D*. Calculate *(i)* current in battery, *(ii)* current in each branch.**

Solution : Ist mesh *BDAB*

$$-4\,(I_1 + I_2) - 3I_2 - 2I_2 = -10$$

$$-4I_1 - 4I_2 - 5I_2 = -10$$

$$-4I_1 - 9I_2 = -10$$

or
$$4I_1 + 9I_2 = 10 \qquad \text{...(i)}$$

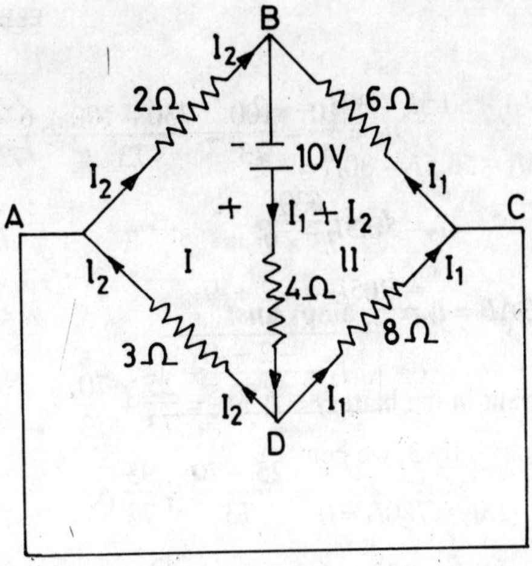

Fig. 1.37.

IInd mesh *BDCB*

$$-4(I_1 + I_2) - 8I_1 - 6I_1 + 10 = 0$$

$$-4I_1 - 4I_2 - 14I_1 + 10 = 0$$

$$-18I_1 - 4I_2 = -10$$

or $$18I_1 + 4I_2 = 10 \quad ...(ii)$$

Multiply eq. *(i)* by 4 and eq. *(ii)* by 9, we get

$$16I_1 + 36I_2 = 40 \quad ...(iii)$$

$$162I_1 + 36I_2 = 90 \quad ...(iv)$$

$$-146I_1 = -50$$

or $$I_1 = \frac{50}{146} = \frac{25}{73}$$

$\therefore$ Current in $DCB = I_1 = \frac{25}{73}$ amp **Ans.**

Now put the value of I_1 in eq. *(i)*, we get

$$4\left(\frac{25}{73}\right) + 9I_2 = 10$$

$$\frac{100}{73} + 9I_2 = 10$$

or $$9I_2 = \frac{10}{1} - \frac{100}{73} = \frac{730 - 100}{73} = \frac{630}{73}$$

$\therefore$ $$I_2 = \frac{630}{73 \times 9}$$

$\therefore$ Current in $DAB = I_2 = \frac{70}{73}$ amp **Ans.**

Now, total current in the battery $= I_1 + I_2 = \frac{25}{73} + \frac{70}{73}$

$$= \frac{25 + 70}{73} = \frac{95}{73} \text{ A}$$

$\therefore$ $$I = I_1 + I_2 = \frac{95}{73} \text{ A} \quad \textbf{Ans.}$$

Example 1.66. An approximately balanced Wheatstone bridge network is constructed as follows :

Two equal resistances of 100 Ω, *AB, BC* are connected in series and another two, *AD, DC* of 200, 201 Ω respectively, connected in series with each other but as a whole in parallel with *ABC*. A 2-volt accumulator of negligible resistance is connected to *A, C* and a galvanometer of resistance 50 Ω across *BD*. Find the current through the galvanometer.

Solution : Find $I_3 = ?$

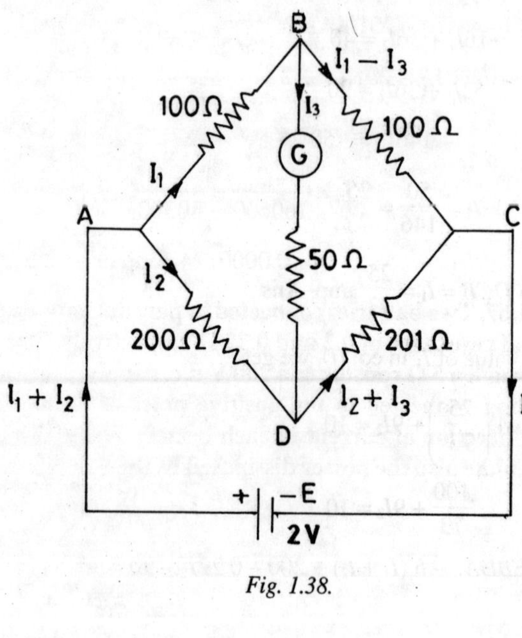

Fig. 1.38.

In mesh *ABDA*,

$$-100I_1 - 50I_3 + 200I_2 = 0 \quad ...(i)$$

In mesh *BCDB*,

$$-100\,(I_1 - I_3) + 201\,(I_2 + I_3) + 50I_3 = 0$$

$$-100I_1 + 100I_3 + 201I_2 + 201I_3 + 50I_3 = 0$$

$$-100I_1 + 201I_2 + 351I_3 = -2 \quad ...(ii)$$

In mesh *ADCEA*,

$$-200I_2 - 201I_2 - 201I_3 + 2 = 0$$

$$-401I_2 - 201I_3 = -2 \quad ...(iii)$$

$$-100I_1 + 200I_2 - 50I_3 = 0$$

$$-100I_1 + 201I_2 + 351I_3 = 0$$

$$-I_2 - 401I_3 = 0$$

$$-I_2 = 401I_3$$

$$\boxed{I_2 = -401\,I_3}$$

Put this value in eqn. *(iii)*, we get

$$-401\,(-401I_3) - 201I_3 = -2$$

$$160801I_3 - 201I_3 = -2$$

$$160600I_3 = -2$$

$$\therefore \quad I_3 = \frac{-2}{160600} = \frac{-1}{80300}$$

$$= -0.0000124 \text{ Amp or } -12.5\ \mu\text{A} \quad \textbf{Ans.}$$

Example 1.67. Two batteries connected in parallel have e.m.f's of 80 and 90 V and internal resistance of 0.2 and 0.22 Ω respectively. The combination is connected through a 5 Ω resistor to a 200 V d.c. supply, the positive poles of the batteries being connected to the positive poles of the supply. Find the magnitude and direction of current in each battery and of the current from the supply. Calculate also the power dissipated in the 5 Ω resistor.

Solution :

In mesh *AEBDA*, $-5\,(I_1 + I_2) + 200 - 0.22I_2 - 90 = 0$

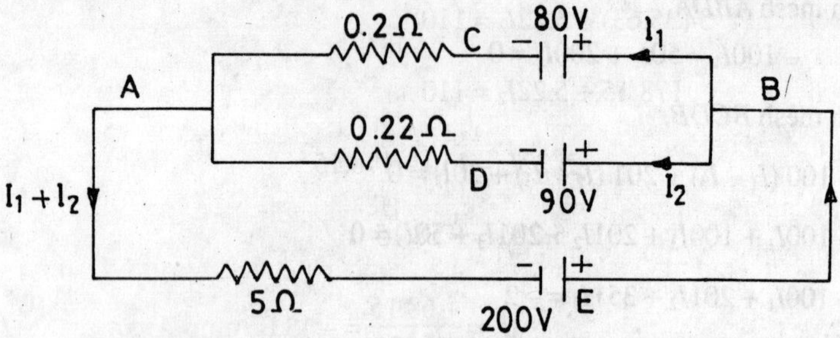

Fig. 1.39.

$$-5I_1 - 5I_2 + 110 - 0.22I_2 = 0$$

$$-5I_1 - 5.22I_2 = -110$$

or $$5I_1 + 5.22I_2 = 110 \quad ...(i)$$

In mesh *AEBCA*, $-5(I_1 + I_2) + 200 - 0.2I_1 - 80 = 0$

$$-5I_1 - 5I_2 + 120 - 0.2I_1 = 0$$

$$-5.2I_1 - 5I_2 = -120$$

or $$5.2I_1 + 5I_2 = 120 \quad ...(ii)$$

$$5I_2 = 120 - 5.2I_1$$

$$\boxed{I_2 = \frac{120 - 5.2I_1}{5}}$$

Put this value in eq. *(i)*, we get

$$5I_1 + 5.22\left(\frac{120 - 5.2I_1}{5}\right) = 110$$

$$\frac{5I_1}{1} + \left(\frac{626.4 - 27.144I_1}{5}\right) = 110$$

$$25I_1 + 626.4 - 27.144I_1 = 550$$

$$-2.144I_1 = 550 - 626.4$$

$$-2.144I_1 = -76.4$$

$$\therefore \quad I_1 = \frac{76.4}{2.144} = 35.63 \text{ amp} \quad \textbf{Ans.}$$

Put this value in eq. *(i)*, we get

$$5\,(35.63) + 5.22 I_2 = 110$$

$$178.15 + 5.22 I_2 = 110$$

$$5.22 I_2 = 110 - 178.15$$

$$5.22 I_2 = -\,68.15$$

$$\therefore \qquad I_2 = \frac{-\,68.15}{5.22} = -\,13.05 \text{ amp } \textbf{Ans.}$$

Example 1.68. Two batteries are connected in parallel. The emf and internal resistance of one are 110 V and 6 Ω respectively and corresponding values for other are 130 V and 4 Ω respectively. A resistance of 20 Ω is connected across parallel combination. Calculate *(i)* the value and direction of current in each battery, *(ii)* the terminal voltage.

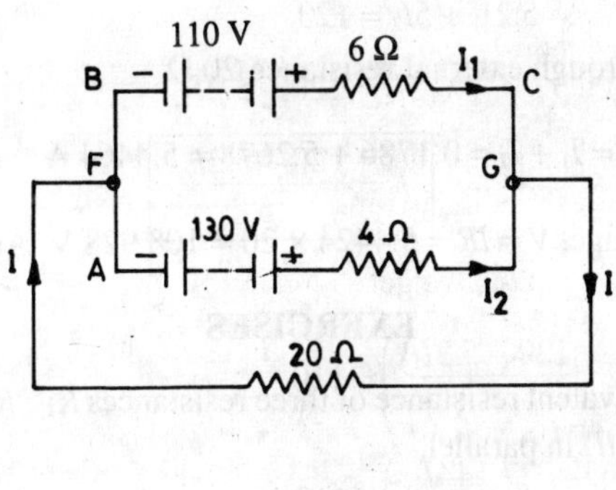

Fig. 1.40.

Solution : Let the current supplied by batteries be I_1 and I_2 respectively.

Current through external resistance $I = I_1 + I_2$

Apply Kirchhoff's second law to meshes *ABCDA* and *FBCGF*, we get

$$6I_1 - 4I_2 = 110 - 130$$

or $\qquad 2I_2 - 3I_1 = 10 \qquad$ ...*(i)*

$$6I_1 + 20\,(I_1 + I_2) = 110$$

or $$13I_1 + 10I_2 = 55 \qquad ...(ii)$$

Multiplying expression *(i)* by 5, we get

$$10I_2 - 15I_1 = 50 \qquad ...(iii)$$

Subtracting expression *(iii)* from expression *(ii)*, we get

(i) $$I_1 = \frac{5}{28} = 0.1786 \text{ A} \quad \textbf{Ans.}$$

$$I_2 = \frac{10 + 3I_1}{2} \qquad \text{From expression } (i)$$

or $$I_2 = \frac{1}{2}\left(10 + 3 \times \frac{5}{28}\right)$$

$$= \frac{295}{56} = 5.2678 \text{ A}$$

(ii) Current through external resistance 20 Ω

$$I = I_1 + I_2 = 0.1786 + 5.2678 = 5.4464 \text{ A}$$

Terminal voltage, $V = IR = 5.4424 \times 20 = 108.928$ V **Ans.**

EXERCISES

1. Compute equivalent resistance of three resistances R_1, R_2 and R_3 connected in *(i)* series ; *(ii)* in parallel.
2. Find the equivalent resistance for the parallel combination of resistances.
3. Disucss with suitable diagram the principle of operation of Wheatstone bridge and determine the condition of balance.
4. State Kirchhoff's laws as applied to an electric circuit.
5. State Kirchhoff's current and voltage laws. Explain them with suitable circu:*

PROBLEMS

1. A circuit is made of a 0.4 Ω wire, a 150 Ω bulb and a 120 Ω rheostat connected in series. Determine the total resistance of the circuit. **[Ans.** 270.4 Ω]

2. A 100 W, 250 V lamp is connected in series with a 100 W, 200 V lamp across 250 V supply mains. Calculate *(i)* the current drawn from the supply mains and *(ii)* voltage across each lamp. Assume lamp resistance constant.

$$\left[\textbf{Ans. } \frac{10}{41}\text{ A} ; \frac{6{,}250}{41}\text{ V , and } \frac{4{,}000}{41}\text{ V}\right]$$

3. The equivalent resistance of four resistors joined in parallel is 30 Ω. The currents flowing through them are 4, 2, $\frac{4}{3}$ and $\frac{2}{3}$ amperes. Determine the value of each resistor. **[Ans.** 60 Ω; 120 Ω; 180 Ω and 360 Ω]

4. A current of 78 A is divided into three branches. The lengths of wire in the three branches are in the ratio of 8, 12 and 16. Th wires are of same material and cross-sectional area. Calculate the current in each branch. **[Ans.** 36 A, 24 A and 18 A]

5. If a resistor is to dissipate energy at the rate of 250 W, find the resistance for terminal voltage of 100 V. **[Ans.** 40 Ω]

6. Four resistors *A, B, C* and *D* are connected in parallel. If *A* is of 56 Ω; *B* is of 50 Ω; *C* is of 40 Ω and takes a current of 2.5A and the total current drawn from the supply is 9.0 A. Determine the resistance of *D* in ohms. **[Ans.** 36.84 Ω]

7. Two resistors of equal values when placed in parallel gives an effective resistance equal to *R*. What will be the effective resistance if the two resistors are connected in series. **[Ans.** 4 *R*]

8. A resistor of value 1.6 Ω is connected in series with two resistors of value 4 Ω and 6 Ω which are connected in parallel. Find the total resistance of the series parallel combination. **[Ans.** 4 Ω]

9. The resistance of two wires is 16 Ω when connected in series and 3 Ω when connected in parallel. Calculate the resistance of each wire. **[Ans.** 12 Ω and 4 Ω]

10. Two resistances of 4 Ω and 6 Ω in parallel are in series with another resistor of 12 Ω. If the current flowing in 12 Ω resistor is 2 A, determine *(i)* the current flowing through 4 Ω and 6 Ω resistors, *(ii)* voltage across the whole circuit and *(iii)* power drawn from the supply. **[Ans.** *(i)* 1.2 A and 0.8 A; *(ii)* 28.8 V; and *(iii)* 57.6 watts]

11. In the circuit shown in Fig. 1.41, find the voltage across and the current in each element.

[**Ans.** Current through 2 Ω and 4 Ω resistors = 2 A and voltage across these resistors are 4 V and 8 V respectively. Current through 4 Ω and 8 Ω resistors = 1 A and voltage across them are 4 V and 8 V respectively]

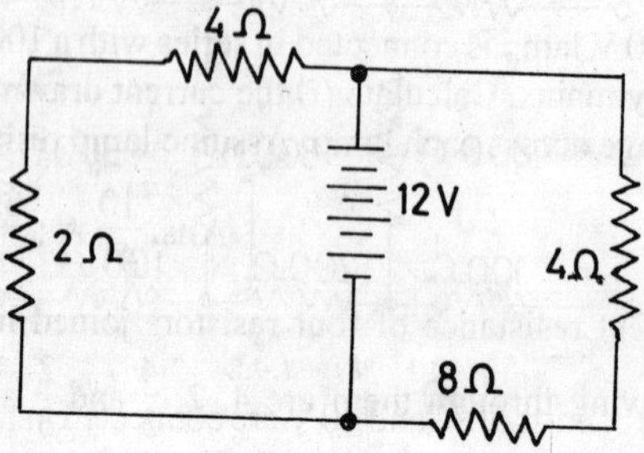

Fig. 1.41.

12. *(a)* Three resistors of 3, 4 and 5 Ω respectively are connected in parallel. This combination is put in series with a 2.5 Ω resistor. Determine the equivalent resistance of the combination.

(b) What is the power absorbed by the circuit if 10 V battery is connected across the group combination of resistor in *(a)* above ? Take internal resistance of battery as 0.5 Ω. Also determine the current flowing in each resistor.

[**Ans.** *(a)* 3.7766 Ω; *(b)* 20.65 watts; 0.995 A, 0.746 A, 0.597 A; 2.338 A]

13. A circuit consisting of three resistances 12 Ω, 18 Ω and 36 Ω respectively joined in parallel is connected in series with a fourth resistance. The whole circuit is supplied at 60 V and it is found that power dissipated in 12 Ω resistance is 36 W. Determine the value of fourth resistance and the total power dissipated in the group. [**Ans.** 11.3 Ω; 208 watts]

14. A resistance of *R* ohms is connected in series with a parallel circuit comprising of two resistances 12 Ω and 8 Ω respectively. The total power dissipated in the circuit is 70 watts when applied voltage is 20 V. Calculate the value of *R*. [**Ans.** 0.914 Ω]

15. How much current a battery of 8 volts will give when it is connected across *AB* in the circuit diagram given in Fig. 1.42. [**Ans.** 1 A]

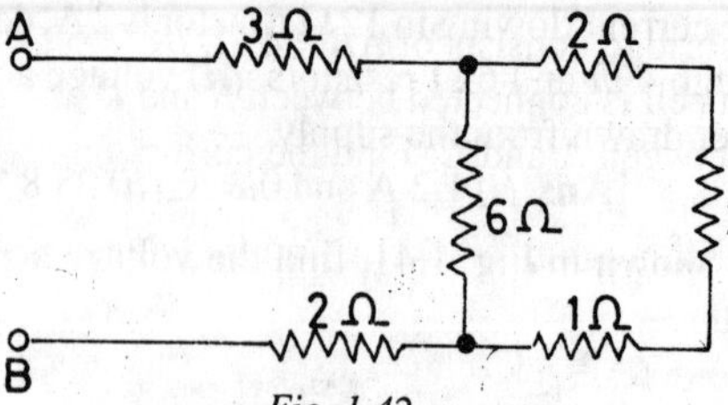

Fig. 1.42.

16. What is the equivalent resistance of the network shown in Fig. 1.43.
[Ans. 600 Ω]

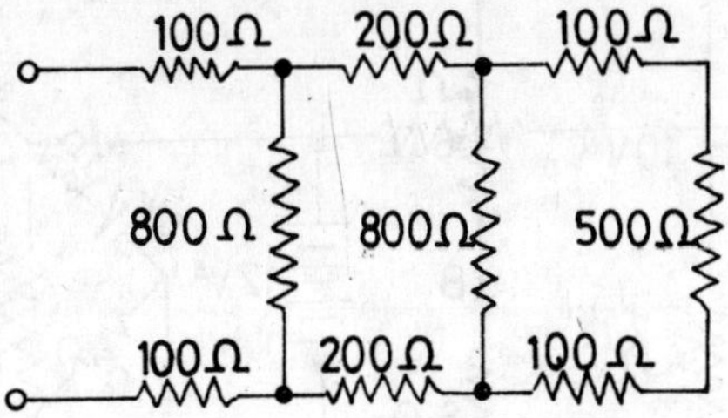

Fig. 1.43.

17. Two batteries of emfs 50 V and 40 V are connected in parallel across a resistor of 10 Ω. Determine the current supplied by each battery and current through 10 Ω resistor if the internal resistance of the batteries are 2.5 Ω and 2 Ω respectively. **[Ans.** 4, zero and 4 A]

18. Calculate the total electrical power consumed in the circuit shown in Fig. 1.44. **[Ans.** 60 watts]

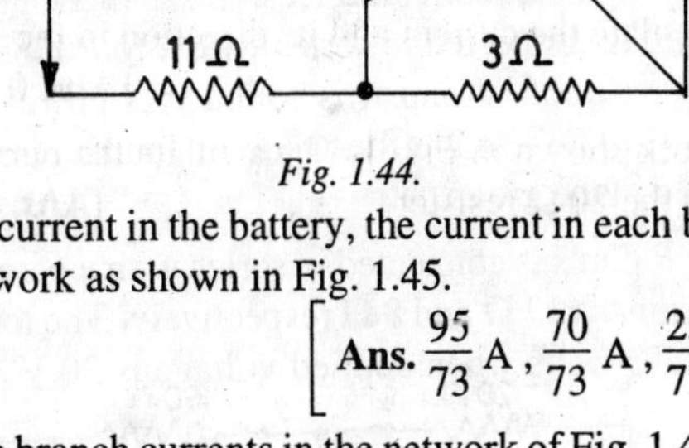

Fig. 1.44.

19. Calculate the current in the battery, the current in each branch and pd across *AB* of the network as shown in Fig. 1.45.
$$\left[\textbf{Ans.}\ \frac{95}{73}\text{ A}, \frac{70}{73}\text{ A}, \frac{25}{73}\text{ A}, \frac{10}{73}\text{ A}, \frac{10}{73}\text{ V}\right]$$

20. Determine the branch currents in the network of Fig. 1.46, when the value of each branch resistance is one ohm.
[Ans. 6.25 A, 6.25 A, 0, 1.25 A, 1.25 A, 7.5 A]

21. A Wheatstone bridge consists of *AB* = 4 Ω, *BC* = 3 Ω, *CD* = 6 Ω anmd *DA* = 5 Ω. A 2 volt cell is connected between *B* and *D* and a galvanometer of 10 Ω resistance between *A* and *C*. Find the current through the galvanometer.
$$\left[\textbf{Ans.}\ \frac{1}{64}\text{ A}\right]$$

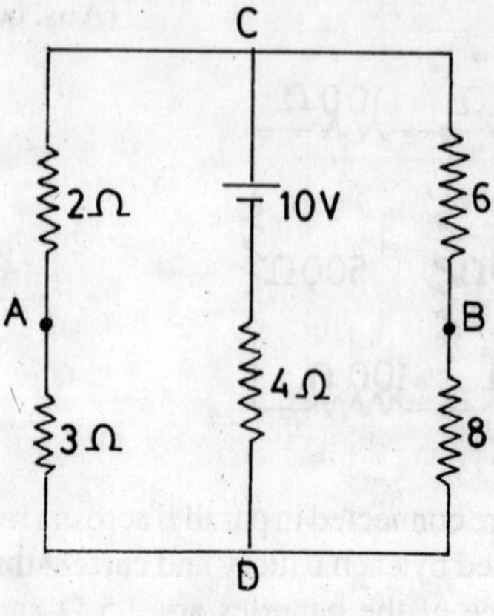

Fig. 1.45.

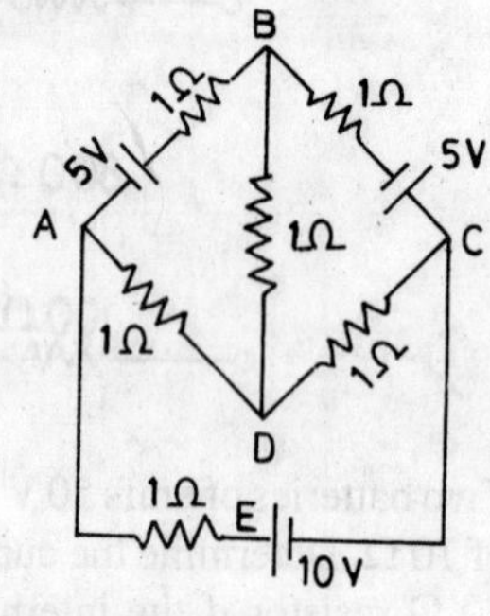

Fig. 1.46.

22. The arms of a Wheatstone bridge have the folloiwng resistances : *AB* = 10 Ω, *BC* = 10 Ω, *CD* = 4 Ω, *DA* = 5 Ω. A galvanometer of 20 Ω resistance is connected across *B* and *D*. Calculate the current through the galvanometer when a pd of 10 V is maintained across *A* and *C*.

$\left[\textbf{Ans. } \frac{1}{49} \text{ A from } B \text{ to } D\right]$

23. The four arms of a Wheatstone bridge have the following resistances : *AB* = 500 Ω, *BC* = 800 Ω, *CD* = 400 Ω and *DA* = 300 Ω. A battery of 6 V is connected across *A* and *C* and a galvanometer of resistance 1,200 Ω across *B* and *D*. Calculate the current and its direction in the galvanometer.

[**Ans.** 0.157 mA from *B* to *D*]

24. For the network shown in Fig. 1.47, calculate the current in 132 Ω resistor and pd across the 20 Ω resistor. [**Ans.** 125/147 A ; 87.75 V]

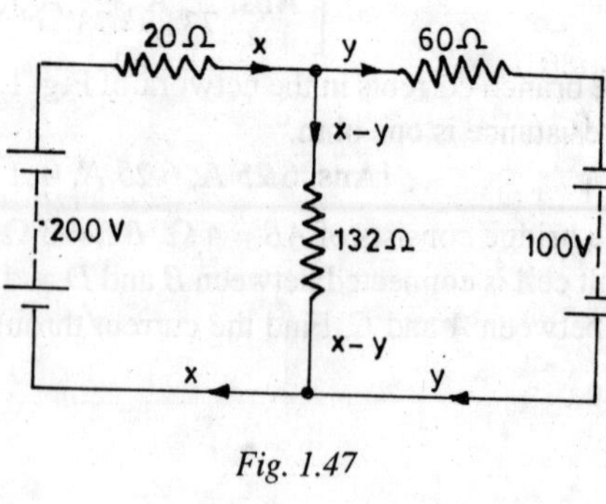

Fig. 1.47

25. Solve the network shown in Fig. 1.48 for the current through 6 Ω resistor.
[Ans. 12.316 A]

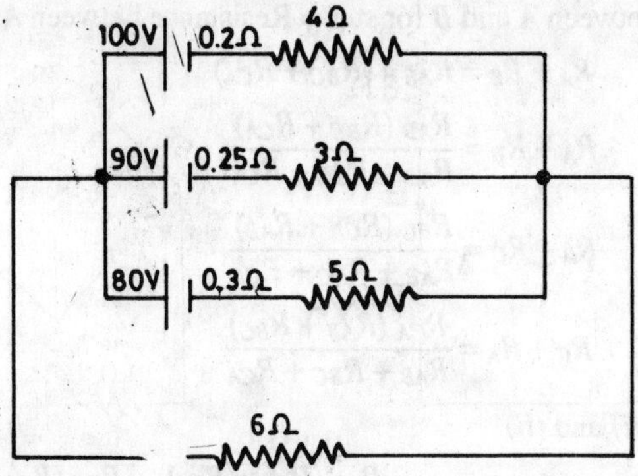

Fig. 1.48.

1.11. DELTA/STAR AND STAR/DELTA TRANSFORMATION

There are some networks in which the resistances are neither in series nor in parallel. A familiar case is a three terminal network *e.g.*, delta network or star network. In such situations, it is not possible to simplify the network by series and parallel circuit rules. However, converting delta network into star and *vice-versa* often simplifies the network and makes it possible to apply series-parallel circuit techniques.

1.11.1. Delta/Star Transformation

Consider three resistor R_{AB}, R_{BC} and R_{CA} connected in delta to three terminals *A*, *B* and *C* as shown in Fig. 1.49 *(i)*. Let the equivalent star- connected network have resistances R_A, R_B and R_C. Since the two arrangements are electrically equivalent, the resistance between any two terminals of one network is equal to the resistance between the corresponding terminals of the other network.

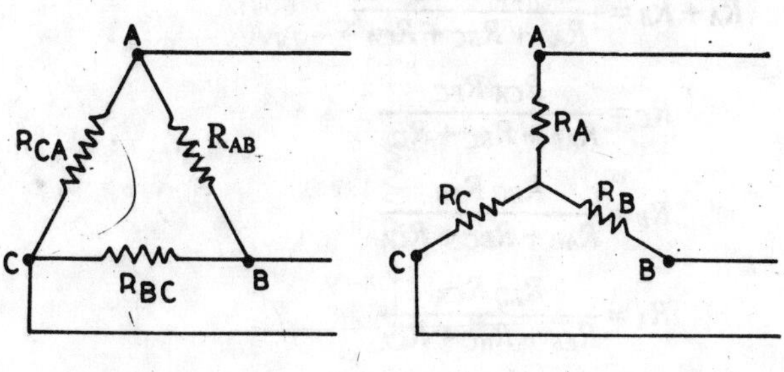

Fig. 1.49.

Let us consider the terminals A and B of the network.

R_{AB}, R_{BC} and R_{CA} are connected in delta.

Equivalent star connected network has resistances R_A, R_B and R_C

Resistance between A and B for star = Resistance between A and B for delta

$$R_A + R_B = R_{AB} \parallel (R_{BC} + R_{CA})$$

or
$$R_A + R_B = \frac{R_{AB}(R_{BC} + R_{CA})}{R_{AB} + R_{BC} + R_{CA}} \qquad ...(i)$$

Similarly,
$$R_B + R_C = \frac{R_{BC}(R_{CA} + R_{AB})}{R_{AB} + R_{BC} + R_{CA}} \qquad ...(ii)$$

$$R_C + R_A = \frac{R_{CA}(R_{AB} + R_{BC})}{R_{AB} + R_{BC} + R_{CA}} \qquad ...(iii)$$

Adding *(i)*, *(ii)* and *(ii)*

$$R_A + R_B + R_B + R_C + R_C + R_A = \frac{R_{AB}(R_{BC} + R_{CA})}{R_{AB} + R_{BC} + R_{CA}} + \frac{R_{BC}(R_{CA} + R_{AB})}{R_{AB} + R_{BC} + R_{CA}} + \frac{R_{CA}(R_{AB} + R_{BC})}{R_{AB} + R_{BC} + R_{CA}}$$

By adding,

$$2(R_A + R_B + R_C) = \frac{R_{AB}R_{BC} + R_{AB}R_{CA} + R_{BC}R_{CA} + R_{BC}R_{AB} + R_{CA}R_{AB} + R_{CA}R_{BC}}{R_{AB} + R_{BC} + R_{CA}}$$

$$2(R_A + R_B + R_C) = \frac{2(R_{AB}R_{BC} + R_{AB}R_{CA} + R_{BC}R_{CA})}{R_{AB} + R_{BC} + R_{CA}}$$

or
$$R_A + R_B + R_C = \frac{R_{AB}R_{BC} + R_{AB}R_{CA} + R_{BC}R_{CA}}{R_{AB} + R_{BC} + R_{CA}} \qquad ...(iv)$$

Subtracting *(i)* from *(iv)*,

$$R_A + R_B + R_C = \frac{R_{AB}R_{BC} + R_{AB}R_{CA} + R_{BC}R_{CA}}{R_{AB} + R_{BC} + R_{CA}}$$

$$R_A + R_B = \frac{R_{AB}R_{BC} + R_{AB}R_{CA}}{R_{AB} + R_{BC} + R_{CA}}$$

$$R_C = \frac{R_{CA}R_{BC}}{R_{AB} + R_{BC} + R_{CA}} \qquad ...(v)$$

$$R_B = \frac{R_{BC}R_{AB}}{R_{AB} + R_{BC} + R_{CA}} \qquad ...(vi)$$

$$R_A = \frac{R_{AB}R_{CA}}{R_{AB} + R_{BC} + R_{CA}} \qquad ...(vii)$$

How to remember ? There is an easy way to remember these relations. Referring to Fig. 1.50, star-connected resistances R_A, R_B and R_C are electrically equivalent to delta-connected resistances R_{AB}, R_{BC} and R_{CA}. What have seen above that :

$$R_A = \frac{R_{AB}\, R_{CA}}{R_{AB} + R_{BC} + R_{CA}}$$

i.e., Any arm of star-connection

$$= \frac{\text{Product of two adjacent arms of } \Delta}{\text{Sum of arms of } \Delta}$$

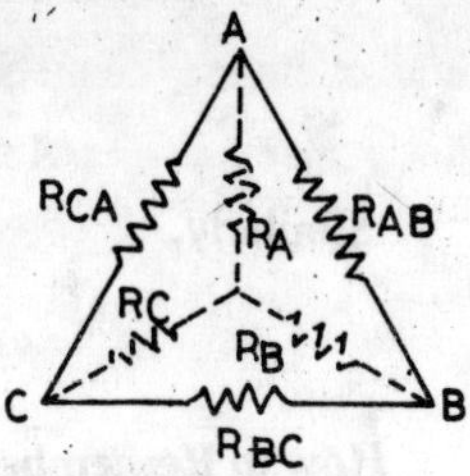

Fig. 1.50.

1.11.2. Star-Delta Transformation

For replacement from star to delta.

Multiply expression *(v)* and *(vi)*, *(vi)* and *(vii)*, *(vii)* and *(v)* and add than taking *(v)* and *(vi)*

$$R_C\, R_B = \frac{(R_{CA}\, R_{BC})\, R_{BC}\, R_{AB}}{(R_{AB} + R_{BC} + R_{CA})^2}$$

$$R_C\, R_B = \frac{R_{CA}\, R_{BC}^2\, R_{AB}}{(R_{AB} + R_{BC} + R_{CA})^2} \qquad \text{...(viii)}$$

Taking *(vi)* and *(vii)*

$$R_B\, R_A = \frac{R_{BC}\, R_{AB}^2\, R_{CA}}{(R_{AB} + R_{BC} + R_{CA})^2} \qquad \text{...(ix)}$$

Taking *(vii)* and *(v)*

$$R_A\, R_C = \frac{R_{AB}\, R_{CA}^2\, R_{BC}}{(R_{AB} + R_{BC} + R_{CA})^2} \qquad \text{...(x)}$$

Add *(viii)*, *(ix)* and *(x)*,

$$R_C\, R_B + R_B\, R_A + R_A\, R_C = \frac{R_{CA}\, R_{BC}^2\, R_{AB} + R_{BC}\, R_{AB}^2\, R_{CA} + R_{AB}\, R_{CA}^2\, R_{BC}}{(R_{AB} + R_{BC} + R_{CA})^2}$$

$$= \frac{R_{AB}\, R_{BC}\, R_{CA}\, (R_{AB} + R_{BC} + R_{CA})}{(R_{AB} + R_{BC} + R_{CA})^2}$$

$$R_C\, R_B + R_B\, R_A + R_A\, R_C = \frac{R_{AB}\, R_{BC}\, R_{CA}}{R_{AB} + R_{BC} + R_{CA}} \qquad \text{...(xi)}$$

Dividing eqn. *(xi)* by *(v)*, *(vi)* and *(vii)* respectively, we get

Eqn. *(v)* is,
$$R_C = \frac{R_{CA} + R_{BC}}{R_{AB} + R_{BC} + R_{CA}}$$

$$R_B + \frac{R_B R_A}{R_C} + R_A = R_{AB}$$

$$R_{AB} = R_A + R_B + \frac{R_A R_B}{R_C} \quad \textbf{Ans.}$$

Similarly,
$$R_{BC} = R_B + R_C + \frac{R_B R_C}{R_A} \quad \textbf{Ans.}$$

$$R_{CA} = R_C + R_A + \frac{R_C R_A}{R_B} \quad \textbf{Ans.}$$

How to Remember ? There is an easy way to remember these relations.

Referring to Fig. 1.51, star-connected resistances R_A, R_B and R_C are electrically equivalent to delta-connected resistances R_{AB}, R_{BC} and R_{CA}. We have seen above that :

$$R_{AB} = R_A + R_B + \frac{R_A R_B}{R_C}$$

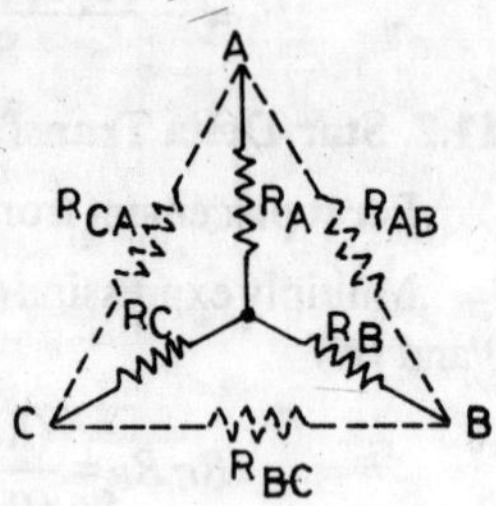

Fig. 1.51.

i.e., Resistance between two terminals of delta = sum of star resistances connected to those terminals *plus* product of same two resistances divided by the third star resistance.

Example 1.69. Three resistances 10 Ω , 20 Ω and 30 Ω are connected in delta as shown in Fig. 1.52 *(a)*. Determine the resistances for an equivalent star-connection.

Solution : As per the condition given the three resistance 10 Ω , 20 Ω and 30 Ω are joined in delta. The equivalent resistance in star is calculated

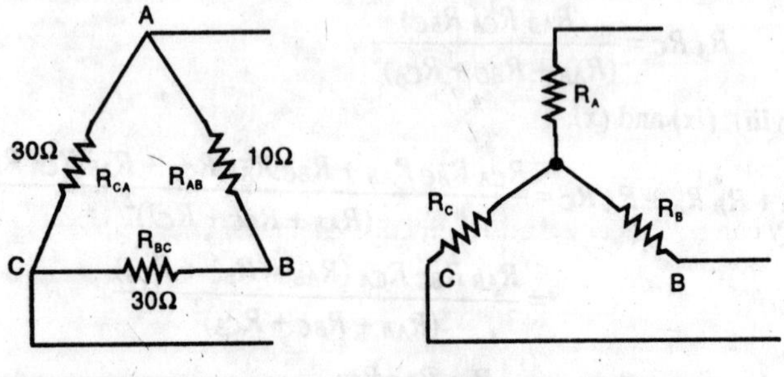

(a) *(b)*

Fig. 1.52.

$$R_A = \frac{10 \times 30}{10 + 20 + 30} = \frac{300}{60} = 5\ \Omega$$

$$R_B = \frac{10 \times 20}{10 + 20 + 30} = \frac{200}{60} = 3.3\ \Omega$$

$$R_C = \frac{20 \times 30}{10 + 20 + 30} = \frac{600}{60} = 10\ \Omega$$

Example 1.70. In Fig. 1.53, 200 volts are applied across line terminals *AB*. Determine the resistance *(i)* between terminals *A* and *B* and *(ii)* circuit current.

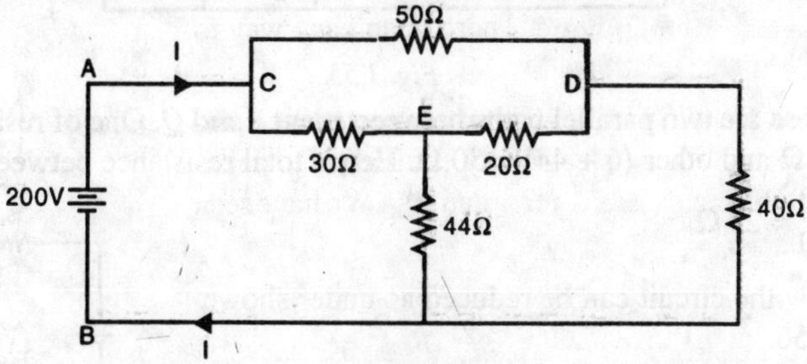

Fig. 1.53.

Solution : Considering the network shown in Fig. 1.54, the three resistances of 50 Ω, 30 Ω and 20 Ω are delta connected between terminal point *C, D* and *E* in Fig. 1.54 *(a)*.

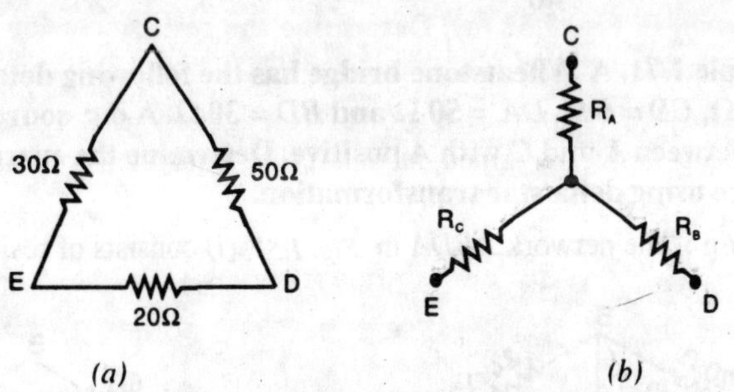

Fig. 1.54.

They can be connected into their equivalent star connections as shown in Fig. 1.54 *(b)*.

$$R_A = \frac{30 \times 50}{50 + 30 + 20} = 15\ \Omega$$

$$R_B = \frac{50 \times 20}{50 + 30 + 20} = 10\ \Omega$$

$$R_C = \frac{20 \times 30}{50 + 30 + 20} = 6\ \Omega$$

The network shown in Fig. 1.55 can be reduced to network as shown under

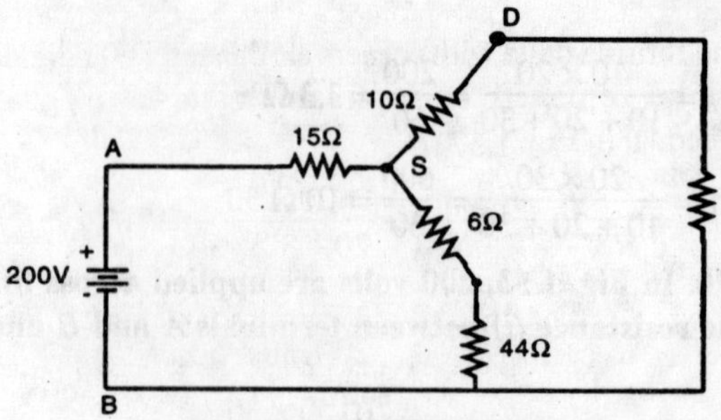

Fig. 1.55.

These are two parallel paths between point S and Q. One of resistance (10 + 40) = 50 Ω and other (6 + 44) = 50 Ω. Hence total resistance between S and Q is

$$= \frac{50 \times 50}{50 + 50} = 25\ \Omega$$

Now the circuit can be reduced as under shown in Fig. 1.56.

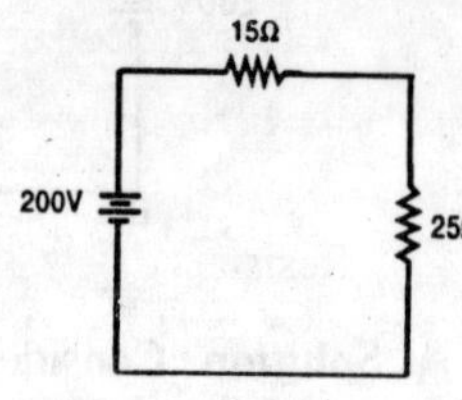

Fig. 1.56.

Resistance between point A and B is

$$15 + 25 = 40\ \Omega \textbf{ Ans.}$$

Circuit current $I = \frac{200}{40} = 5$ A **Ans.**

Example 1.71. A Wheatstone bridge has the following details : AB = 20 Ω, BC = 24 Ω, CD = 5 Ω, DA = 50 Ω and BD = 30 Ω. A d.c. source of 220 V is connected between A and C with A positive. Determine the current delivered by the source using delta/star transformation.

Solution : The network $ABDA$ in *Fig. 1.57. (i)* consists of resistances 20 Ω,

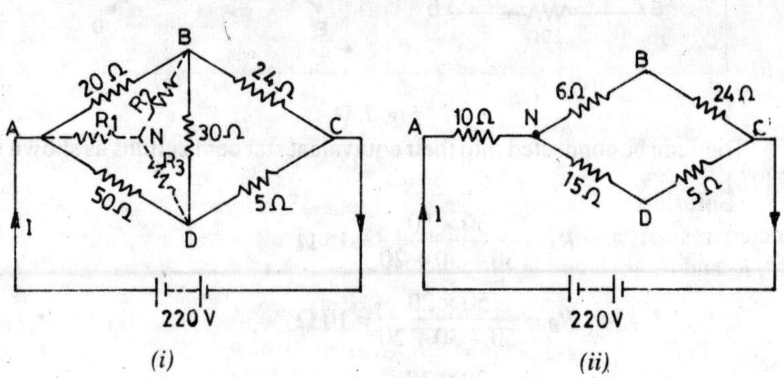

Fig. 1.57

30 Ω and 50 Ω and forms a delta with corners at *A*, *B* and *D*. These delta-connected resistances can be replaced by equivalent star-connected resistances R_1, R_2 and R_3 as shown in Fig. 1.57 *(i)*.

$$R_1 = \frac{R_{AB} R_{DA}}{R_{AB} + R_{BD} + R_{DA}} = \frac{20 \times 50}{20 + 30 + 50} = 10\ \Omega$$

$$R_2 = \frac{R_{AB} R_{BD}}{R_{AB} + R_{BD} + R_{DA}} = \frac{20 \times 30}{20 + 30 + 50} = 6\ \Omega$$

$$R_3 = \frac{R_{AD} R_{BD}}{R_{AB} + R_{BD} + R_{DA}} = \frac{50 \times 30}{20 + 30 + 50} = 15\ \Omega$$

Thus the network shown in Fig. 1.52 *(i)* reduces to the network shown in Fig. 1.52 *(ii)*. As seen from Fig. 1.57 *(ii)*, there are two parallel paths between *N* and *C*.

∴ Resistance betwene *N* and *C*

$$= (6 + 24)\ \Omega \parallel (15 + 5)\ \Omega = \frac{30 \times 20}{30 + 30} = 12\ \Omega$$

Resistance between *A* and *C* = 10 + 12 = 22 Ω

∴ Current delivered by source, *I* = 220/22 = 10 A **Ans.**

Example 1.72. Using delta/star transformation, find the galvanometer current in the Wheatstone bridge shown in Fig. 1.58 *(i)*.

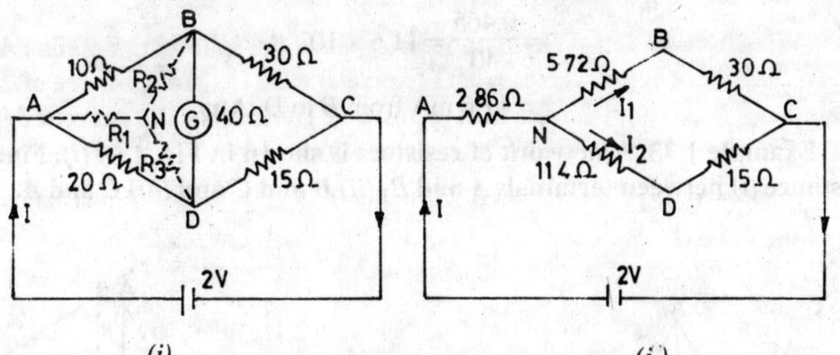

Fig. 1.58.

Solution : The network *ABDA* in Fig. 1.53 *(i)* forms a delta. These delta-connected resistances can be replaced by equivalent star-connected resistances R_1, R_2 and R_3 as shown in Fig. 1.58 *(i)*.

$$R_1 = \frac{R_{AB} R_{DA}}{R_{AB} + R_{BD} + R_{DA}} = \frac{10 \times 20}{10 + 40 + 20} = 2.86\ \Omega$$

$$R_2 = \frac{R_{AB} R_{BD}}{R_{AB} + R_{BD} + R_{DA}} = \frac{10 \times 40}{10 + 40 + 20} = 5.72\ \Omega$$

$$R_3 = \frac{R_{DA}\, R_{BD}}{R_{AB} + R_{BD} + R_{DA}} = \frac{20 \times 40}{10 + 40 + 20} = 11.4\ \Omega$$

Thus the network shown in Fig. 1.53 *(i)* reduces to the network shown in Fig. 1.58 *(ii)*.

$$R_{AC} = 2.86 + \frac{(30 + 5.72)\,(15 + 11.4)}{(30 + 5.72) + (15 + 11.4)} = 18.04\ \Omega$$

Battery current, $I = 2/18.04 = 0.11$ A

The battery current divides at N into two parallel paths.

$\therefore$ Current in branch NBC, $I_1 = 0.11 \times \dfrac{26.4}{26.4 + 35.72} = 0.047$ A

Current in branch NDC, $I_2 = 0.11 \times \dfrac{35.72}{26.4 + 35.72} = 0.063$ A

Potential of B. w.r.t. $C = 30 \times 0.047 = 1.41$ V

Potential of D. w.r.t. $C = 15 \times 0.063 = 0.945$ V

Clealry, point B is at higher potential than point D by

$$1.41 - 0.945 = 0.465 \text{ A}$$

$$\therefore \text{Galvanometer current} = \frac{\text{P.D. between } B \text{ and } D}{\text{Galvanometer resistance}}$$

$$= \frac{0.465}{40} = 11.6 \times 10^{-3} \text{ A}$$

$$= 11.6 \text{ mA from } B \text{ to } D$$ **Ans.**

Example 1.73. A network of resistors is shown in Fig. 1.59 *(i)*. Find the resistance *(i)* between terminals *A* and *B*, *(ii)* *B* and *C* and *(iii)* *C* and *A*.

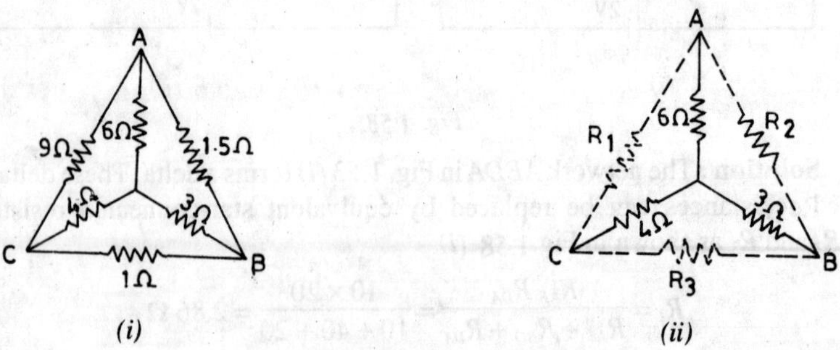

Fig. 1.59.

Solution : The star-connected resistances 6 Ω, 3 Ω and 4 Ω in Fig. 1.59 *(i)* are shown separately in Fig. 1.59 *(ii)*. These start-connected resistances can be convered into equivalent delta-connected resistances R_1, R_2 and R_3 as shown in Fig. 1.59 *(ii)*.

$$R_1 = 4 + 6 + \left(4 \times \frac{6}{3}\right) = 18\ \Omega$$

$$R_2 = 6 + 3 + \left(6 \times \frac{3}{4}\right) = 13.5\ \Omega$$

$$R_3 = 4 + 3 + \left(4 \times \frac{3}{6}\right) = 9\ \Omega$$

These delta-connected resistances R_1, R_2 and R_3 come in parellel with the original delta-connected resistances. The circuit shown in Fig. 1.59 *(i)* reduces to the circuit shown in Fig. 1.59 *(i)*.

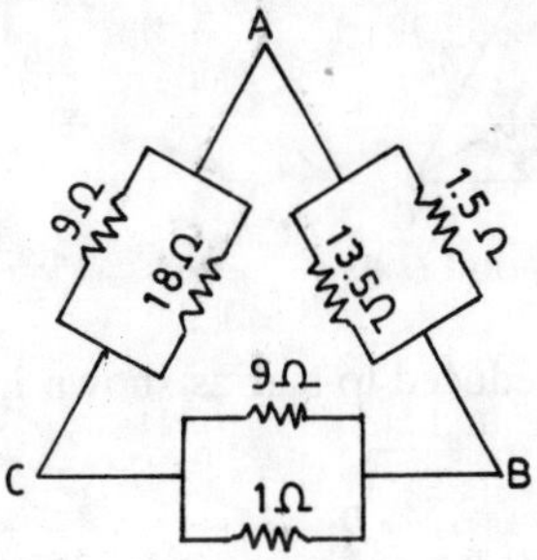

(i)

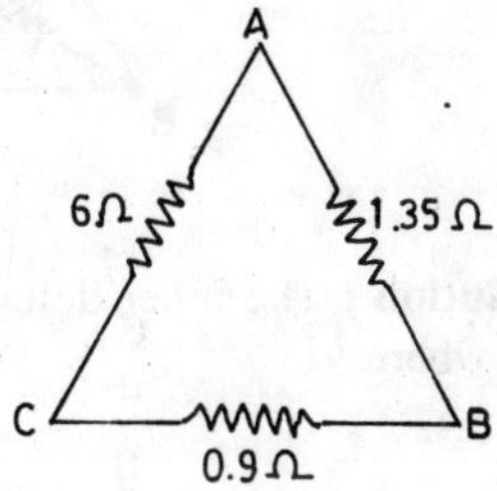

(ii)

Fig. 1.60.

The parallel resistances in each leg of delta in Fig. 1.60 *(i)* can be replaced by a single resistor as shown in Fig. 1.60 *(ii)* where

$$R_{AC} = 9 \times \frac{18}{27} = 6\ \Omega$$

$$R_{BC} = 9 \times \frac{1}{10} = 0.9\ \Omega$$

$$R_{AB} = 1.5 \times \frac{13.5}{15} = 1.35\ \Omega$$

Resistance between A and $C = 6\ \Omega \parallel (1.35 + 0.9)\ \Omega$

$$= 6 \times \frac{2.25}{8.25} = 1.636\ \Omega \ \textbf{Ans.}$$

Resistance between B and $C = 0.9\ \Omega \parallel (6 + 1.35)\ \Omega$

$$= 0.9 \times \frac{7.35}{8.25} = 0.8\ \Omega \ \textbf{Ans.}$$

Resistance between A and $B = 1.35\ \Omega \parallel (6+0.9)\ \Omega$

$$= 1.35 \times \frac{6.9}{8.25} = 1.13\ \Omega \ \textbf{Ans.}$$

Example 1.74. In the network shown in Fig. 1.61, determine the resistance between *A* and *B*. The numbers represent the respective resistances in ohms.

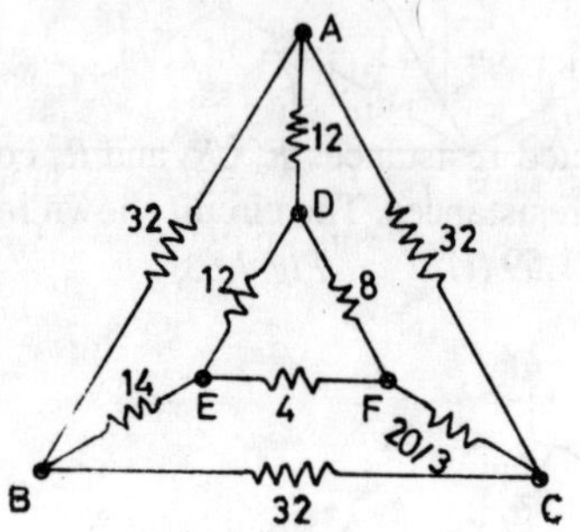

Fig. 1.61

Solution : The inner delta *DEF* can be reduced to star as shown in Fig. 1.62 *(b)*; where

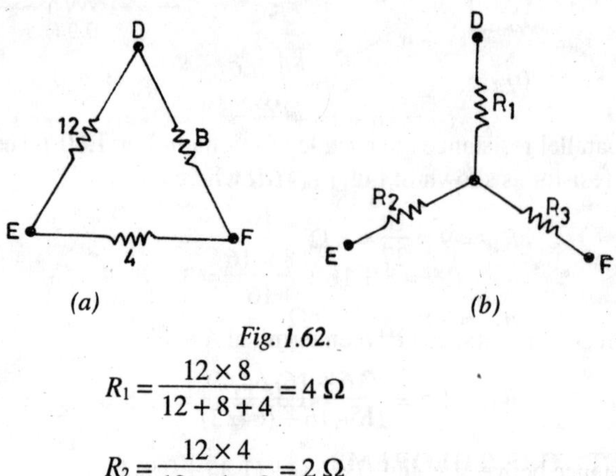

(a) *(b)*

Fig. 1.62.

$$R_1 = \frac{12 \times 8}{12+8+4} = 4\ \Omega$$

$$R_2 = \frac{12 \times 4}{12+4+8} = 2\ \Omega$$

$$R_3 = \frac{8 \times 4}{8+4+12} = \frac{4}{3}\ \Omega$$

The original network is reduced to the shape shown in Fig. 1.63. The resultant internal star is shown in Fig. 1.64 *(a)*, which can be further reduced to delta connections shown in Fig. 1.64 *(b)*.

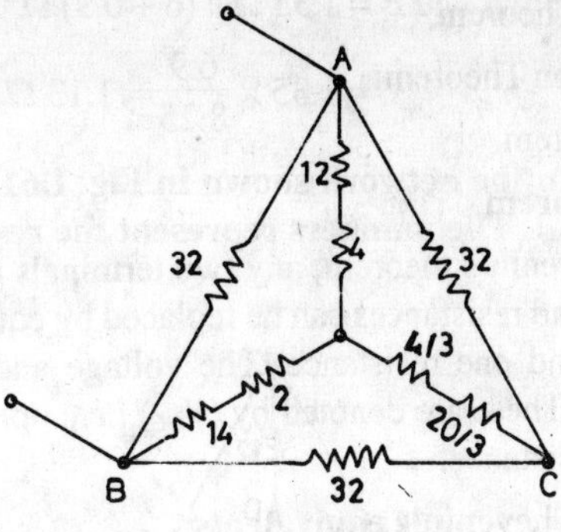

Fig. 1.63.

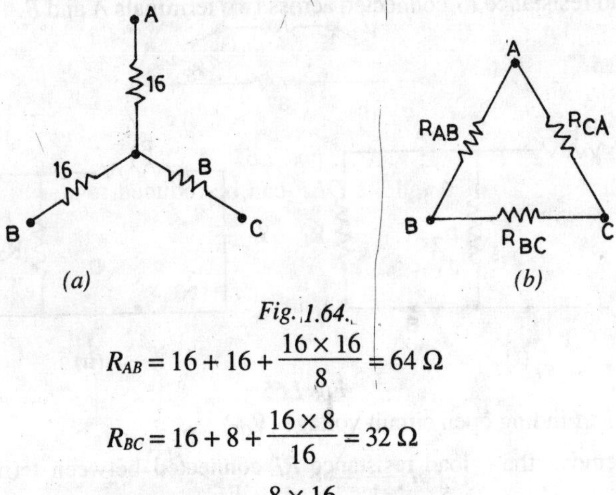

Fig. 1.64.

$$R_{AB} = 16 + 16 + \frac{16 \times 16}{8} = 64\ \Omega$$

$$R_{BC} = 16 + 8 + \frac{16 \times 8}{16} = 32\ \Omega$$

$$R_{CA} = 8 + 16 + \frac{8 \times 16}{16} = 32\ \Omega$$

Effective resistance between terminal *A* and *B*

$$= \frac{(16 + 16)\ 64/3}{16 + 16 + (64/3)} = 12.8\ \Omega \quad \textbf{Ans.}$$

1.12. NETWORK THEOREMS

The arrangement by which various sources of electrical energy and resistances are connected together is called an electrical network. To solve simple and complex networks many laws and theorems were developed. The main advantage of these theorems is that they save time and reduce the complicated mathematical calculations.

Various theorems discussed are :

(i) Thevenin's Theorem.

(ii) Super-position Theorem.

(iii) Norton Theorem.

1.12.1. Thevenin's Theorem

According to Thevenin's theorem, any two terminals network containing a number of emf sources and resistances can be replaced by equivalent simple series circuit of one voltage and one resistance. The voltage and resistance is called Thevenin's equivalents. These are denoted by (V_{OC}),*i.e.*, open circuit votlage and R_{th} is called Thevenin resistance.

Procedure for finding Thevenin's equivalents

Consider a network as shown in Fig. 1.65. It is desired to find current I_L through load resistance R_L connected across two terminals *A* and *B*.

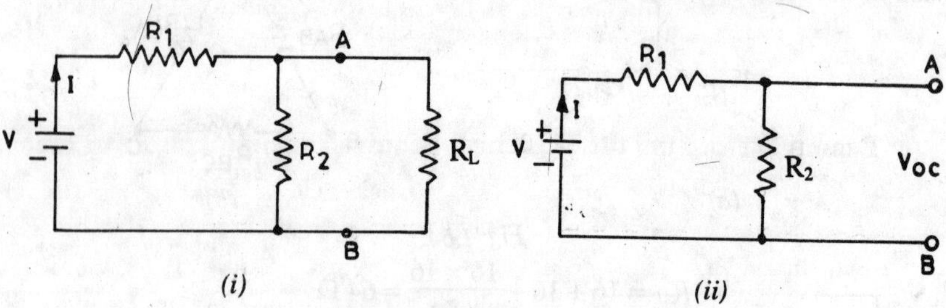

Fig. 1.65.

Step I : Finding open circuit voltage (V_{OC})

(i) Remove the load resistance R_L connected between terminal *AB* of network.

$$I = \frac{V}{R_1 + R_2}$$

$$V_{OC} = IR_2$$

or

$$V_{OC} = \left(\frac{V}{R_1 + R_2}\right) \times R_2$$

Step II : Finding Thevenin's equivalent resistance

The Thevenin equivalent resistance is the resistance between the terminals *AB* with the conditions

(a) Load R_L is removed.

(b) Emf reduced to zero and the circuit can be redrawn as shown in Fig. 1.65 *(iii)*.

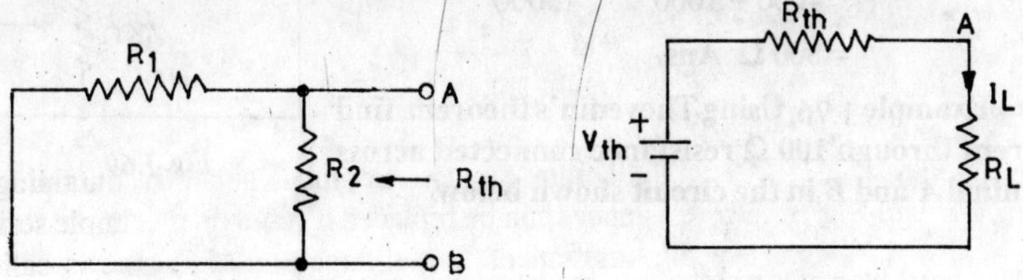

(iii) (iv)

Fig. 1.66

$$R_{th} = \frac{R_1 \times R_2}{R_1 + R_2}$$

Thus the value of the R_{th} is determined. Once the values of V_{th} and R_{th} are determined, then the current through load resistance R_L is determined by connecting back as shown in Fig. 1.66 *(iv)*.

$$I_L = \frac{V_{th}}{R_{th} + R_L}$$

Example 1.75. Find the Thevenin's requivalent circuit for Fig. 1.67.

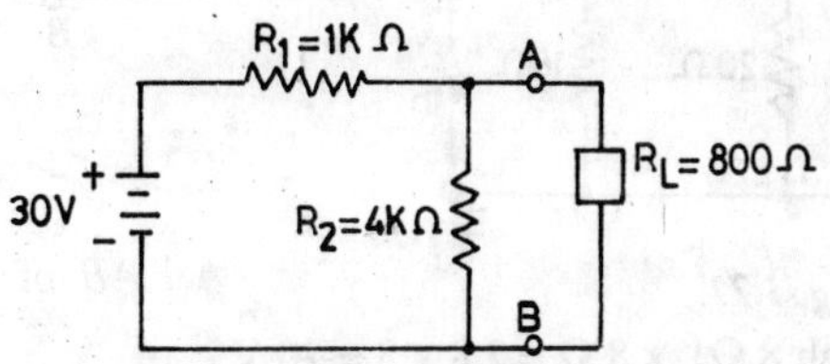

Fig. 1.67

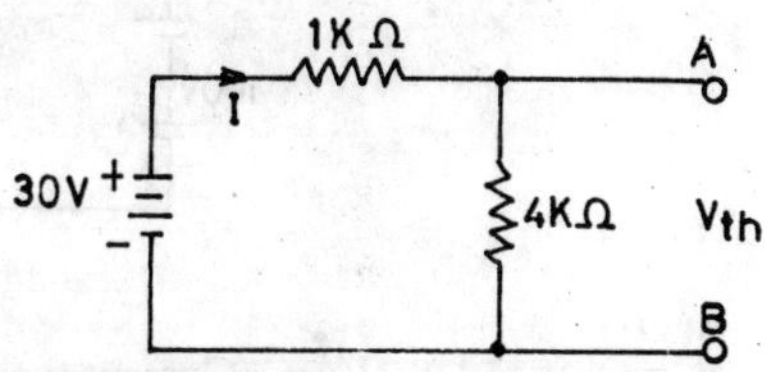

Fig. 1.68.

Solution : Step I : Remove the load resistance $R_L = 800\ \Omega$ and redraw the circuit (Fig. 1.68).

or Now current through 4 KΩ resistor

$$I = \frac{V}{R_1 + R_1} = \frac{30}{1000 + 4000} = \frac{30}{5000} = \frac{3}{500}\ \text{A}$$

Now Thevenin's voltage $V_{th} = I \times R_2 = \frac{3}{500} \times 4000 = 24\ \text{V}$

Step II : For finding Thevenin's equivalent R_{th} remove the battery of 30 V and redraw the circuit (Fig. 1.69).

$$R_{th} = \frac{4000 \times 1000}{4000 + 1000} = \frac{4000 \times 1000}{5000}$$

$$= 800\ \Omega \ \textbf{Ans.}$$

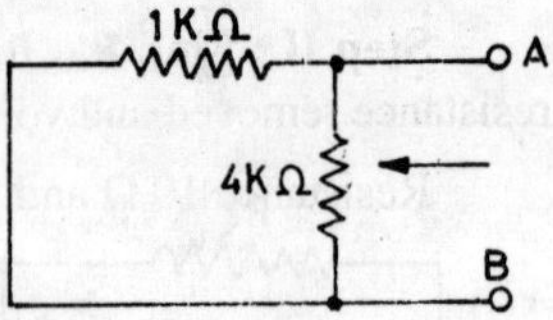

Fig. 1.69.

Example 1.76. Using Thevenin's theorem, find current through 100 Ω resistance connected across terminal *A* and *B* in the circuit shown below.

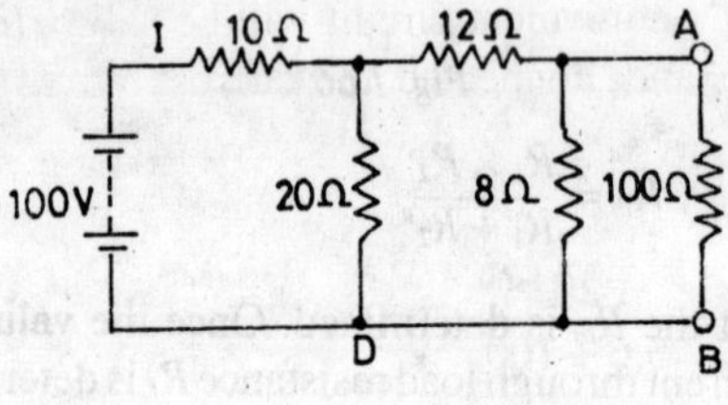

Fig. 1.70.

Solution : Finding V_{th} : It is the voltage across terminals *A* and *B* with 100 Ω resistance removed as shown in Fig. 1.71.

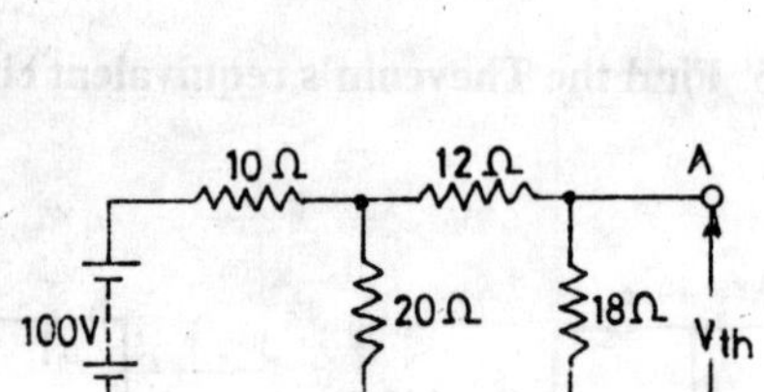

Fig. 1.71.

$$V_{th} = (\text{Current through } 8\ \Omega) \times 8\ \Omega = 2.8 \times 8 = 20\ \text{V}$$

By solving this series–parellel circuit.

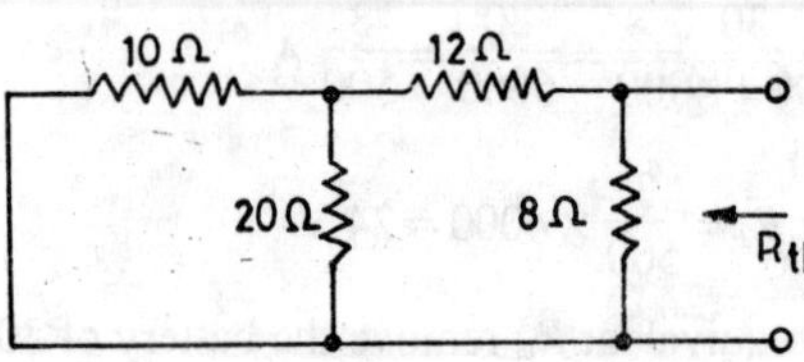

Fig. 1.72.

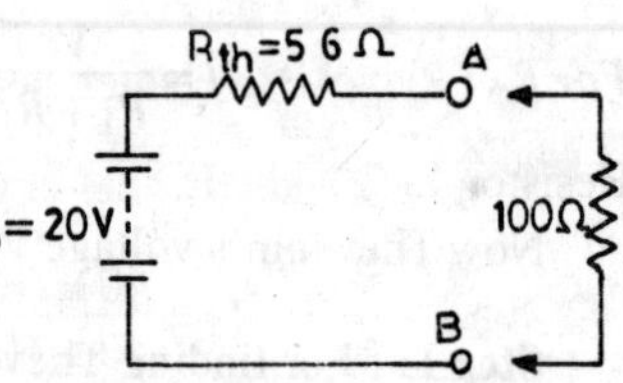

Fig. 1.73.

Step II : Find R_{th}. It is the resistance between terminals A and B with 100 resistance removed and voltage source short circuited as shown in Fig. 1.72.

Resistance 10 Ω and 20 are in parellel.

$$= \frac{\left(\frac{10 \times 20}{10 + 20} + 12\right) \times 8}{\left(\frac{10 \times 20}{10 + 20} + 12\right) + 8} = 5.6\ \Omega$$

Therefore, Thevenin's equivalent circuit will be shown in Fig. 1.73. Now current through 100 Ω resistance connected back across terminals A and B can be found by using Ohm's law.

Current through 100 Ω

$$\frac{V_{th}}{R_{th} + R_L} = \frac{20}{5.6 + 100} = 0.19 \text{ A } \textbf{Ans.}$$

Example 1.77. In the circuit shown in Fig. 1.74, find the current in R_2 when $R_2 = 11$ ohms. Using Thevenin's theorem.

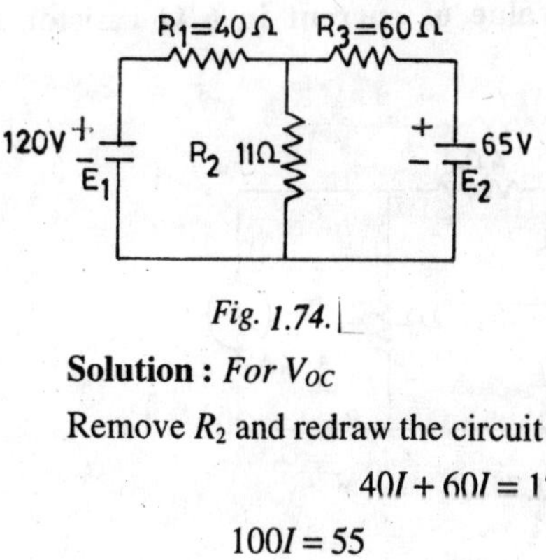

Fig. 1.74.

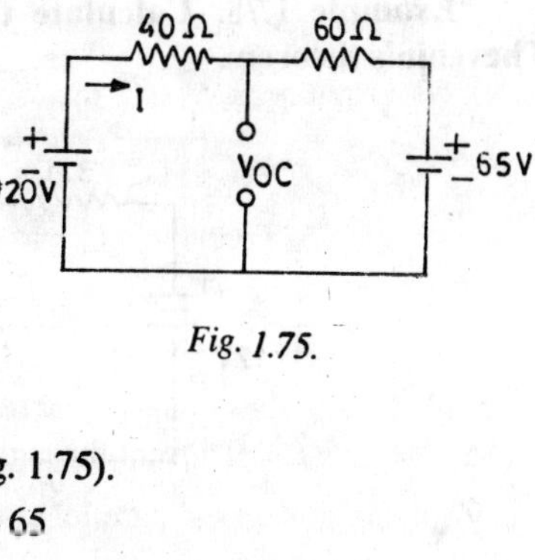

Fig. 1.75.

Solution : *For V_{OC}*

Remove R_2 and redraw the circuit **(Fig. 1.75).**

$$40I + 60I = 120 - 65$$

$$100I = 55$$

$$I = \frac{55}{100} + 0.55 \text{ Amp}$$

$$V_{OC} = E_1 - IR_1 = 120 - 0.55 \times 40 = 120 - 22 = 98 \text{ V}$$

For R_{th} : Remove the batteries E_1 and E_2 redraw the circuit (Fig. 1.76).

Resistor 40 Ω and 60 Ω are in parallel.

$$R_{th} = \frac{40 \times 60}{40 + 60} = \frac{2400}{100} = 24\ \Omega$$

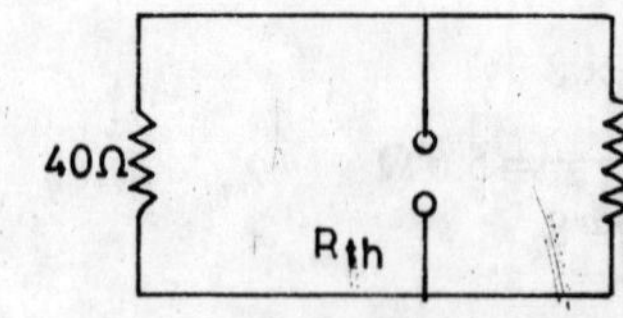

Fig. 1.76.

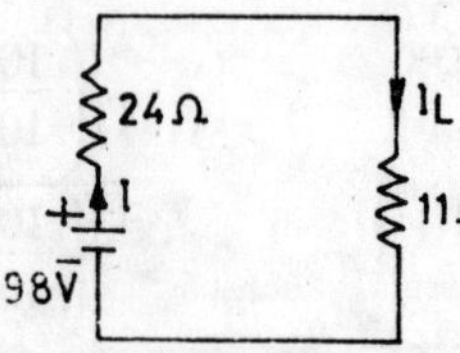

Fig. 1.77.

By taking open circuit voltage V_{OC} and Thevenin resistance R_{th}, draw the circuit (Fig. 1.77) and connecte the 11 Ω resistance back.

I_L is the current in the 11 Ω resistance

$$I_L = \frac{V_{OC}}{R_{th} + R_L} = \frac{98}{24+11} = \frac{98}{35} = 2.8 \text{ A}$$

Hence current in R_2 is 2.8 A. **Ans.**

Example 1.78. Calculate the value of current in 6 Ω resistor by Thevenin's theorem.

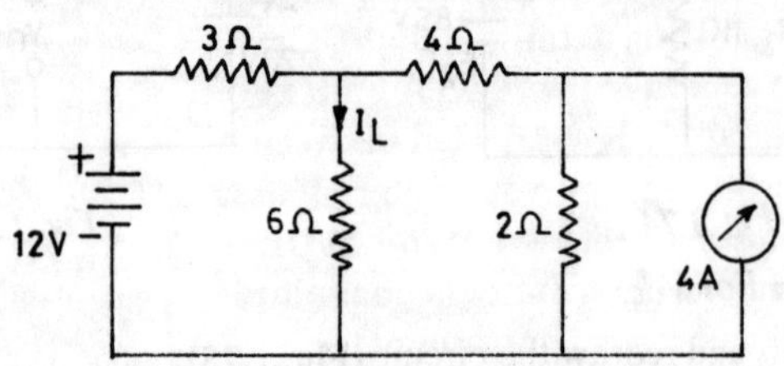

Fig. 1.78.

Solution : *For* V_{OC}

Step I : *(i)* Remove the 6 Ω resistance.

(ii) Convert 4 A current source into voltage source and redraw the circuit (Fig. 1.79).

$$3I + 6I = 12 - 8$$

$$9I = 4$$

$$I = \frac{4}{9} \text{ A}$$

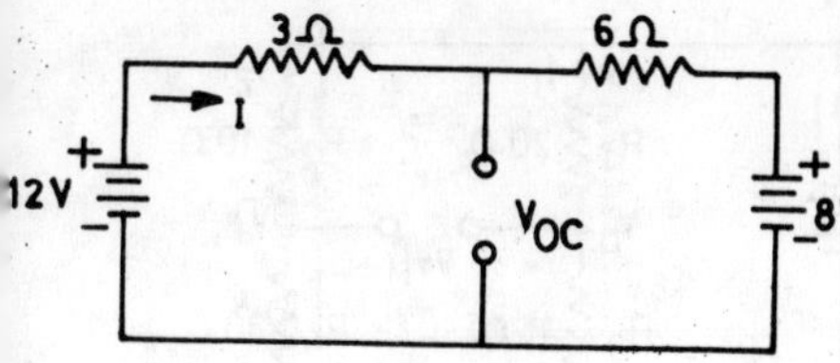

Fig. 1.79.

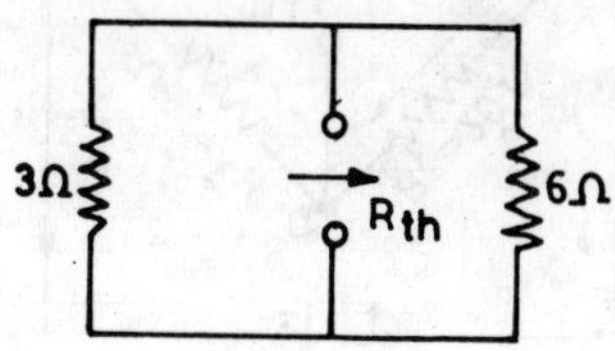

Fig. 1.80.

$$V_{OC} = E_1 - IR_1 = 12 - \frac{4}{9} \times 3 = 12 - \frac{4}{3} = \frac{32}{3} = 10.06 \text{ V}$$

Step II : ***For*** R_{th}

Short circuit the batteries and redraw the circuit (Fig. 1.80).

$$R_{th} = \frac{3 \times 6}{3+6} = \frac{18}{9} = 2\ \Omega$$

$$\therefore \quad I_L = \frac{V_{OC}}{R_{th} + R_L} = \frac{10.06}{2+6} = \frac{10.06}{8} = 1.25 \text{ A } \textbf{Ans.}$$

Example 1.79. A Wheatstone bridge *ABCD* has the following details : *AB* = 10 Ω, *BC* = 30 Ω, *CD* = 15 Ω and *DA* = 20 Ω. A battery of emf 2 V and negligible is connected between *A* and *C* with *A* positive. A galvanometer of 40 Ω resistance is connected between *B* and *D*. Using Thevenin's theorem, determine the magnitude and direction of current in galvanometer.

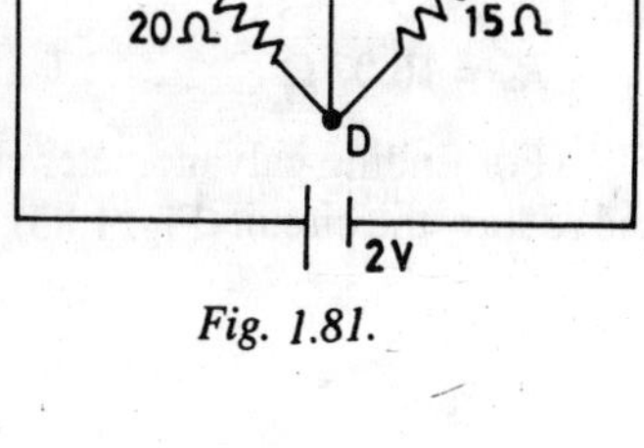

Fig. 1.81.

Solution : We shall find Thevenin's equivalent circuit of terminal *BD* in Fig. 1.81.

Step I : Finding V_{th} **:** To find V_{th}, at terminals *BD*, remove the load (*i.e.*, 40 Ω galvanometer) as shown in Fig. 1.82.

The voltage between terminals *B* and *D* is equal to V_{th}.

$$I_1 \text{ current in branch } R_1 \text{ and } R_3 = \frac{E}{R_1 + R_3} = \frac{2}{20+15} = \frac{2}{35} = 0.0571 \text{ A}$$

$$I_2 \text{ current in } R_2 \text{ and } R_4, = \frac{E}{R_2 + R_4} = \frac{2}{10+30} = \frac{1}{20} = 0.05 \text{ A}$$

$$\text{P.D. across } R_1,\ V_{oa} = \frac{2}{35} \times 20 = 1.142 \text{ A}$$

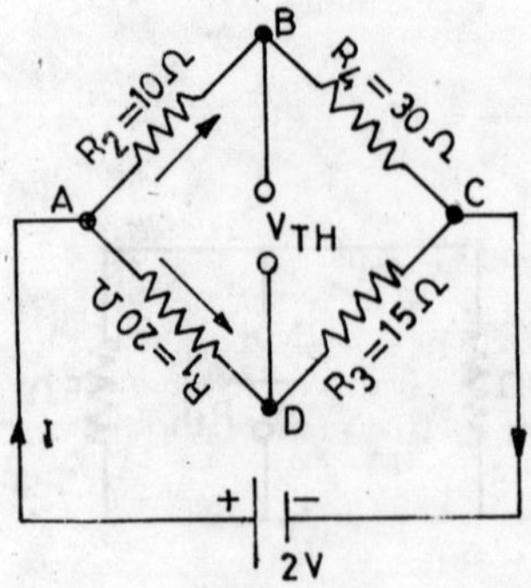

Fig. 1.82.

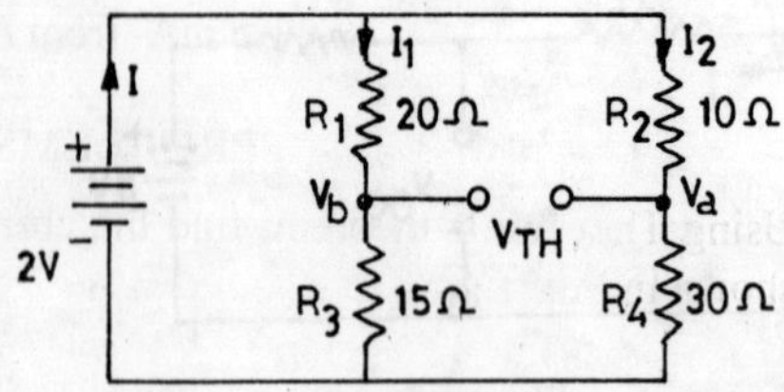

Fig. 1.83.

$$\text{P.D. across } R_2,\ V_{ob} = \frac{1}{20} \times 10 = 0.5 \text{ V}$$

P.D. between *B* and *D*, V_{th}

$$V_{th} = V_{oa} - V_{ob} = 1.412 - 0.5 = 0.642 \text{ V} \quad \textbf{Ans.}$$

Finding R_{th} : In order to find R_{th}, remove the load and replace the battery by a short as shown in Fig. 1.84. Then resistance measured beteen terminals *B* and *D* is equal to R_{th}.

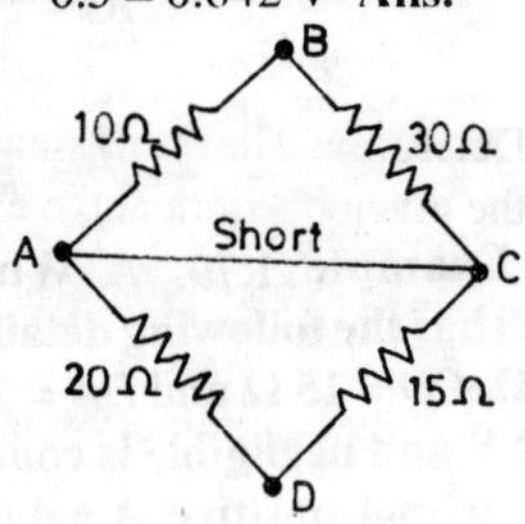

Fig. 1.84.

R_{th} = Resistance at terminals BD in Fig. 1.84.

$$= \frac{10 \times 30}{10 + 30} + \frac{20 \times 15}{20 + 15} = 7.5 + 8.57 = 16.07\ \Omega$$

$$R_{th} = 16.07\ \Omega$$

For finding galvanometer current connect the galvanometer resistance back and redraw the circuit (Fig. 1.85).

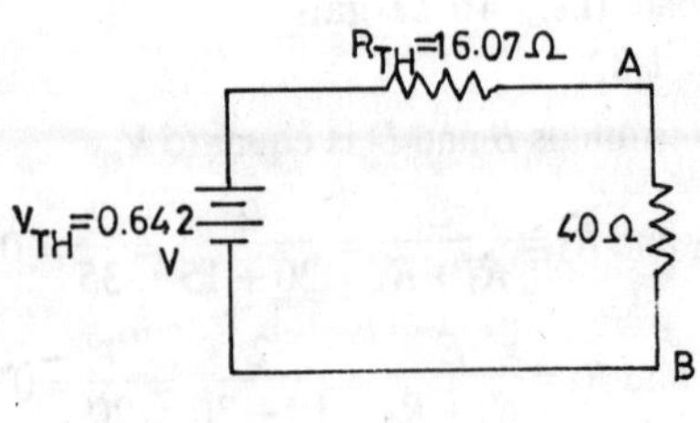

Fig. 1.85.

$$\text{Galvanometer current} = \frac{V_{th}}{R_{th} + 40} = \frac{0.642}{16.07 + 40} = 11.5 \times 10^{3} \text{ A}$$

$$= 11.5 \text{ mA from } B \text{ to } D \quad \textbf{Ans.}$$

PROBLEMS

1. Using Thevenin's theorem, find the current in 10 Ω resistor in the circuit shown in Fig. 1.86. **[Ans.** 0.481 A]

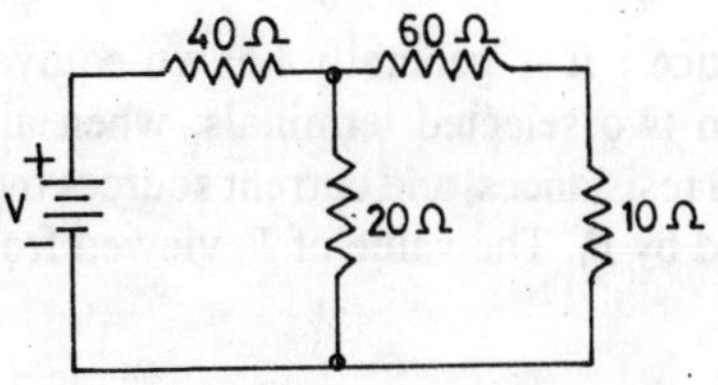

Fig. 1.86.

2. Determine Thevenin's equivalents circuit which may be used to represent the given network at the terminals *AB*. **[Ans.** $V_{th} = 5.63$ V ; $R_{th} = 2.21$ Ω]

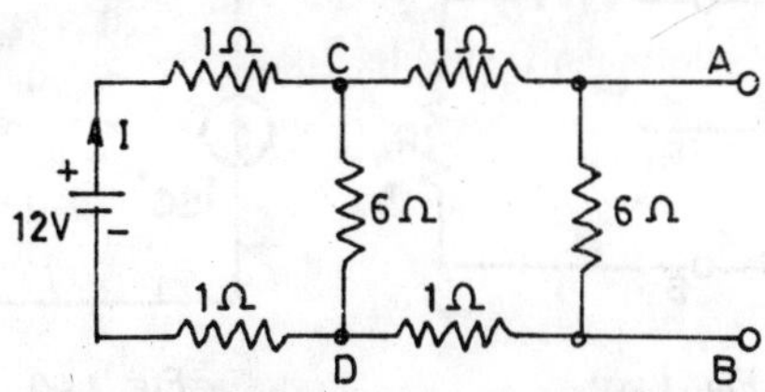

Fig. 1.87.

1.21.2. Norton's Theorem

According to this theorem, the current passing through a resistance between any two terminals of the network can be determined by replacing the entire network by an equivalent constant current source and a parallel resistance.

- **Norton's Current :** It is supplied by the source which will pass through two selected terminals, when they are short circuited. It is denoted by I_{SC}. Its value is $I_{SC} = \frac{V}{R_1}$ A.

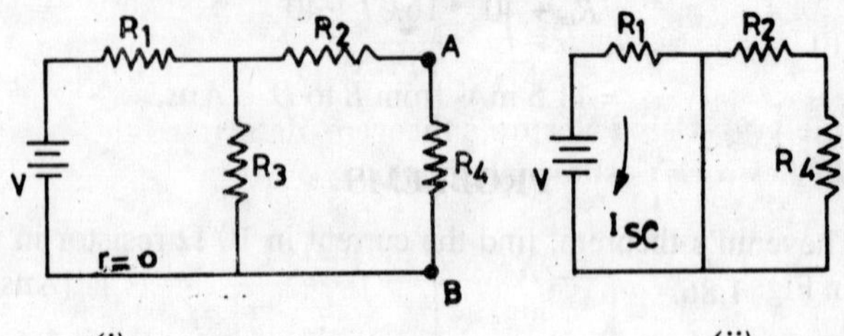

(i) (ii)

Fig. 1.88.

Norton's Resistance : It is generally known equivalent resistance of the network as viewed from two selected terminals, when all the emf sources are replaced by their internal resistances, and current sources by open circuit as shown in Fig. 1.89. It is denoted by R_i. The value of R_i viewed from the terminals *A* and *B* is given :

$$R_i = \frac{R_1 \times (R_2 + R_4)}{R_1 + (R_2 + R_4)}$$

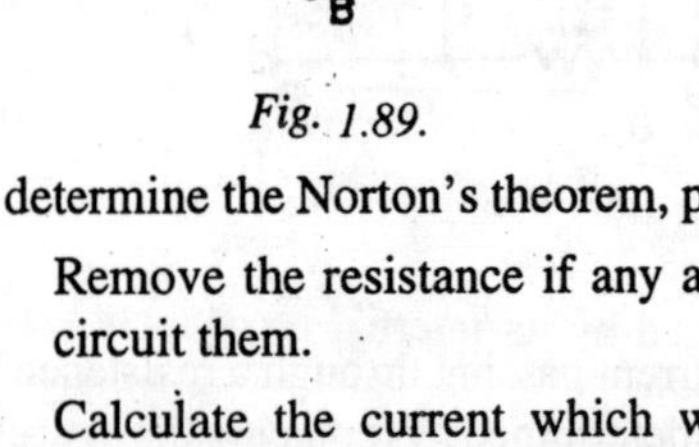

Fig. 1.89.

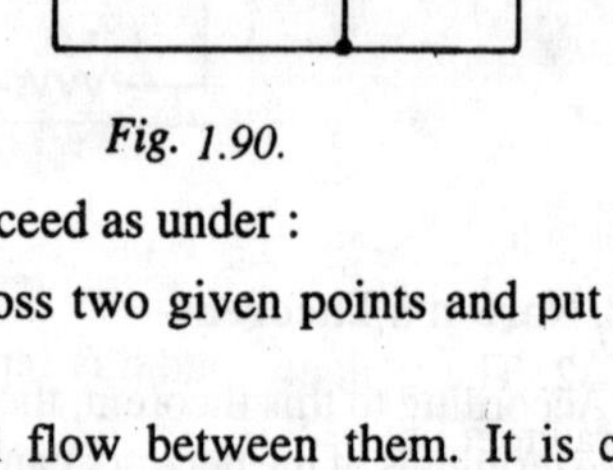

Fig. 1.90.

To determine the Norton's theorem, proceed as under :

(i) Remove the resistance if any across two given points and put short circuit them.

(ii) Calculate the current which will flow between them. It is called Norton's current and is written as I_{SC}.

(iii) Find the resitance R_i of the whole network as viewed from terminals *AB*, after all the voltage sources have been removed by short circuits and replaced by their internal resistances.

(iv) The Norton's resistance and load resistance R_3 are connected parellel across current sources as showin in Fig. 1.90.

(v) Then the load current I_L is determined from following relation :

$$I_L = I_{SC} \times \frac{R_i}{R_i + R_3}$$

Example 1.80. Using Norton's theorem, determine the current in 20 Ω resistor in the network as shown in Fig. 1.91.

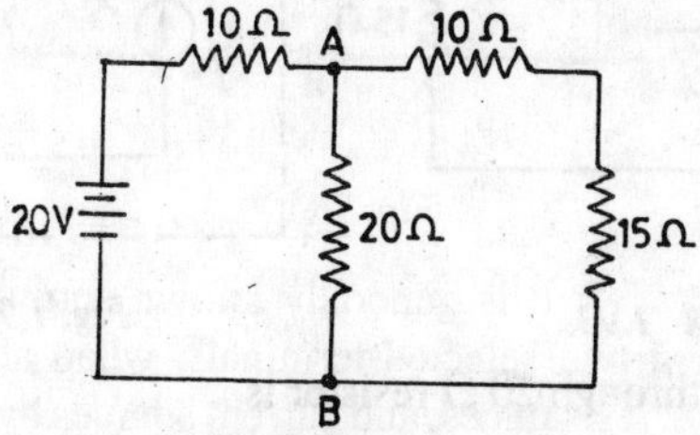

Fig. 1.91.

Solution : To determine the current in 20 Ω resistor the following steps should be followed :

1. Remove the 20 Ω resistor and short circuit the two terminals as shown in Fig. 1.92. Determine the current I_{SC}, which is limited by only 10 Ω resistor.

$$I_{SC} = \frac{20}{10} = 2 \text{ A}$$

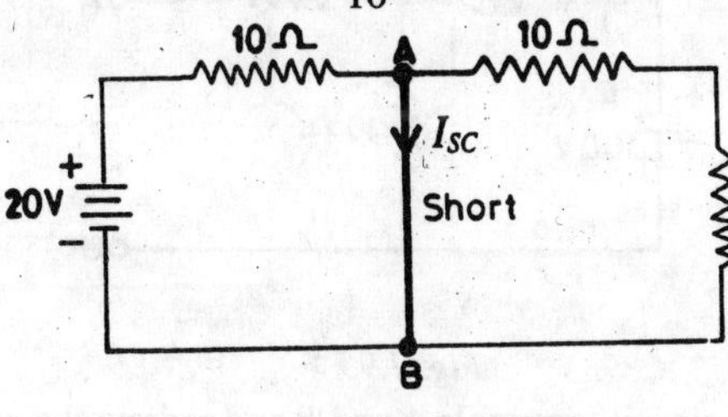

Fig. 1.92.

2. The voltage source is replaced by its internal resistance. The circuit is re-drawn in Fig. 1.93.

Keeping the selected terminals A and B open circuited R_i. Norton's resistance or resistance of the network as seen from terminals A and B is under. Resistance 10 Ω and 15 Ω are in series and this series combination is in parallel with 10 Ω resistor

$$\therefore \quad R_i = \frac{25 \times 10}{25 + 10} = \frac{250}{35} = 7.14 \ \Omega$$

3. The Norton's current source is shown in Fig. 1.94.

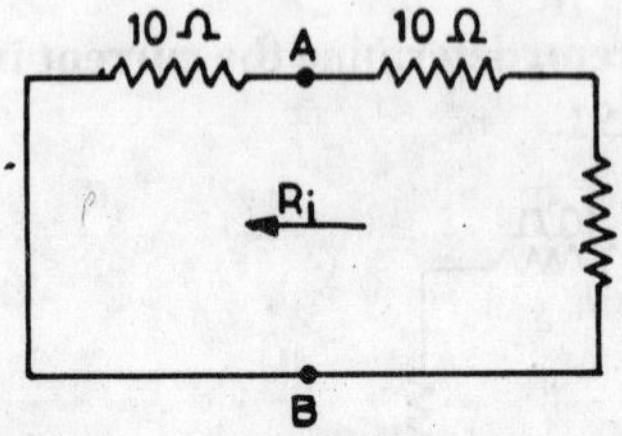

Fig. 1.93.

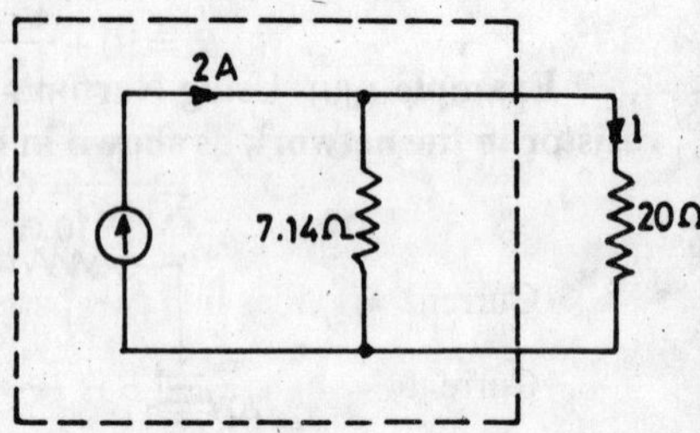

Fig. 1.94.

Current flowing through 20 Ω resistor is

$$I = I_{SC} \times \frac{R_i}{R_i + R_L} = 2 \times \frac{7.14}{7.14 + 20} = 0.526 \text{ A}$$

Example 1.81. Using Norton's theorem find the constant current equivalent of circuit shown in Fig. 1.95.

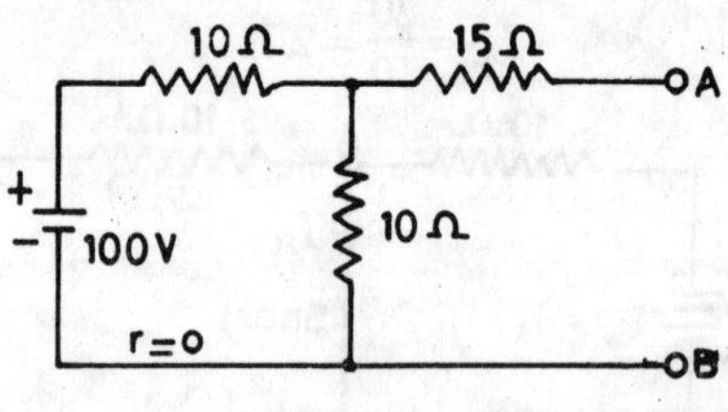

Fig. 1.95.

Solution : Short circuit terminals *A* and *B* and redraw the circuit (Fig. 1.96).

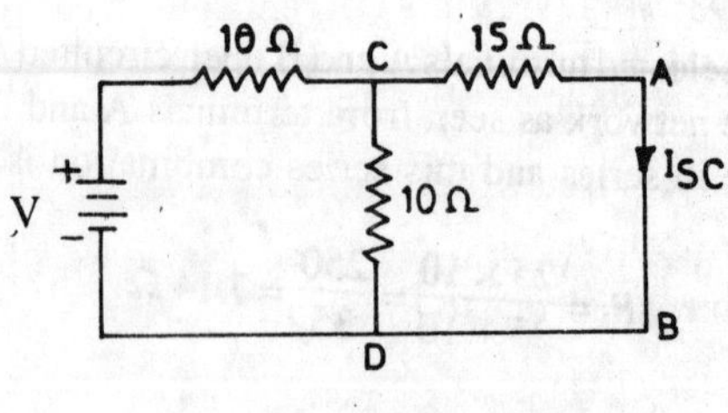

Fig. 1.96.

Resistor 10 Ω and 15 Ω are connected in parallel and this combination is connected in series with 10 Ω.

$$R_i = 10 + \frac{15 \times 10}{15 + 10} = 16\ \Omega$$

$$I = \frac{100}{16} = 6.25\ \text{A}$$

Current I divides into two parts at point C.

Current through A and B is I_{SC}

$$I_{SC} = \frac{I \times 10}{10 + 15} = \frac{6.25 \times 10}{10 + 15} = 2.5\ \text{A}$$

Now remove the battery source and redraw the circuit (Fig. 1.97).

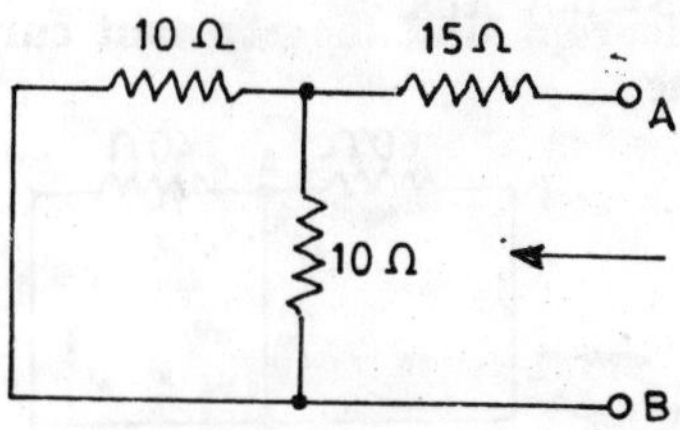

Fig. 1.97.

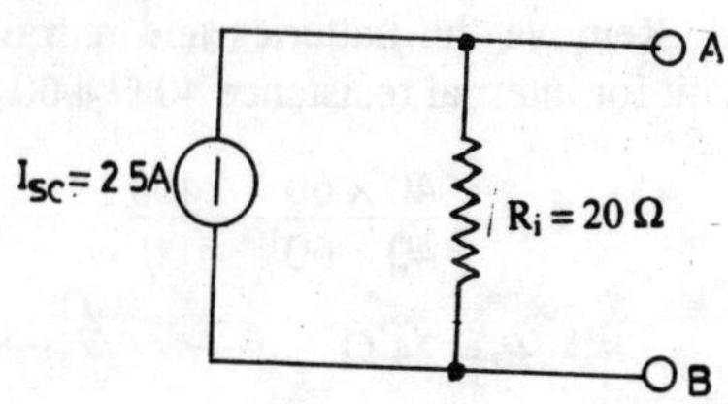

Fig. 1.98.

$$R_i = \frac{10 \times 10}{10 + 10} + 15 = \frac{100}{20} + 15 = 5 + 15 = 20\ \Omega$$

Now R_i and I_{SC} are connected in parallel.

Hence Norton's equivalents

$$I_{SC} = 2.5\ \text{A}$$

$$R_i = 20\ \Omega \quad \textbf{Ans.}$$

Example 1.82. Using Norton's current source equivalent circuit of network shown in Fig. 1.99. Find the current flowing when it takes the value of 12 Ω and 24 Ω.

Solution : Short circuit the terminals A and B and redraw the circuit **(Fig. 1.100).**

Current in shorted path due to E_1

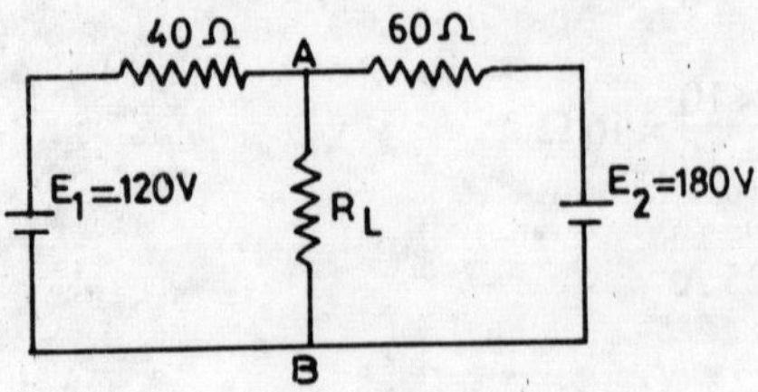

Fig. 1.99.

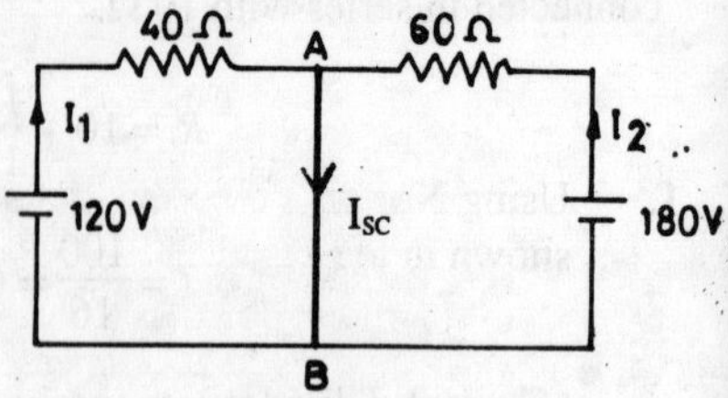

Fig. 1.100

$$I_1 = \frac{120}{40} = 3 \text{ A From } A \text{ to } B$$

Current due to battery E_2

$$I_2 = \frac{180}{60} = 3 \text{ A from } A \text{ to } B$$

Hence $\qquad I_{SC} = I_1 + I_2 = 3 + 3 = 6 \text{ A}$ **Ans.**

Remove the batteries and redraw the circuit for internal resistance 40 Ω || 60 Ω.

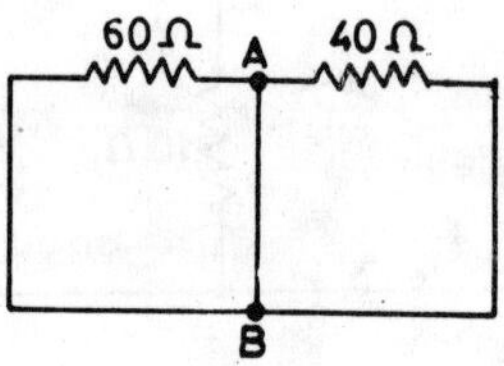

Fig. 1.101

$$R_i = \frac{40 \times 60}{40 + 60} = \frac{2400}{100}$$

$$R_i = 24 \, \Omega$$

Now the circuit can be drawn again with Norton's equivalents

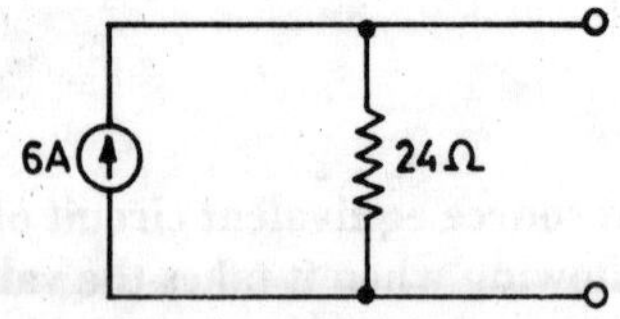

Fig. 1.102.

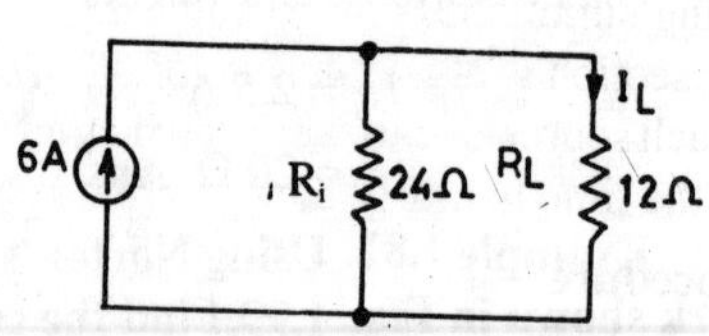

Fig. 1.103.

Now connect $R_L = 12 \, \Omega$

$$I_L = \frac{I_{SC} \times R_i}{R_i + R_L} = \frac{6 \times 24}{24 + 12} = 4 \text{ A}$$

When $R_L = 24$

$$\therefore \quad I_L = \frac{6 \times 24}{24 + 24} = 3\text{ A} \quad \textbf{An.}$$

PROBLEMS

1. Using Norton's theorem, determine current in 10 Ω resistor of the network shown in the Fig. 1.98. **[Ans.** 1.887 A**]**

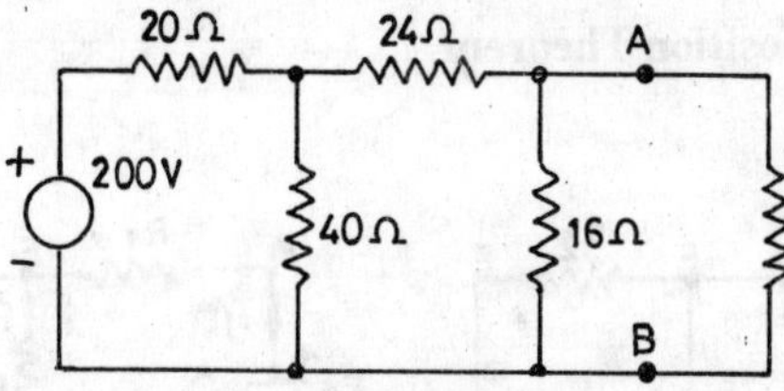

Fig. 1.104.

2. Apply Norton's theorem to calculate current flowing through 5 Ω resistor as shown in Fig. 1.99. **[Ans.** $I_{AB} = 0.5$ A**]**

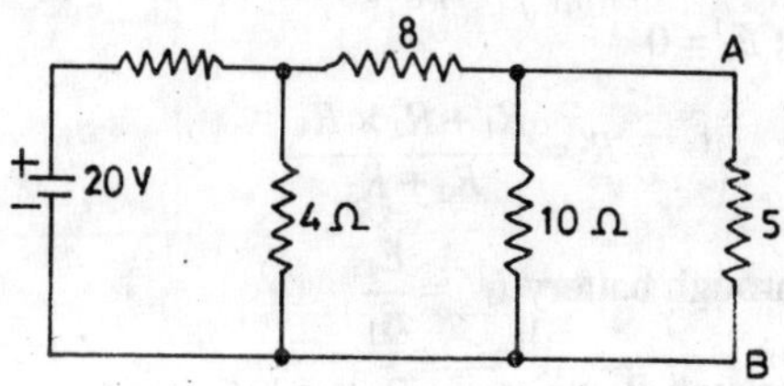

Fig 1.105.

1.12.3. Superposition Theorem

According to this theorem, if there are two or more than two sources of emf's acting simultaneously in a linear bilateral network. The current flowing through any section is algebraic sum of all the currents, which would flow in that section. If each source of emf were considered separately and all other sources are removed for the time being by their internal resistance.

Procedure

(i) Mark the direction of currents in various sections I_1, I_2, I_3.

(ii) Now consider only one source of emf say E_1. Redraw the circuit and determine the current in those section as I_1', I_2' and I_3' respectively.

(iii) Now consider the second source E_2 and short circuit E_1 and redraw the circuit.

(*iv*) Again determine the current, I_1'', I_2'' and I_3''. By superposition theorem, the actual flow of current in various sections can be determined by superimposing the current, *i.e.*,

$$I_1 = I_1' - I_1''$$

$$I_2 = I_2' - I_2''$$

$$I_3 = I_3' + I_3''$$

Solution of Superposition Theorem

Case I :

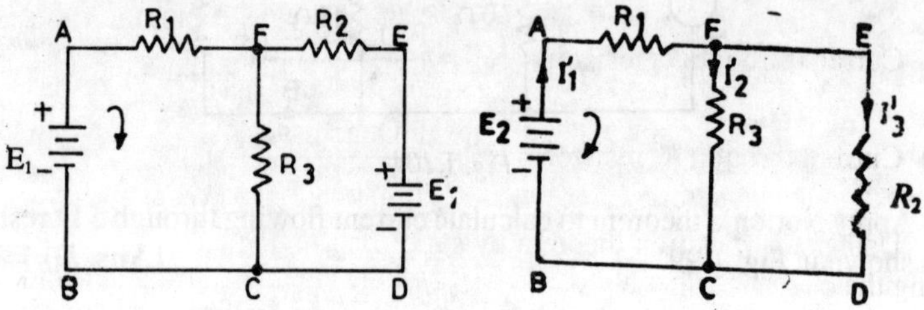

Fig. 1.106.

Short circuit, $E_2 = 0$

$$R' = \frac{R_1 + R_2 \times R_3}{R_2 + R_3}$$

I_1' current through battery $I_1' = \dfrac{E_1}{R_1}$

I_1' current through R_1 resistor = Battery current I_1

P.D. across FC and $ED = E_1 - I_1' R_1$

Current through R_2 and resistor $= I_2' = \dfrac{\text{P.D. across } ED}{R_2}$

Current through R_3 and resistor $= I_3' = \dfrac{\text{P.D. across } FC}{R_3}$

Case II :

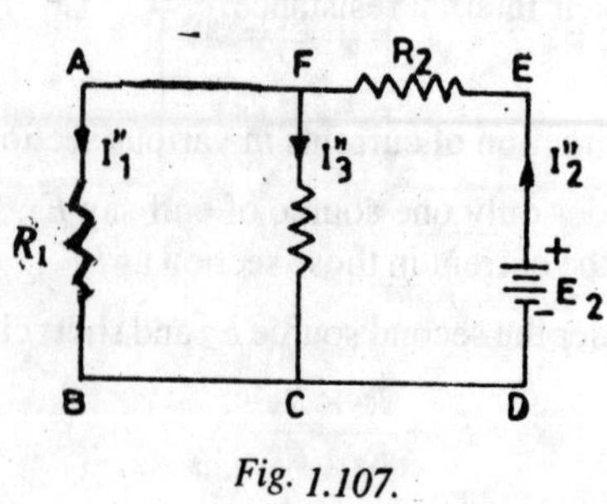

Fig. 1.107.

Short circuit, $E_1 = 0$

Circuit resistance $R'' = R_2 + \dfrac{R_1 R_3}{R_1 + R_3}$

I_2 current through battery $= \dfrac{E_2}{R''}$ amp

I_2'' current through R_2 resistor = Battery current

P.D. across FC and $AB = E_2 - I_2'' R_2$

Current through R_3 resistor $= I_3'' = \dfrac{\text{P.D. across } FC}{R_3}$

Current through R_1 resistor $= I_1'' = \dfrac{\text{P.D. across } AB}{R_1}$

Using superposition theorem by superposing the circuit I & II.

Current in $R_1\ \Omega$ resistor, $I_1 = I_1' - I_1''$

Current in $R_2\ \Omega$ resistor, $I_2 = I_2' + I_2''$

Current in $R_3\ \Omega$ resistor, $I_3 = I_3' + I_3''$

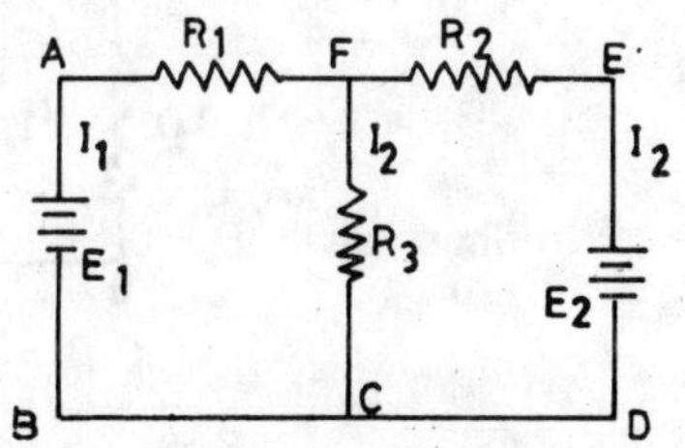

Fig. 1.108.

Problem : Verify superposition theorem in the following network for current I.

Note : In order to make use of superposition theorem each source, once at a time the remaining source is represented by their internal resistor thus short circuiting E_2 voltage source, we get the following circuit shown in Fig. 1.104.

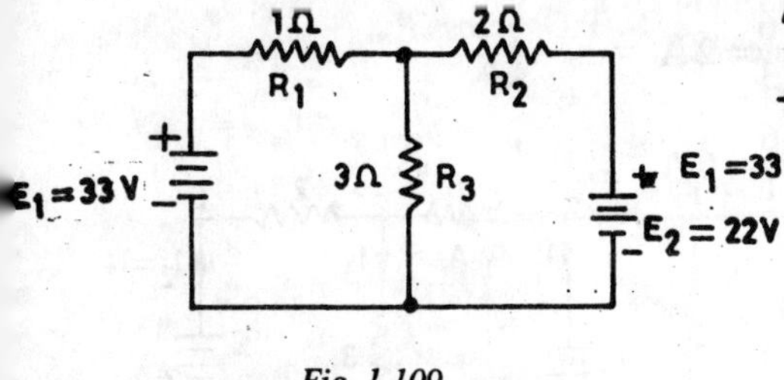

Fig. 1.109.

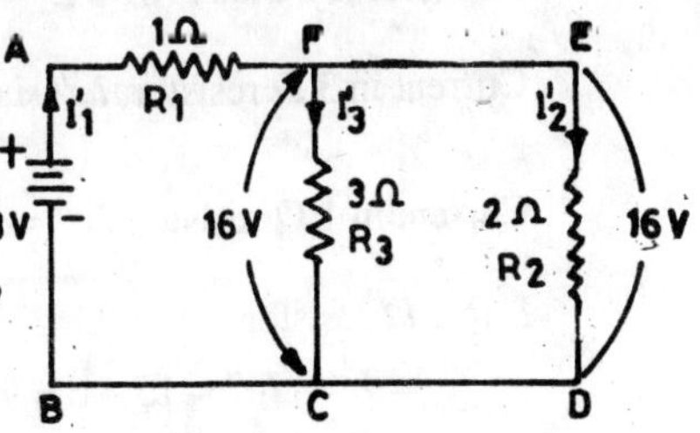

Fig. 1.110.

Total circuit resistance

$$R' = R_1 + \frac{R_2 \times R_3}{R_2 + R_3}$$

$$R' = 1 + \frac{2 \times 3}{2+3} = 1 + \frac{6}{5} = \frac{11}{5} = 2.2\ \Omega$$

$$I_1 \text{ current through battery} = \frac{E_1}{R'} = \frac{33}{11} \times 5 = 15\ \text{A}$$

$$\text{Current through } 1\ \Omega \text{ resistor} = I_1{}' = \text{Battery current} = 15\ \text{A}$$

$$\text{P.D. across } FC \text{ and } ED = E_1 - I_1{}' R_1 = 33 - 15 \times 1 = 18\ \text{V}$$

$$I_2{}' \text{ current through } 2\ \Omega \text{ resistor} = \frac{18}{2} = 9\ \text{A}$$

$$I_3{}' \text{ current through } 1\ \Omega \text{ resistor} = \frac{18}{3} = 6\ \text{A}$$

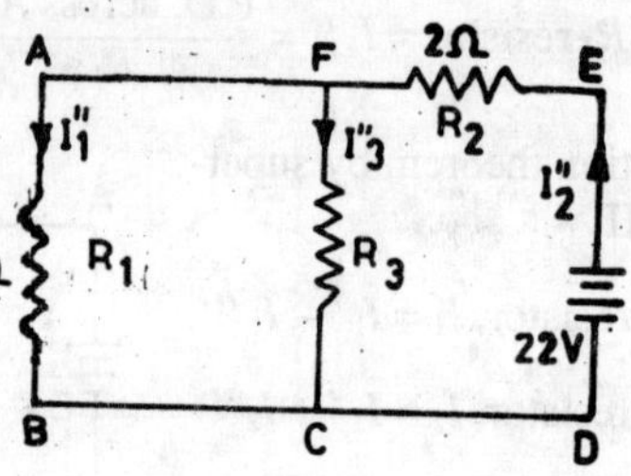

Fig. 1.111.

$$\text{Circuit resistor } R'' = R_2 + \frac{R_1 R_3}{R_1 + R_3} = 2 + \frac{1 \times 3}{1+3} = 2 + \frac{3}{4} = \frac{11}{4}$$

$$\text{Battery current } I_2 = \frac{E_2}{R''} = \frac{22}{11} \times 4 = 8\ \text{A}$$

$I_2{}''$ current through 2 Ω resistor = Battery current 8 A

P.D. across FC and $AB = E_2 - I_2{}'' R_2 = 22 - 8 \times 2 = 6$ V

$$\text{Current in } 3\ \Omega \text{ resistor, } I_3{}'' = \frac{6}{3} = 2\ \text{A}$$

$$\text{Current in } 1\ \Omega \text{ resistor, } I_1{}'' = \frac{6}{1} = 6\ \text{A}$$

I_1 in 1 Ω resistor

$= I_1{}' - I_1{}'' = 15 - 6 = 9$ A

I_3 in 3 Ω resistor

$= I_3{}' + I_3{}'' = 6 + 2 = 8$ A

I_2 in 2 Ω resistor

$= I_2{}' - I_2{}'' = 9 - 8 = 1$ A

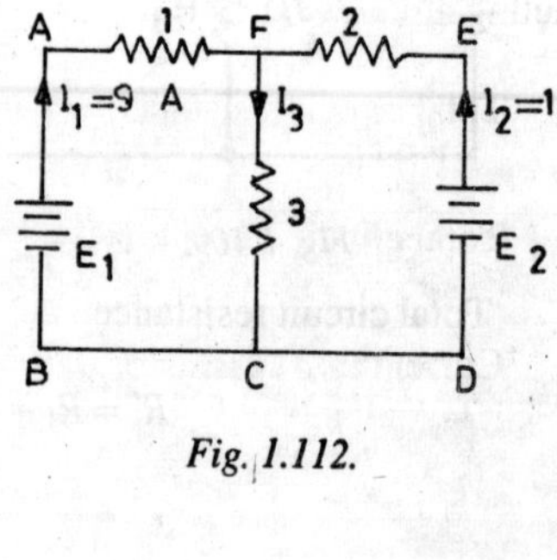

Fig. 1.112.

Example 1.83. In the network shown in Fig. 1.113 *(i)*, find the different branch currents by superposition theorem.

Solution : In Fig. 1.113 *(ii)*, 40 V source is replaced by a short so that e.m.f. of 35 V is acting alone. The various branch currents due to 35 V alone are [Refer to Fig. 1.113 *(ii)*] :

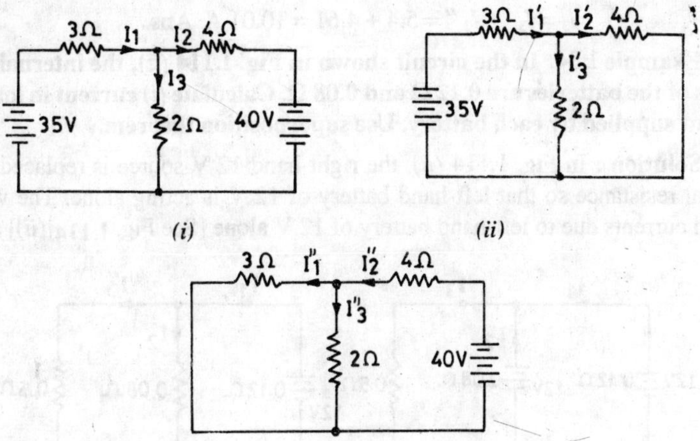

(i) *(ii)* *(iii)*

Fig. 1.113.

$$\text{Total circuit resistance} = 3 + \frac{2 \times 4}{2+4} = 4.33\ \Omega$$

$$\text{Total circuit current, } I_1{}' = \frac{35}{4.33} = 8.08 \text{ A}$$

$$\text{Current in 4 } \Omega \text{ resisotr, } I_2{}' = 8.08 \times \frac{2}{(2=4)} = 2.69 \text{ A}$$

$$\text{Current in 2 } \Omega \text{ resistor, } I_3{}' = 8.08 \times \frac{4}{(2+4)} = 5.39 \text{ A}$$

In Fig. 1.113 *(iii)*, 35 V source is replaced by a short so that now 40 V source is acting alone.

$$\text{Total circuit resistance} = 4 + \frac{3 \times 2}{3+2} = 5.2\ \Omega$$

$$\text{Total circuit current, } I_2{}'' = \frac{40}{5.2} = 7.69 \text{ A}$$

$$\text{Current in 3 } \Omega \text{ resistor, } I_1{}'' = 7.69 \times \frac{2}{2+3)} = 3.08 \text{ A}$$

Current in 2 Ω resistor, $I_3'' = 7.69 \times \frac{3}{(2+3)} = 4.61$ A

The actual current values I_1, I_2 and I_3 in Fig. 1.113 *(i)* can be found by algebraically adding the component values.

$$\therefore \quad I_1 = I_1' - I_1'' = 8.08 - 3.08 = 5 \text{ A} \textbf{ Ans.}$$

$$I_2 = I_2' - I_2'' = 2.69 - 7.69 = -5 \text{ A} \textbf{ Ans.}$$

$$I_3 = I_3' + I_3'' = 5.4 + 4.61 = 10.01 \text{ A} \textbf{ Ans.}$$

Example 1.84. In the circuit shown in Fig. 1.114 *(ii)*, the internal resistances of the batteries are 0.12 Ω and 0.08 Ω. Calculate *(i)* current in load, *(ii)* current supplied by each battery. Use superposition theorem.

Solution : In Fig. 1.114 *(ii)*, the right-hand 12 V source is replaced by its internal resistance so that left-hand battery of 12 V is acting alone. The various branch currents due to left-hand battery of 12 V alone [See Fig. 1.114 *(ii)*] are :

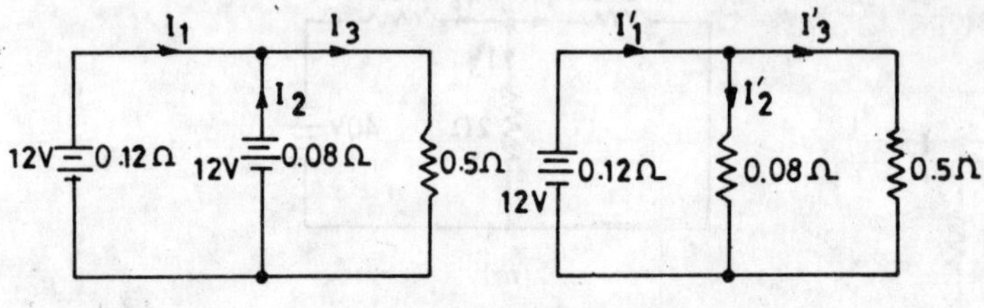

(i) *(ii)*

Fig. 1.114.

Total circuit resistance $= 0.12 + \frac{0.08 \times 0.5}{0.08 + 0.5} = 0.189$ Ω

Total circuit current, $I_1' = \frac{12}{0.189} = 63.5$ A

Current in 0.08 Ω, $I_2' = 63.5 \times \frac{0.5}{0.08 + 0.5} = 54.74$ A

Current in 0.5 Ω, $I_3' = 63.5 \times \frac{0.08}{0.08 + 0.5} = 8.76$ A

In Fig. 1.114 *(iii)*, left-hand 12 V source is replaced by its internal resistance so that now right-hand 12 V source is acting alone.

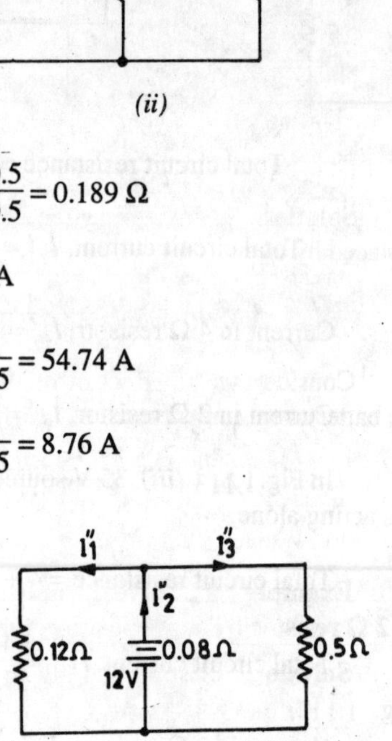

Fig. 1.114 (iii).

Total circuit resistance

$$= 0.08 = \frac{0.12 \times 0.5}{0.12 + 0.5} = 0.177 \text{ Ω}$$

Total circuit current,

$$I_2'' = \frac{12}{0.177} = 67.8 \text{ A}$$

Current in 0.12 Ω, $I_1'' = 67.8 \times \frac{0.5}{0.12 + 0.5} = 54.6 \text{ A}$

Current in 0.5 Ω, $I_3'' = 67.8 \times \frac{0.12}{0.12 + 0.5} = 13.12 \text{ A}$

The actual current values of I_1 (current in first battery), I_2 (current in second battery) and I_3 (load current) can be found by algebraically adding the component values.

$$I_1 = I_1' - I_1'' = 63.5 - 54.6 = 8.9 \text{ A} \quad \textbf{Ans.}$$

$$I_2 = I_2'' - I_2' = 67.8 - 54.74 = 13.06 \text{ A} \quad \textbf{Ans.}$$

$$I_3 = I_3' + I_3'' = 8.76 + 13.12 = 21.88 \text{ A} \quad \textbf{Ans.}$$

Example 1.85. Using superposition theorem find the current through the 6 Ω resistor of the network.

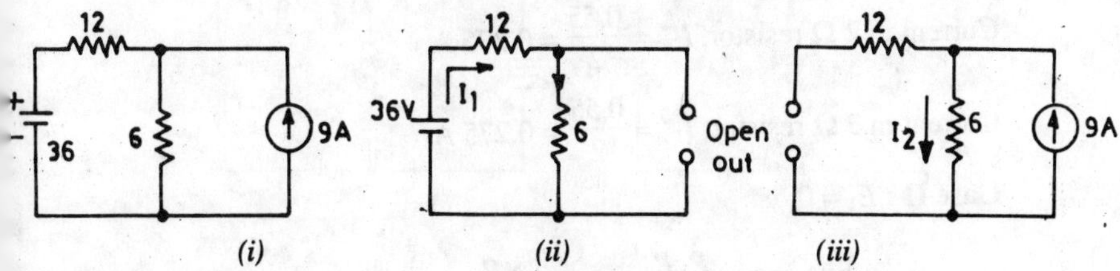

Fig. 1.115.

Solution : Considering the effect of 36 V source, *i.e.*, current source is replaced by circuit [Fig. 1.109 *(ii)*].

$$I_1 = \frac{E}{R_T} = \frac{35}{12 + 6} = 2 \text{ A}$$

Considering the effect of the 9 A source byu current divider rule. Replace the battery by circuit [Fig. 1.115 *(iii)*].

$$I_2 = \frac{9 \times 12}{12 + 6} = \frac{108}{18} = 6 \text{ A}$$

$$I_{6\,\Omega} = 2 + 6 = 8 \text{ A} \quad \textbf{Ans.}$$

Example 1.86. Solve the network shown in Fig. 1.116, for finding current in 2 Ω resistor by superposition theorem.

Solution : Case I : Taking $E_2 = 0$, then the circuit is reduced as shown in Fig. 1.117.

$$R' = R_1 + \frac{R_2 R_3}{R_2 + R_3} = 1 + \frac{2 \times 2}{2 + 2} = 1 + 1 = 2\ \Omega$$

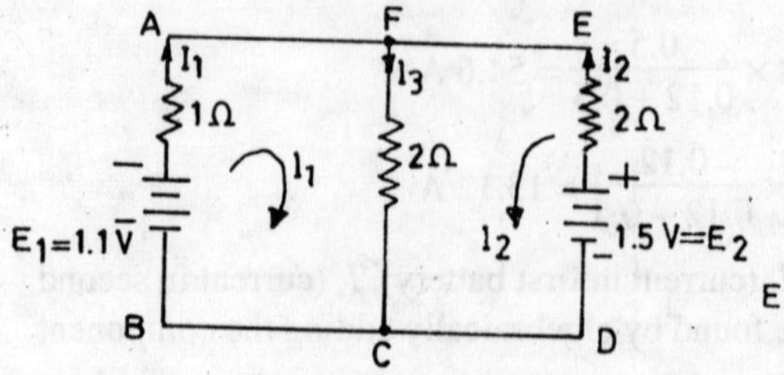

Fig. 1.116.

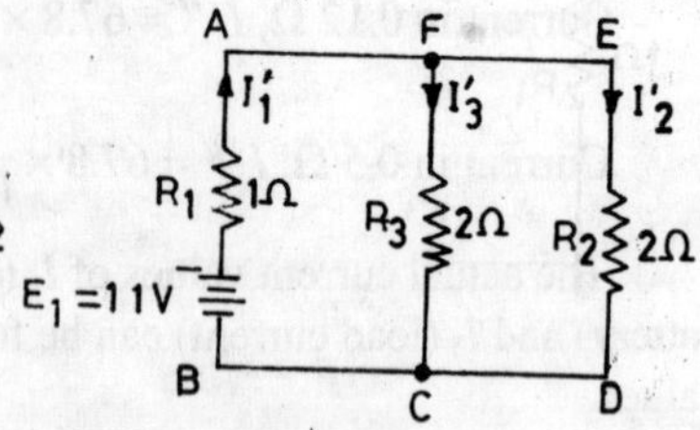

Fig. 1.117.

Current supplied by battery E_1, $I_1{}' = \dfrac{E_1}{R'} = \dfrac{1.1}{2} = 0.55$ A

Current through 1 Ω resistor, $I_1{}' = 0.55$ A

P.D. across FC and $ED = E_1 - I_1{}' R_1 = 1.1 - 0.55 \times 1 = 0.55$ V

Current in 2 Ω resistor, $I_2{}' = \dfrac{0.55}{2} = 0.275$ A

Current in 2 Ω resistor, $I_3{}' = \dfrac{0.55}{2} = 0.275$ A

Case II : $E_1 = 0$

$$R'' = R_2 + \frac{R_1 R_3}{R_1 + R_3} = 2 + \frac{1 \times 2}{1 + 2} = 2 + 0.66 = 2.66\ \Omega$$

Current supplied by battery E_2, $I_2{}'' = \dfrac{E_2}{R''} = \dfrac{1.5}{2.66} = 0.563$ A

$I_2{}''$ current in 2 Ω resistor = 0.563 A

P.D. across branch FC and AB,

$$= E_2 - I_2{}'' R_2 = 1.5 - 0.563 \times 2 = 1.5 - 1.126 = 0.374 \text{ V}$$

$I_3{}''$ current in 2 Ω resistor $= \dfrac{0.374}{2} = 0.187$ A

$I_1{}''$ current in 1 Ω resistor $= \dfrac{0.374}{1} = 0.374$ A

Superposing the circuit shown in Fig. 1.118 and Fig. 1.119.

Current in 1 Ω resitor $= I_1{}' + (-I_1{}'') = 0.55 - 0.374 = 0.176$ A

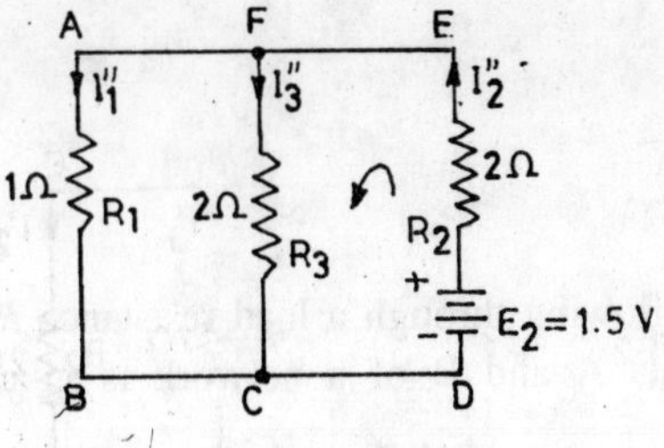

Fig. 1.118.

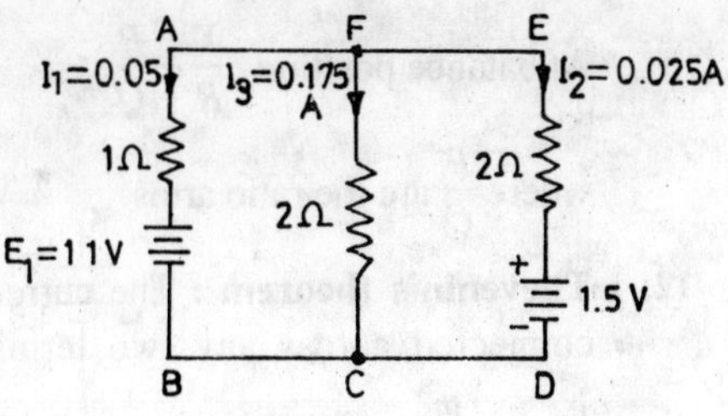

Fig. 1.119.

Current in 2 Ω resistor = $I_2' + (-I_2'') = 0.275 - 0.562 = -0.287$ A

Current in 2 Ω resistor = $I_3' + (I_3'') = 0.275 + 0.187$

∴ $I_3 = 0.462$ A **Ans.**

HIGHLIGHTS

1. **Electricity :** The invisible energy which constitute the flow of electrons in a closed circuit to do work is called electricity.
2. **Electric current :** The rate of movement of electrons is called electric current.
3. **Electric potential :** The capacity of a charged body to do work is called electric potential.
4. Law of resistance are summed up as $R = \rho\, l/A$ ohm.
5. **Resistivity :** It is defined as the resistance between the opposite faces of a metre cube of that material.
6. **Conductance :** Reciprocal of resistance. Its unit is Siemens (S).
7. **Temperature coefficient of resistance :** It is defined as the change in resistance per ohm original resistance per degree change in temperature.

$$\alpha = \frac{R_t - R_0}{R_o\, t} /^{\circ}\text{C}$$

8. Ohm's law can be written as $I = \dfrac{V}{R}$. This law is applicable both to d.c. as well as a.c. circuits.
9. In series circuit : $R = R_1 + R_2 + \ldots + R_n$

In parallel circuit : $= \dfrac{1}{R} = \dfrac{1}{R_1} + \dfrac{1}{R_2} + \ldots + \dfrac{1}{R_n}$

10. Two laws of Kirchhoff's can be briefly stated as below :

(i) current law $\Sigma I = 0$ at a point or junction.

(ii) voltage law : $\Sigma IR + \Sigma$ emf = 0 round a mesh.

11. Wheatstone bridge :

At balance position : $\frac{X}{R} = \frac{P}{Q}$

where $\frac{P}{Q}$ are the ratio arms.

12. **Thevenin's theorem :** The current flowing through a load resistance R_L connected across any two terminals A and B of a network is given $I = \frac{V_{th}}{R_{th} + R_L}$ where V_{th} is open circuit voltage and R_{th} is the internal resistance.

13. **Superposition theorem :** According to this theorem; if there are two or more than two sources of emf's acting simultaneously in a linear bilateral network. The current flowing through any section is algebraic sum of all the current, which would flow in that section. If each source considered separately and all other sources are removed by their internal resistance.

14. **Norton's theorem :** According to Norton's theorem, the current passing through a resistance between any two terminals of the network can be determined by replacing the entire network by an equivalent constant current source and a parallel resistance.

15. **Delta-star transformation :** Resistance of each arm of the star is given by the product of the resistance of the two delta sides that meet as its end divided by the sum of the three delta resistances.

16. **Star-delta transformation :** The equivlanet delta resistance between any two terminals is given by the sum of star resistances between these terminals plus the product of these two star resistances divided by the third star resistance.

SHORT QUESTIONS

1. The resistivity of different materials is different. Why ?
2. Why carbon and silicon has negative temperature coefficient resistance ?
3. When load is connected across the cell terminals, there is fall in p.d.
4. A 100 W bulb has more resistance than 200 W bulb. Why ?
5. Why the resistances are connected in series ?
6. Why the cells are connected in series ?
7. Why the electrical equipments are connected in parallel ?
8. What is the difference between loop and mesh ?
9. Why the different theorems are used in solving electrical network ?

MULTIPLE CHOICE QUESTIONS

1. According to KCl as applied to a junction in a network :
   (a) total sum of currents meeting at a junction is zero.
   (b) no current can leave the junction without some entering it.
   (c) algebraic sum of currents meeting at a junction is zero.
2. Kwh is the unit of :
   (a) Power (b) Energy (c) Work
3. The specific resistance of a conductor depends upon
   (a) length of conductor.
   (b) nature of conductor material.
   (c) diameter of conductor.
4. Kirchhoff's voltage law is concerned with :
   (a) IR drops
   (b) junction voltage
   (c) battery emf's
   (d) both (a) and (b).
5. The area of cross section is doubled and length of conductor is also doubled, the resistance will be
   (a) increase four times
   (b) remain unchanged
   (c) decrease to four times.
6. The temperature coefficient of resistance is negative in case of
   (a) Insulators (b) Conductor (c) Electrolytes
   (d) both in (a) and (c)
7. The unit of electric energy
   (a) watt-sec (b) Kwh (c) joule
   (d) all in (a), (b) & (c)
8. While Thevenizing a circuit between two terminals, V_{th} equals :
   (a) short circuit terminal voltage.
   (b) open circuit terminal voltage
   (c) net voltage available in the circuit.

9. Norton's equivalent of a circuit is
   (*a*) constant-current source with a conductance in parallel.
   (*b*) constant voltage source in parallel with high resistance.
   (*c*) single current source and single voltage source.

10. A 100 watt, 220 volts bulb takes a current of
   (*a*) .455 A (*b*) 2.2 A (*c*) 0.22 A

ANSWERS

1. (*c*)	2. (*b*)	3. (*b*)	4. (*d*)	5. (*b*)
6. (*d*)	7. (*d*)	8. (*b*)	9. (*a*)	10. (*a*)

EXERCISE

1. State and explain Kirchhoff's current and voltage laws.
2. State and explain superposition theorem how is it applied for solving a network ?
3. State Thevenin's theorem illustrate the application of the theorem with reference to an appropriate electric network. Can Thevenin's theorem be applied to AC ?
4. State and explain Norton's theorem. Show that this theorem is just the converse to Thevenin's theorem.
5. Derive the necessary equations for converting a delta network into an equivalent star network.
6. Write short note on Star-delta transforamtion.
7. By using star-delta transformation for Fig. 1.120. Find out current supplied by battery. **[Ans.** 0.8422 A]
8 In the network in Fig. 1.121. Find the resistance between the point A and B. **[Ans.** 2.5 Ω]

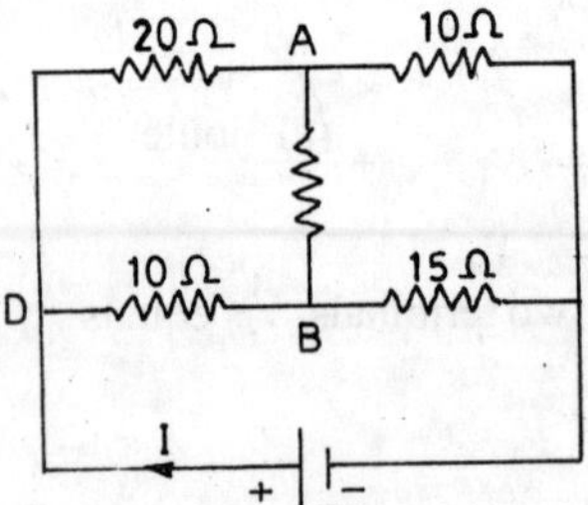

Fig. 1.120.

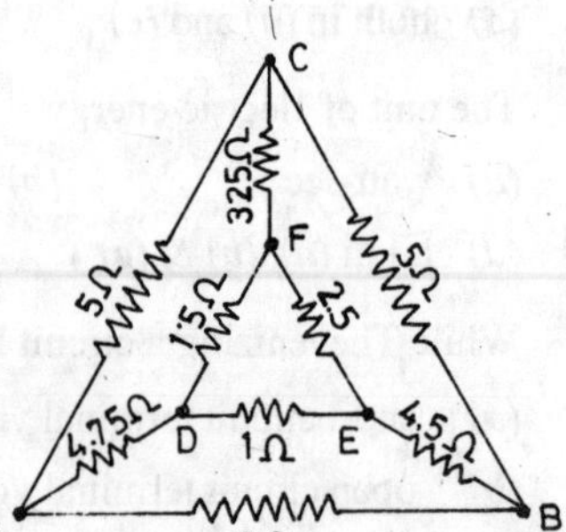

Fig. 1.121.

9. Solve the given circuit (Fig. 1.122) by applying superposition theorem and determine the current in 1 Ω resistor, the internal resistance of 1.6 V battery is 1.0 Ω and internal resistance of 1.2 V battery is 0.5 Ω. [**Ans.** 1.0 A]

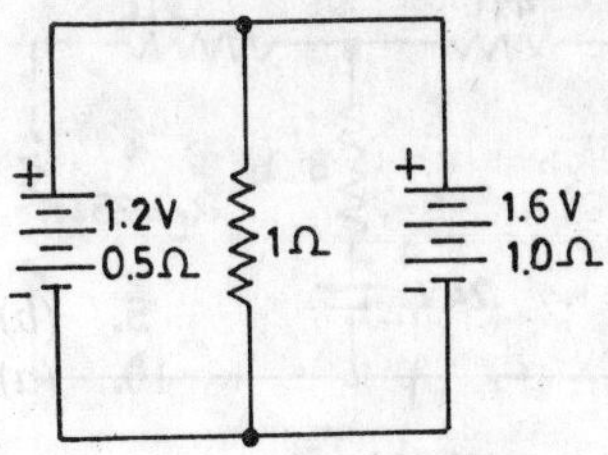

Fig. 1.122.

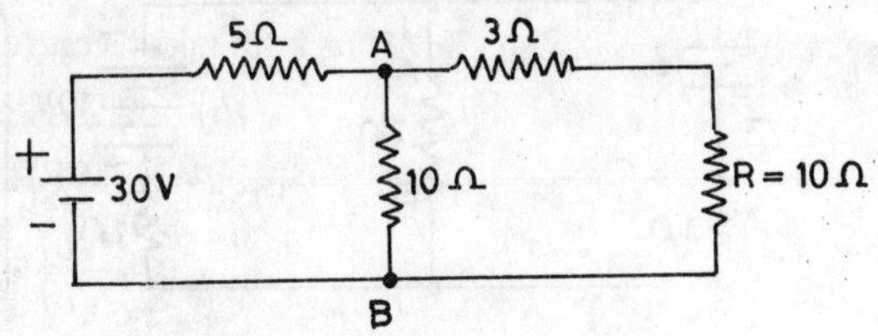

Fig. 1.123.

10. By applying Thevenin's theorem or otherwise, find the current through the resistance R and the voltage across it when connected, as shown in Fig. 1.123.

$$\left[\textbf{Ans. } \frac{60}{49}\text{ A}\,;\ \frac{600}{49}\text{ V}\right]$$

11. Find the current flowing in the 7 Ω resistance branch with the help. Thevenin's theorem. Also find power dissipated in the 7 Ω resistance branch (Fig. 1.124). [**Ans.** 1.43 A; 14.3 watts]

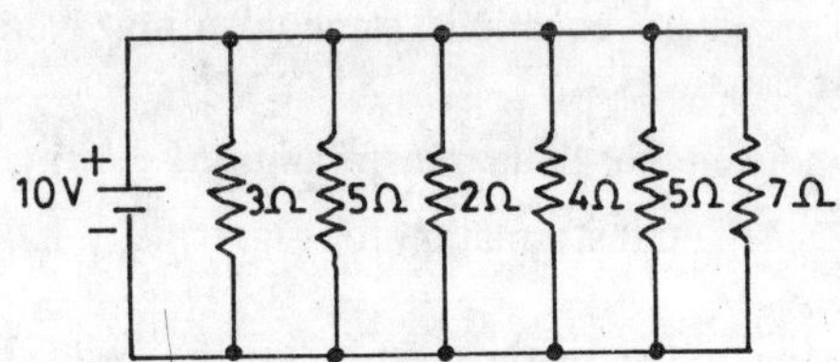

Fig. 1.124.

12. Determine Thevenin's equivalent circuit which may be used to represent the given network at the terminals *A-B*. (Fig. 1.125).

[**Ans.** 5.68 V in series with a resistor of 2.21 Ω]

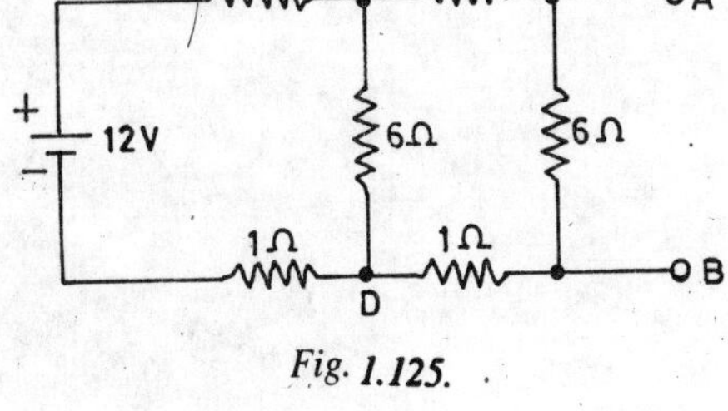

Fig. 1.125.

13. In the circuit shown in Fig. 1.126, find the current in 2 Ω resistance by superposition or Thevenin's or Norton theorem. **[Ans. –1 A]**

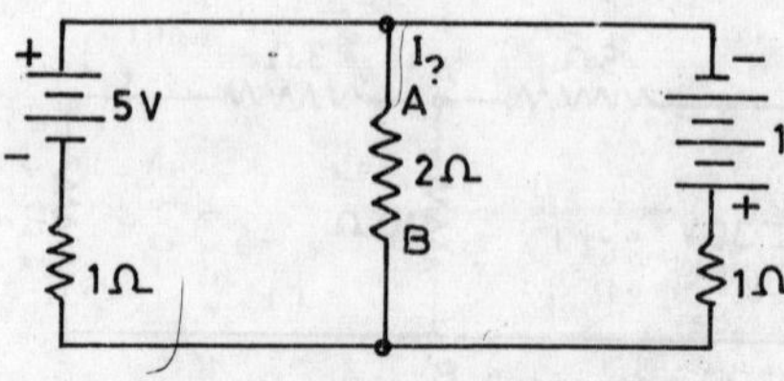

Fig. 1.127.

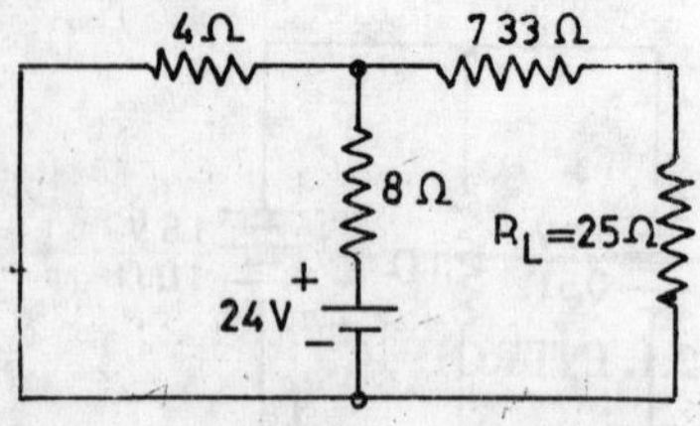

Fig 1.128.

14. Use Norton's theorem to develop an equivalent circuit to supply R_L in Fig. 1.127. Determine current and power delivered to R_L.

[**Ans.** 0.8 A in parallel with a resistance of 10 Ω; 0.2286 A; 1.3 watts]

2

Batteries

2.1. INTRODUCTION

A cell is a source of d.c. supply. The e.m.f. and current obtained from a single cell are generally small. For example, an ordinary dry cell has an e.m.f. of 1.5 volts and can deliver about 1/8 ampere continuously. Such a cell can, therefore, supply electrical energy to a circuit requiring 1.5 volts and not more than 1/8 ampere. Many occasions arise where higher voltage or current or both are required. To meet these needs, a number of cells are suitably connected or grouped. The combination of cells thus obtained is called a *battery*. Batteries are playing important role in many industrial and commercial applications, *e.g.*, in transistor radios, telephone exchanges etc. In this chapter, we shall focus our attention on various types of cells with special reference to their practical applications.

2.2. ELECTRO CHEMICAL CELL

Electrochemical cell is an *arrangement in which chemical reactions take place at a steady rate so as to convert chemical energy into electrical energy*. The amount of energy which can be provided by a electrochemical cell depends upon the amount of chemical reactions taking place within the cell.

2.2.1. Types of Electrochemical Cells

Electrochemical cells are of two types :

1. Primary Cells

Voltaic cell, Daniel cell, Laclanche cell and Dry cells are examples of primary cells.

Main feature of primary cells is that these cells cannot be recharged. The chemicals within the cells are to be replaced when their entire chemical energy is converted into electrical energy. Chemical reactions taking place within the primary cells are irreversible.

2. Secondary Cells

Secondary cells are those cells in which chemical reactions taking place are reversible. They do not possess inherent chemical energy. Electrical energy is first converted into chemical energy by passing electrical current through the cell. The electrical energy which is supplied to the cell is stored into the cell as chemical energy. The chemical energy stored into the cells can be reconverted into the electrical energy by reversing the chemical reaction taking place within the cell. Secondary cells therefore, act as storehouse for electrical energy which can be used whenever required. Secondary cells are sometimes also called accumulators.

It is assumed that students are already familiar with construction and working of primary cell. In this chapter, the secondary cells will be studied in details.

Types of Secondary Cells

Secondary cells are of following types :

1. Lead Acid Cells.
2. Alkaline Cells.

Several cells are usually connected in series to make a storage battery.

2.3. LEAD ACID CELLS

2.3.1. Construction of a Lead Acid Battery

A battery consists of a number of cells and each cell of a battery consists of following parts.

1. Plates

Plates of a lead acid cell are made of antimonial lead alloy grid. Grids for the positive and negative plates are of same design. Plates support the active materials which take part in the chemical reactions within the cell. The plates are framed in following two ways.

***(a)* Formed Plates or Plante Process**

In the Plante Process, the active material on the plates is formed by the process of charging and discharging the battery a number of times. By this process the lead on the surface of the plates gets changed into acting material. The plates are grooved or made in the form of grille so as to have a large exposed area.

***(b)* Pasted or Faure Plates**

These plates are formed by spreading a paste of the active material on supporting grid of antimonial lead alloy. These plates are lighter in weight and cheap in cost as compared to plante plates; and are usually employed for batteries used in automobiles, whereas the plante plates are used for stationary batteries.

It is customary to use positive plates of plante type and negative plates of faure types. A cell consists of a number of plates connected in parallel to form a group. The two sets of plates are then sandwiched together, adjoining plates being separated from each other by perforated plastic, hard rubber or wooden separators. The –ve group of plates contains one more plate than that of the +ve group. The arrangement of plates in a cell is illustrate in Fig. 2.1.

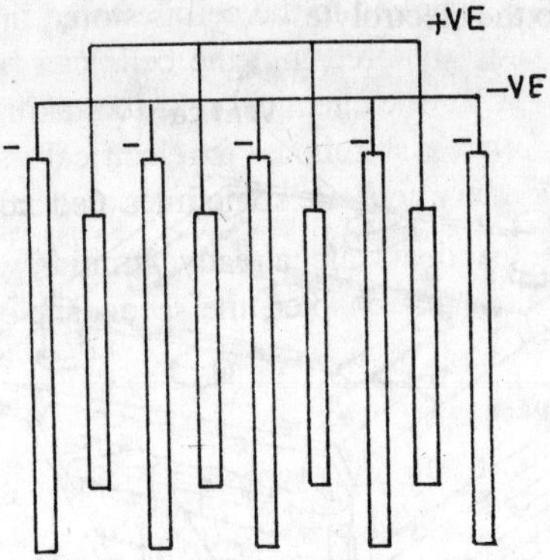

Fig. 2.1.

2. Separators

The separators are thin sheets of a porous material which are placed between +ve and –ve plates to prevent internal short circuits of the +ve and –ve plates. The plates are made porous so that electrolyte can diffuse freely between the plates. The separators are made of insulating material which may be chemically treated wood, rubber or glass fibre mats.

3. Electrolyte

Electrolyte used in lead acid cells is dilute sulphuric acid which is approximately three parts water and one part sulphuric acid. The +ve and –ve plates are immersed completely in this solution.

4. Container

Plates and electrolyte are placed in a container which may be made of vulcanised rubber or moulded hard rubber, ceramics, glass or celluloid. Container is sealed at the top. Three holes are left at top, *i.e.*, one at each end for the terminal posts and one in the middle for the vent plug. The vent plug is made of polystyrene or rubber and is screwed in the cover. Electrolyte is poured into the container through the vent plug. It allows free escape of gases but prevent spilling of electrolyte. Four ribs are fitted at the bottom of the container. Their function is to support the plates and hold them in position and at the same time protect the plates from being short circuited due to fall of active material from plates on the bottom of container. These are also called support blocks.

5. Connecting Bar

It is used to join the cells together in series connecting the +ve terminals of one cell to –ve terminal of the other cell.

6. Cell Terminals

Each cell has two terminals. The terminals are generally made of lead as it does not corrode due to the electrolyte.

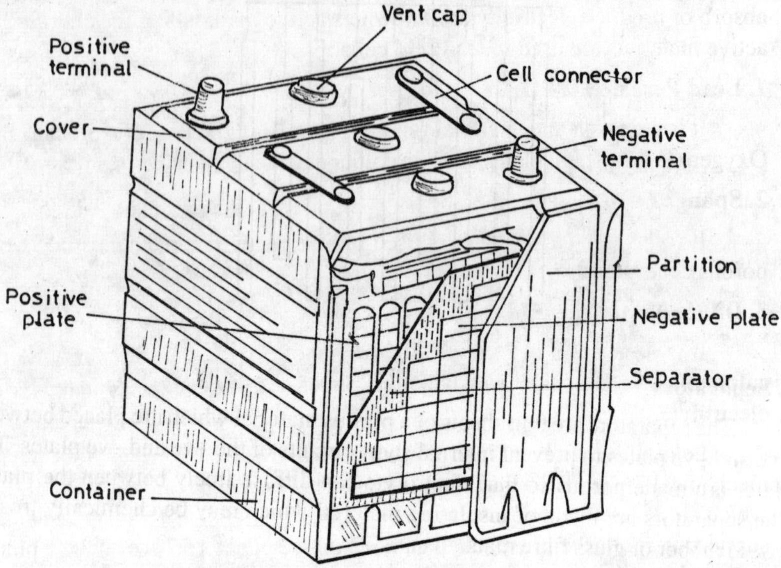

Fig. 2.2. Lead acid battery

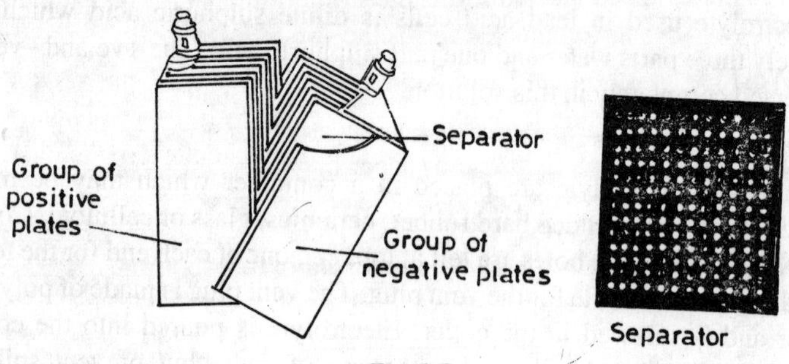

Fig. 2.3.

7. Cell Cover

Each cell compartment has a cover, usually made of molded hard rubber. Openings are provided in these covers for the two terminal posts and for a vent. The joints between covers and container are sealed with an acid-resistance material.

8. Vent Caps

Each cell cover has a hole into which is fitted the vent cap. This cap has a vent hole to allow the free exit of the gas formed in the cell. The vent caps can be easily removed for adding water or taking hydrometer reading.

2.3.2. Active Materials used in Lead Acid Cells

The substances of the cell which take part in chemical reactions and hence absorb or produce electricity are known as active materials of the cell. Following active materials are used in lead acid cells.

1. Lead Peroxide (PbO_2)

It is used as active material for +ve plates. It is combination of lead (Pb) and Oxygen (O_2). It is chocolate brown in colour and is a quite hard but brittle substance.

2. Sponge Lead

It forms the active material used on –ve plate. It is pure lead in Sponge or porous condition.

3. Dilute Sulphuric Acid

It is used as an electrolyte. It consists of 3 parts of water and one part of sulphuric acid. Positive and negative plates are immersed completely in the electrolyte.

Lead in the form of lead peroxide (PbO_2) and sponge lead has very little mechanical strength. Therefore, it is supported by plates of pure lead. The plates which are covered with lead peroxide are known as +ve plates or anodes and those supporting sponge lead are called cathode or –ve plates. the +ve and –ve plates are arranged alternately and are connected to two common terminals. The terminal to which +ve plates are connected forms the +ve terminal of the cell and the terminal to which –ve plates are connected forms the –ve terminal.

2.3.3. Working Principle of Lead Acid Cells

When sulphuric acid is dissolved in water its molecules break up into +ve Hydrogen ions and –ve Sulphate ions according to reaction.

$$H_2SO_4 \rightarrow 2H^+ + SO_4^-$$

Now if two lead electrodes are immersed in this solution and are connected

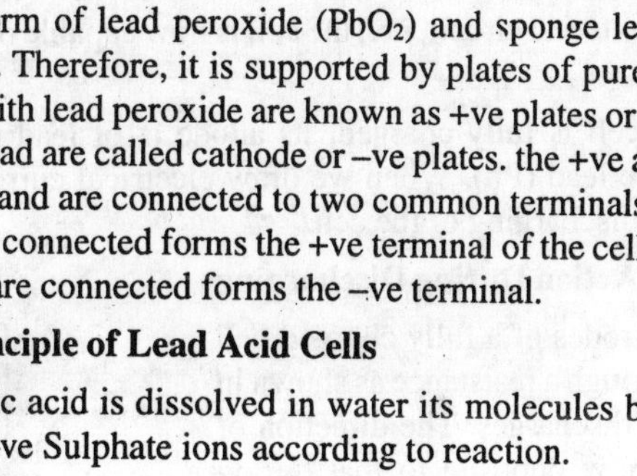

Fig. 2.4.

to DC supply mains, the electrode which is connected to +ve terminal of DC supply is called Anode and the electrode which is connected to –ve terminal of supply is called cathode as shown in Fig. 2.4.

The hydrogen ions being +vely charged move towards the cathode and sulphate ions being –vely charged move towards the anode. Each Hydrogen ion (H^+) on reaching the cathode gets one electron from it, becomes hydrogen atom and escapes out as hydrogen gas. Each negatively charged sulphate ion (SO_4^{--}) on reaching the anode, gives its two electrons to the anode, reacts with water to form sulphuric acid and oxygen according to chemical reaction.

$$SO_4 + H_2O \rightarrow H_2SO_4 + O_2$$

The oxygen produced attacks the lead anode to form lead peroxide, *i.e.*,

$$Pb + O_2 \rightarrow PbO_2$$

Now, if the cell is disconnected from DC source, and a voltmeter is connected between the electrodes, it will indicate potential difference between the electrodes. If these electrodes are connected by a wire, the current will flow from anode to cathode through external circuit, *i.e.*, the cell is now capable of supplying electrical energy.

When the cell is fully charged, its anode is of lead peroxide (PbO_2) and cathode is of pure lead (Pb). When we draw electrical current from the cell, the process is called discharging of the cell.

2.3.4. Chemical Action During Discharging

When electrodes of a fully charged cell are connected through a resistance as shown in Fig. 2.5, the cell discharges. The direction of current flow is now opposite to that during charging. The +ve hydrogen ions now move to anode and –ve sulphate ions move to cathode. Each hydrogen ion on reaching the anode, receives one electron from it (which comes to anode from cathode through external circuit) and becomes hydrogen atom. The hydrogen atom attacks the anode to form lead sulphate (which is whitish in colour) and water according to chemical reactions.

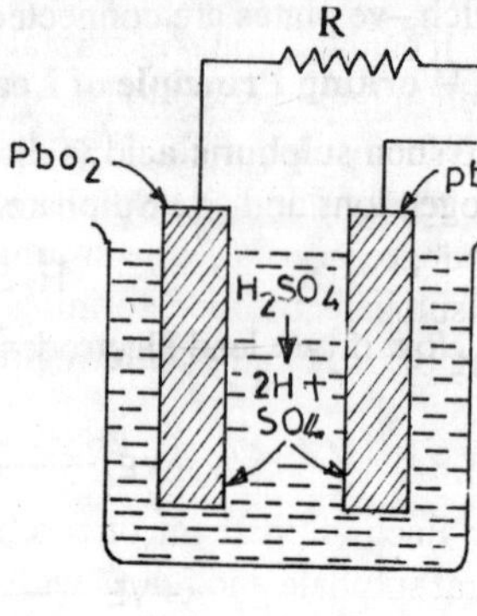

Fig. 2.5.

$$PbO_2 + H_2SO_4 + 2H \rightarrow PbSO_4 + 2H_2O$$

Each sulphate ion (SO_4) on reacting the cathode gives up its two electrons (which move to anode through external circuit), becomes neutral sulphate and attacks the metallic lead cathode to form lead sulphate. The chemical reaction taking place at cathode is

$$Pb + SO_4 \rightarrow PbSO_4$$

From above discussion it is noted that during discharging :

(i) Both anode and cathode becomes lead sulphate ($PbSO_4$) which is whitish in colour.

(ii) Due to formation of water, specific gravity of electrolyte falls.

(iii) Voltage of cell decreases.

(iv) Chemical energy stored in the cell is converted into electrical energy.

2.3.5. Recharging of Lead Acid Cell

When a lead acid cell is fully discharged, both the electrodes are of lead sulphate ($PbSO_4$) as shown in Fig. 2.6.

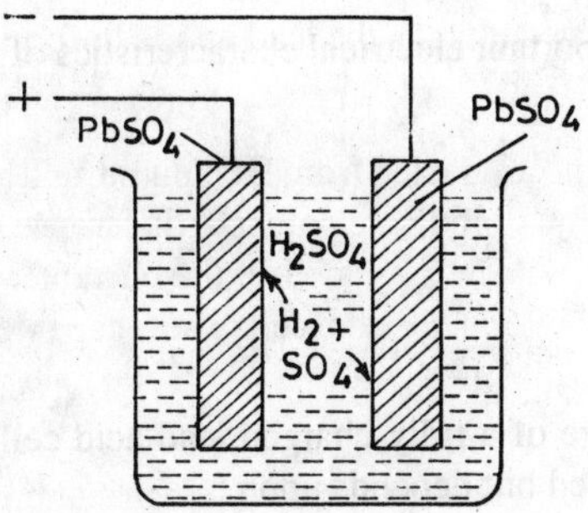

Fig. 2.6.

It cannot supply electrical energy unless it is recharged. In order to recharge the cell, it is connected to DC supply agains. Its anode is connected to +ve terminal of the supply and cathode to –ve terminal of the supply. The hydrogen ions now move to cathode and sulphate ions to anode. Each hydrogen ion on reaching the cathode, receives the electron from there, form hydrogen atom which react with lead sulphate cathode forming lead and sulphuric acid. The chemical reaction taking place at cathode during recharging is

$$PbSO_4 + H_2 \rightarrow Pb + H_2SO_4$$

Each sulphate ion on reaching the anode, gives up its two Electrons, becomes neutral sulphate and reacts with lead sulphate to form lead peroxide and sulphuric acid. The chemical reaction taking place at anode during recharging is

$$PbSO_4 + SO_4 + 2H_2O \rightarrow PbO_2 + 2H_2SO_4$$

It is noted that during re-charging.

(i) Anode becomes dark chocolate brown in colour (PbO_2) and cathode becomes grey metallic lead (Pb).

(ii) Specific gravity of electrolyte increases.

(iii) Terminal voltage of the cell rises.

(iv) Electrical energy supplied to the cell gets stored in the cell in the form of chemical energy.

It is found that chemical reactions taking place during discharging are just opposite to that during recharging. Such reactions are called reversible reactions. The charging and discharging of a lead acid cell can be replaced by single reversible reaction given below :

$$\underset{\text{+ve Plate}}{PbO_2} + 2H_2SO_4 + \underset{\text{−ve plate}}{Pb} \xrightarrow[\text{charge}]{\text{Discharge}} \underset{\text{+ve Plate}}{PbSO_4} + 2H_2O + \underset{\text{−ve plate}}{PbSO_4}$$

2.3.6. Electrical Characteristic of Lead Acid Cell

There are three important electrical characteristics of a lead acid accumulation. They are :

1. Voltage.
2. Capacity.
3. Efficiency.

1. Voltage

Open circuit voltage of a fully charged lead acid cell is approximately 2.2 volt. This value is not fixed but depends upon :

(a) **Length of time since the cell was last charged :** Terminal voltage of cell falls with time due to local reactions taking place within the cell even if cell is not in use.

(b) **Specific gravity of electrolyte :** The voltage increases with increase in specific gravity and vice versa. If specific gravity of electrolyte comes near to density of water, the voltage of the cell is reduced to zero. Specific gravity of electrolyte should not be increased beyond. 1.23 as it will increase the internal resistance of the cell.

(c) **Temperature of electrolyte :** Terminal voltage of the cell increases slightly with increase in temperature.

When the cell is discharged, the terminal voltage falls. The fall of voltage depend upon the rate of discharge. The rate of discharge is generally specified by the number of hours during which the cell will sustain the rate in question, before the terminal voltage falls to 1.8 volt. The cell voltage rapidly falls in the beginning to about 2 volt, then very slowly upto 1.85 volt and again suddenly to 1.8 volt.

The cell voltage should not be allowed to fall below 1.8 volt, otherwise hard insoluble lead sulphate is formed which increases the internal resistance of the cell.

2. Capacity

Capacity of the cell is given by product of current in amperes and time in hours during which the cell can supply current until its e.m.f. falls to 1.8 volt. It is measured in amp hours. The capacity is always given at a specified rate of discharge. In U.S.A. capacity is specified at 8-hour discharge rate and in U.K. at 10 hour discharge rate. Capacity depends upon following factors :

(a) **Rate of discharge :** The capacity of a cell decreases with increase of discharge rate. It is due to the fact that if rate of discharge is increased, there is rapid fall in terminal voltage of the cell, and the chemical reactions taking place within the cell become very brisk which weakens the plates and reduces their capacity. It is found that a cell having 100 Ah capacity at 10 hour discharge rate has 82.5 Ah capacity at 5 hour rate and 50 Ah at 1 hour rate. The variation in capacity with discharge rate is shown in Fig. 2.7.

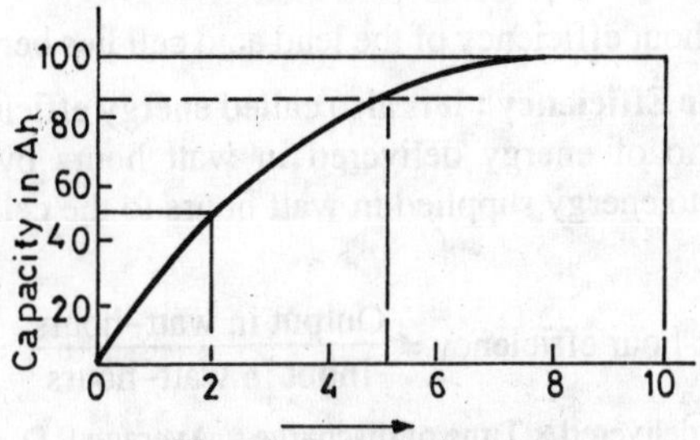

Fig. 2.7.

(b) **Density of Electrolyte :** Increase in density of electrolyte increases the capacity of cell and vice-versa.

(c) **Temperature :** Capacity of cell increases with increase in temperature, because at high temperature.

(i) Chemical reactions taking place with the cell take place more vigorously.

(ii) Internal resistance of electrolyte decreases.

(iii) There is a better diffusion of electrolyte.

However, at high temperature the acid attacks the antimony lead alloy grid, terminal posts and wooden separators. Also the paste of active material is rapidly changed into lead sulphate which is always accompanied by expansion of paste particularly at the +ve plate, resulting in buckling and cracking of the grid. Therefore, it is not advisable to work batteries above 40°C.

Capacity of the cell decreases with decrease in temperature. It was found that capacity of cell is reduced to zero at freezing point, *i.e.*, (0°C) even though the cell is fully charged.

(d) **Amount of Active Material :** Greater is the amount of active material on the plates, larger is the capacity.

3. Efficiency

The efficiency is defined as the rate of output to input for any device. Efficiency of a cell can be defined in two ways :

(i) **The Quantity or Ampere-Hour (AH) Efficiency :** The quantity or ampere-hour efficiency of the cell does not take into account the variation of terminal voltage of the cell during charge and discharge. It is defined as the rate of ampere-hours of discharge to ampere hours of charge, *i.e.*,

$$\text{Ah efficiency} = \frac{\text{Amp–hour of discharge}}{\text{Amp–hour of charge}}$$

The Amp hour efficiency of the lead acid cell lies between 90 to 95 %.

(ii) **Watt-hour Efficiency :** It is also called energy efficiency and is given by the ratio of energy delivered in watt hours by the cell during discharge to energy supplied in watt hours to the cell while charging, *i.e.*,

$$\text{Watt-hour efficiency} = \frac{\text{Output in watt–hours}}{\text{Input in watt–hours}}$$

$$= \frac{\text{Current delivered} \times \text{Time of discharge} \times \text{Average P.D. during discharge}}{\text{Current supplied} \times \text{Time of charge} \times \text{Average P.D. during charge}}$$

$$= \text{Ah efficiency} \times \frac{\text{Average P.D. during discharge}}{\text{Average P.D. during charge}}$$

It is found that during discharge the e.m.f. of the cell falls from about 2.1 volt to 1.8 volt whereas during charge, it rises from 1.8 volt to about 2.6 volt. As the average voltage during charge is more than that during discharge, the watt hour efficiency is always found to less than Ah efficiency. The wh efficiency varies between 72 to 80 %.

2.3.7. Internal Resistance of a Cell

Internal resistance of the cell is the resistance offered by the cell itself to the flow of current. Due to internal resistance of the cell some voltage drop occurs within the cell when internal resistance of the cell some voltage drop occurs within the cell when current flows through it. Therefore, the internal resistance of the cell should be kept minimum. Internal resistance of the cell can be reduced by :

(i) increasing the size of plates, but there is a limit to it because it will make the cell too big to handle.

(ii) Increasing the number of plates inside the cell and by joining all the –ve plates together and all the +ve plates together. Its effect is equivalent to joining a number of cells in parallel. At the same time the length of electrolyte between the electrodes is reduced with consequent decreases in internal resistance of the cell. There is always one more

negative plate than the positive plates, *i.e.*, there is a negative plate at both ends, which assures that both sides of a +ve plate are used.

2.3.8. Charging Methods used for Lead Acid Batteries or Accumulator

The batteries are usually never discharged beyond 1.8 volt, otherwise excessive sulphation is formed on the plates and it becomes difficult to reconvert it. Therefore, when cell voltage of a lead acid battery falls to 1.8 volt it needs to be recharged.

To recharge a lead acid battery, the cells are wiped dry, the terminal posts and cable connectors are cleaned of oxides, and the plugs in the cell covers are removed. The level of electrolyte in the cells is checked and if necessary the cell is filled upto top with distilled water. Vent caps are removed from the vent holes.

For initial charging of a new battery, the new battery is filled with distilled water and is left as such for 2 to 3 hours so that the plates are soaked in water. Then water is removed and the battery is filled with the electrolyte and is kept for 24 hours. The new battery is charged initially at low rate, *i.e.*, ⅓rd of rated charging rate and is overcharged 20 % more than its rated capacity. The battery is then discharged at relatively higher rate, *i.e.*, 2 to 3 times than the rated charge rate. The battery is then kept idle for 2 to 3 hours and then recharged at normal rate.

Charging Method

For charging a lead acid battery DC supply is required. if the available source is AC, then it is converted into DC by rectifiers. Depending upon the source of DC power employed, the charging methods are of three types. They are :

1. Normal voltage DC supply method.
2. A low voltage motor-generator method.
3. A rectifier unit method.

1. Normal Voltage DC Supply Method

In this method, +ve terminal of battery is connected to the –ve terminal of DC supply mains and –ve terminal of battery is connected to the –ve terminal of DC supply mains as shown in Fig. 2.8.

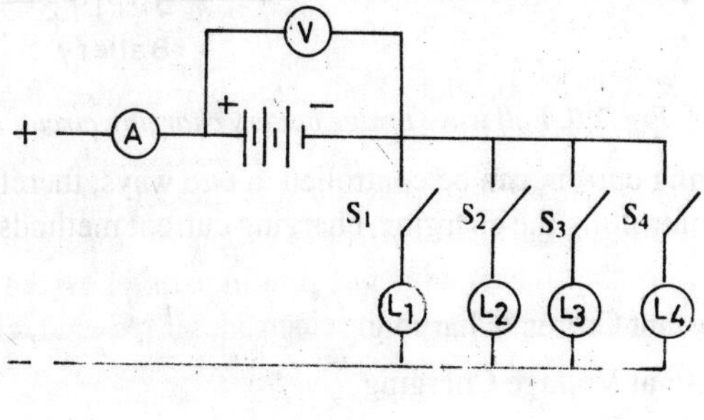

Fig. 2.8.

L_1, L_2, L_3 and L_4 are high wattage lamps connected in parallel. These lamps are used as resistance to control the charging current drawn by the battery from the supply mains. Potential across the battery shown by the volt meter must be about 5% higher than the normal terminal voltage of the battery. The charging current should be adjusted to such a value that temperature of electrolyte may not increase beyond 43°C. The charging is continued until the gases evolve freely, indicating that the batteries have been fully charged.

The advantage of this method is that no additional apparatus except the lamps or variable resistances is required, but it has a great disadvantage of low efficiency because of the wastage of considerable amount of energy in the lamps or in controlling resistances.

2. Low Voltage Motor-Generator Set Method

When available power supply is AC, a low voltage DC generator may be used as source of power for charging of batteries. In this method, charging current can be easily controlled simply by regulating the generator voltage by means of field regulators.

3. Charging of Battery from AC Power Source or Rectifier Method

When available power supply is AC, this method is normally used. Schematic diagram of this method is shown in Fig. 2.9. Normal AC supply if first stepped down by means of multiple tap step down transformer and then rectified by a rectifier. Rectified DC voltage is then applied to battery.

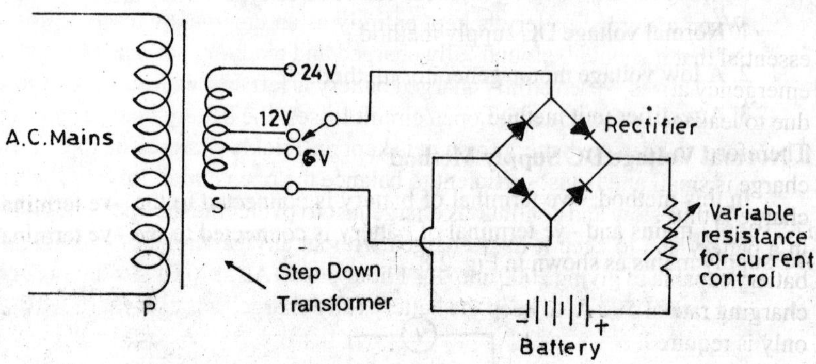

Fig. 2.9. Full wave bridge battery charging circuit

The charging current can be controlled in two ways, therefore, according to the method of controlling the charging, charging current methods are of two types. They are :

(i) Constant Current Charging

(ii) Constant Voltage Charging

(i) Constant Current Charging

In this method, the charging current is kept constant throughout the charging period, by reducing the resistance in the circuit as the voltage of battery goes up. This method is usually employed for initial charging of lead acid batteries and for charging portable batteries in general. In this method the charging current is kept 1/8th of its ampere hour rating. To limit the charging current a variable resistance is connected in series, which is gradually reduced in its value as the battery voltage builds up. The current carrying capacity of the series resistance should be equal to or greater than the charging current required otherwise the resistance will overheat and burnout.

(ii) Constant Voltage Charging

In this method, the charging voltage is kept constant throughout the charging process. Charging current is high in the beginning when the battery is in discharged condition and gradually decreases as the terminal voltage of battery increases. The supply voltage is kept slightly higher than the rated battery voltage. At the end of the charge the charging current drops to almost zero as the battery voltage becomes nearly equal to voltage of supply circuit.

By this method, cells with different capacities and at different degrees of discharge can be charged. The large charging current in the beginning is of short duration and will not harm the cell. This method is however, not suitable for badly sulphated and old batteries which require prolonged charging at slow rate.

2.3.9. Trickle Charging

When a storage battery is kept entirely as an emergency reserve, it is very essential that it should be found fully charged and ready for service whenever, the emergency arises. When a fully charged battery is left unused for a long time, then due to leakage action and other open circuit losses, the battery voltage falls down. Therefore, to keep the battery fresh, it is kept on Trickle charge. The rate of trickle charge is small and is just sufficient to balance the open circuit losses. The trickle charge is thus slow and continuous charge just to overcome the open circuit losses in a battery and to keep it fully charged without any gassing. For example, for a battery capable of giving 2000 amp for 1 hour or 400 Ah at 10 Hr rate having normal charging rate of 500 A, to keep the battery fully charged a trickle charge of 1 Amp only is required.

When during emergency, the battery gets discharged, it is recharged at normal charging rate and then kept on a continuous trickle charge.

2.3.10. Floating Batteries

In various applications the batteries are kept floating on the line and are so connected that they are being charged when load demands are light and automatically discharge during peak periods when load demands are heavy or when the usual power supply fails or is disconnected. Such batteries are called floating batteries.

2.3.11. Indications of a Fully Charged Cell

During the charging process, it is very essential that the battery is taken out from the charging circuit as soon as it is fully charged. Overcharging as well as undercharging are undesirable and should always be avoided. The indications of a fully charged cell (or battery) are :

(i) Voltage

(ii) Specific gravity of electrolyte

(iii) Gassing

(iv) Colour of plates.

(i) **Voltage :** During charging, the terminal potential of a cell increases and provides an indication to state of charge. A fully charged lead–acid cell has a terminal voltage of about 2.1 volts.

(ii) **Specific gravity :** During the charging process, the specific gravity of the electrolyte (H_2SO_4) increases and provides an important indication to the state of charge of the cell. The specific gravity of the electrolyte of a fully charged lead-acid cell is about 1.28. This can be measured by means of a *hydrometer* (See Fig. 2.10). It consists of a long glass tube fitted with a thin rubber hose at its lower end and a rubber bulb at the upper end. Inside the tube is sealed glass float weighted

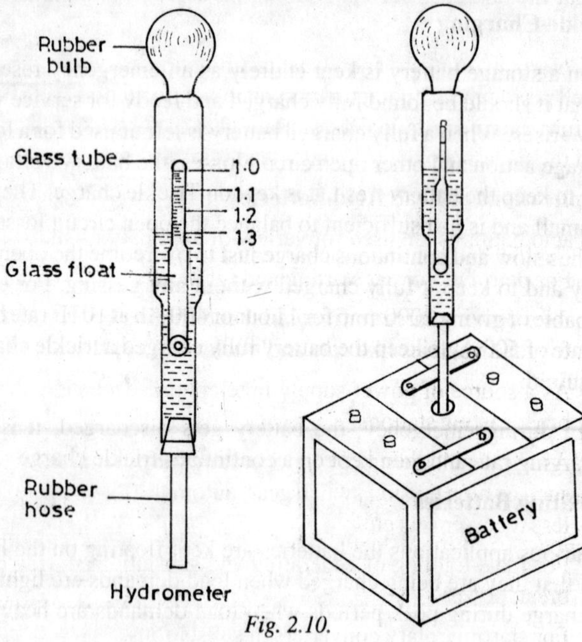

Fig. 2.10.

at one end to make it float upright. By inserting the hose into the cell and then operating the bulb, a quantity of electrolyte (H_2SO_4) may be drawn into the glass tube. The depth to which the float sinks in the drawn liquid indicates the specific gravity of the electrolyte. If the float sinks low, the specific gravity of the liquid (*i.e.*, electrolyte H_2SO_4), is low. If it floats high, the specific gravity is high. Hydrometer float is marked with a scale starting with 1 and extending to 1.3. When the cell is fully charged, the hydrometer reading approaches 1.28 mark.

In practice, the state of charge of a lead-acid cell (or battery) is determined from the specific gravity of the electrolyte (H_2SO_4). The following table may be useful in this regard :

Table 2.1.

Sp. Gr. of H_2SO_4	*State of charge*
1.13	Discharged
1.19	25% charged
1.22	50% charged
1.25	75% charged
1.28	100% charged

(iii) **Gassing :** When the cell is fully charged, the charging current starts electrolysis of water. The result is that hydrogen is given off at the cathode and oxygen at the anode; the process being known as *gassing*. Gassing at both the plates indicates that the charging current is doing no useful work and hence should be stopped.

(iv) **Colour of Plates :** The visual examination of colour of the plates of a lead-acid cell provides yet another important indication to its state of charge. When the cell is fully charged, the positive plate gets converted into PbO_2 (chocolate brown) and the negative plate to spongy lead (grey).

2.3.12. Applications of Lead Acid Batteries

Lead acid batteries are used for variety of applications such as :

(i) For petrol motor car starting and ignition.

(ii) As a power source for industrial and mining battery locomotives, and for road vehicles like cars and trucks.

(iii) As a source of power supply in telephone exchange, laboratories and broadcasting stations.

(iv) As a low voltage supply for operating purpose in many different ways such as high tension switch gear, automatic telephone exchange, wireless receiver sets etc.

(v) For local lighting of generating and substations during odd times and break downs.

(vi) For starting rotary converters in substations.

(vii) For starting of diesel generating sets.

(viii) For compensating the feeder drops at the time of heavy loads.

(ix) As a source of supply for submarines when submerged.

(x) Hospital and other places, *e.g.*, banks, theaters etc. where continuous supply is absolutely essential often use battery as an emergency supply.

2.3.13. Maintenance and Care of Lead Acid Batteries

In order to keep the battery in good condition for long period, the following instructions must be followed.

(i) The cells must always be charged in right direction with direct current.

(ii) The cells should be charged at proper rate.

(iii) Overcharge or discharge of the cells should be avoided.

(iv) Cells should not be discharged below 1.8 volt and never allow them to completely discharge. It will cause permanent layer of lead sulphate to be formed on the plates. The cells are then said to be sulphated.

(v) Sulphated cells can be cured by giving them successive over charges. The sulphated cells are cut out of the battery during discharge so that they can get two successive charges with no intervening discharge.

(vi) The temperature of the cell should not be allowed to exceed $43^{o}C$.

(vii) The level of electrolyte must be maintained at least 10 mm to 15 mm above the plates and should not be exposed to air.

(viii) Specific gravity of the electrolyte must be checked after and before charging the cell.

(ix) Evaporation of electrolyte should be made up by adding distilled water only. Only pure and colourless sulphuric acid must be added whenever necessary.

(x) When cells are not in use or are kept as stand by, keep them on trickle charge, which is just sufficient to balance the open circuit losses.

(xi) Vent openings in the filling plugs should be kept open to prevent the gasses formed within the cell, from building a high pressure.

(xii) The acid and corrosion on the battery top should be washed off with a cloth moistened with baking soda or ammonia and water.

(xiii) The battery terminals and metal supports should be cleaned down to bare metal and covered with petroleum jelly.

2.4. ALKALINE BATTERIES

Like lead acid cells, alkaline batteries also consist of two electrodes, *i.e.*, anode and cathode, immersed in an electrolyte. The plates and electrolyte are placed in a suitable container. There are two types of alkaline batteries which are in common use. These are :

(i) Nickel Cadmium Type

(ii) Nickel Iron Type.

2.4.1. Nickel-Cadmium Batteries

Active material used for positive plate and the electrolyte used in nickel cadmium cells are the same as that used for nickel iron cell. For negative plate a mixture of cadmium and iron is used.

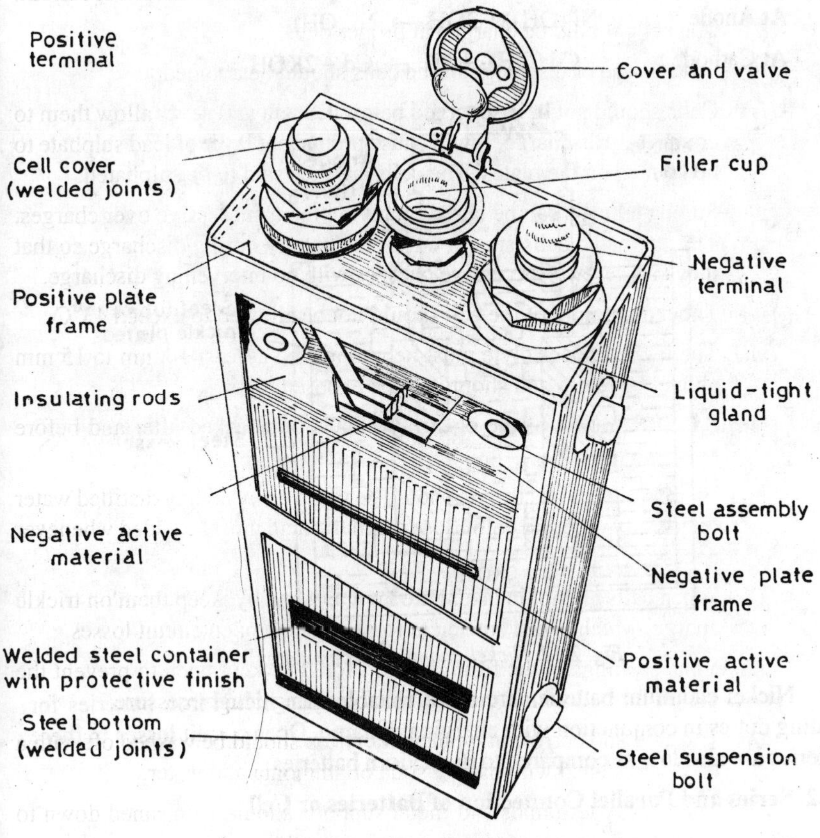

Fig. 2.11. Nickel-cadmium alkaline accumulator

The number of positive paltes used in nickel-cadmium cell is one more than the number of negative plates.

The working of nickel-cadmium cell is almost similar to nickel iron cell.

When current passes through the cell, the electrolyte KOH breaks up into K^+ and OH^- ions. During discharging the K^+ ions move to anode and OH^- ions move to cathode. Following chemical changes take place when the cell is supplying current to the external circuit.

At Anode : $Ni(OH)_4 + 2K \rightarrow NI(OH)_2 + 2KOH$

and at Cathode : $Cd + 2OH \rightarrow Cd(OH)_2$

When the cell is charged by connecting the anode to the positive terminal of supply and cathode with the negative terminal of supply the direction of current flow is opposite to that during discharging. Therefore, K^+ ions move to cathode and OH^- ions move to anode. Following chemical reactions take place during charging.

At Anode : $Ni(OH)_2 + 2OH \rightarrow Ni(OH)_4$

At Cathode : $Cd(OH)_2 + 2K \rightarrow Cd + 2KOH$

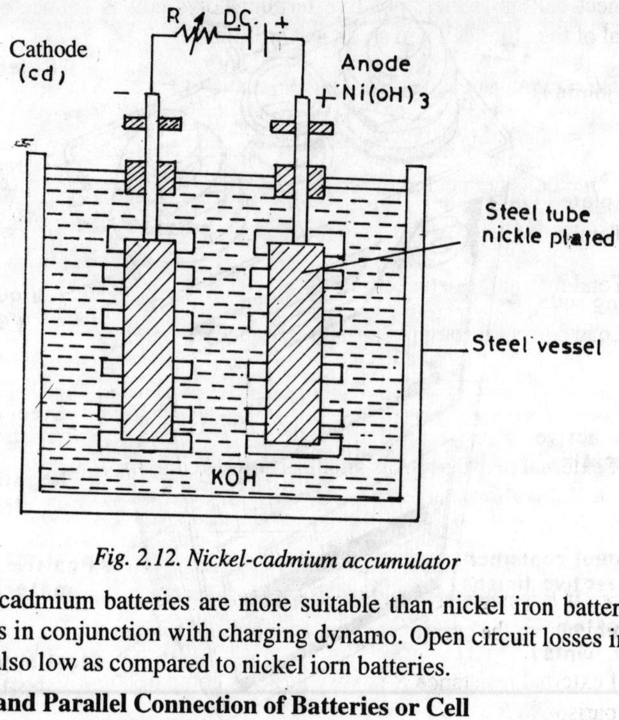

Fig. 2.12. Nickel-cadmium accumulator

Nickel cadmium batteries are more suitable than nickel iron batteries for floating duties in conjunction with charging dynamo. Open circuit losses in these batteries are also low as compared to nickel iorn batteries.

2.4.2. Series and Parallel Connection of Batteries or Cell

We have already studied in this chapter that the cell is a device which converts chemical energy into electrical energy. A cell consists of two electrodes immersed in an electrolyte. Due to chemical actions opposite charges are developed at the electrodes. The electrodes serve as +ve and –ve terminals of the cell. Terminal voltage of a cell is 1.5 volt in case of dry cells and 2.1 volt in case of lead acid cells and 1.2 volt in nickel ion cells. In order to increase the terminal voltage or current, a number of cells are required to be connected in series or parallel. A cell is represented by the symbol as shown in Fig. 2.13.

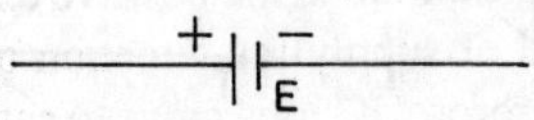

Fig. 2.13.

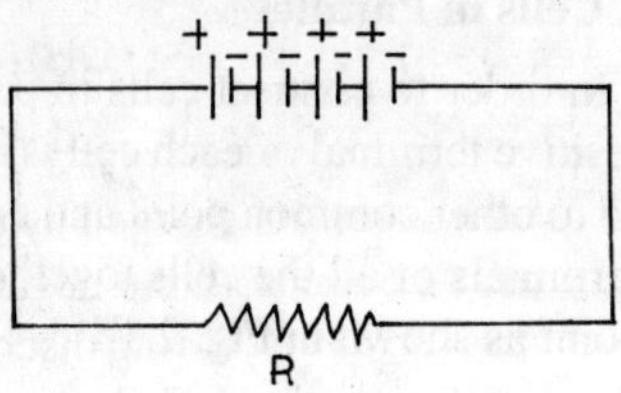

Fig. 2.14.

2.4.3. Grouping of Cells in Series

The cells are connected in series when higher voltages are required. In order to connect cells in series, positive terminal of a cell is connected to negative terminal of the other cell and so on as shown in Fig. 2.14.

If E = e.m.f. of each cell,

r = internal resistance of each cell,

n = cells connected in series.

Total e.m.f. = nE

Total internal resistance = nr

Load current through external resistance R,

$$I = \frac{nE}{R + nr}$$

If external resistance R is small as compared to nr, R may be neglected and

$$I = \frac{nE}{nr} = \frac{E}{r}$$

i.e., If R is negligible as compared to total internal resistance of the cells in the combination, the current is same as supplied by an individual cell.

If external resistance R is very large as compared to nr, then neglecting nR in comparison to R

$$I = \frac{nE}{r} = n \times \text{curernt supplied by single cell}$$

Therefore, if internal resistance of the cell is negligible, then the cells must be connected in series in order to get maximum current through load resistance.

Applications of Series Grouping : This arrangement of cells is used where higher voltage is required but the current required is not more than the current supplied by a single cell.

2.4.4. Cells in Parallel

In order to connect cells in parallel, positive terminal of each cell is connected to other common point and negative terminals of all the cells together at one point as shown in Fig. 2.15.

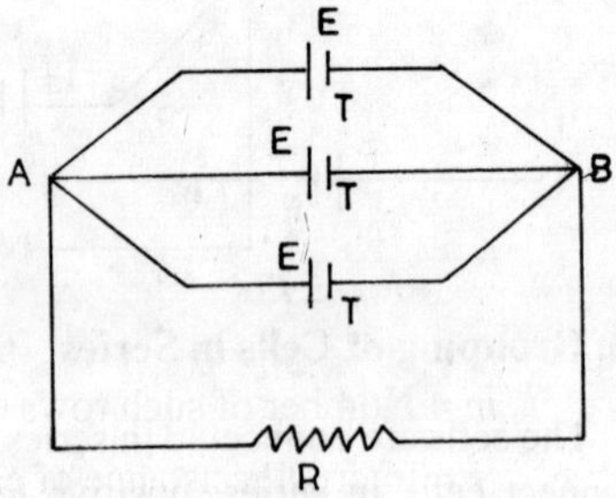

Fig. 2.15.

Let there are n cells each of emf E and internal resistance r connected in parallel. In such a case voltage across the output terminals A and B will be the same as the voltage of each cell. Total internal resistance of the source will now be $= r/n$

$$\text{Total circuit resistance} = R + \frac{r}{n}$$

$$\text{Load current, } I = \frac{E}{R + \frac{r}{n}}$$

If r is very small as compared to R, neglecting $\frac{r}{n}$,

$$I = \frac{E}{R} = \text{current due to a single cell}$$

when R is very small as compared to $\frac{r}{n}$, the

$$I = \frac{E}{\frac{r}{n}} = n \cdot \frac{E}{r} = n \times \text{current due to each cell}$$

Therefore, from above discussion it is clear that parallel grouping of cells is advisable when internal resistance of each cell is high as compared to external resistance in order to get maximum current through load resistance.

Applications of Parallel Grouping : This arrangement of cells is used where higher current is required to operate the circuit but the voltage required is not more than voltage produced by a single cell.

Mixed Grouping of cells : In mixed grouping some cells are connected in series and a number of such series sets are connected in parallel as shown in Fig. 2.16.

Let n = Number of cells connected in series

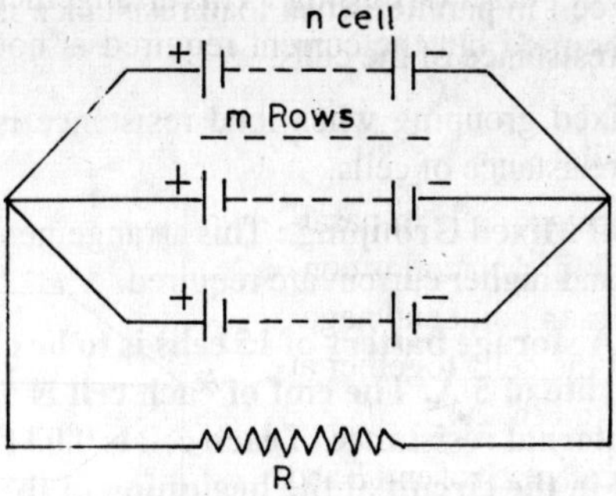

Fig. 2.16.

m = Number of such rows connected in parallel

r = Internal resistance of each cell

E = e.m.f. of each cell

R = Load resistance

Internal resistance of n cell in series = nr

Internal resistance of m such rows in parallel = $\frac{nr}{m}$

Total resistance of the circuit = $R + nr/m$

Potential across load resistance = nE

Current through the circuit,

$$I = \frac{nE}{R + \frac{nr}{m}} = \frac{nE}{\frac{n^2 r}{N} + R}$$

where $N = m \times n$ = Total no. of cells

For maximum current, $\frac{dI}{dn} = 0$

Differentiating the above equation for I wrt n and equating the result obtained to zero, it can be proved that for maximum current

$$R = \frac{nr}{m}$$

i.e., external load resistance must be equal to internal resistance of the series parallel group in order to get maximum load current.

From above discussion it is concluded that in order to get maximum load current it is advisable to

(*i*) Connect cells in series when load resistance is high as compared to internal resistance of the cells.

(ii) Connect cell in parallel when load resistance is small as compared to internal resistance of the cells.

(iii) Have mixed grouping when load resistance is comparable with the internal resistance of cells.

Applications of Mixed Grouping : This arrangement of cells is used where both higher voltage and higher current are required.

Example 2.1. A storage battery of 15 cells is to be charged from a 40 volt supply at constant rate of 5 A. The emf of each cell is 1.2 V at the beginning of charge and the internal resistance of each cell is 0.03 Ω. Calculate the series resistance required in the circuit at the beginning of the charging.

Solution : Supply voltage, $V = 40$ volts

Battery emf, $E_b = 15 \times 1.2 = 18$ V

Internal resistance of the battery $= n \times r = 15 \times 0.03 = 0.45\ \Omega$

Charging current at start of charging, $I_c = 5$ A

$$\text{Total resistance in the circuit} = \frac{V - E_b}{I_c}$$

or
$$R + nr = \frac{40 - 18}{5} = 4.4\ \Omega$$

or
$$R = 4.4 - 0.45 = 3.95\ \Omega \quad \textbf{Ans.}$$

Example 2.2. A battery of 60 cells is charged from a supply of 250 V. Each cell has an emf of 2 V at the start of charge and 2.5 V at the end. If the internal resistance of each cell is 0.1 Ω and if there is an external resistance of 19 Ω in the circuit, calculate *(a)* the initial charging current; *(b)* the final charging current; and *(c)* the additional resistance which must be added to give a finishing charge of 2 A rate.

Solution : Supply voltage, $V = 250$ V

EMF of battery at the start of charging, $E_{b1} = 60 \times 2 = 120$ V

EMF of battery at the end of charging, $E_{b2} = 60 \times 2.5 = 150$ V

Internal resistance of battery, $r = 60 \times 0.1 = 6\ \Omega$

External resistance, $R = 19\ \Omega$

(a) Initial charging current, $I_{C1} = \dfrac{V - E_{b1}}{R + r} = \dfrac{250 - 120}{19 + 6} = 5.2$ A **Ans.**

(b) Final charging current, $I_{C2} = \dfrac{V - E_{b2}}{R + r} = \dfrac{250 - 150}{19 + 6} = 4$ A **Ans.**

(c) Total resistance in the circuit to given charging current of 2 A at the end of charging.

$$R' + R + r = \frac{V - E_{b2}}{I_{C2}'} = \frac{250 - 150}{2} = 50\ \Omega$$

or additional resistance required,

$$R' = 50 - (R + r) = 50 - (19 + 6) = 25\ \Omega \quad \textbf{Ans.}$$

Example 2.3. A battery has taken a charging current of 5.2 A for 24 hours at a voltage of 2.25 V, while discharging it gave a current of 4.5 A for 24 hours at an average voltage of 1.85 V. Calculate the quantity efficiency and the energy efficiency of the battery.

Solution : Charging current, $I_C = 5.2$ A

Charging mean voltage, $V_C = 2.25$ V

Charging period, $T_C = 24$ hours

Discharging current, $I_d = 4.5$ A

Discharging mean voltage, $V_d = 1.85$ V

Discharging period, $T_d = 24$ hours

Quantity efficiency, $\eta_{AH} = \dfrac{I_d\, T_d}{I_C\, T_C} \times 100 = \dfrac{4.5 \times 24}{5.2 \times 24} \times 100 = 86.54\ \%$ **Ans.**

Energy efficiency,

$$\eta_{WH} = \frac{I_d\, T_d}{I_c\, T_c} \times \frac{V_d}{V_c} \times 100 = \frac{4.5 \times 24}{5.2 \times 24} \times \frac{1.85}{2.25} \times 100 = 71.15\ \% \quad \textbf{Ans.}$$

Example 2.4. An alkaline cell is discharged at a steady current of 4 A for 12 hours, the average terminal voltage being 1.2 V. To restore it to original state of voltage, a steady current of 3 A for 20 hours is required, the average terminal voltage being 1.44 V. Calculate the ampere-hour and watt-hour efficiencies in this particular case.

Solution : Charging current, $I_C = 3$A

Charging average voltage, $V_C = 1.44$ V

Charging period, $T_C = 20$ hours

Discharging current, $I_d = 4$ A

Discharging average voltage, $V_d = 1.2$ V

Discharging period, $T_d = 12$ hours

Ampere-hour efficiency, $\eta_{AH} = \dfrac{I_d\, T_d}{I_c\, T_c} \times 100 = \dfrac{4 \times 12}{3 \times 20} \times 100 = 80\ \%$ **Ans.**

Watt-hour efficiency, $\eta_{WH} = \eta_{AH} \times \dfrac{V_d}{V_c} = 80 \times \dfrac{1.2}{1.44} = 66.7\ \%$ **Ans.**

Example 2.5. A discharged battery is put on charge at 5 A for 3 ½ hours at a mean charging voltage of 13.5 V. It is then discharged in 6 hours at a constant terminal voltage of 12 V through a resistance of *R* ohms. Determine *(i)* value of *R* for an ampere hour efficiency of 85% *(ii)* watt- hour efficiency of the battery.

Solution : Charging current, $I_c = 5$ A

Charging period, $T_C = 3\frac{1}{2}$ hours

Charging voltage, $V_c = 13.5$ V

Discharging period, $T_d = 6$ hours

Discharging voltage, $V_d = 12$ V

Ampere-hour efficiency, $\eta_{AH} = 85\% = 0.85$

$$\because \quad \eta_{AH} = \frac{I_d T_d}{I_c T_c}$$

$$\therefore \text{ Discharging current, } I_d = \eta_{AH} \times \frac{I_C T_C}{T_d} = 0.85 \times \frac{5 \times 3\frac{1}{2}}{6} = 2.48 \text{ A}$$

$$\text{Discharging resistance, } R = \frac{V_d}{I_d} = \frac{12}{2.48} = 4.84\ \Omega \ \textbf{ Ans.}$$

$$\text{Watt-hour efficiency, } \eta_{WH} = \eta_{AH} \times \frac{V_d}{V_c} = 85 \times \frac{12}{13.5} = 75.55\ \% \ \textbf{ Ans.}$$

Example 6. Four cells each of emf 2 volt and internal resistance of 0.1 ohm supply maximum current to a 10 ohm resistance. How should these cells be connected ? Also find out the maximum value of current.

Solution : Let the cells are connected in mixed grouping so that there are *n* cells in series and *m* such groups in parallel.

For maximum current, $R = \frac{nr}{m}$

$$\therefore \quad 10 = \frac{n \times 0.1}{m}$$

$$\text{or} \quad 10m = n \times 0.1$$

$$\text{or} \quad n = 100\,m \quad ...(i)$$

as total number of cells, $m \times n = 4$

$$\therefore \quad n = \frac{4}{m} \quad ...(ii)$$

or $$\frac{4}{m} = 100\,m$$

or $$4 = 100\,m^2$$

or $$m^2 = \frac{4}{100}$$

i.e., *m* is less than 1, therefore, the cells are connected in series alone. The maximum current,

$$I = \frac{nE}{R+nr} = \frac{4\times 2}{10+4\times .1} = \frac{8}{10.4} = 769 \text{ Amp} \textbf{ Ans.}$$

Example 7. 125 cells of 2 volt each are to supply maximum current to a external resistance of 0.5 ohm. Give the arrangement of cells and find out the maximum current if internal resistance of each cell is 0.1.

Solution : Let n = no. of cells in series

m = no. of such groups in paralell

Total number of cells, $m \times n = 100$

For maximum current, $R = \frac{nr}{m}$

$$0.5 = \frac{n\times 0.1}{m}$$

or $$0.5\,m = 0.1\,n$$

As $$m \times n = 125 \textbf{ Ans.}$$

Example 8. Twelve cells, each of emf 2 volts and internal resistance 0.5 Ω, are connected in series across an external resistance of 4.5 Ω. Determine *(i)* the current supplied by the battery *(ii)* terminal voltage of the battery and *(iii)* fall in voltage per cell.

Solution : E.M.F of each cell, $E = 2$ V

Internal resistance of each cell, $r = 0.5\ \Omega$

Number of cells, $n = 12$

External resistance, $R = 4.5\ \Omega$

Total e.m.f. of battery $= nE = 12 \times 2 = 24$ V

Internal resistance of battery $= nr = 12 \times 0.5 = 6\ \Omega$

Total circuit resistance $= R + nr = 4.5 + 6 = 10.5\ \Omega$

(i) Current supplied by battery, $I = \dfrac{24}{10.5} = 2.286$ A **Ans.**

(ii) Terminal voltage of battery $= IR = 2.286 \times 4.5 = 10.287$ V **Ans.**

(iii) Fall in voltage per cell $= Ir = 2.286 \times 0.5 = 1.143$ V **Ans.**

Example 9. How many cells, each having an e.m.f. of 1.5 volts and internal resistance of 0.2 Ω, would be required to pass a current of 1.5 A through a resistance of 40 Ω ?

Solution : $E = 1.5$ V ; $r = 0.2\ \Omega$; $R = 40\ \Omega$; $I = 1.5$ A ; $n = ?$

For series grouping of cells,

Circuit current, $I = \dfrac{nE}{R + nr}$

or $$1.5 = \frac{1.5n}{40 + 0.2n}$$

$\therefore$ $n = 50$ **Ans.**

Therefore, 50 cells should be connected in series to satisfy the required condition.

Example 10. A battery consisting of 10 cells in series has two of the cells reversed. E.M.F. of each cells is 1.5 V and internal resistance 0.1 Ω. The resistance of the external circuit is 4 Ω. Find the reduction in current due to the two cells being reverse connected.

Solution : When 10 cells are properly connected in series,

Total e.m.f. of battery $= nE = 10 \times 1.5 = 15$ V

Total circuit resistance $= R + nr = 4 + 10 \times 0.1 = 5\ \Omega$

$\therefore$ Circuit current $= \dfrac{15}{5} = 3$ A

When two of the cells are reverse connected,

Total e.m.f. of battery $= 8 \times 1.5 - 2 \times 1.5 = 9$ V

Total circuit resistance = 5 Ω, *i.e.*, same as before

$\therefore$ Circuit current $= \dfrac{9}{5} = 1.8$ A

$\therefore$ Reduction in current $= 3 - 1.8 = 1.2$ A **Ans.**

Example 11. A battery is formed by five cells in series. When the external resistance is 4 Ω, the current is 1.5 A and when it is 9 Ω, the current falls to 0.75 A. Find the e.m.f. and internal resistance of each cell.

Solution : In series grouping of cells,

$$I = \frac{nE}{R + nr}$$

or $\quad nE = I\,(R + nr)$

When $\quad R = 4\,\Omega\ ,\ I = 1.5\text{ A}$

$\therefore \quad 5E = 1.5\,(4 + 5 \times r)$

or $\quad 5E - 7.5r = 6 \qquad \ldots(i)$

When $\quad R = 9\,\Omega\ ,\ I = 0.75\text{ A}$

$\therefore \quad 5E = 0.75\,(9 + 5 \times r)$

or $\quad 5E - 3.75\,r = 6.75 \qquad \ldots(ii)$

Solving eq. *(i)* and eq. *(ii)*, we get,

$$E = 1.4\text{ V and } r = 0.2\,\Omega \ \textbf{Ans.}$$

Example 12. Four cells, each of e.m.f. 1.5 V and internal resistance 0.5 Ω, are connected in parallel across an external resistance of 2.5 Ω. Determine *(i)* the current supplied by the battery *(ii)* current supplied by each cell and *(iii)* the terminal voltage of the battery.

Solution : Fig. 2.17 shows the arrangement of cells in parallel.

E.M.F. of each cell, $E = 1.5$ V

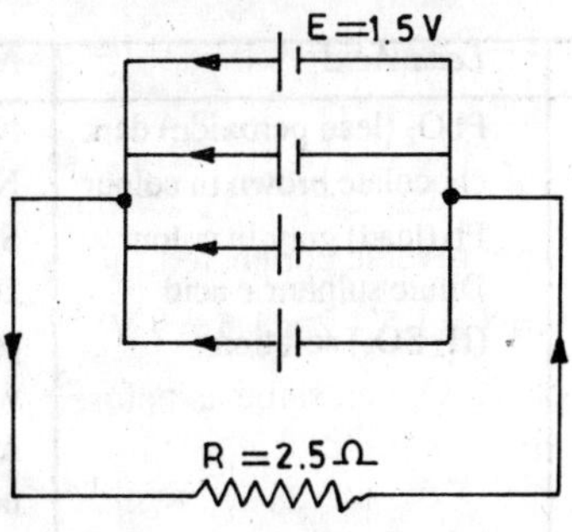

Fig. 2.17.

Resistance of each cell, $r = 0.5\ \Omega$

No. of rows of cells, $m = 4$

Battery e.m.f. = 1.5 V

$$\text{Resistance of battery} = \frac{r}{m} = \frac{0.5}{4} = 0.125\ \Omega$$

$$\text{Total circuit resistance} = R + \frac{r}{m} = 2.5 + 0.125 = 2.625\ \Omega$$

(i) Current supplied by batery, $I = \frac{1.5}{2.625} = 0.571$ A **Ans.**

(ii) Current supply by each cell, $= \frac{0571}{4} = 0.1427$ A **Ans.**

(iii) Terminal voltage of battery $= IR = 0.571 \times 2.5 = 1.4275$ V **Ans.**

Example 13. Sixteen cells, each having an internal resistance of 4 Ω, when joined in parallel send a current of 2.72 A through an external resistance of 0.3 Ω. Find the e.m.f. of each cell.

Solution : $m = 16$; $r = 4\ \Omega$; $I = 2.72$ A

$R = 0.3\ \Omega$; $E = ?$

Battery e.m.f. = E (to be found out)

$$\text{Total circuit resistance} = R + \left(\frac{r}{m}\right) = 0.3 + \left(\frac{4}{16}\right) = 0.55\ \Omega$$

$$I = \frac{E}{0.55} = 2.72 = \frac{E}{0.55}$$

Battery e.m.f. $E = 2.72 \times 0.55 = 1.496$ V **Ans.**

2.5. COMPARISON BETWEEN LEAD ACID AND NICKEL-CADMIUM ACCUMULATORS

Particulars	*Lead-Acid*	*Nickel-Cadmium*
1. Active materials		
(a) Positive plate	PbO_2 (lead peroxide) dark chocolate brown in colour	Nickel hydroxide Ni $(OH)_3$
(b) Negative plate	Pb (lead) grey in colour	Spongy cadmium
(c) Electrolyte	Dilute sulphuric acid (H_2SO_4) solution	20% of water solution of potassium hydroxide (KOH) with a samll adition of lithium monohydrate containing not less than 50% lithium hydroxide (LiOH)

2. Mechanical Properties (for same A-H rating)		
(a) Weight	Moderate	Heavy
(b) Volume	Varies	Large
(c) Number of plates per cell	Few	Many
(d) Mechanical strength	Poor	Good
(d) High temperature damage	Yes	No
(f) Life	1,250 charges and dis-charges.	Long life
3. Maintenance Factors		
(a) Damage likely due to freezing	Yes	No
(b) Rate of self dis-charge	High	Low
(c) Trickle Charge Use	Yes	Yes
4. Electrical Properties		
(a) Average emf	2.0 V per cell	1.2 V per cell
(b) Internal resistance	Comparateively low (about 0.001 to 0.002 Ω)	
(c) Efficiency		
(i) Quantity	90 to 95%	80%
(ii) Energy	75 to 85%	65%
(d) Voltage regulator control	Easy	Easy
(e) Gas production during charge	Yes	Slight
(f) Ampere-hour capacity		
(i) Reduced by over-charge	Yes	No
(ii) Reduced by oc-casional over-discharge	No	No
(iii) Reduced by frequent over discharge	Yes	Yes

2.6. SILVER OXIDE CELLS

The active material of positive plates is silver oxide which is pressed into the plate and then subjected to heat treatment. The active material of the negative plates is a mixture of zinc power and zinc oxide.

The chemical changes taking place within the cell can be represented by the following single equation

$$\underset{+\text{ve plate}}{Ag_2O} + \underset{-\text{ve plate}}{Zn} \underset{\text{charge}}{\overset{\text{Discharge}}{\rightleftharpoons}} \underset{+\text{ve plate}}{2Ag} + \underset{-\text{ve plate}}{ZnO}$$

A silver oxide cell has specific capacity 4 to 5 times greater than other types of cells. It has ground application such as military, *i.e.*, communication equipments portable radar set and nightvision equipment. Moreover the efficiency is high and self-discharge is small.

Silver zinc batteries can withstand much heavier discharge current as compared to other types. The operating temperature range is between –20°C to +60°C.

It is also used for submarine propulsion. Its life time in wet condition is 1-2 years and dry storage life is upto 5 years. However, the only disadvantage is of its higher cost.

HIGHLIGHTS

1. **Cell :** A source of emf in which chemical energy is converted with electric energy.
2. **Primary cell :** The chemical action is not reversible.
3. **Secondary cell :** The cell in which chemical action is reversible.
4. **Battery :** When number of cells are connected in series, the combination is called battery.
5. The two efficiencies of a cell are :

   (a) Ampere-hour efficiency.

   (b) Watt-hour efficiency.

$$\eta_{A-H} = \frac{\text{Amp–hours during discharge}}{\text{Amp–hours during charge}}$$

$$\eta_{W-H} = \frac{\text{Watt–hours on discharge}}{\text{Watt–hour on charge}}$$

6. Charge indications of lead acid battery

   (a) **Specific gravity :** 1280 to 1290-100% charged.

   (b) **Voltage :** Above 2 volts fully charged; below 1.8 V fully discharged.

(c) **Colour of plate :** anode-chocolate brown and cathode-grey fully charge.

7. **Nickel-Cadmium cell :** Anode — Ni $(OH)_4$

Cathode — Cd (cadmium)

Electrolyte hydroxide — KOH (Pottassium)

Electrical characteristics :

E.M.F. — fully charged cell – 1.4 V

E.M.F. — fully discharged – 1.0 V.

SHORT QUESTIONS

1. Terminal voltage or cell is less than the emf. Why ?
2. Primary cell cannot be recharged. Why ?
3. Why the separator is perforated ?
4. Positive & negative terminals of battery should be connected to positive & negative terminals of supply. Why ?
5. The battery should never be left standing in a discharged condition.
6. The battery terminals should never be short circuited. Why ?
7. Large space is left at the bottom of battery container. Why ?
8. W-H efficiency is less than A-H efficiency. Why ?

MULTIPLE CHOICE QUESTIONS

1. The emf of a fully charged lead acid cell is ________ volts.

(a) 1.21 V (b) 1.8 V (c) 2.1 V

(d) 1.4 V

2. Capacity of storage cell is measured in ________.

(a) joule (b) watt-hour (c) ampere hour

(d) coulomb.

3. The indications of fully charged cells are given by

(a) voltage

(b) gassing

(c) specific gravity of electrolyte

(d) all of the above.

4. To check the specific gravity of electrolyte, the instrument used is

(a) voltmeter (b) lectometer (c) hydrometer

(d) barometer.

5. The cell in which the chemical action is reversible is known as
   (*a*) Secondary cell
   (*b*) Primary cell
   (*c*) Nickel cadmium cell.

ANSWER

1. (*c*) 2. (*c*) 3. (*d*) 4. (*c*) 5. (*a*)

EXERCISES

1. Distinguish between a primary and a seconary cell. In what form is the energy stored in the cell ? **(B.T.E. Ph. June 1975)**
2. Describe the construction of Weston cadmium standard cell and explain why this type is suitable for use a standard.
3. Draw and explain the equivalent circuit for a cell. Explain the term 'internal resistance' of a cell and discuss its significance. **(B.T.E. Pb. July 1981)**
4. Derive expression for the current when a number of cells are connected in (*i*) series and (*ii*) parallel.
5. Describe the construction of lead-acid battery. Write down the chemical equations during the charge and discharge of lead-acid battery. **(B.T.E. Pb. Feb. 1991)**
6. (*a*) Explain with connection diagrams the two charging methods of a lead-acid cell by constant current.
   (*b*) What are the indications of fully charged lead-acid cells ?
   (*c*) State the precautions to be taken during charging.
7. State the indications of a fully charged lead-acid cell.
8. Describe the constructional details and operating characteristics of lead-acid cells.
9. Write down the active metarials of a lead acid cell and the relation between watt-hour efficiency and amphere-hour efficiency. **(B.T.E. Pb. January 1992)**
10. (*a*) List out the different parts of lead acid battery and mention the materials used for each part.
    (*b*) What are the indications of a fully charged cell ?
    (*c*) List factors affecting the capacity of a cell.
11. State the factors which affect the capacity of battery mentioning the nature of effects.
12. Write the care and maintenance of lead acid batteries.

13. Draw a sketch of nickel cadmium cell and name each component giving its material.

14. Give the construction, characteristics, merits, demerits and applications of nickel-cadmium accumulators.

15. Differentiate between

(i) Primary and secondary cell.

(ii) Battery and coil.

(iii) EMF and terminal potential of a cell.

(iv) Ampere-hour efficiency and watt-hour efficiency.

16. Write short notes on the following :

(i) Dry cell.

(ii) Grouping of cells.

(iii) Battery ratings.

(iv) Uses of lead acid accumulators.

(v) Care and maintenance of lead-acid batteries.

17. Answer the following to the points :

(i) When a battery is not supplying any load, a voltmeter connected across the terminals gives a certain reading. If a load is now connected across the battery will the voltmeter reading increase or decrease ? Give reasons for your answer.

(ii) When are the cells connected in series, parallel and mixed grouping. **(B.T.E. Pb. June 1994)**

(iii) In what form energy is stored in a secondary battery ?

(iv) How will you make sure that lead acid accumulator is *(a)* fully charged *(b)* fully discharged ? **(B.T.E. June 1975)**

(v) How will you know that a battery has completely discharged. **(B.T.E. Haryana June 1974)**

(vi) Do we use ac or dc to charge a battery ? **(B.T.E. Haryana June 1994)**

(vii) What are the factors which affect the capacity of a battery ? **(B.T.E. Rajasthan Nov., 1974)**

(viii) Which type of storage cells are in commertical use ? **(B.T.E. Haryana Dec., 1974)**

(ix) If the level of the electrolyte in a given battery is low, what liquid will you use to fill it up ? **(B.T.E. Haryana June 1974)**

PROBLEMS

1. Ten cells, each of e.m.f. 1.5 V and internal resistance 0.2 Ω are connected in series. What current will they send through an external resistance of 4 Ω ? **(Ans.** 2.5 A)
2. Three cells of e.m.f 1.1, 1.3 and 1.5 volts are connected in series to supply current to an external resistance of 1.6 Ω. The internal resistances of three cells are 0.25, 0.35 and 0.4 Ω respectively. Determine the current through the external resistance. **(Ans.** 1.5 A)
3. A battery consists of 5 cells, each of e.m.f. 1.5 V and internal resistance 0.25 Ω, connected in series. Through what external resistance will it send a current of 2A ? **(Ans.** 2.5 Ω)
4. Six cells, each of e.m.f. 1.5 volts and internal resistance of 2 Ω, are joined in parallel and connected to an external resistance of 5 Ω. What current will flow ? **(Ans.** 0.5294 A)
5. A batter consists of 4 cells, each of e.m.f. 1.46 volts. The cells are connected in parallel to an external resistance of 0.525 Ω. If the current supplied by the battery is 0.8 A, find the internal resistance of each cell. **(Ans.** 5.2 Ω)
6. 24 cells each having an e.m.f. of 2.1 V and internal resistance of 2 Ω are to supply the maximum current to a 12 Ω external resistance. Show how the cells are to be connected and what is the value of current to external resistance ? **(Ans.** Series-parallel, $m = 3$; $n = 8$; 1.625 A)

3

Magnetism and Electromagnetism

3.1. INTRODUCTION TO ELECTROMAGNETISM

The branch of Electrical Engineering which deals with the magnetic effects of electric current by which mangetism is produced is called *electromagnetism.*

The close relationship between electric current and magnetism has opened a new era.

The study of electromagnetism is of utmost importance to electrical engineer as the operation of all electrical machinery is due to the application of the magnetic effects of electric current in one form or the other.

Magnets used in electrical machines/equipments are of artificial type and generally horse shoe or U-shaped because they can retain their magnetism for a very long time.

3.2. MAGNETIC FIELD

The area around a magnetic pole or a magnet within which its influence is perceptible, is called its magnetic field. A small compass needle can be used to map a magnetic field. Iron filings may also be used to demonstrate the shape and distribution of the magnetic field in any horizontal plane. These filings set themselves in the form of curved chains between the poles as shown in Fig. 3.1. Such continuous curves in a magnetic field are normally termed as magnetic lines of force. These lines of force travel from the north to the south pole as shown in Fig. 3.1. Thus each line of force forms a closed path.

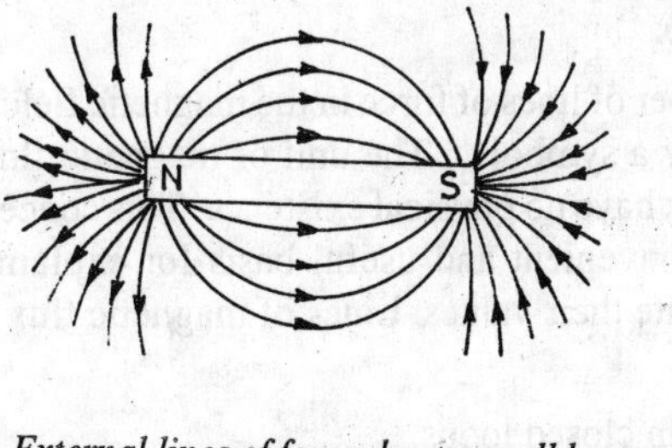

Fig. 3.1. External lines of force due to small bar magnet

3.2.1. Magnetic Effect of Electric Current

When an electric current flows through a conductor, magnetic field is set up all along the length of the conductor. Fig. 3.2 shows the magnetic field produced by the current flowing in a straight wire. The magnetic lines of force are in the form of concentric circles around the conductor.

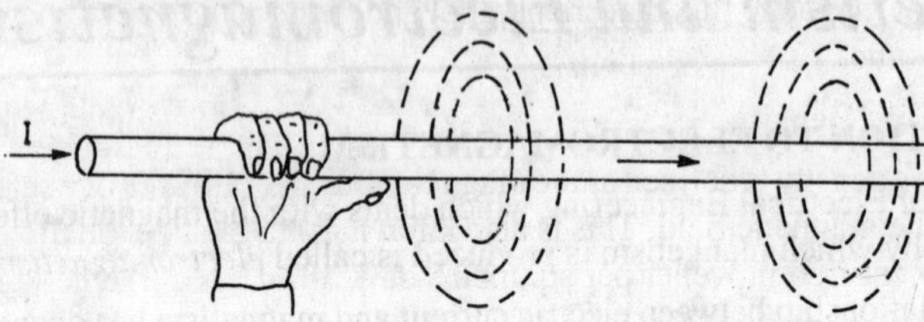

Fig. 3.2.

The direction of lines of force depends upon the direction of current and may be determined by **right-hand rule**. *Hold the conductor with the right hand with the thumb pointing in the direction of current (See Fig. 3.2). Then the fingers will point in the direction of magnetic field around the conductor.* Applying this rule to Fig. 3.2, it is clear that when viewed from left hand side, the direction of magnetic lines of force will be clockwise.

The following points may be noted about the magnetic effect of electric current :

(i) The greater the current through the conductor, the stronger the magentic field and vice-versa.

(ii) The magnetic field near the conductor is stronger and becomes weaker and weaker as we move away from the conductor.

(iii) The magnetic lines of force round the conductor will be either clockwise or anticlockwise, depending upon the direction of current. One may be *right-hand rule* to determine the direction of magnetic fieid round the conductor.

3.2.2. Magnetic Flux

The total number of lines of force in the magnetic field is called the magnetic flux. It is denoted by a symbol ϕ. The unit of magnetic flux is weber (Wb). The lines of magnetic flux have no physical existence. This concept of lines of magnetic flux forms a very convenient and useful basis for explaining various magnetic effects and to calculate their values. Lines of magnetic flux possess the following properties :

(i) They form closed loops.

(ii) They always start from the north pole and end in the south pole **and are** then continuous through the body of the magnet.

(iii) They never intersect each other.

(iv) Lines of force are like stretched elastic cords, tending to contract length wise.

(v) Lines of force exert lateral pressure, *i.e.*, they tend to bulge out side ways.

(vi) Lines of magnetic flux that are parallel and in the same direction, repel one another.

3.2.3. Magnetic Flux Density

Magnetic flux density is defined as the magnetic flux per unit area of a surface at right angles to the magnetic field. This is also known as magnetic induction. Its symbol is B and measured in weber per sq. metre or Tesla. Hence magnetic flux

ϕ = flux density $B \times$ surface area A, or flux density, $B = \dfrac{\phi}{A}$, wb/m^2 or Tesla.

3.2.4. Magnetic Field around a Straight Current Carrying Conductor

When a conductor carries an electric current, a magnetic field is produced all along its length, the lines of magnetic flux being concentric circles in planes at right angles of the conductor. The magnetic field associated with a current carrying conductor depends upon the magnitude of current and also on the direction of flow of current. Fig. 3.3 shows the magnetic field set up around a straight current carrying conductor. The two conductors shown in Fig. 3.3 carry the same current in opposite directions and their fields so produced have different directions. If these two conductors are laid side by side, the two fields so produced will cancel each other thereby resulting in zero magnetic effect. Thus the magnetic effect produced by two conductors laid side by side and carrying the same current in opposite directions is nil. However, if they carry the same current in the same direction, the magnetic effect will be twice compared to one such conductor alone.

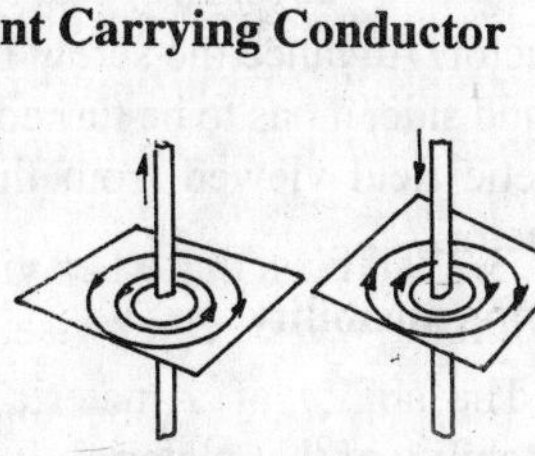

Fig. 3.3. Magnetic field around a current carrying conductor

3.2.5. Method to find the Direction of Magnetic Field

The direction of magnetic field can be found out by any one of the following rules :

Right Hand Rule

Grip the current carrying conductor in the right hand, with the thumb pointing in the direction of current as shown in Fig. 3.4, then the fingers encircling the wire point in the direction of the magnetic field or lines of magnetic flux.

Cork Screw Rule

Place a cork screw or the wood screw along the side of the current carrying

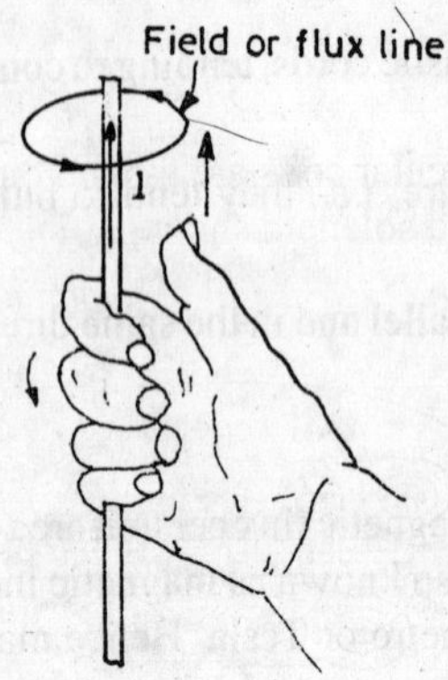

Fig. 3.4. Right hand rule showing direction of field

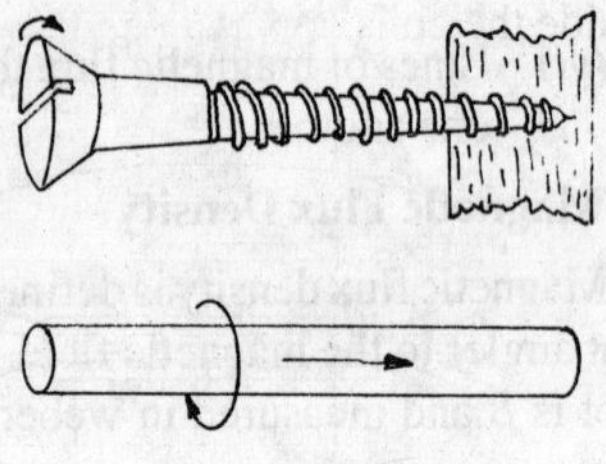

Fig. 3.5. Right hand screw rule

conductor. Advance the screw in the direction of current. When viewed from the left hand side, it has to be turned clock-wise as shown in Fig. 3.5. The direction of magnetic field viewed from the same side will then be clockwise around the conductor.

3.2.6. Permeability

The ability of a material to conduct magnetic flux through it is called permeability of that material. It is denoted as Greek letter μ (mu).

The permeability of all nonmagnetic materials including air is represented by μ_0, equal to $4\pi \times 10^{-7}$ H/m.

The absolute permeability μ of a medium or magnetic material can also be expressed in terms of its relative permeability μ_r and the permeability of free space or air μ_0, *i.e.*,

Relative permeability : $\mu = \mu_0 \times \mu_r$

Relative Permeability : The relative permeability of a magnetic material may be defined as the ratio of the flux density produced in the material to the flux density produced in vacuum or in a nonmagnetic core, provided the magnetic field strength is same in both the cases. It is denoted by μ_r. Relative permeability varies with the value of flux density.

3.2.7. Solenoid

When a number of circular coils are arranged coaxially in such a way that same current flows in the same direction in the coils, then this arrangement is called a *solenoid*.

Solenoid is a Greek word meaning tube-like.

The solenoids are of two types :

(i) Iron-cored solenoid.

(ii) Air-cored solenoid.

In the iron-cored solenoid, the set of circular coils are wound on an iron bar as shown in Fig. 3.6. However, in the air-cored solenoid shown in Fig. 3.7 there is air inside the coils.

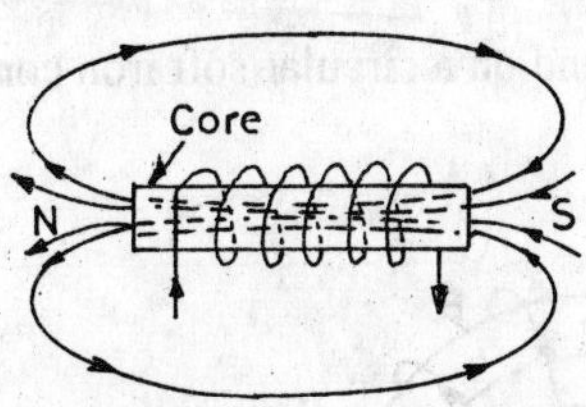

Fig. 3.6.

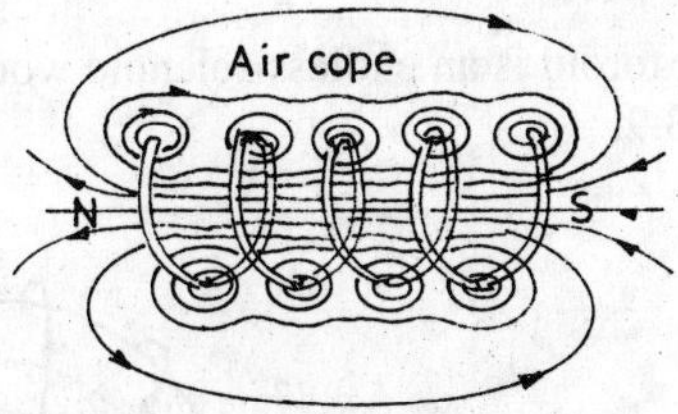

Fig. 3.7.

When current is passed through a solenoid (iron-cored or air- cored), it behaves like a magnet with its ends forming opposite poles.

The polarity of a solenoid can be determined by *Right-hand rule.*

Right-Hand Rule

Hold the solenoid in your right hand so that the fingers point in the direction in which the current flows in coils. Then thumb stretched longitudinally points towards the north pole or the direction of magnetic lines of force (See Fig. 3.8).

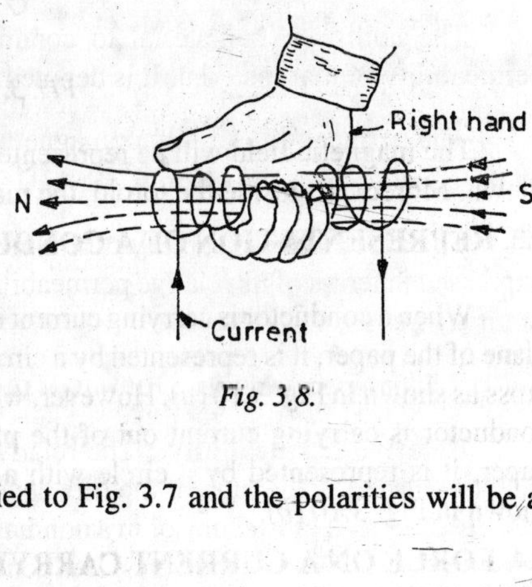

Fig. 3.8.

Applying this rule to Fig. 3.8, the left hand face of the solenoid will become *N*-pole and the right hand face *S*-pole.

Similarly, this rule can be applied to Fig. 3.7 and the polarities will be as indicated.

Shape of Magnetic Field

The magnetic field due to solenoid is shown in Fig. 3.8.

The following important points are noteworthy regarding the magnetic field due to a solenoid :

(i) The magnetic field is practically identical with that of a bar magnet of the same dimensions.

(ii) Inside the coils (*i.e.*, in Iron-core or Air-core), the magnetic field is uniform and is represented by a set of straight, parallel and equally spaced magnetic lines of force.

(iii) Outside the solenoid, the magnetic field is represented by lines of force spreading out in all directions from the ends of the solenoid.

(iv) By increasing the number of turns of current, strength of magnetic field can be increased.

3.2.8. Toroid

A toroid is an endless solenoid wound on a circular soft iron core as shown in Fig. 3.9.

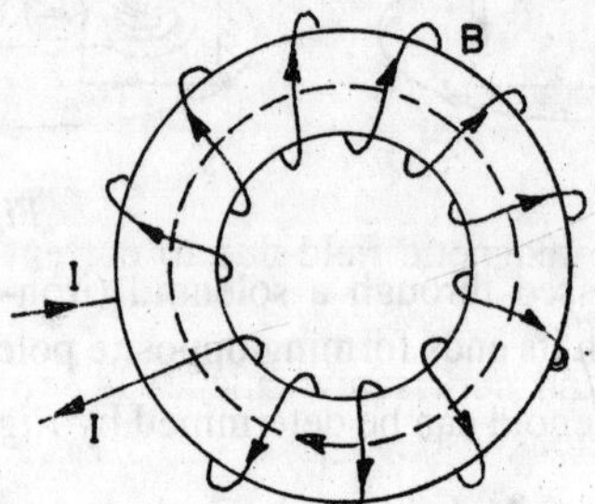

Fig. 3.9.

The magnetic field will be represented by lines of force which are circles as shown. Moreover, inside the toroid, the magnetic field is uniform.

3.3. REPRESENTATION OF A CONDUCTOR CARRYING CURRENT

When a conductor is carrying current into the plane of the paper, it is represented by a circle with cross as shown in Fig. 3.10 *(a)*. However, when the conductor is carrying current out of the plane of paper, it is represented by a circle with a dot as shown in Fig. 3.10 *(b)*.

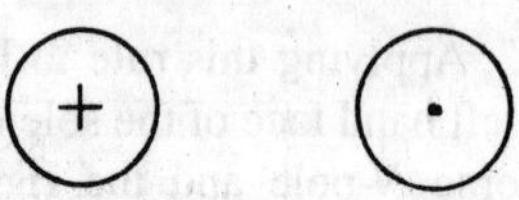

(a) *(b)*

Fig. 3.10.

3.4. FORCE ON A CURRENT CARRYING CONDUCTOR PLACED IN MAGNETIC FIELD

When a current carrying conductor is placed in a magnetic field, mechnical force acts on it.

Consider a conductor carrying current into the plane of paper and placed in a uniform magnetic field as shown in Fig. 3.11. Then, the force F will act on the conductor tending to move it downwards. The fact that force acts on the conductor and also conductor moves downwards can be explained as under :

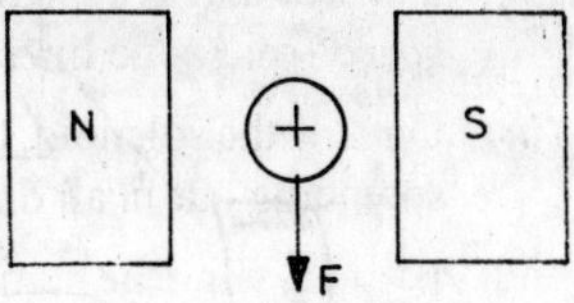

Fig. 3.11.

Consider Fig. 3.12, in which only the field due to the magnets is shown. It is seen that magnetic field is uniform and acts from *N*-pole to *S*-pole.

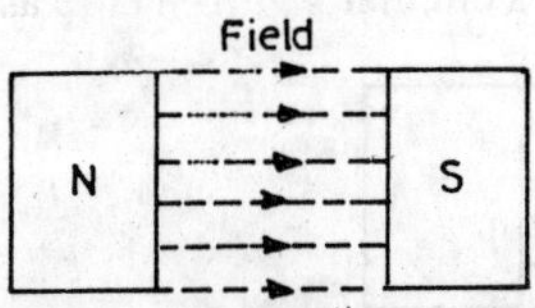

Fig. 3.12.

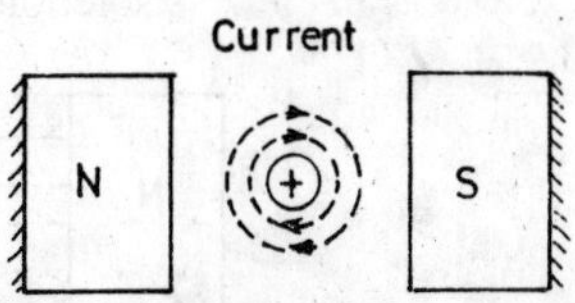

Fig. 3.13.

In Fig. 3.13 only the magnetic field due to current carrying conductor is shown.

By applying right hand cork screw rule, it is seen that the direction of magnetic lines of force is clockwise.

In Fig. 3.14, magnetic field due to poles and conductor are shown together. It is seen that at the top of the conductor, magnetic lines due to poles are in the same direction as that due to conductor. In other words, the two fields are helping each other at the top of the conductor. However, at the bottom of the conductor, field due to lines is in opposite direction to the field due to the conductor (*i.e.*, the two fields are opposing each other). The result is that magnetic lines of the force are crowded at the top of the conductor and thinned at the bottom as shown in Fig. 3.15.

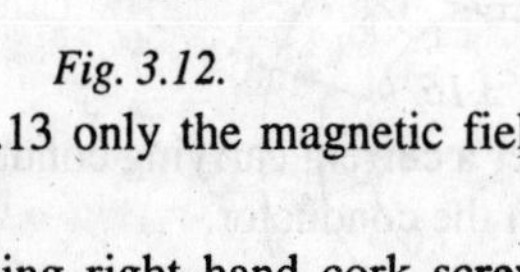

Fig. 3.14.

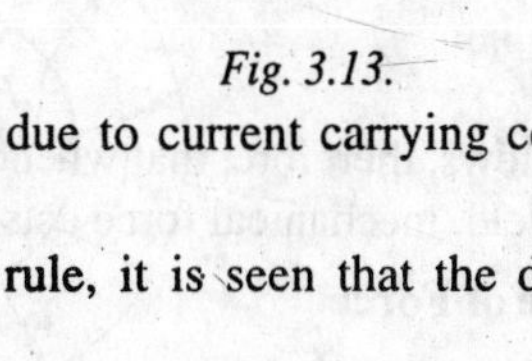

Fig. 3.15.

As the magnetic lines of force behave like rubber bands (property of magnetic lines of force), tending to contract, therefore, a force *F* acts on the conductor from top towards bottom, *i.e.*, the force moves the conductor downwards.

If the direction of current in the conductor is reversed, the direction of force is also reversed (see Fig. 3.16).

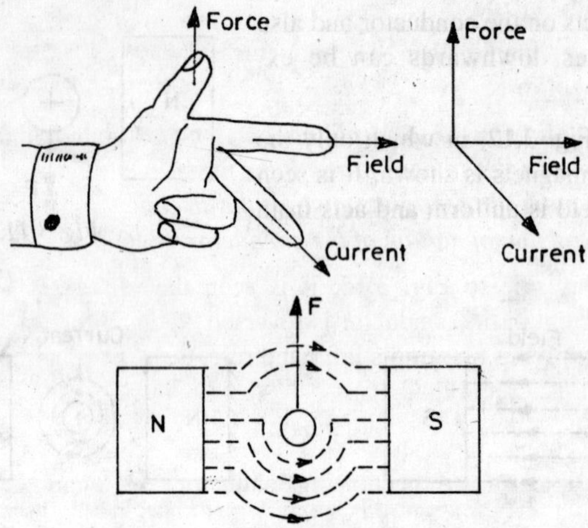

Fig. 3.16.

If follows, therefore, that whenever a current carrying conductor placed in a magnetic field, mechanical force acts on the conductor.

Magnitude of Force

The magnitude of force acting on a current carrying conductor placed in a magnetic field depends upon the following factors :

(i) *Flux density (B) due to poles.* Greater the flux density, greater the force acting on the conductor.

(ii) *Magnitude of current (I).* Larger the current in the conductor, greater the force on the conductor.

(iii) *Length of conductor (l).* Greater the length of conductor, greater the force.

It can be proved mathematically that the magnitude of the force acting on a current carrying conductor placed at *right angles* to magnetic field is given as :

$$F = BIl \text{ newtons}$$

where, B = flux density in Wb/m^2

I = conductor current in amperes

l = conductor length in metres.

If the conductor is making an angle θ with the field, then

$$F = BI\,l \sin\theta \text{ newtons}$$

Direction of Force

The direction of force acting on a current carrying condcutor placed in a magnetic field can be found by *Fleming's left hand rule*.

Fleming's Left Hand Rule

This rule is applicable when the conductor is at right angles to the magnetic field and is stated below :

Stretch out the first finger, central finger and thumb of your hand so that they are mutually at right angles to each other. If the first finger points in the direction of magnetic field, central finger points towards the direction of current, then thumb will point in the direction of motion of the conductor. (See Fig. 3.16).

Applying this rule to Fig. 3.15, it is seen that direction of force F is downwards. If current in the conductor is reversed, keeping the direction of field unchanged, then applying Fleming's left hand rule, the direction of motion of the conductor is upwards. (See. Fig. 3.16).

Applications

The force acting on a current carrying conductor placed in magnetic field has many useful applications, *e.g.*, in motors, electrical instruments etc.

Exmaple 3.1. A conductor 100 cm long carries a current of 100 amperes at right angles to a uniform magnetic field of 1 Wb/m². Calculate the force and power required to move the conductor at a speed of 10 m/s in a plane at right angles to the magnetic field.

Solution : Flux density, $B = 1\ \text{Wb/m}^2$

Conductor current, $I = 100$ A

Conductor length, $l = 100$ cm $= 1$ m

Conductor speed, $v = 10$ m/s

$\therefore$ Force on the conductor, $F = BIl = 1 \times 100 \times 1 = 100$ newtons **Ans.**

Power required = Work done/s.

= Force × distance moved/second = $F \times v$

$= 100\ \text{N} \times 10\ \text{ms}^{-1} = 1000$ watts **Ans.**

Example 3.2. A conductor 0.1 metre long and carrying a current of 50 A, lies perpendicular to the field of strength 1000 AT/m. Calculate :

(i) **Force acting on the conductor,**

(ii) **Mechanical power required to move the conductor against this force with a speed of 1 m/s.**

(iii) **e.m.f. induced in the conductor.**

Solution : Length of the conductor, $l = 0.1$ m

Current in the conductor, $I = 50$ A

Field intensity, $H = 1000$ AT/m

Flux density, $B = \mu_0 H$ ($\mu_r = 1$ for air)

$$= 4\pi \times 10^{-7} \times 1000 = 12.57 \times 10^{-4} \text{ Wb/m}^2$$

Force acting on the conductor, $F = BIl \sin\theta$ $(\because \theta = 90^\circ)$

$\therefore$ $F = 12.57 \times 10^{-4} \times 50 \times 0.1 = 62.85 \times 10^{-4}$ newtons **Ans.**

Mechanical power required to move the conductor against this force at a velocity of 1 m/s.

$\therefore$ Power, P = Force × velocity

$$= 62.85 \times 10^{-4} \times 1 \text{ Nm/s} = 62.85 \times 10^{-4} \text{ watts}$$ **Ans.**

E.M.F. induced in the conductor

$$e = \frac{P}{I} = \frac{62.85 \times 10^{-4}}{50} = 1.257 \times 10^{-4} \text{ volts}$$ **Ans.**

3.5. FORCE BETWEEN TWO PARALLEL CURRENT CARRYING CONDUCTORS

Mechanical force exists between two conductors carrying current. In most of the cases, this force is small but if the conductors are close together and the currents are large, the magnitude of this force is appreciable.

Magnitude of Force

Let there be two parallel conductors *a* and *b* carrying current and placed close together as shown in Fig. 3.17.

Let l = length of each conductor in metres

d = distance between conductors in metres

I_a = current carried by conductor *a*

I_b = current carried by conductor *b*

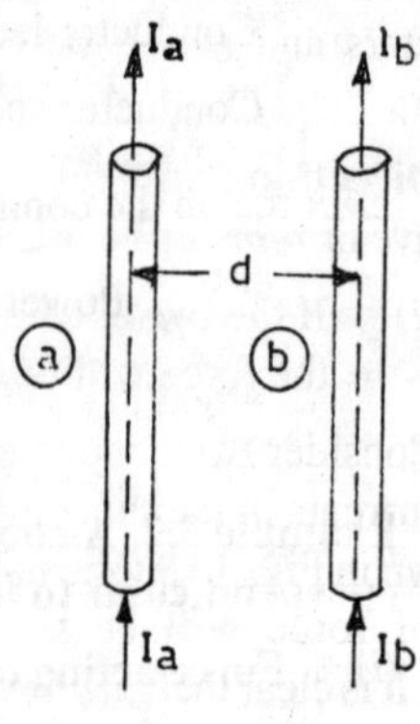

Fig. 3.17.

Let us consider one of the conductors producing magnetic field and the other conductor carrying current placed in this magnetic field. Suppose we consider the magnetic field due to conductor *a*. Then conductor *b* carrying current is placed in the magnetic field produced by conductor *a*. Therefore, as already seen, mechanical force acts on the conductor *b*. The magnitude of this force is given by

$$F = BI_b\, l \text{ newtons} \quad ...(i)$$

B = flux density produced by the conductor *a* at the centre of conductor *b*.

Now, the field strength H due to current carrying conductor *a* at the centre of condcutor *b* is given by

$$H = \frac{I_a}{2\pi d}$$

But $$B = \mu_0 \mu_r H$$

where $\mu_0 = 4\pi \times 10^{-7}$ H/m

μ_r = Relative permeability of the medium in which conductors are placed.

$$\therefore \quad B = (\mu_0 \mu_r) \frac{I_a}{2\pi d}$$

Putting the value of B in expression (i), we get,

$$F = \left[(\mu_0 \mu_r) \frac{I_a}{2\pi d} \right] I_a\, l \text{ newtons}$$

$$= \frac{I_a I_b \mu_0 \mu_r l}{2\pi d} \text{ newtons} = \frac{4\pi \times 10^{-7} \mu_r I_a I_b}{2\pi d} \times l \text{ newtons}$$

$$= \frac{2 \times 10^{-7} I_a I_b}{d} \times l \text{ newtons}$$

$$\therefore \quad F = \frac{2 I_a I_b l}{d} \times 10^{-7} \text{ newtons } (\mu_r = 1, \text{ for air})$$

From the above expression, it is clear that larger the currents carried by the conductors and closer the conductors, greater is the mechanical force between them.

Direction (Nature) : The force between the two conductors will either be *attractive* or *repulsive* depending upon the direction of currents in the conductors:

(i) If the two conductors are carrying currents in the same direction, then the force will be attractive.

Consider two conductors carrying currents in the same direction as shown in Fig. 3.18. The magnetic lines of force will be set up as shown. It is clear that magnetic lines are in the opposite direction. In the space between the conductor. Since magnetic lines of force in the opposite direction attract (property of magnetic lines of force), therefore, attractive force exists between the conductors.

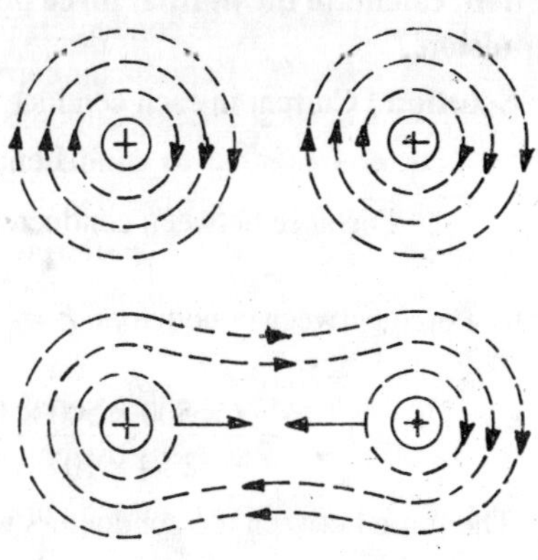

Fig. 3.18.

(ii) If the two conductors are carrying currents in the opposite direction then repulsive force exists between them. Consider two conductors carrying currents in the opposite directions. The magnetic lines of force due to conductors are shown in Fig. 3.19. It is clear that magnetic lines are in the same direction in the space between the conductors. Since lines in the same direction repel, therefore, repulsive force exists between the conductors.

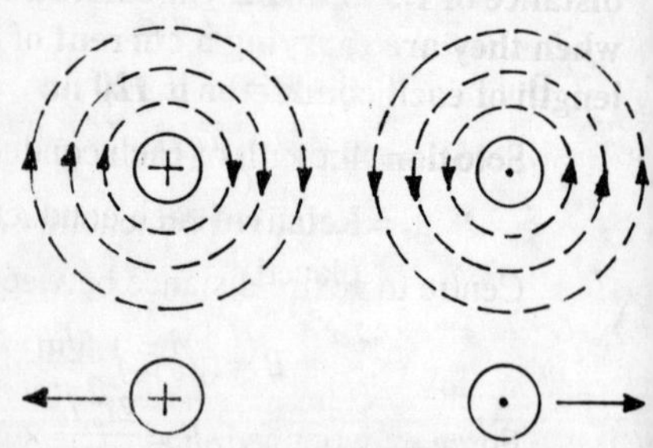

Fig. 3.19.

Definition of one amphere : We know,

$$F = \frac{2\,I_a\,I_b \times l}{d} \times 10^{-7} \text{ newtons}$$

If $I_a = I_b = 1$ ampere

$l = 1$ m , $d = 1$ m

Then, $F = 2 \times 10^{-7}$ newtons

Hence we can define one ampere on this basis as under :

One ampere is that much current which when flowing in each of the two infinitely long parallel conductors situated in vacuum and separated 1 m apart between centres, produces on each conductor a force of 2×10^{-7} newtons per metre length.

Example 3.3. Two long straight parallel conductors have their centres 2.54×10^{-2} m apart. If each carries a current of 500 amperes in the same direction, calculate the mutual force per metre run. Is it one of attraction or of repulsion ?

Solution : Current in each conductor, $I_a = I_b = 500$ A

Length, $l = 1$ m

Distance between conductors, $d = 2.54 \times 10^{-2}$ m

$\therefore$ Force between conductors, $F = \dfrac{2\,I_a\,I_b \times l}{d} \times 10^{-7}$

$$= \frac{2 \times 500 \times 500 \times 1}{2.54 \times 10^{-2}} \times 10^{-7} = 1.96 \text{ N } \textbf{Ans.}$$

The force between the conductors will be of attraction.

Example 3.4. Two long, parallel conductors are laid at a centre to centre distance of 1.5 m in air. Calculate and state the nature of force between them when they are carrying a current of 1,000 A each in opposite directions. The length of each conductor is 120 m.

Solution : Length of each conductor, $l = 120$ m

Current in each conductor, $I_a = I_b = 1{,}000$ A

Centre to centre distance between conductors,

$$d = 1.5 \text{ m}$$

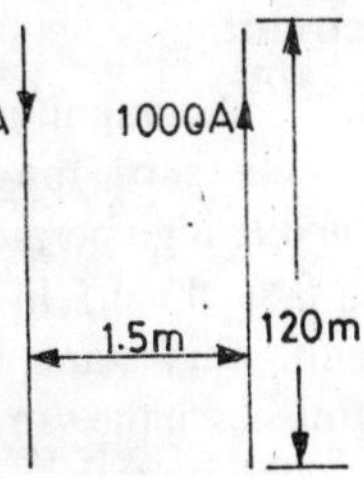

Fig. 3.20.

$$\text{Force experienced, } F = \frac{2 I_a I_b}{d} \times l \times 10^{-7} \text{ newtons}$$

$$= \frac{2 \times 1{,}000 \times 1{,}000}{1.5} \times 120 \times 10^{-7}$$

$$= 16 \text{ newtons } \textbf{Ans.}$$

Since the conductors carry current in the opposite directions, therefore, conductors will be pulled apart (*i.e.*, repulsive force).

Example 3.5. Two long parallel wires A and B, 12 cm apart carry currents of 750 A and 500 A respectively in opposite directions. Determine the flux density at the mid point of the perpendicular line between the wires.

Solution : Current $I_a = 750$ A

Current $I_B = 500$ A

Flux density at mid point between two conductors, due to conductor A,

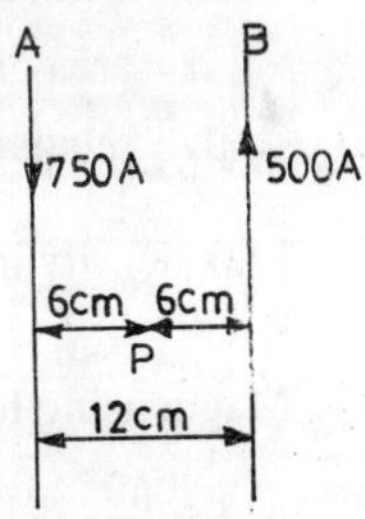

Fig. 3.21.

$$B_1 = \frac{\mu_0 \mu_r I_A}{2 \pi r} = \frac{4 \pi \times 10^{-7} \times 1 \times 750}{2 \pi \times 0.06} = 2.5 \times 10^{-3} \text{ T}$$

Flux density at mid point between two conductors due to conductor B.

$$B_2 = \frac{\mu_0 \mu_r I_B}{2 \pi r} = \frac{4 \pi \times 10^{-7} \times 1 \times 500}{2 \pi \times 0.06} = 1.667 \times 10^{-3} \text{ T}$$

When the two conductors carry current in opposite directions, the fields are additive, so

Net flux density, $B = B_1 = B_2 = 2.5 + 1.667 = 4.167$ T **Ans.**

Example 3.6. Force between two wires carrying currents in opposite directions is 20.4 kg/m, when they are placed parallel with axis 5 cm apart. Calculate the current in one conductor when the current flowing through the other conductor is 5,000 A.

Solution : Since force experienced, $F = 2I_a \times I_b \times \frac{l}{d} 10^{-7}$ newtons

$$\therefore \quad I_a = \frac{F \times d}{2I_b\, l \times 10^{-7}} = \frac{20.4 \times 9.81 \times 0.05}{2 \times 5{,}000 \times 1 \times 10^{-7}} = 10{,}000 \text{ A} \quad \textbf{Ans.}$$

3.6. MAGNETIC CIRCUIT AND ITS ANALYSIS

The closed path followed by magnetic flux is called a **magnetic circuit**.

A magnetic circuit usually consists of magnetic materials having high permeability (*e.g.*, iron, soft steel etc.). In this circuit, magnetic flux starts from a point and finishes at the same point after completing its path.

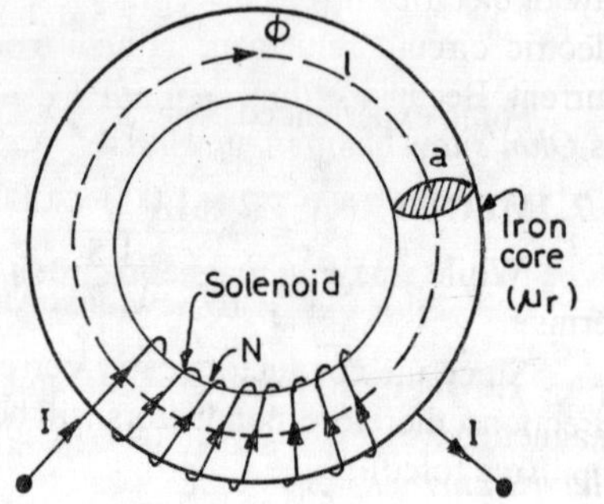

Fig. 3.22.

Fig. 3.22 shows a solenoid having N turns wound on an iron core (ring). When current I ampere is passed through the solenoid, magnetic flux ϕ wb is set up in the core.

Let l = mean length of magnetic circuit in m ;

a = area of cross-section of core in m^2 ;

μ_r = relative permeability of core material.

Flux density in the core material, $B = \frac{\phi}{a}$ wb/m^2

Magnetising force in the core material, $H = \frac{B}{\mu_0 \mu_r} = \frac{\phi}{a \mu_0 \mu_r}$ AT/m

According to work law, the work done in moving a unit pole once round the magnetic circuit (or path) is equal to the ampere-turns enclosed by the magnetic circuit.

i.e., $$Hl = NI \quad \text{or} \quad \frac{\phi}{a \mu_0 \mu_r} \times l = NI$$

or $$\phi = \frac{NI}{(l/a \mu_0 \mu_r)} \text{ wb}$$

The above expression reveals that the amount of flux set-up in the core is

(*i*) directly proportional to N and I, *i.e.*, NI called *magnetomotive force (m.m.f.)*. It shows that the flux increases if either of the two increases and vice-versa.

(ii) inversely proportional to $l/a\ \mu_0\ \mu_r$ called *reluctance* of the magnetic path. In fact, reluctance is the opposition offered to the magnetic flux by the magnetic path. The lower is the reluctance, the higher will be the flux and vice- versa.

Thus, $$\text{Flux} = \frac{\text{m.m.f.}}{\text{reluctance}}$$

It may be noted that the above expression has a strong resemblance to Ohm's law of electric current (I = e.m.f./resistance). The m.m.f. is analogous to e.m.f. in electric circuit, reluctance is analogous to resistance and flux is analogous to current. Because of this similarity, the above expression in sometimes referred to as *Ohm's law of magnetic circuits.*

3.7. IMPORTANT TERMS

While studying magnetic circuits, generally, we come across the following terms :

1. Magnetic flux (ϕ) : The amount of magnetic lines of force set-up in a magnetic circuit is called *magnetic flux.* Its unit is weber (wb). It is analogous to *electric current I* in electric circuit.

2. Magnetomotive force (m.m.f.) : The magnetic pressure which sets-up or tends to set-up magnetic flux in a magnetic circuit is called *magnetomotive force.* As per work law it may be defined as under :

The work done in moving a unit magnetic pole (1 wb) once round the magnetic circuit is called *magnetomotive force.* In general

$$\text{m.m.f.} = NI \text{ ampere-turns (or AT)}$$

It is analogous to e.m.f. in an electric circuit.

3. Reluctance (S) : The opposition offered to the magnetic flux by a magnetic circuit is called its *reluctance.*

It depends upon length (l), area of cross section (a) and permeability ($\mu = \mu_0\ \mu_r$) of the material that makes up the magnetic circuit. It is measured in AT/wb.

$$\text{Reluctance, } S = \frac{1}{a\ \mu_0\ \mu_r}$$

It is analogous to *resistance* in an electric circuit.

4. Permeance : It is a measure of the ease with which flux can be set-up in a material. It is just reciprocal of reluctance of the material and is measured in wb/AT or henry.

$$\text{Permeance} = \frac{1}{\text{reluctance}} = \frac{a\,\mu_0\,\mu_r}{l}\ \text{wb/AT or H}$$

It is analogous to *conductance* in an electric circuit.

3.8. COMPARISON BETWEEN MAGNETIC AND ELECTRIC CIRCUITS

Although magnetic and electric circuits have many points of similarity but still they are not analogous in all respects. A comparison of the two circuits is given below :

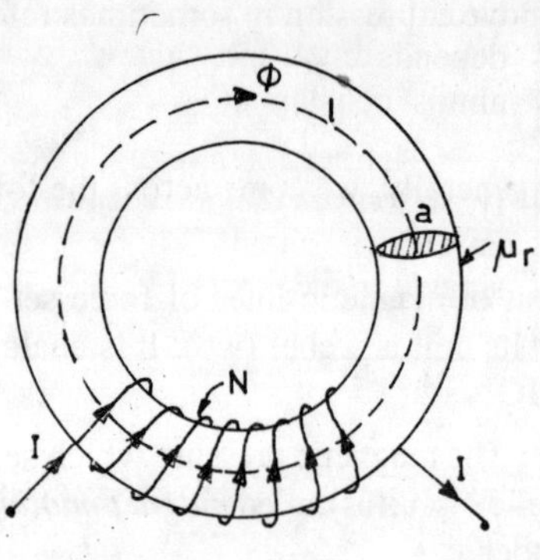

Fig. 3.23.

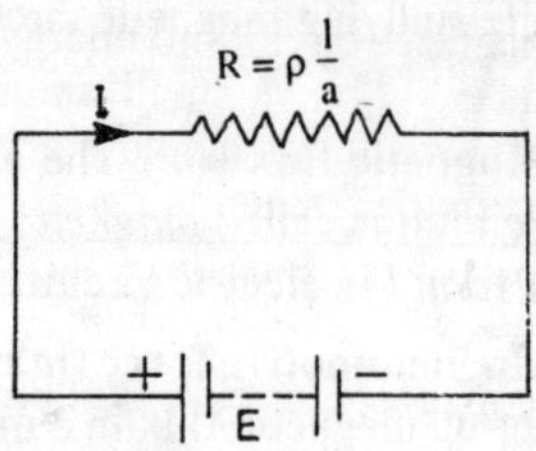

Fig. 3.24.

Similarities

1. The closed path for magnetic flux is called magnetic circuit.	1. The closed path for electric current is called electric circuit.
2. Flux = m.m.f./reluctance	2. Current = e.m.f./resistance
3. Flux, ϕ in wb.	3. Current, I in amperes.
4. m.m.f. in AT.	4. e.m.f. in V.
5. Reluctance, $S = \frac{1}{a\mu} = \frac{1}{a\,\mu_0\,\mu_r}$ AT/wb	5. Resistance, $R = \rho\frac{l}{a}\,\Omega$ or $R = \frac{1}{\sigma}\frac{l}{a}\,\Omega$
6. Permeance = 1/reluctance	6. Conductance = 1/resistance
7. Permeability, μ	7. Conductivity, $\sigma = 1/\rho$
8. Flux density, $B = \frac{\phi}{a}$ wb/m^2	8. Current density, $J = \frac{I}{a}$ A/m^2
9. Magnetic intensity, $H = NI/l$	9. Electric intensity, $E = V/d$

Dissimilarities

1. In fact, the magnetic flux does not flow but it sets-up in the magnetic circuit.	1. The electric current (electrons) actually flow in an electric circuit.
2. For magnetic flux, there is no perfect insulator. It can set- up even in the non-magnetic materials like air, rubber, glass etc. with reasonable m.m.f.	2. For electric current, there are large number of perfect insulators like glass, air, rubber etc. which do not allow it to flow through them under normal conditions.
3. The reluctance (S) of a magnetic circuit is not constant rather it varies with the value of B.	3. The resistance (R) of an electric circuit is almost constant as its value depends upon the value of ρ which is almost constant.
4. Once the magnetic flux is set-up in a magnetic circuit, no energy is expanded. However, a small amount of energy is required at the start to create flux in the circuit.	4. Energy is expanded continuously, so long as the current flows through an electric circuit. This energy is dissipated in the form of heat.

3.9. AMPERE-TURNS CALCULATIONS

In a magnetic circuit, flux produced, $\phi = \dfrac{\text{m.m.f.}}{\text{reluctance}} = \dfrac{NI}{l/a\,\mu_0\,\mu_r}$

or AT required, $NI = \dfrac{\phi\, l}{a\,\mu_0\,\mu_r} = \dfrac{B}{\mu_0\,\mu_r}\, l = Hl$

3.10. SERIES MAGNETIC CIRCUIT

A magnetic circuit that has a number of parts of different dimensions and materials carrying the same magnetic field is called a *series magnetic circuit.* Such a series magnetic circuit (composite circuit) is shown in Fig. 3.25.

Total reluctance of the magnetic circuit,

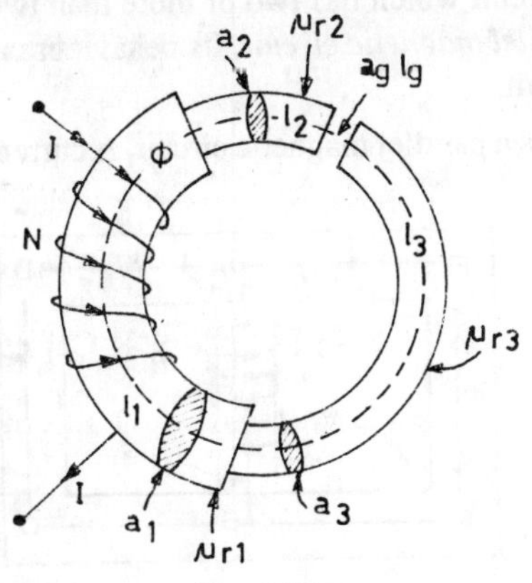

Fig. 3.25.

$$S = S_1 + S_2 + S_3 + S_g = \frac{l_1}{a_1\,\mu_0\,\mu_{r1}} + \frac{l_2}{a_2\,\mu_0\,\mu_{r2}} + \frac{l_3}{a_3\,\mu_0\,\mu_{r3}} + \frac{l_g}{a_g\,\mu_0}$$

$$\text{Total m.m.f} = \phi\,.\,S = \phi\left(\frac{l_1}{a_1\,\mu_0\,\mu_{r1}} + \frac{l_2}{a_2\,\mu_0\,\mu_{r2}} + \frac{l_3}{a_3\,\mu_0\,\mu_{r3}} + \frac{l_g}{a_g\,\mu_0}\right)$$

$$= \frac{B_1\,l_1}{\mu_0\,\mu_{r1}} + \frac{B_2\,l_2}{\mu_0\,\mu_{r2}} + \frac{B_3\,l_3}{\mu_0\,\mu_{r3}} + \frac{B_g\,l_g}{\mu_0}$$

$$= H_1\,l_1 + H_2\,l_2 + H_3\,l_3 + H_g\,l_g$$

3.11. IMPORTANCE OF AIR GAP

Air gap play very important role in electrical machinery. The magnetic circuits of most of the electrical machinery and instruments have air gap, *i.e.,* magnetic flux made to flow through certain distance in air. Air being non-magnetic offers much more reluctance than iron for the same length and area of *X*-section. The reluctance of air gap is only few hundred times greater than reluctance of iron. The width of air gap is one by forty times of mili-metre.

The air gap is also one of the main sources of low power factor at which the induction motors operate. The air gap increases the magnetising current which is responsible for setting the air gap flux. Small air gap means decreased value of magnetising current.

If the air gap is decreased to extremly small value, it causes mechanical difficulties. The shaft and frame of the machinery must be rigid enough to prevent scraping of rotor. For small air gap the stator as well as rotor must be perfectly circular.

Too much decreased in the air gap may increase the motor noise. It also prevent the machine from accelerating to its rated speed.

3.12. PARALLEL MAGNETIC CIRCUIT

A magnetic circuit which has two or more than two parts for the magnetic flux is called a *parallel magnetic circuit.* Its behaviour can be just compared to a parallel electric circuit.

Fig. 3.26 shows a parallel magnetic circuit. A current carrying coil is wound

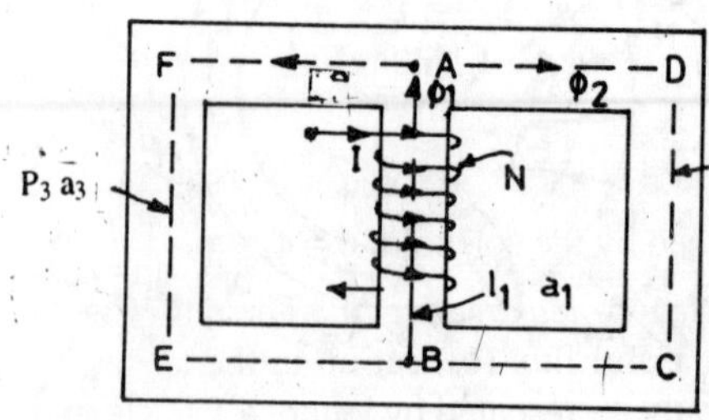

Fig. 3.26.

on the central limb *AB*. This coil sets-up a magnetic flux ϕ_1 in the central limb which is further divided into two parts *i.e., (i)* path *ADCB* which carries flux ϕ_2 and *(ii)* path *AFEB* which carries flux ϕ_3.

It is clear that $\phi_1 = \phi_2 + \phi_3$

The two magnetic paths *ADCB* and *AFEB* are in parallel. The ATs required for this parallel circuit is equal to the ATs required for any one of the paths.

If S_1 = reluctance of path *BA, i.e.,* $l_1/a_1 \mu_0 \mu_{r1}$

S_2 = reluctance of path *ADCB, i.e.,* $l_2/a_2 \mu_0 \mu_{r2}$

S_3 = reluctance of path *AFEB, i.e.,* $l_3/a_3 \mu_0 \mu_{r3}$

∴ Total m.m.f. required = m.m.f. required for path *BA* + m.m.f. required for path *ADCB* or path *AFEB*.

i.e., Total m.m.f. or AT $= \phi_1 S_1 + \phi_2 S_2 + \phi_3 S_3$

3.13. LEAKAGE FLUX

The magnetic flux which does not follow the intended path in a magnetic circuit is called *leakage flux.*

When a current is passed through a solenoid, as shown in Fig. 3.27, magnetic flux is produced by it. Most of this flux is set- up in the magnetic core and passes through the air gap. This flux is known as *useful flux* ϕ_u. However, some of the flux is just set-up around the coil and is not utilised for any work. This flux is called *leakage flux* ϕ_l

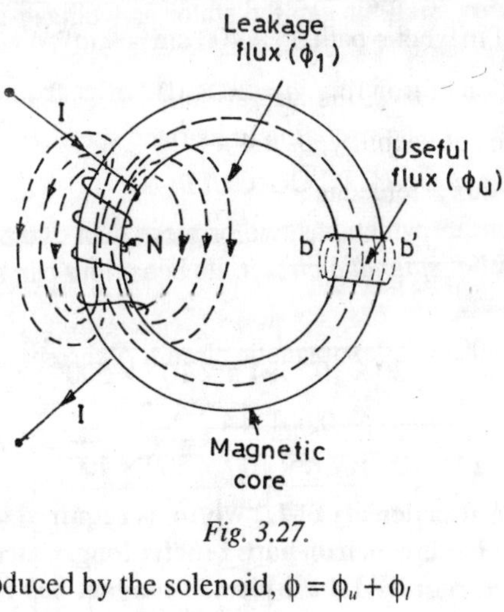

Fig. 3.27.

Total flux produced by the solenoid, $\phi = \phi_u + \phi_l$

Leakage co-efficient or leakage factor : The ratio of total flux (ϕ) produced by the solenoid to the useful flux (ϕ_u) set-up in the air gap is known as *leakage co-efficient.* It is generally represented by letter 'λ'.

∴ Leakage co-efficient, $\lambda = \phi/\phi_u$

Fringing : It may be seen in Fig. 3.27 that the useful flux when sets-up in the air gap, it tends to bulge outwards at b and b'. This increases the effective area in the air gap and decreases the flux density. This effect is known as *fringing*. The fringing is directly proportional to the length of the air gap.

Example 3.7. An iron ring of 400 cm mean circumference is made from round iron of cross section 20 cm². Its permeability is 500. If it is wound with 400 turns, what current would be required to produce a flux of 0.001 wb ?

Solution : The magnetic circuit is shown in Fig. 3.28.

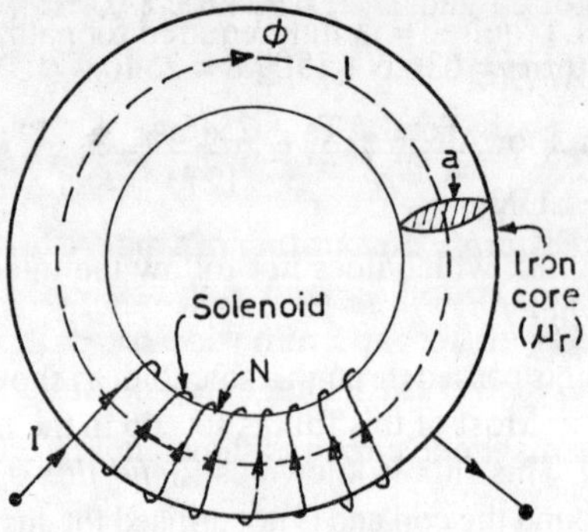

Fig. 3.28.

Mean length of magnetic path, $l_m = 400$ cm $= 4$ m

Area of X-section of iron ring, $a = 20 \times 10^{-4}$ m^2

Absolute permeability, $\mu_0 = 4\pi \times 10^{-7}$

Now m.m.f. = flux × reluctance

$$NI = \phi \times \frac{l_m}{a\,\mu_0\,\mu_r}$$

$$400 \times I = 0.001 \times \frac{4}{20 \times 10^{-4} \times 4\pi \times 10^{-7} \times 500}$$

$$\therefore \text{Current, } I = \frac{0.001 \times 4}{20 \times 10^{-4} \times 4\pi \times 10^{-7} \times 500 \times 400} = 7.958 \quad \textbf{Ans.}$$

Example 3.8. A flux density of 1.2 wb/m² is required in the 2 mm air gap of an electro-magnet having an iron path 1 metre long. Calculate the magnetising force and current required if the electro-magnet has 1273 turns. Assume relative permeability of iron to be 1500.

Solution : Flux density, $B = 1.2$ wb/m^2

Relative permeability of iron, $\mu_r = 1500$

No. of turns, $N = 1273$

Length of iron path, $l_i = 1$ m

Length of air gap, $l_g = 2$ mm = 0.002 m

Magnetising force for iron, $H_i = \dfrac{B}{\mu_0 \mu_r} = \dfrac{1.2}{4\pi \times 10^{-7} \times 1500} = 636.6$ AT/m

Magnetising force for air gap, $H_g = \dfrac{B}{\mu_0} = \dfrac{1.2}{4\pi \times 10^{-7}} = 954900$ AT/m

ATs required for iron path, $H_i\, l_i = 636.6 \times 1 = 636.6$

ATs required for air gap, $H_g\, l_g = 954900 \times 0.002 = 1909.8$

Total ampere turns = 636.6 + 1909.8 = 2546.4

Current required, $I = \dfrac{\text{Total ATs}}{N} = \dfrac{2546.4}{1273} = 2$ A **Ans.**

Example 3.9. Estimate the number of ampere turns necessary to produce a flux of 1,00,000 lines round an iron ring of 6 cm^2 cross-section and 20 cm mean diameter having an air gap 2 mm wide across it. Permeability of the iron may be taken 1200. Neglect the leakage flux outside the 2 mm air gap.

Solution : The magnetic circuit is shown in Fig. 3.29.

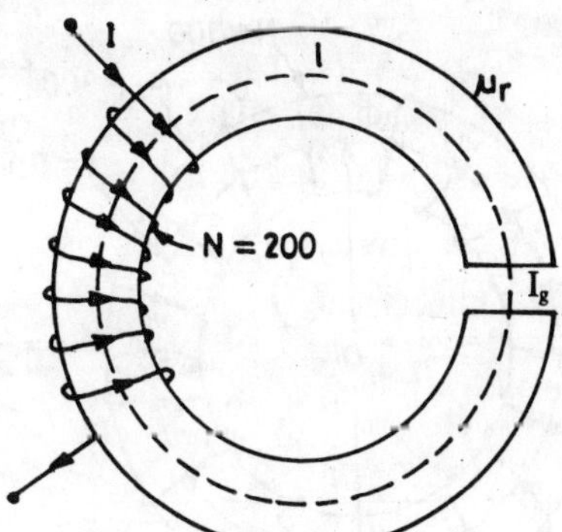

Fig. 3.29.

Area of cross section of the ring, $a = 6\text{ cm}^2 = 6 \times 10^{-4}\text{ m}^2$

Mean diameter of the ring, $D_m = 20$ cm = 0.2 m

Length of air gap, $l_g = 2\text{ mm} = 2 \times 10^{-3}$ m

Flux set up the ring, $\phi = 100{,}000\text{ lines} = 100{,}000 \times 10^{-8} = 0.001$ wb

Relative permeability of iron, $\mu_r = 1200$

Mean length of ring, $l_m = \pi D = \pi \times 0.2 = 0.6283$ m

Length of air gap, $l_g = 0.002$ m

Length of iron path, $l_i = 0.6283 - 0.002 = 0.6263$ m

Now, m.m.f. = flux × reluctance

Ampere turns required for iron path,

$$AT_i = \phi \times \frac{l_i}{a\,\mu_0\,\mu_r} = 0.001 \times \frac{0.6263}{6 \times 10^{-4} \times 4\,\pi \times 10^{-7} \times 1200} = 692.21 \text{ AT}$$

Ampere turns required for air gap,

$$AT_g = \phi \times \frac{l_g}{a\,\mu_0} = 0.001 \times \frac{0.002}{6 \times 10^{-4} \times 4\,\pi \times 10^{-7}} = 2652.58 \text{ AT}$$

Total ampere turns required to produce the given flux

$$= AT_i + AT_g = 692.21 + 2652.58 = 3344.79 \text{ AT} \quad \textbf{Ans.}$$

Example 3.10. A coil is wound uniformly with 300 turns over a steel ring of relative permeability 900 having a mean diameter of 20 cm. The steel ring is made of a bar having circular cross- section of diameter 2 cm. If the coil has a resistance of 50 ohm and is connected to 250 V d.c. supply, calculate *(i)* the m.m.f. of the coil, *(ii)* the field intensity in the ring, *(iii)* reluctance of the magnetic path, *(iv)* total flux and *(v)* permeance of the ring.

Solution : The magnetic circuit is shown in Fig. 3.30.

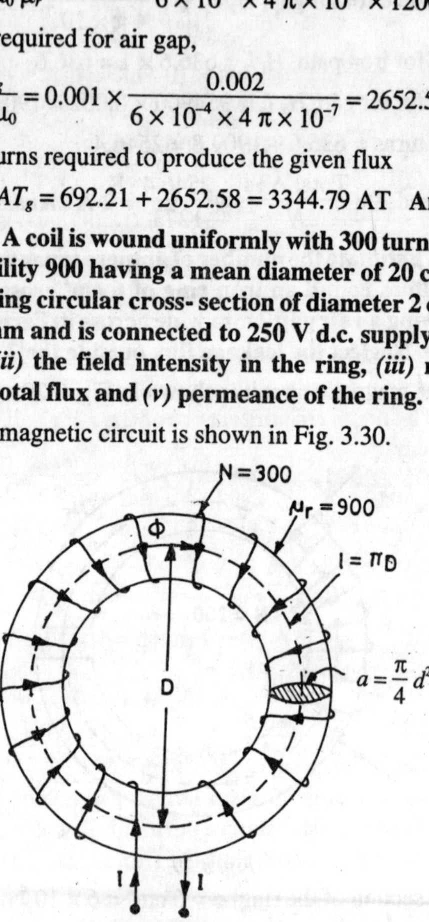

Fig. 3.30.

Current through the coil, $I = \frac{V}{R} = \frac{250}{50} = 5$ A

(i) m.m.f. of the coil, $= NI = 300 \times 5 = 1500$ T **Ans.**

(ii) Field intensity, $H = \frac{NI}{l}$

where, $l = \pi D = 0.2\,\pi$ metres

$$\therefore \quad H = \frac{1500}{0.2\,\pi} = 2387.3 \text{ AT/m} \quad \textbf{Ans.}$$

(iii) Reluctance of the magnetic path, $S = \frac{l}{a\,\mu_0\,\mu_r}$

$$\text{where, } a = \frac{\pi d^2}{4} = \frac{\pi \times (2)^2 \times 10^{-4}}{4} = \pi \times 10^{-4} \text{ m}^2 \;;\; \mu_r = 900$$

$$\therefore \quad S = \frac{0.2\,\pi}{\pi} \times 10^{-4} \times 4\,\pi \times 10^{-7} \times 900 = 17.684 \times 10^5 \text{ AT/wb} \quad \textbf{Ans.}$$

$$\textit{(iv)} \text{ Total flux, } \phi = \frac{\text{m.m.f.}}{S} = \frac{1500}{17.684 \times 10^5} = 0.848 \text{ m wb} \quad \textbf{Ans.}$$

$$\textit{(v)} \text{ Permeance} = \frac{1}{S} = \frac{1}{17.684 \times 10^5} = 5.655 \times 10^{-7} \text{ wb/AT} \quad \textbf{Ans.}$$

Example 3.11. Calculate the relative permeability of an iron ring when the exciting current taken by the 600 turn coil is 1.2 A and the total flux produced is 1 m wb. The mean circumference of the ring is 0.5 m and the area of cross-section is 10 cm^2.

Solution : $NI = \frac{\phi \times l}{a\,\mu_0\,\mu_r}$

$$\therefore \quad \mu_r = \frac{\phi \times l}{a\,\mu_0\,NI}$$

Where $N = 600$ turns ; $I = 1.2$ A ; $\phi = 1$ m wb $= 1 \times 10^{-3}$ wb ; $l = 0.5$ m ;

$a = 10 \text{ cm}^2 = 10 \times 10^{-4} \text{ m}^2$

$$\therefore \quad \mu_r = \frac{1 \times 10^{-3} \times 0.5}{10 \times 10^{-4} \times 4\,\pi \times 10^{-7} \times 600 \times 1.2} = 552.6 \quad \textbf{Ans.}$$

Example 3.12. An iron ring of mean length 1 m has an air gap of 1 mm and a winding of 200 turns. If the relative permeability of iron is 500 when a current of 1 A flows through the coil, find the flux density.

Solution : The magnetic circuit is shown in Fig. 3.31.

Now, m.m.f. = flux × reluctance

$$\text{i.e.,} \quad NI = \phi \left(\frac{l_1}{a\,\mu_0\,\mu_r} + \frac{l_g}{a\,\mu_0} \right)$$

$$\text{or} \quad NI = B \left(\frac{l_1}{\mu_0\,\mu_r} + \frac{l_g}{\mu_0} \right)$$

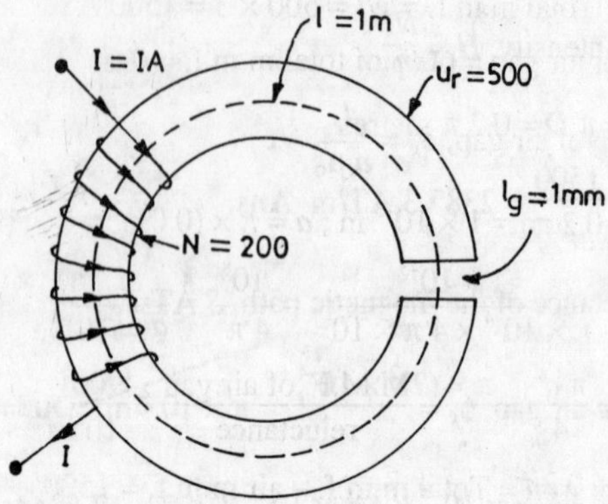

Fig. 3.31.

where $N = 200$ turns ; $I = 1$ A ; $\mu_r = 500$

$$l_g = 1 \text{ mm} = 0.001 \text{ m} ; l_i = (1 - 0.001) = 0.999 \text{ m}$$

$$\therefore \quad 200 \times 1 = B\left(\frac{0.999}{4\pi \times 10^{-7} \times 500} + \frac{0.001}{4\pi \times 10^{-7}}\right)$$

or $$B = \frac{200}{2385.73} = 0.0838 \text{ wb/m}^2 \textbf{ Ans.}$$

Example 3.13. An iron ring of 10 cm² area has a mean circumference of 100 cm. It has a saw cut of 0.2 cm wide. A flux of one m wb is required in the air gap. The leakage factor is 1.2. The flux density of iron for relative permeability 400 is 1.2 wb/m². Calculate the number of ampere-turns required.

Solution : Flux density in air gap, $B_g = \phi_g ; a = \frac{1 \times 10^{-3}}{10 \times 10^{-4}} = 1 \text{ wb/m}^2$

Flux in iron ring, $\phi_i = \lambda \times \phi_g$ (where λ is leakage factor)

$$= 1.2 \times 1 \times 10^{-3} = 1.2 \times 10^{-3} \text{ wb}$$

Flux density in iron ring, $B = \frac{\phi_i}{a} = \frac{1.2 \times 10^{-3}}{10 \times 10^{-4}} = 1.2 \text{ wb/m}^2$

Total ampere-turn required, $= H_g \, l_g + H_i \, l_i = \frac{B_g}{\mu_0} l_g + \frac{B_i}{\mu_0 \mu_r} l_i$

$$= \frac{1}{4\pi \times 10^{-7}} \times 0.2 \times 10^{-2} + \frac{1.2}{4\pi \times 10^{-7} \times 400} \times 1 = 3978.87 \quad \textbf{Ans.}$$

Example 3.14. A steel ring of 10 cm mean radius and of circular cross-section 1 cm in radius has an air gap of 1 mm length. It is wound uniformly with 500 turns of wire carrying current of 3 A. Neglect magnetic leakage. The air gap takes 60% of the total m.m.f. Find the total reluctance.

Solution : Total m.m.f. $= NI = 500 \times 3 = 1500$ ATs

M.M.F. for air gap = 60% of total m.m.f. $= 0.6 \times 1500 = 900$ ATs

Reluctance of air gap, $S_g = \dfrac{l_g}{a\,\mu_0}$

where $l_g = 1$ mm $= 1 \times 10^{-3}$ m ; $a = \pi \times (0.01)^2 = \pi \times 10^{-4}$ m^2 ;

$$\therefore \quad S_g = \frac{1 \times 10^{-3}}{\pi \times 10^{-4} \times 4\pi \times 10^{-7}} = \frac{10^8}{4\pi^2} \text{ ATs/wb}$$

Flux in the air gap, $\phi_g = \dfrac{\text{M.M.F. of air gap}}{\text{reluctance}} = \dfrac{900}{10^8} \times 4\pi^2 = 36\pi^2 \times 10^{-6}$ wb

M.M.F. for iron = Total m.m.f. – air m.m.f. $= 1500 - 900 = 600$ ATs

Flux in the iron ring, $\phi_i = \phi_g = 36\pi^2 \times 10^{-6}$ wb

(since there is no magnetic leakage)

Reluctance of iron ring, $S_i = \dfrac{\text{M.M.F. of iron}}{\phi_i} = \dfrac{600}{36\pi^2 \times 10^{-6}} = \dfrac{10^8}{6\pi^2}$ AT/wb

Total reluctance, $S = S_g + S_i = \dfrac{10^8}{4\pi^2} + \dfrac{10^8}{6\pi^2} = \dfrac{10^8}{\pi^2}\left(\dfrac{1}{4} + \dfrac{1}{6}\right)$

$= 4.22 \times 10^6$ ATs/wb **Ans.**

Example 3.15. Determine magnetomotive force, magnetic flux, reluctance and flux density in case of a steel ring 30 cm mean diameter and a circular cross-section 2 cm in diameter has an air gap 1 mm long. It is wound uniformly with 600 turns of wire carrying a current of 2.5 A. Neglect magnetic leakage. The iron path takes 40% of the total magnetomotive force.

Solution : M.M.F. of the magnetic circuit

$= NI = 600 \times 2.5 = 1500$ ATs **Ans.**

As iron path takes 40% of the total m.m.f., the reluctance of iron in 40% and the rest of the reluctance (60%) is of air path.

$$\therefore \quad \frac{S_a}{S_i} = \frac{60}{40} = \frac{3}{2} = 1.5$$

Reluctance of air path, $S_a = \dfrac{l_a}{a\,\mu_0}$

where $l_a = 1 \times 10^{-3}$ m , $a = \dfrac{\pi}{4}(2)^2 \times 10^{-4} = \pi \times 10^{-4}$ m^2

$$\therefore \quad S_a = \frac{1 \times 10^{-3}}{\pi \times 10^{-4} \times 4\pi \times 10^{-7}} = 2.533 \times 10^6 \text{ ATs/wb}$$

Reluctance of iron path, $S_i = \frac{S_a}{1.5} = \frac{2.533 \times 10^6}{1.5} = 1.688 \times 10^6$ ATs/wb

Total reluctance $= S_a + S_i = (2.533 + 1.688) \times 10^6$

$= 4.221 \times 10^6$ ATs/wb **Ans.**

Magnetic flux, $\phi = \frac{\text{m.m.f.}}{\text{reluctance}} = \frac{1500}{4.221 \times 10^6} = 0.3554$ m wb **Ans.**

Flux density, $B = \frac{\phi}{a} = \frac{0.3554 \times 10^{-3}}{\pi \times 10^{-4}} = 1.131$ wb/m² **Ans.**

Example 3.16. The ring shaped core shown in Fig. 3.32 is made of a material having a relative permeability of 1000. The flux density in the smallest area of cross-section is 2 T. If the current through the coil is not to exceed 1.5 A, compute the number of turns of the coil.

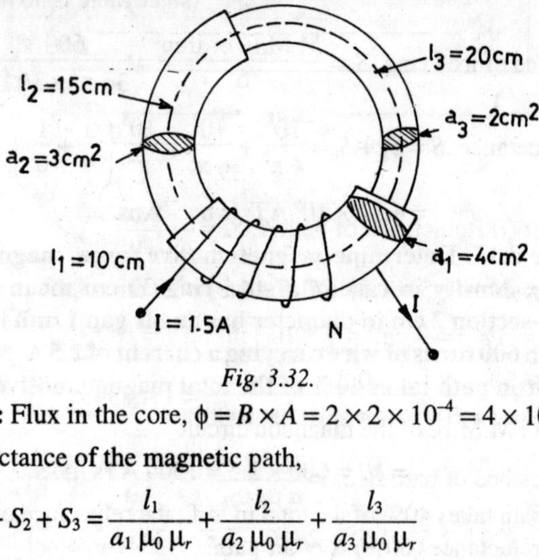

Fig. 3.32.

Solution : Flux in the core, $\phi = B \times A = 2 \times 2 \times 10^{-4} = 4 \times 10^{-4}$ wb

Total reluctance of the magnetic path,

$$S = S_1 + S_2 + S_3 = \frac{l_1}{a_1 \mu_0 \mu_r} + \frac{l_2}{a_2 \mu_0 \mu_r} + \frac{l_3}{a_3 \mu_0 \mu_r}$$

$$= \frac{1}{\mu_0 \mu_r}\left(\frac{l_1}{a_1} + \frac{l_2}{a_2} + \frac{l_3}{a_3}\right) = \frac{1}{4\pi \times 10^{-7} \times 1000}\left(\frac{0.1}{4 \times 10^{-4}} + \frac{0.14}{3 \times 10^{-4}} + \frac{0.2}{2 \times 10^{-4}}\right)$$

$= 13.926 \times 10^5$ AT/wb

Total m.m.f. required, $NI = \phi S$

or $\quad N \times 1.5 = 4 \times 10^{-4} \times 13.926 \times 10^5$

$\therefore$ No. of turns, $N = 371.36$ **Ans.**

Example 3.17. The magnetic frame shown in Fig. 3.33 is built up of iron of square cross-section, 3 cm side. Each air gap is 2 mm wide. Each of the coils is wound with 1000 turns and the exciting current is 1.0 A. The relative

permeability of part A and part B may be taken as 1000 and 1200 respectively. Calculate *(i)* reluctance of part A, *(ii)* reluctance of part B, *(iii)* reluctance of two air gaps, *(iv)* total reluctance of the complete magnetic circuit, *(v)* m.m.f. produced and *(vi)* flux set-up in the circuit.

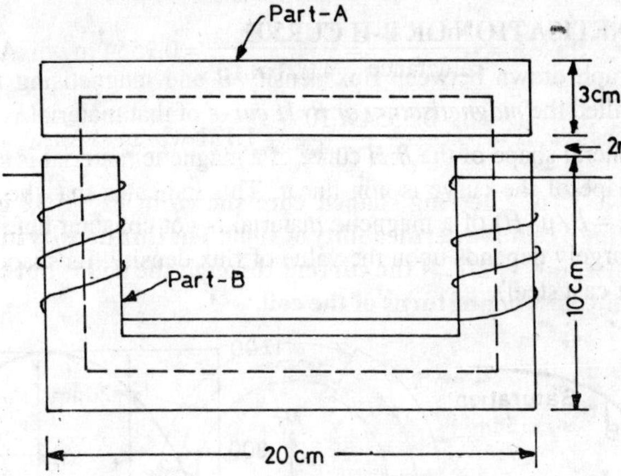

Fig. 3.33.

Solution : *(i)* Reluctance of path A, $S_A = \dfrac{l_A}{a\,\mu_0\,\mu_{rA}}$

where $l_A = 20 - (1.5 + 1.5) + (1.5 + 1.5) = 20$ cm $= 0.2$ m ;

$$a = 3 \times 3 = 9 \text{ cm}^2 = 9 \times 10^{-4} \text{ m}^2 \text{ ; } \mu_{rA} = 1000$$

$$\therefore \quad S_A = \frac{0.2}{9 \times 10^{-4} \times 4\,\pi \times 10^{-7} \times 1000} = 176839 \text{ AT/wb} \quad \textbf{Ans.}$$

(ii) Reluctance of path B, $S_B = \dfrac{l_B}{a\,\mu_0\,\mu_{rB}}$

where $l_B = (20 - 1.5 - 1.5) + 2\,(10 - 1.5) = 34$ cm $= 0.34$ m ; $\mu_{rB} = 1200$

$$\therefore \quad S_B = \frac{0.34}{9 \times 10^{-4} \times 4\,\pi \times 10^{-7} \times 1200} = 250521 \text{ AT/wb} \quad \textbf{Ans.}$$

(iii) Reluctance of two air gaps, $S_g = \dfrac{l_g}{a\,\mu_0}$

where $l_g = 2 + 2 = 4$ mm $= 4 \times 10^{-3}$ m

$$\therefore \quad S_g = \frac{4 \times 10^{-3}}{9 \times 10^{-4} \times 4\,\pi \times 10^{-7}} = 3536776 \text{ AT/m} \quad \textbf{Ans.}$$

(iv) Total reluctance of the composite magnetic circuit,

$$S = S_A + S_B + S_g = 176839 + 250521 + 3536776 = 3964136 \text{ AT/wb} \quad \textbf{Ans.}$$

(*v*) Total m.m.f. $= NI = (2 \times 1000) \times 1 = 2000$ AT **Ans.**

(*vi*) Flux set-up in the circuit, $\phi = \dfrac{\text{m.m.f}}{\text{reluctance}} = \dfrac{2000}{3964136}$

$= 0.5045$ m wb **Ans.**

3.14. MAGNETISATION OR B-H CURVE

The graph drawn between flux density *B* and magnetising force *H* of a material is called the *magnetisation or B- H curve* of that material.

The general shape of the *B-H* curve of a magnetic material is shown in Fig. 3.34. The shape of the curve is non-linear. This indicates that the relative permeability ($\mu_r = B/\mu_0 H$) of a magnetic material is not constant but it varies. The value of μ_r largely depends upon the value of flux density. Its shape is shown in Fig. 3.35 (for cast steel).

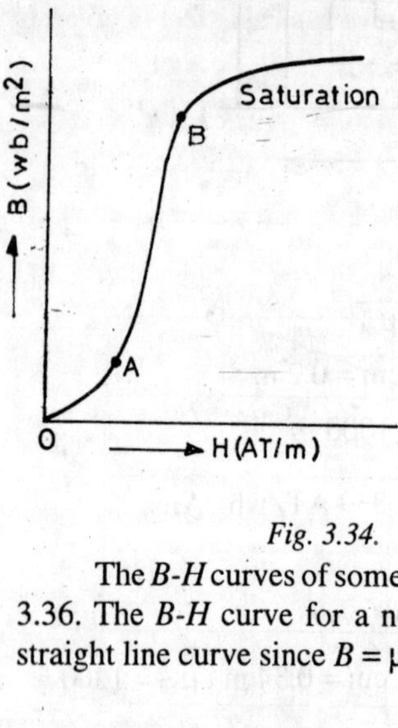

Fig. 3.34.

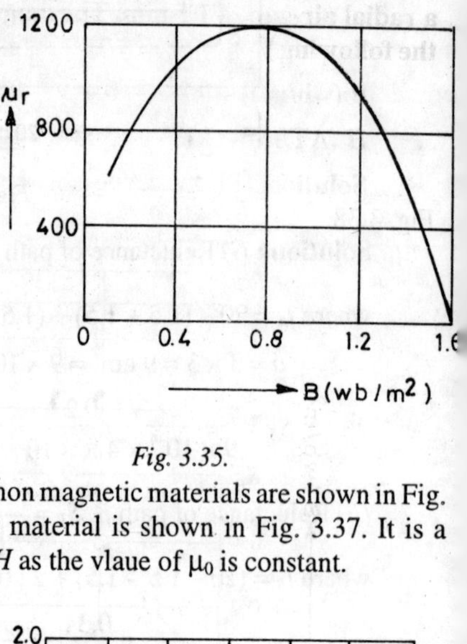

Fig. 3.35.

The *B-H* curves of some of the common magnetic materials are shown in Fig. 3.36. The *B-H* curve for a non-magnetic material is shown in Fig. 3.37. It is a straight line curve since $B = \mu_0 H$ or $B \propto H$ as the vlaue of μ_0 is constant.

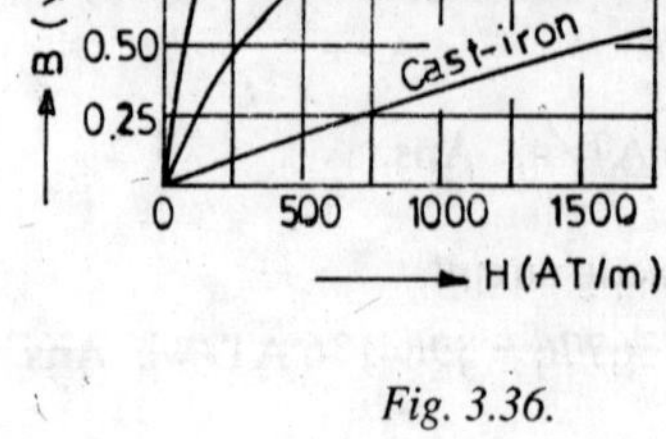

Fig. 3.36.

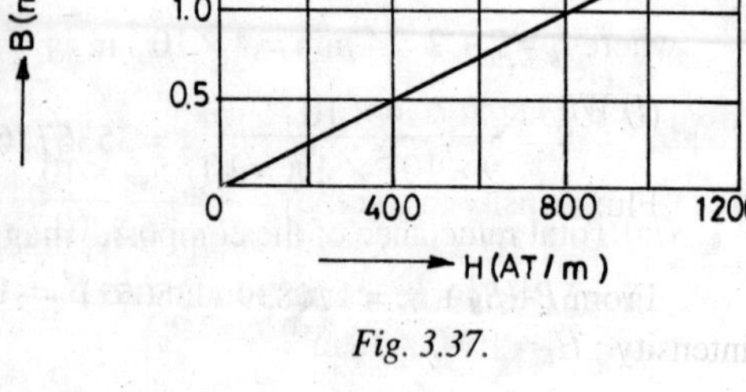

Fig. 3.37.

Magnetic Calculations from B-H Curves

B-H curves are very useful in solving the problems of magnetic circuits. While solving the problems, the following steps are followed :

(i) Find the value of flux density in the magnetic circuit if not known.

(ii) Determine the magnetising force *H* from the *B-H* curve corresponding to flux density in the magnetic circuit.

(iii) Compute the length of magnetic path *l*.

(iv) Determine m.m.f. required (m.m.f. = *Hl*).

Example 3.18. A cast steel ring of mean diameter 50 cm having a circular cross-section of 5 cm² is uniformly wound with 500 turns. Determine the current required to establish a flux of 5.5×10^{-4} wb *(i)* with no air-gap, *(ii)* with a radial air-gap of 1.5 mm. The magnetisation curve for cast steel is given by the following :

B (wb/m²) –	**0.2**	**0.4**	**0.6**	**0.8**	**1.0**	**1.2**
H (AT/m) –	**175**	**300**	**400**	**600**	**850**	**1250**

Solution : Plot a curve between *B* and *H* on a graph paper as shown in Fig. 3.38.

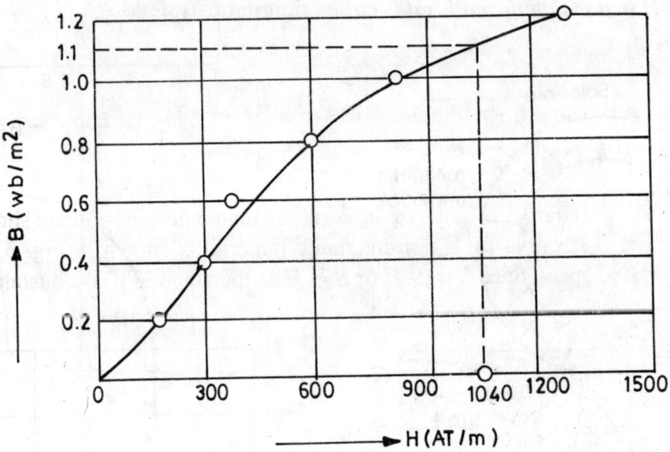

Fig. 3.38.

(i) With no air gap,

$$\text{Flux density in steel, } B_1 = \frac{\phi}{a} = \frac{5.5 \times 10^{-4}}{5 \times 10^{-4}} = 1.1 \text{ wb/m}^2$$

From *B-H* curve, corresponding to $B_i = 1.1$ wb/m², value of magnetic field intensity ; $H_i = 1040$ AT/m.

Mean length of steel ring, $l_i = \pi D = 0.5\,\pi$ m

$\therefore$ m.m.f. required, $= H_i\,l_i = 1040 \times 0.5\,\pi = 1633.63$ AT

$$\text{Magnetising current, } I = \frac{1633.63}{500} = 3.627 \text{ A} \quad \textbf{Ans.}$$

(ii) With an air gap of 1.5 mm

$$\text{Field intensity in the air gap, } H_g = \frac{B_g}{\mu_0} = \frac{1.1}{4\,\pi \times 10^{-7}} = 8.753 \times 10^5 \text{ AT/m}$$

m.m.f. required for air gap $= H_g\,l_g = 8.753 \times 10^5 \times 1.5 \times 10^{-3} = 1313$ AT

Total m.m.f. required, $NI = 1633.63 + 1313 = 2946.63$ AT

$$\therefore \text{Magnetising current, } I = \frac{NI}{N} = \frac{2946.63}{500} = 5.893 \text{ A} \quad \textbf{Ans.}$$

3.15. MAGNETIC HYSTERESIS

When a magentic material is magnetised first in one direction and then in the other, it is found that flux density B in the material lags behind the applied magnetising force H. This phenomenon is known as *magnetic hysteresis*.

Hence, *the phenomenon of flux density B lagging behind the magnetising force H in a magnetic material is called* **magnetic hysteresis.**

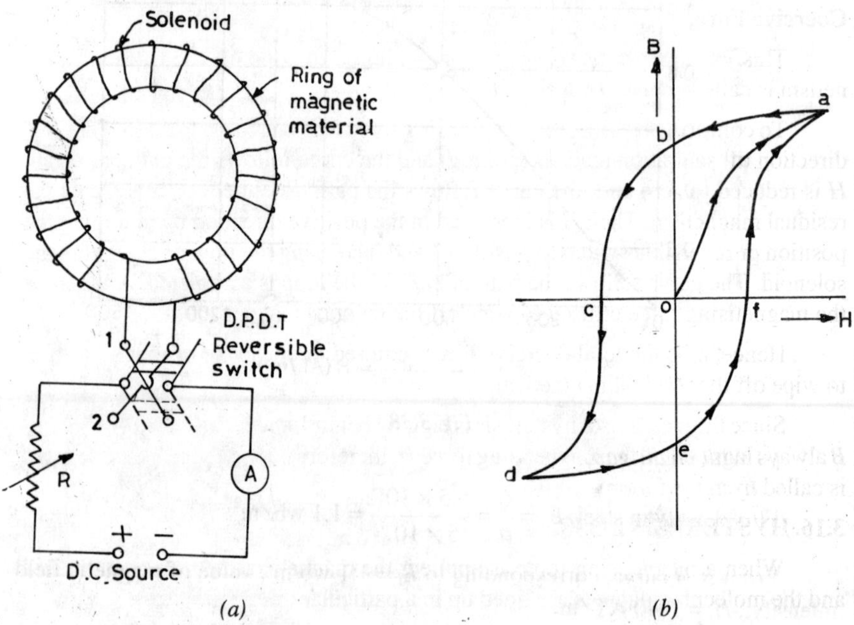

Fig. 3.39.

'Hysteresis' is the term derived from the Greek word *hysterein* meaning to lag behind. To understand the complete phenomenon of magnetic hysteresis, consider a ring of magnetic material on which a solenoid is wound uniformly as shown in Fig. 3.39 *(a)*. The solenoid is connected to d.c. source through a double pole double throw reversible switch (position 1).

When the field intensity *H* is increased gradually by increasing current in the solenoid (by decreasing the value of *R*), the flux density *B* also increases untill saturation point *a* is reached and curve so obtained is *oa*. If now the magnetising force is gradually reduced to zero by decreasing current in the solenoid to zero, the flux density does not become zero and the curve so obtained is *ab* as shown in Fig. 3.39 *(b)*. When magnetising force *H* is zero, the flux density still has value *ob*.

Residual Magnetism and Retentivity

This value of flux density *'ob'* retained by the magnetic material is called **residual magnetism** and the power of retaining this residual magnetism is called **retentivity** of the material.

To demagnetise the magnetic ring, the magnetising force *H* is reversed by reversing the direction of flow of current in the solenoid. This is achieved by changing the position of double pole, double throw switch (*i.e.*, position 2). When *H* is increased in reverse direction, the flux density starts decreasing and becomes zero and curve follows the path *bc*. Thus residual magnetism of the magnetic material is wiped off by applying magnetising force *oc* in opposite direction.

Coercive Force

This value of magnetising force *oc* required to wipe off the residual magnetism is called **coercive force**.

To complete the loop, the magnetising force *H* is increased further in reverse direction till saturation reaches (point *d*) and the curve follows the path *cd*. Again *H* is reduced to zero and the curve follows the path *de*. Where *oe* represents the residual magnetism. Then *H* is increased in the positive direction by changing the position or reversible switch to position 1 and increasing the flow of current in the solenoid. The curve follows the path of *efa* and the loop is completed. Again *of* is the magnetising force utilised to wipe off the residual magnetism *oe*.

Hence, *cf* is the total coercive force required in one cycle of magnetisation to wipe off the residual magnetism.

Since the meaning of hysteresis is lagging behind, and in this case flux density *B* always lags behind the magnetising force *H*, therefore, loop (*abcdefa*) so obtained is called *hysteresis loop*.

3.16. HYSTERESIS LOSS

When a magnetising force is applied, the magnetic material is magnetised and the molecular magnets are lined up in a particular direction.

However, when the magnetising force in a magnetic material is reversed, the internal friction of the molecular magnets opposes the reversal of magnetism, resulting in hysteresis. To overcome this internal friction of the molecular magnets, a part of the magnetising force is used. The work done by the magnetising force against this internal friction of molecular magnets produces heat. This energy, which is wasted in the form of heat due to hysteresis, is called *hysteresis loss*.

Hysteresis loss occurs in all the magnetic parts of electrical machines where there is reversal of magnetisation. This loss results in wastage of energy in the form of heat. Consequently, it increases the temperature of the machine which is undesirable. Therefore, a suitable magnetic material is selected for the construction of such parts, *e.g.*, silicon steel in which hysteresis loss is minimum.

3.17. IMPORTANCE OF HYSTERESIS LOOP

The shape and size of the hysteresis loop *largely depends upon the nature of the material. The choice of a magnetic material for a particular application often depends upon the shape and size of the hysteresis loop. A few cases are discussed below by way of illustration :

(i) The smaller the hysteresis loop area of a magnetic material, the less is the hysteresis loss. The hysteresis loop for silicon steel has a very small area [See Fig. 3.40 *(i)*]. For this reason, silicon steel is widely used for making transformer cores and rotating machines which are subjected to rapid reversals of magnetisation.

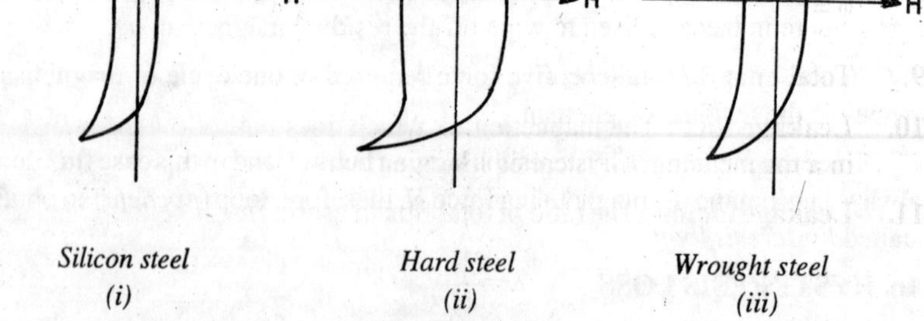

Silicon steel *(i)* *Hard steel* *(ii)* *Wrought steel* *(iii)*

Fig. 3.40.

*It is also depends upon (i) the maximum value of flux density established and (ii) the initial magnetic state of the material.

(ii) The hysteresis loop for hard steel [See Fig. 3.40 *(ii)*] indicates that this material has high retentivity and coercivity. Therefore, hard steel is quite suitable for making permanent magnets. But due to the large area of the loop, there is greater hysteresis loss. For this reason, hard steel is not suitable for the construction of electrical machines.

(iii) The hysteresis loop for wrought iron [See Fig. 3.40 *(iii)*] shows that this material has fairly good residual magnetism and coercivity. Hence it is suitable for making cores of electromagnets.

HIGHLIGHTS

1. **Magnetic field:** The area around a magnet within which its influence is pereptible is called magnetic field.
2. **Magnetic flux :** The total number of lines of force in the magnetic field is called magnetic flux. It is denoted by ϕ.
3. **Magnetic circuit :** Path followed by magnetic flux.
4. **Magnetic flux density :** The magnetic flux per unit area of a surface at right angle to the magnetic field.

$$\text{Magnetic flux density, } B = \frac{\phi}{A} \text{ Wb/m}^2.$$

5. **Permeability :** The ability of a material to conduct magnetic flux through it is called permeability of that material. It is represented by Greek letter μ (mu).
6. **MMF :** The magnetic pressure which sets up magnetic flux in a magnetic circuit is called MMF.

$$\text{mmf} = NI \text{ amp-turns}$$

7. **Reluctance :** The opposition offered to magnetic flux by a magnetic circuit is called its reluctance

$$\text{Reluctance, } S = \frac{1}{a\,\mu_0\,\mu_r}$$

8. $\phi = \dfrac{\text{mmf}}{\text{reluctance}}$
9. Total mmf $= H_1\, l_1 + H_2\, l_2$
10. **Leakage flux :** The magnetic flux which does not follow the intended path in a magnetic circuit is called leakage flux.
11. **Leakage factor :** The ratio of total flux to useful flux is called leakage factor.

$$\lambda = \frac{\phi}{\phi_u}$$

12. **Fringing :** The useful flux in the air gap increases the effective area of air gap and decrease the flux density is called fringing.

13. **B–H. curve :** The graph drawn between the flux density B and magnetising force H of a material is called B-H curve.
14. **Magnetic hysteresis :** The phenomenon of flux density B lagging behind the magnetic force H in a magnetic material is called magnetic hysteresis.
15. **Hysteresis loss :** The continuous reversal in a magnetic material causes hysteresis loss. It depends upon the area of hysteresis loop.

SHORT QUESTIONS

1. Why flux in air gap is useful flux ?
2. Why alloy sheet steel is most suited for making transformer cores ?
3. Why air gap essential ?
4. Solid core has more losses than laminated core why ?

MULTIPLE CHOICE QUESTIONS

1. The materials, which are not strongly attracted by magnet are known as
   (*a*) Ferro-magnetic materials
   (*b*) Non-magnetic materials
   (*c*) Paramagnetic materials.
2. Permeability is expressed in
   (*a*) H/M (*b*) H (*c*) H/M^2
3. The unit of flux density in SI system is
   (*a*) AT/m (*b*) coulombs (*c*) weber/m.
4. Hysteresis loss in magnetic material is due to its
   (*a*) retentivity (*b*) coercivity (*c*) flux density
5. The mmf of a magnetic circuit is analogous to
   (*a*) emf (*b*) current (*c*) resistance
   (*d*) power.

ANSWERS

1. (*b*) 2. (*a*) 3. (*c*) 4. (*c*) 5. (*a*)

NUMERICALS

1. An iron ring has a cross-sectional area of 400 mm^2 and a mean diameter of 25 cm. It is wound with 500 turns. If the value of relative permeability is 500, find the total flux set-up in the ring. The coil resistance is 400 Ω and the supply voltage is 200 V. (**Ans.** 0.08 m wb)
2. Estimate the total m.m.f. required to produce a flux of 1 milli weber round an iron ring of 6 cm^2 cross-section and 90 cm mean diameter having an air gap of 9 mm wide across it. Relative permeability of iron is 1200. (**Ans.** 15052 ATs)

3. A steel ring 15 cm mean radius and of circular section 1 cm in radius has an air gap of 1 mm length. It is wound uniformly with 500 turns of wire carrying a current of 3 A. Neglect magnetic leakage. The air gap takes 20% of the total m.m.f. Find reluctance. (**Ans.** 12.6×10^6 ATs/wb)

4. A circular iron ring has a mean circumference of 1.5 m and a cross-sectional area of 100 cm^2. A saw-cut of 0.4 cm wide is made in the ring. Calculate the magnetising current required to produce a flux of 0.8 m wb in the air gap if the ring is wound with a coil of 350 turns. Assuming relative permeability of iron as 400 and leakage factor 1.25. (**Ans.** 1.58 A)

5. A cast steel ring of mean circumference 50 cm has a cross-section of 0.52 cm^2. It has a saw-cut of 1 mm at one place. Given the following data :

B (wb/m²) :	1.0	1.25	1.46	1.60
μ_r :	714	520	360	247

Calculate how many ampere-turns are required to produce a flux of 5200 max wells. If leakage factor is 1.2.

[**Hint :** $B_2 = 1$ T , $B_1 = 1.2$ T and $\mu_r = 558$ from graph which is a straight line] (**Ans.** 1651.44 AT)

6. A steel ring 50 cm mean diameter and of circular section 3 cm in diameter has an air gap 2 mm long. It is wound uniformly with 1000 turns of wire carrying a current of 4 A. Calculate *(i)* m.m.f., *(ii)* magnetic flux, *(iii)* reluctance and *(iv)* flux density. Neglect magnetic leakage. Iron path takes about 40% of total m.m.f.

(**Ans.** 4000 AT ; 1.066 m wb ; 3.753×10^6 AT/wb ; 1.508 T)

7. In the magnetic circuit shown in Fig. 3.41, a coil of 500 turns in wound on the central limb. The magnetic path from A to B by way of outer limbs have a mean length of 100 cm each and an effective cross-sectional area of 2.5 cm^2. The central limb is 25 cm long and 5 cm^2 cross-sectional area. The air gap is 0.8 cm long. Calculate the current flowing through the coil to produce a flux of 0.3 m wb in the air gap. The relative permeability of the core material is 800 (neglect leakage and fringing).

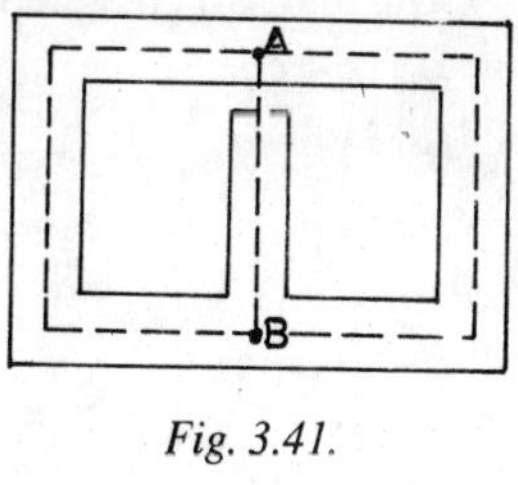

Fig. 3.41.

PROBLEMS

1. Define the term m.m.f., magnetic flux and magnetic reluctance and establish the relation between these quantities for a magnetic circuit.
2. Define the terms permeance and leakage flux.
3. Make comparison between magnetic and electric circuits.

4. Explain what do you mean by magnetic hysteresis, residual magnetism and coercive force.
5. Explain hysteresis and hysteresis loop of magnetic material.
6. What happens to force between two current carrying conductors if distance between them is double and current in them is halved.
7. Find the percentage change in stored energy in a coil if current through it is doubled and its inductance halved ?
8. State the utility of Lenz's law and Fleming right hand and left hand rules.

□□

4

Electro-magnetic Induction

4.1. INTRODUCTION

In the begining of nineteenth century, Oersted discovered that magnetism can be produced by means of electric current soon after he found that the electric current can be produced by the action of magnetic field.

In 1831, Michael Faraday's discovered that an emf is induced in a conductor, when the conductor is cut the magnetic flux.

This phenomenon is known as Electro-magnetic Induction. Today most of the electrical devices are based on this principle.

4.2. ELECTROMAGNETIC INDUCTION

The process by which an emf and hence current is induced in a conductor when magnetic flux linking with it changes is called electromagnetic induction.

The emf produced in this way is called induced emf and current is also called induced current.

Illustration

In order to demonstrate the phenomenon of electro-magnetic induction, consider a coil *'C'* having large number of turns and also connected galvanometer *G*.

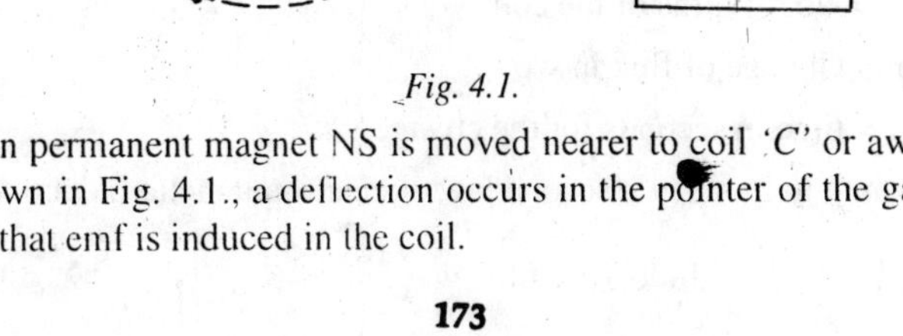

Fig. 4.1.

When permanent magnet NS is moved nearer to coil *'C'* or away from the coil as shown in Fig. 4.1., a deflection occurs in the pointer of the galvanometer indicating that emf is induced in the coil.

On the other hand, if the magnet is kept stationary and the coil is brought nearer to the magnet or away from the magnet as shown in Fig. 4.2. There is again deflection occurs in the pointer of the galvanometer.

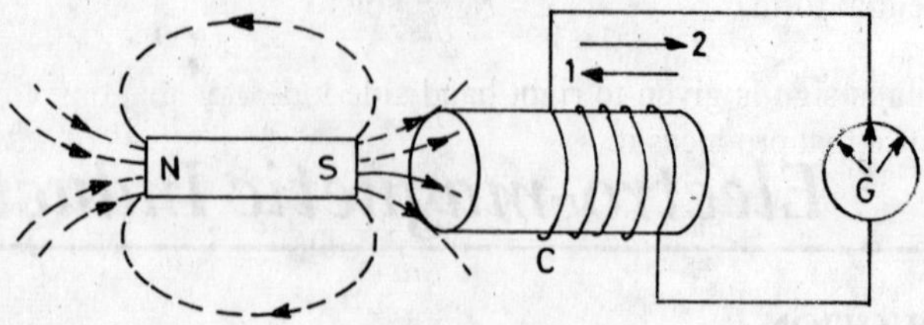

Fig. 4.2.

However when the coil and magnet both are kept stationary, that means no flux linking the coil then pointer will show zero deflection.

The magnitude of induced emf depends upon the rate of change of flux linking the coil. Greater speed of magnet or coil greater emf is induced in the coil. The following points are worth noting.

(i) Whenever flux linking with the circuit changes either coil or magnet is in motion, the galvanometer pointer shows deflection. This indicates that emf is induced in the coil.

(ii) The direction of induced emf depends upon the direction of magnetic field and direction of movement of coil.

(iii) The magnitude of induced emf depends upon the rate of change of flux linking with the coil.

4.3. FARADAY'S LAWS OF ELECTRO-MAGNETIC INDUCTION

Faraday's First Law : This law states that when the flux linking with the coil or circuit changes an emf is induced in it or whenever the magnetic flux is cut by the conductor an emf is induced in the conductor.

Faraday's Second Law : This law states that the magnitude of emf induced is directly proportional to the rate of change of flux linkage or to the product of number of turns and rate of change of flux linking the coil.

$$\text{Rate of change of flux linkages} = \frac{N(\phi_2 - \phi_1)}{t} \text{ wb-turn/s}$$

where N = No. of turns of the coil

$\phi_2 - \phi_1$ = Change of flux in wb.

t = time in seconds for the change.

According to Faraday's second law of electromagnetic Induction

$$\text{Induced emf, } e \propto \frac{N(\phi_2 - \phi_1)}{t}$$

$$e = \frac{N(\phi_2 - \phi_1)}{t}$$

In differential form, $e = N\frac{d\phi}{dt}$ volts

Using minus sign is given to right hand side indicates that emf is induced opposes the cause that produces it.

$$e = -N\frac{d\phi}{dt} \text{ volts}$$

Example 4.1. A coil of 2000 turns surrounds a flux of 5 mwb produced by a permanent magnet. The magnet is suddenly drawn away causing the flux inside the coil to drop to 2 mwb in 0.1 second. What is the average emf induced?

Solution :

Change in flux, $d\phi = 5 - 2 = 3$ mwb

$= 3 \times 10^{-3}$ **wb**

Time taken for the change, $dt = 0.1$ sec

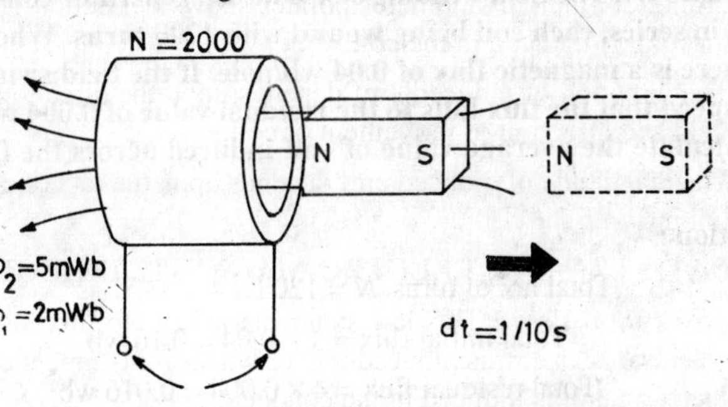

Fig. 4.3.

$$\therefore \quad e = N\frac{d\phi}{dt} = 2000 \times \frac{3 \times 10^{-3}}{0.1} = 60 \text{ V}$$

Example 4.2. A coil of 100 turns is linked by a flux of 20 mwb. If this flux is reversed in a time of 2 ms, calculate the average emf induced in the coil.

Solution :

change in flux, $d\phi = 20 - (-20) = 40 \text{ mwb} = 40 \times 10^{-3}$ wb

Time taken for the change, $dt = 2 \text{ ms} = 2 \times 10^{-3}$ s

$$\therefore \quad e = N\frac{d\phi}{dt} = 100 \times \frac{40 \times 10^{-3}}{2 \times 10^{-3}} = 2000 \text{ V } \textbf{Ans.}$$

Example 4.3. A coil of 200 turns of wire is wound on a magnetic circuit of reluctance 2000 AT/wb. If a current of 1 A flowing in the coil is reversed in 10 ms, find the average emf induced in the coil.

Solution :

$$\text{Flux in the coil} = \frac{\text{m.m.f.}}{\text{reluctance}} = \frac{200 \times 1}{2000} = 0.1 \text{ wb}$$

When the current (*i.e.*, 1 A) in the coil is reversed, flux through the coil is also reversed.

$$e = N \frac{d\phi}{dt}$$

Here, $N = 200$

$$d\phi = 0.1 - (-0.1) = 0.2 \text{ mwb}$$

$$dt = 10 \times 10^{-3} \text{ s}$$

$$\therefore \quad e = 200 \times \frac{0.2 \times 10^{-3}}{10 \times 10^{-3}} = 4 \text{ V } \textbf{Ans.}$$

Example 4.4. The field winding of 4-pole d.c. generator consists of 4 coils connected in series, each coil being wound with 1200 turns. When the field is excited, there is a magnetic flux of 0.04 wb/pole. If the field switch is opened at such a speed that the flux falls to the residual value of 0.004 wb/pole in 0.1 second, calculate the average value of emf induced across the field winding terminals.

Solution :

$$\text{Total no. of turns, } N = 1200 \times 4 = 4800$$

$$\text{Total initial flux} = 4 \times 0.04 = 0.16 \text{ wb}$$

$$\text{Total residual flux} = 4 \times 0.004 = 0.016 \text{ wb}$$

$$\text{Change in flux, } d\phi = 0.16 - 0.016 = 0.144 \text{ wb}$$

$$\text{Time taken } dt = 0.1 \text{ second}$$

$$\therefore \quad \text{Induced emf, } e = N\frac{d\phi}{dt} = 4800 \times \frac{0.144}{0.1} = 6912 \text{ V } \textbf{Ans.}$$

4.4 . LENZ'S LAW AND FLEMING'S RIGHT AND LEFT HAND RULE

The direction of induced emf and hence current in a conductor or coil can be determined by the following method.

1. Lenz's Law

This law states that the direction of induced emf is such that the current produced by it sets up a magnetic field opposing the motion or change producting it.

The above statement will be clarified from the following experiment.

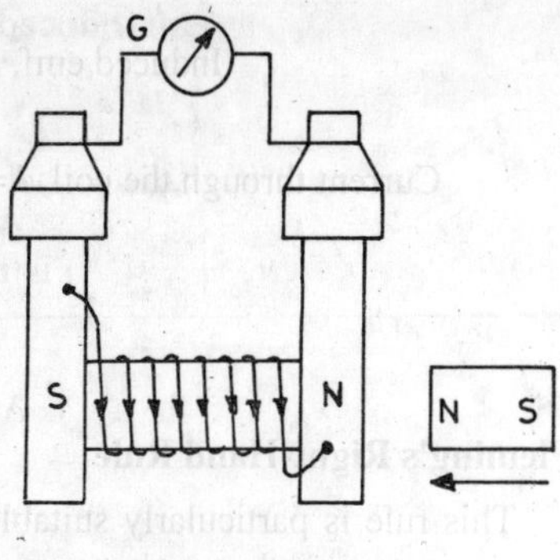

Fig. 4.4.

When the bar magnet is moved towards the coil, as shown in Fig. 4.4, an emf is induced in the coil, which causes an electric current to flow through the coil in the direction shown in Fig. 4.4. Due to flow of this electric current through the coil magnetic field is pruduced, north pole adjacent to north pole of the bar magnet and south pole on the other end, which can be determined from helix rule explained in Art 5.17. Now since similar poles repel each other, therefore, the movement of the bar magnet towards the coil is opposed by the field produced. Mechanical energy is expended in moving the bar magnet towards the coil against this force of repulsion and this expended energy is converted into electrical energy.

The same results will be observed when the bar magnet is moved away from the coil.

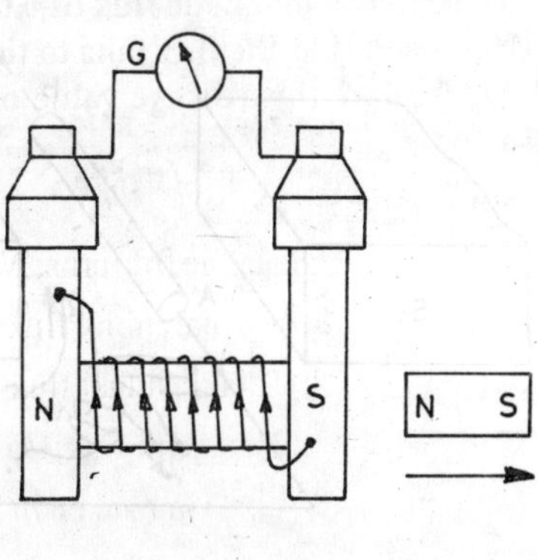

Fig. 4.5.

When the bar magnet is moved away from the coil, an emf is induced in the coil, current flows through it in the direction opposite to that in which it was flowing when the magnet was moved towards the coil, so magnetic field will be produced of opposite polarity, *i.e.*, south pole adjacent to north pole of the bar magnet. Therefore, force of attraction will be exerted between the coil and bar magnet, which will oppose the movement of the bar magnet away from the coil. Mechanical energy is expended in this case also and is converted into electrical energy.

Example 4.5. A coil of resistance 100 Ω is placed in a magnetic field of 1 mwb. The coil has 100 turns and a galvanometer of 400 Ω resistance is connected in series with it. Find the average emf and the current if the coil is moved from given field to a field of 0.2 mwb in 0.1 second.

Solution :

Number of turns on the coil, N = 100

Change of flux in 0.1 s, $d\phi = 0.001 - 0.0002 = 0.0008$ wb

$$\text{Rate of change of flux, } \frac{d\phi}{dt} = \frac{0.0008}{0.1} = 0.008 \text{ wb/s}$$

$$\text{Induced emf, } e = N\frac{d\phi}{dt} = 100 \times 0.008 = 0.8 \text{ V } \textbf{Ans.}$$

$$\text{Current through the coil, } I = \frac{e}{\text{coil resistance + galvanometer resistance}}$$

$$= \frac{e}{100+400} = \frac{0.8}{500}$$

$$= 0.0016 \text{ A or } 1.6 \text{ mA } \textbf{Ans.}$$

2. Fleming's Right-Hand Rule

This rule is particularly suitable to find the direction of induced emf and hence current when the conductor moves at right angles to a stationary magnetic field. It may be stated as under :

Stretch out the forefinger, middle finger and thumb of your right hand so that they are at right angles to one another. If the forefinger points in the direction of magnetic field, thumb in the direction of motion of the conductor, then the middle finger will point in the direction of induced current.

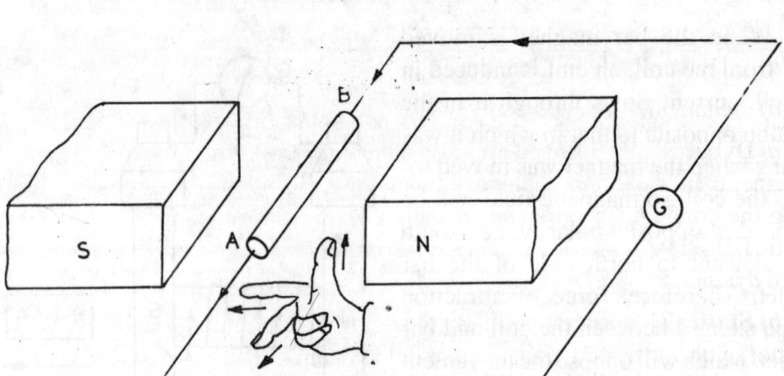

Fig. 4.6.

Consider a conductor *AB* moving upwards at right angles to a uniform magnetic field as shown in Fig. 4.6. Applying Fleming's right hand rule, it is clear that the direction of induced current is from *B* to *A*. If the motion of the conductor is downward, keeping the direction of magnetic field unchanged, then the direction of induced current will be from *A* to *B*.

3. Fleming Left Hand Rule

The direction of the force on a current carrying conductor, the following rule, called the Left Hand Rule, can be used (See. Fig. 4.7.)

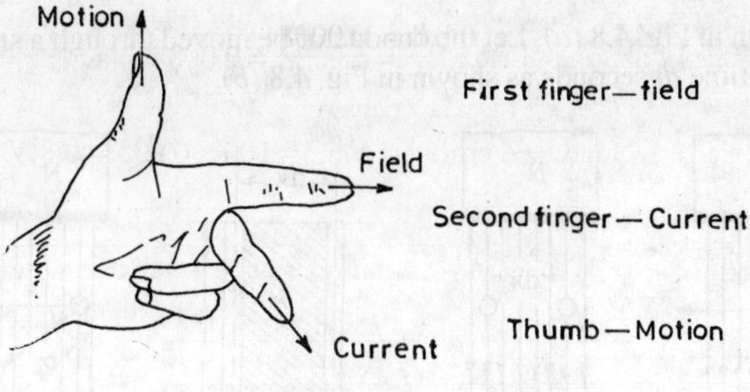

Fig. 4.7.

It may be noted that in the Fig. 4.7, the direction of flux, the current in the conductor and the resulting force are mutually at right angles. The left hand rule can be stated as follows : Keep the first and second fingers and the thumb of the left hand at right angles to one another; point the first finger in the direction of the flux and the second finger, in the direction of the current, then, the thumb will point in the direction of the force on the conductor.

4.5. INDUCED EMF

When flux linking with a conductor (or coil) changes, an emf is induced in it.

Types of Induced emf

(*a*) Dynamically induced emf

(*b*) Statically induced emf

(*a*) **Dynamically induced emf :** By either moving the conductor and keeping the magnetic field system stationary or moving the magnetic field system and keeping the conductor stationary, in such a way that conductor cuts across the magnetic field (as in case of d.c. and a.c. generators). The emf induced in this way is called *dynamically induced emf.*

(*b*) **Statically induced emf :** By changing the flux linking with the coil (or conductor) without moving either coil or field system. However, the change of flux produced by the field system linking with the coil is obtained by changing the current in the field system (solenoid), as in transformers. The emf induced in this way is called *statically induced emf.*

4.5.1. Dynamically Induced emf

By either moving the conductor keeping the magnetic field system stationary or moving the field system keeping the conductor stationary so that flux is cut by the conductor, the emf thus induced in the conductor is called *dynamically induced emf.*

Mathematical Expression

Considering a conductor of length l metres placed in the magnetic field of flux density B wb/m^2 is moving at right angle to the field at a velocity v metres/second as shown in Fig. 4.8 *(a)*. Let the conductor be moved through a small distance dx metresin time dt seconds as shown in Fig. 4.8 *(b)*.

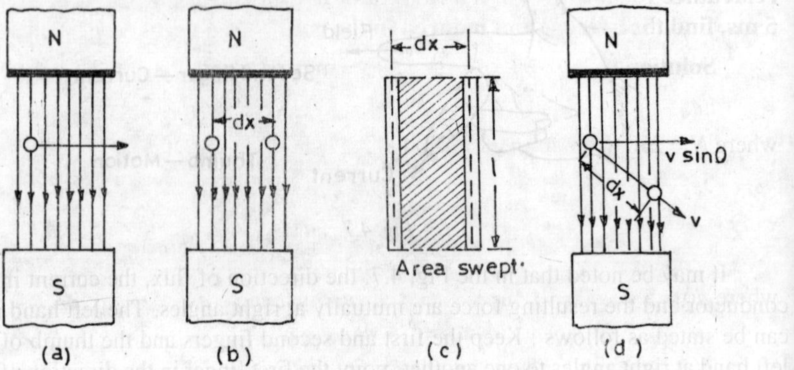

Fig. 4.8.

Area swept by the conductor, $A = l \times dx$

Flux cut by the conductor, $\phi = B \times A = B\,l\,dx$

According to Faraday's Law of electro-magnetic induction ;

$$\text{Induced emf, } e = \frac{\text{flux cut}}{\text{time}} = \frac{\phi}{dt} = \frac{B\,l\,dx}{dt}$$

$$= B\,l\,v \qquad [\text{Since } dx/dt = v \text{ (velocity)}]$$

Now, if the conductor is moved at an angle θ with the direction of magnetic field at a velocity v metres/second as shown in Fig. 4.8 *(d)*. A small distance covered by the conductor in that direction is dx in time dt seconds. Then the component of distance perpendicular to the magnetic field, which produces emf, is $dx \sin\theta$.

∴ Area swept by the conductor, $A = l \times dx \sin\theta$

Flux cut by the conductor, $\phi = B \times A = B\,l\,dx \sin\theta$

$$\text{Induced emf, } e = \frac{B\,l\,dx \sin\theta}{dt} = B\,l\,v \sin\theta$$

Example 4.6. A coil of 500 turns is linked with a flux of 2 mwb. If this flux is reversed in 4 ms, calculate the average emf induced in the coil.

Solution : Average induced emf, $e = N\dfrac{d\phi}{dt}$

where $N = 500$ turns;

$d\phi = 2 - (-2) = 4$ mwb

$dt = 4 \times 10^{3}$ s

∴ $e = 500 \times 4 \times 10^{-3} / 4 \times 10^{-3} = 500$ V **Ans.**

Example 4.7. A coil of 250 turns is wound on a magnetic circuit of reluctance 100000 AT/wb. If a current of 2 A flowing in the coil is reversed in 5 ms, find the average emf induced in the coil.

Solution : ϕ = m.m.f./reluctance,

i.e., $\phi = NI/S$

where $N = 250$, $I = 2$ A and $S = 100000$ AT/wb

$$\therefore \quad \phi = 250 \times 2/100000 = 5 \text{ mwb}$$

$$\text{Average induced emf, } e = N\frac{d\phi}{dt}$$

where $d\phi = 5 - (-5) = 10$ mwb (since current is reversed)

$$e = 250 \times 10 \times 10^{-3}/5 \times 10^{-3} = 500 \text{ V } \textbf{Ans.}$$

Example 4.8. The field winding of a 6 pole d.c. generator contains 6 bunches of coils connected in series, each bunch of coil carries 1000 turns wound on each pole. When the field is excited, there is a magnetic flux of 0.03 wb/pole. If the field switch is opened, the magnetic flux fall to the residual flux value 0.003 wb/pole in 0.1 second. What will be the average value of emf induced across the field winding terminals?

Solution : Average emf induced in one bunch of coil,

$$e_1 = N\frac{d\phi}{dt}$$

where, $N = 1000$ turns, $d\phi = 0.03 - 0.003 = 0.027$ wb, $dt = 0.1$ s

$$\therefore \quad e_1 = 1000 \times 0.027/0.1$$
$$= 270 \text{ V}$$

Total emf induced across the field winding terminals,

$$e = 6\,e_1 = 6 \times 270$$
$$= 1620 \text{ V } \textbf{Ans.}$$

Example 4.9. A wire of length 50 cm moves at right angles to its length at 40 m per second in a uniform magnetic field of density 1 wb/m². Calculate the electromotive force induced in the conductor when the direction of motion is *(i)* perpendicular to the field and *(ii)* inclined at 30° to the direction of field.

Solution :

Induced emf in a conductor, $e = B\,l\,v \sin\theta$

where $B = 1$ wb/m², $l = 0.5$ m, $v = 40$ m/s

(i) When $\theta = 90°$, as shown in Fig. 4.9 *(a)*

$$\therefore \quad e = 1 \times 0.5 \times 40 \times \sin 90°$$
$$= 20 \text{ V } \textbf{Ans.}$$

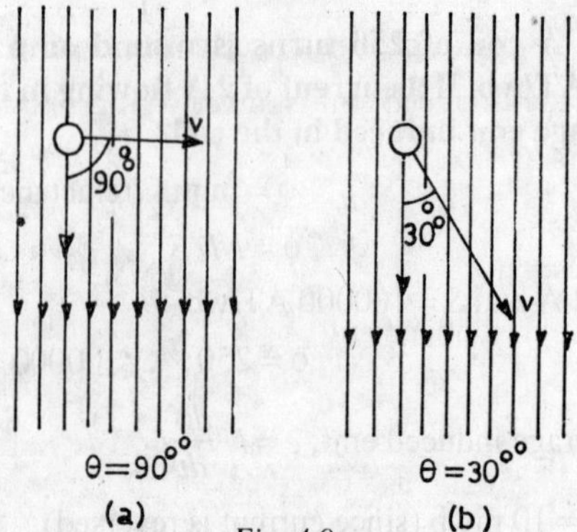

Fig. 4.9.

(ii) When $\theta = 30°$, as shown in Fig. 4.9 *(b)*.

$$e = 1 \times 0.5 \times 40 \times \sin 30°$$
$$= 10 \text{ V } \textbf{Ans.}$$

4.5.2. Statically Induced emf

When the coil and magnetic field system both are stationary but the magnetic field linking with the coil changes (by changing the current producing the field), the emf thus induced in the coil is called *statically induced emf.*

The statically induced emf may be

(i) Self induced emf

(ii) Mutually induced emf

(i) **Self induced emf :** The emf induced in a coil due to the change of flux produced by it linking with its own turns is called *self induced emf.*

Consider a coil having *N* turns as shown in Fig. 4.10 when switch *S* is closed and current *I* flows through the coil, it produces flus ϕ which also links with its own turns. If the current flowing through the coil is changed by changing the value of variable resistance *R*, it changes the flux linking with the coil and hence, an emf is induced in the coil. This induced emf is called *self induced emf.* Although it can not be measured directly but its effect can be seen.

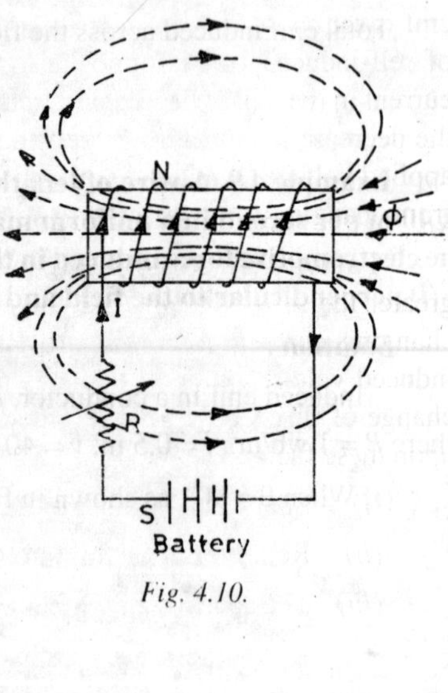

Fig. 4.10.

The direction of this induced emf is such that it opposes the cause which produces it (Lenz's law), *i.e.*, change of current in the coil.

Magnitude of self induced emf is directly proportional to the rate of change of current in the coil, *i.e.*,

$$e \propto \frac{dI}{dt} \quad \text{or} \quad e = L\frac{dI}{dt}$$

$$\text{Self induced emf, } e = -N\frac{d\phi}{dt} = -N\frac{d}{dt}\left[\frac{Ni}{l/\mu_r\,\mu_o\,a}\right]$$

$$= -N\frac{N}{l/\mu_r\,\mu_o\,a}\cdot\frac{di}{dt}$$

$$= -\frac{N^2\,\mu_r\,\mu_o\,a}{l}\cdot\frac{di}{dt}$$

The quantity $\frac{N^2\,\mu_r\,\mu_o\,a}{l}$ is a constant for any given coil or circuit and is called *coefficient of inductance.* It is represented by symbol L and measured in henries.

Hence self induced emf, $e = L\frac{di}{dt}$

where $L = \frac{N^2\,\mu_o\,\mu_r\,a}{l}$ henrys

4.6. SELF-INDUCTANCE (L)

The property of a coil that opposes any change in the amount of current flowing through it is called its *self-inductance* or *inductance.*

This property (*i.e.*, inductance) is due to the self- inductance emf in the coil itself by the changing current. If the current in the coil is increasing, the self-induced emf is set up in such a direction so as to oppose the rise of current, *i.e.*, direction of self-induced emf is opposite to that of the applied voltage. Similarly, if the current in the coil is decreasing, self-induced voltage will be such so as to oppose the decrease in current, *i.e.*, self-induced emf will be in the same direcion as the applied voltage. It may be noted that self-inductance does not prevent the current from changing; it serves only to delay the change.

Factors Affecting Inductance : The greater the self-induced voltage, the greater the self-inductance of the coil and hence larger is the opposition to the changing current. According to Faraday's laws of electromagnetic induction, induced voltage in a coil depends upon the number of turns (*N*) and the rate of change of flux ($d\phi/dt$) linking the coil. Hence, the inductance of a coil depends upon these factors, *viz.*

(*i*) Shape and number of turns.

(*ii*) Relative permeability of the material surrounding the coil.

(*iii*) The speed with which the magnetic field changes.

If fact, anything that affects the magnetic field also affects the inductance of the coil. Thus, increasing the number of turns of a coil increases its inductance. Similarly, substituting an iron core for air core increases its inductance.

4.7. MAGNITUDE OF SELF-INDUCED EMF

Consider a coil of N turns carrying a current of I amperes. If current in the coil changes, the flux linkages of the coil will also change. This will set up a self-induced emf e in the coil given by

$$e = N\frac{d\phi}{dt} = \frac{d}{dt}(N\,\phi)$$

Since flux is due to current in the coil, it follows that flux linkages $(= N\,\phi)$ will be proportional to I.

$$\therefore \qquad e = \frac{d}{dt} = (N\,\phi)$$

$$\propto \frac{dI}{dt}$$

$$= \text{Constant} \times \frac{dI}{dt}$$

or $$e = L\frac{dI}{dt} \qquad ...(i)$$

where L is a constant called *self-inductance* or *inductance* of the coil. The unit of inductance is henry (H). If in eqn. *(i)* above, $e = 1$ volt, $dI/dt = 1$ A/second,

then $L = 1$ H.

Hence a coil (or circuit) has an inductance of 1 henry if an emf of 1 volt is induced in it when current through it changes at the rate of 1 ampere per second.

4.8. COEFFICIENT OF SELF INDUCTION

The coefficient of self induction (L) can be determined from any one of the following three relations.

First Method : In case the dimensions of the solenoid are given, the coefficient of self induction may be determined from the relation

$$L = \frac{N^2\,a\,\mu_r\,\mu_0}{l} \text{ henrys}$$

Second Method : In case the magnitude of induced emf in a coil for a given rate of change of current in the coil is known, self inductance of the coil may be determined from the following relation.

$$e = L\frac{di}{dt}$$

or $$L = \frac{e}{di/dt}$$

Fig. 4.11.

Third Method : In case the number of turns of the coil and flux produced per ampere of current in the coil is known, the self inductance of the coil may be determined from the following relation.

$$L = \frac{N\phi}{i}$$

The above relation can be derived as follows.

Magnetic flux produced in a coil of N turns, length l metres, area of x-section a metres^2 and relative permeability μ_r when carrying a current of i amperes is given by

$$\phi = \frac{Ni}{l/\mu_r \mu_0 a} = \frac{N a \mu_r \mu_0}{l} i$$

and self inductance of the coil $L = \frac{N^2 a \mu_r \mu_0}{l} = \frac{N}{i} \frac{N a \mu_r \mu_0}{l} i = \frac{N\phi}{i}$

From the above relation it is obvious that the self inductance of a coil or circuit is equal to weber turns per ampere in the coil or circuit.

In the above relation if $N\phi = 1$ wb-turn and $i = 1$ A then $L = 1$ H

Hence a coil is said to have a self inductance of one henry if a current of 1 A, when flowing through it, produces flux linkage of 1 wb-turn in it.

Example 4.10. A wooden toroid of mean diameter 400 mm and cross-sectional area 400 mm^2 is uniformly wound with a coil of 1000 turns which carries a current of 2 A. Determine *(i)* self- inductance of the coil and *(ii)* the emf induced in the coil when the current is uniformly reduced to zero in 10 ms.

Solution :

(i) $$L = \frac{N^2 a \mu_0 \mu_r}{l}$$

Here $N = 1000$; $a = 400 \times 10^{-6}$ m^2 ; $l = 0.4\pi$ m ; $\mu_r = 1$

$\therefore$ $$L = \frac{(1000)^2 \times (400 \times 10^{-6}) \times (4\pi \times 10^{-7}) \times 1}{0.4\pi}$$

$$= 0.4 \times 10^{-3} \text{ H} = 0.4 \text{ mH } \textbf{Ans.}$$

(ii) $$e = L\frac{dI}{dt} = 0.4 \times 10^{-3} \times \frac{2-0}{10 \times 10^{-3}} = 0.08 \text{ V } \textbf{Ans.}$$

Example 4.11. A solenoid has 1,200 turns and carries a current of 2 A. The iron core has a length of 0.4 m and cross-section of 80 cm^2, the relative

permeability is 1,000. Calculate the self induced emf in the solenoid, if the current is switched off in 0.01 second.

Solution :

Neumber of turns on solenoid, $N = 1{,}200$

Length of iron core, $l = 0.4$ m

Area of x-section of iron core, $a = 80 \text{ cm}^2 = 0.008 \text{ m}^2$

Relative permeability of iron, $\mu_r = 1{,}000$

Self inductance of coil, $L = \dfrac{N^2 a \mu_0 \mu_r}{l}$

$$= \frac{(1200)^2 \times 0.008 \times 1000 \times 4\pi \times 10^{-7}}{0.4}$$

$$= 36.191 \text{ H}$$

Rate of change of current, $\dfrac{di}{dt} = \dfrac{2-0}{0.01} = 200$ A/s

Self induced emf, $e = L\dfrac{di}{dt}$

$$= 36.191 \times 200 = 7238 \text{ V } \textbf{Ans.}$$

Example 4.12. A toroid has a core of square x-section 2500 m^2 in area and a mean diameter of 250 mm. The core material has a relative permeability 1000. Calculate the number of turns wound on the core to obtain an inductance of 1 H.

Solution : Length of core, $l = \pi \times D = \pi \times 0.25 = 0.25\pi$ metres

Area of x-section of core, $a = 2500 \text{ mm}^2 = 0.0025 \text{ m}^2$

Relative permeability of core material,

$$\mu_r = 1000$$

$$\mu_0 = 4\pi \times 10^{-7} \text{ H/m}$$

Self inductance of coil, $L = 1$ H

$$\because \quad L = \frac{N^2 a \mu_0 \mu_r}{l}$$

$\therefore$ Number of turns of the coil, $N = \sqrt{\dfrac{N \times l}{\mu_0 \mu_r a}}$

$$= \sqrt{\frac{1 \times 0.25\pi}{4\pi \times 10^{-7} \times 1000 \times 0.0025}}$$

$$= 500 \textbf{ Ans.}$$

Example 4.13. The resistance and inductance of a coil are 3 Ω and 0.1 mH respectively. What potential difference exists at the terminals of the solenoid at the instant when the current is 1 A, but increasing at the rate of 10000 A per second?

Solution :

Resistance of the coil, $R = 3\ \Omega$

Inductance of the coil, $L = 0.1\text{ mH} = 1 \times 10^{-4}\text{ H}$

Current flowing through the coi, $I = 1.0$ A

Rate of increase of current, $\frac{di}{dt} = 10000$ A/s

Potential difference across the solenoid,

V = Voltage drop in resistance + induced voltage in the coil

$$= i\,R + L\frac{di}{dt} = 1 \times 3 + 1 \times 10^{-4} \times 10000 = 4\text{ V } \textbf{Ans.}$$

Example 4.14. An iron ring of 0.15 metre diameter and 0.001 m² in cross-section with a saw cut 2 mm wide is wound with 300 turns of wire. The gap flux density is 1 T. The relative permeability of the iron is 500. Determine the exciting current and inductance. Ignore leakage and fringing.

Solution : Flux density, $B = 1$ T

Relative permeability for iron, $\mu_r = 500$

Length of iron path $= \pi \times 0.15 - 0.002 = 0.469$ m

Total ampere-turns required, $AT = H_g\, l_g + H_i\, l_i$

$$= \frac{B}{\mu_0}\, l_g + \frac{B}{\mu_0\, \mu_r}\, l_i$$

$$= \frac{1}{4\pi \times 10^{-7}} \times 0.002 + \frac{1}{4\pi \times 10^{-7} \times 500} \times 0.469$$

$$= 1591 + 747 + 2338$$

Exciting current, $i = \frac{AT}{N} = \frac{2338}{300} = 7.76$ A **Ans.**

Flux, $\phi = B \times A = 1 \times 0.001 = 0.001$ wb

Inductance, $L = \frac{N\phi}{i} = \frac{300 \times 0.001}{7.76} = 38.66$ mH **Ans.**

Example 4.15. A circuit has 1000 turns enclosing a magnet circuit 20 cm² in cross-section. With 4 A, the flux density is 1.0 wb/m² and with 9 A it is 1.4 wb/m². Find the mean value of the inductance between these current limits and the induced emf if the current falls from 9 A to 4 A in 0.05 second.

Solution : Flux corresponding to the current of 4 A,

$$\phi_1 = 1 \times 20 \times 10^{-4} = 0.002 \text{ wb}$$

Self inductance corresponding to the current of 4 A,

$$L_1 = \frac{N\phi_1}{i_1} = \frac{1000 \times 0.002}{4} = 0.5 \text{ H}$$

Flux corresponding to the current of 9 A,

$$\phi_2 = 1.4 \times 20 \times 10^{-4} = 0.0028 \text{ wb}$$

Self inductance corresponding to the current of 9 A,

$$L_2 = \frac{N\phi_2}{i_2} = \frac{1000 \times 0.0028}{9} = 0.31 \text{ H}$$

Mean value of self inductance,

$$L = \frac{L_1 + L_2}{2} = \frac{0.5 + 0.31}{2} = 0.405 \text{ H } \textbf{Ans.}$$

Rate of change of current,

$$\frac{di}{dt} = \frac{9-4}{0.05} = 100 \text{ A/s}$$

$$\text{emf induced, } e = L\frac{di}{dt} = 0.405 \times 100 = 40.5 \text{ V } \textbf{Ans.}$$

Example 4.16. A coil of 300 turns; wound on a core of non- magnetic material, has an inductance of 10 mH. Calculate *(i)* the flux produced by a current of 5 A and *(ii)* the average value of emf induced when a current of 5 A is reversed in 8 milliseconds.

Solution :

Inductance of coil, $L = 10 \text{ mH} = 0.01 \text{ H}$

Number of turns on the coil, $N = 300$

Current through coil, $i = 5$ A

(i) Flux produced, $\phi = \dfrac{iL}{N} = \dfrac{5 \times 0.01}{300}$

$= 0.1667$ m wb **Ans.** $\quad \because \quad L = \dfrac{N\phi}{i}$

Rate of change of current, $\dfrac{di}{dt} = \dfrac{5-(-5)}{0.008} = 1.250$ A/s

Induced emf, $e = L\dfrac{di}{dt}$

$= 0.01 \times 1250 = 12.5$ V **Ans.**

4.9. MUTUALLY INDUCED EMF

Consider two coils *A* and *B* placed closed together so that the flux created by one coil completely links with the other coil. Let coil *A* have a battery and switch *S* and coil *B* be connected to the galvanometer *G* as shown in Fig. 4.12.

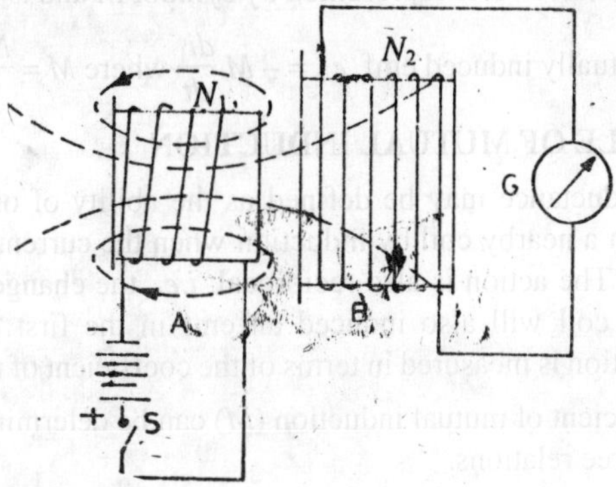

Fig. 4.12.

When switch *S* is opened, no current flows through coil *A*, so no flux is created in coil *A*, *i.e.*, no flux links with coil *B*, therefore, no emf is induced across coil *B*, the fact is indicated by galvanometer zero deflection. Now when the switch *S* is closed current in coil *A* starts rising from zero value to a finite value, the flux is produced during this period and increases with the increase in current of coil *A*, therefore, flux linking with the coil *B* increases and an emf, known as *mutually induced emf*, is produced in coil *B*, the fact is indicated by galvanometer deflection. As soon as the current in coil *A* reaches its finite value, the flux produced or flux linking with coil *B* becomes constant, so no emf is induced in coil *B*, and galvanometer pointer returns back to zero position. Now if the switch *S* is opened, current will start decreasing, resulting in decrease in flux linking with coil *B*, an emf will be again induced but in direction opposite to previous one, this fact will be shown by the galvanometer deflection in opposite direction.

Hence whenever the current in coil *A* changes, the flux linking with coil *B* changes and an emf, known as mutually induced emf, is induced in coil *B*.

Consider coil *A* of turns N_1 wound on a core of length l metres, area of cross-section a square metres and relative permeability μ_r. When the current of i_1 amperes flows through it, a flux of $\dfrac{N_1\, i_1}{l/\mu_0\, \mu_r\, a}$ is set up around the coil. Let whole of the flux produced due to flow of current in coil *A* be linked with the coil *B* having N_2 turns and placed near by coil *A*.

Mutually induced emf, $e_m = -$ Rate of change of flux linkage of coil *B*

$= - N_2 \times$ rate of change of flux in coil *A*

$$= -N_2 \frac{d}{dt}\left[\frac{N_1 i_1}{l/\mu_0 \mu_r a}\right] = \frac{N_1 N_2 a \mu_0 \mu_r}{l} \cdot \frac{di_1}{dt}$$

The quantity $\frac{N_1 N_2 a \mu_0 \mu_r}{l}$ is called the *coefficient of mutual induction* of coil *B* with respect to coil *A*. It is represented by symbol M and is measured in henrys.

Hence mutually induced emf, $e_m = -M \frac{di_1}{dt}$ where $M = \frac{N_1 N_2 a \mu_0 \mu_r}{l}$ henrys.

4.10. PRINCIPLE OF MUTUAL INDUCTION

Mutual inductance may be defined as the ability of one coil or circuit to induce an emf in a nearby coil by induction when the current flowing in the first coil is changed. The action is also reciprocal, *i.e.*, the change in current flowing through second coil will also induced an emf in the first coil. The ability of reciprocal induction is measured in terms of the coefficient of mutual induction *M*.

The coefficient of mutual induction (*M*) can be determined from any one of the following three relations.

First Method : In case the dimensions of the coils are given, the coefficient of mutual induction may be determined from the relation

$$M = \frac{N_1 N_2 a \mu_0 \mu_r}{l} \text{ henrys}$$

Second Method : In case the magnitude of induced emf in the second coil for a given rate of change of current in the first coil is known, mutual inductance between the coil may be determined from the following relation

$$e_m = M \frac{di_1}{dl}$$

or
$$M = \frac{e_m}{di_1/dt}$$

Third Method : In case the number of turns of the coil and flux linking with this coil per ampere of current in another coil is known, the mutual inductance of the coil may be determined from the following relation

$$M = N_2 \frac{\phi_2}{i_1} \text{ Henry}$$

Example 4.17. A wooden ring has a mean diameter of 150 mm and a cross-sectional area of 250 mm². It is wound with 1500 turns of insulated wire. A second coil of 900 turns is then wound on the top of the first. Assuming that all the flux produced by the first coil links with the second, calculate the mutual inductance.

Solution : Number of turns on coil *A*, $N_1 = 1500$

Number of turns on coil *B*, $N_2 = 900$

Relative permeability of wood, $\mu_r = 1.0$

$$\mu_0 = 4\pi \times 10^{-7}$$

Area of x-section of wooden ring,

$$a = 250 \text{ mm}^2 = 2.5 \times 10^{-4} \text{ m}^2$$

Length of wooden ring, $l = 150\ \pi$ mm $= 0.15\ \pi$ m

$$\text{Mutual inductance, } M = \frac{N_1 N_2 \mu_0 \mu_r a}{l}$$

$$= \frac{1500 \times 900 \times 4\pi \times 10^{-7} \times 1 \times 2.5 \times 10^{-4}}{0.15\pi} = 0.9 \text{ mH } \textbf{Ans.}$$

Example 4.18. Two coils having 100 and 150 turns respectively are wound side by side on a closed iron circuit of section 125 cm² and mean length 200 cm. If the permeability of iron is 2000, calculate *(a)* self inductance of each coil, *(b)* mutual inductance between them, *(c)* the emf indcued in the second coil if current in first coil changes from 0 to 5 A in 0.02 s.

Solution :

Number of turns on first coil, $N_1 = 100$

Number of turns on second coil, $N_2 = 150$

Relative permeability of iron, $\mu_r = 2000$

Area of x-section of iron circuit, $a = 125 \text{ cm}^2 = 0.0125 \text{ m}^2$

Length of iron circuit, $l = 200$ cm $= 2.0$ m

(a) Self inductance of first coil,

$$L_1 = \frac{N_1^2 a \mu_0 \mu_r}{l}$$

$$= \frac{(100)^2 \times 4\pi \times 10^{-7} \times 2000 \times 0.0125}{2.0}$$

$$= 157.1 \text{ mH } \textbf{Ans.}$$

$$\text{Self inductance of second coil, } L_2 = \frac{N_2^2 a \mu_0 \mu_r}{l}$$

$$= \frac{(150)^2 \times 4\pi \times 10^{-7} \times 2000 \times 0.0125}{2.0}$$

$$= 353.4 \text{ mH } \textbf{Ans.}$$

(b) Mutual inductance between coil,

$$M = \frac{N_1 N_2 a \mu_0 \mu_r}{l}$$

$$= \frac{100 \times 150 \times 4\pi \times 10^{-7} \times 2000 \times 0.0125}{2.0}$$

$$= 235.6 \text{ mH } \textbf{Ans.}$$

Rate of change of current in first coil,

$$\frac{di_1}{dt} = \frac{5-0}{0.02} = 250 \text{ A/s}$$

emf induced in second coil,

$$e_2 = M\frac{di_1}{dt}$$

$$= 235.6 \times 10^{-3} \times 250 = 58.9 \text{ V } \textbf{Ans.}$$

Example 4.19. Two coils *A* and *B* are wound on the same iron core. There are 500 turns on *A* and 3000 turns on *B*. A current of 5 amperes through coil *A* produces a flux of 600×10^{-6} wb in the core. If this current is reversed in 0.02 s, calculate the average emf induced in coils *A* and *B*.

Solution :

Number of turns on coil A, $N_1 = 500$

Number of turns on coil B, $N_2 = 3000$

Flux per ampere in coil A, $\frac{\phi_1}{i_1} = \frac{600 \times 10^{-6}}{5} = 120 \times 10^{-6}$ wb/A

Self inductance of coil A, $L_1 = N_1 \frac{\phi_1}{i_1} = 500 \times 120 \times 10^{-6} = 0.06$ H **Ans.**

Flux linking the coil B per ampere of current in coil A,

$$\frac{\phi_2}{i_1} = 120 \times 10^{-6} \text{ wb}$$

(Assuming flux created in coil A links the coil B completely, *i.e.*, $\phi_2 = \phi_1$)

Mutual inductance between the coils,

$$M = N_2 \frac{\phi_2}{i_1} = 3000 \times 120 \times 10^{-6} = 0.36 \text{ H}$$

Rate of change of current in coil A,

$$\frac{di_1}{dt} = \frac{5-(-5)}{0.02} = 500 \text{ A/s}$$

Average emf induced in coil A,

$$e_1 = L_1 \frac{di_1}{dt} = 0.06 \times 500 = 30 \text{ V } \textbf{Ans.}$$

Average emf induced in coil B,

$$e_2 = M\frac{di_1}{dt} = 0.36 \times 500 = 180 \text{ V } \textbf{Ans.}$$

Example 4.20. Two coils have a mutual inductance of 0.3 H. If the current in one coil is varied from 5 A to 2 A in 0.4 s, calculate *(a)* the average emf induced in the second coil, *(b)* the change of flux linked with the second coil assuming that it is wound with 200 turns.

Solution : Mutual inductance $M = 0.3$ H

Rate of change of current in one coil,

$$\frac{di_1}{dt} = \frac{5-2}{0.4} = 7.5 \text{ A/s}$$

(i) emf induced in second coil, $e_2 = M\frac{di_1}{dt} = 0.3 \times 7.5 = 2.25$ V **Ans.**

(ii) Since mutually induced emf, $e_2 = N_2\frac{d\phi_2}{dt}$

∴ Change of flux linked with the second coil,

$$\frac{d\phi_2}{dt} = \frac{e_2}{N_2} = \frac{2.25}{200} = 0.01125 \text{ wb/s } \textbf{Ans.}$$

Example 4.21. The self-inductance of a coil of 500 turns is 0.25 H. If 60% of the flux is linked with a second coil of 10000 turns, calculate *(a)* The mut'al inductance of the two coils, *(b)* emf induced in the second coil when curr the first coil changes at the rate of 100 A/s.

Solution : Self-inductance of the first coil, $L_1 = 0.25$ H

Number of turns on first coil, $N_1 = 500$

Flux per ampere of current in first coil,

$$\frac{\phi_1}{i_1} = \frac{L}{N_1} = \frac{0.25}{500} = 0.5 \text{ m wb} \qquad \because \; L$$

Flux linked with second coil per ampere of current in first coil,

$$\frac{\phi_2}{i_1} = \frac{0.6\,\phi}{i_1} = 0.6 \times 0.5 = 0.3 \text{ m wb}$$

(i) The mutual inductance of the two coils,

$$M = \frac{N_2\,\phi_2}{i_1} = 10000 \times 0.3 \times 10^{-3} = 3 \text{ H } \textbf{Ans.}$$

(ii) emf induced in the second coil,

$$e_2 = M\frac{di_1}{dt} = 3 \times 100 = 300 \text{ V } \textbf{Ans.}$$

Example 4.22. Two identical coils *A* and *B* consisting of 1500 turns each lie in parallel planes such that 70% of the flux produced by the current in coil *A* link the coil *B*. Current of 4A flowing in the coil *A* produces in it ia flux of 0.04 mWb. Calculate *(i)* self-inductance of each coil and *(ii)* mutual inductance between them.

If the current in coil *A* changes from 4A to –4A in 0.02s, what will be the emf induced in coil *B*.

Solution : *(i)* Self-inductance of a coil is given by

$$L = \frac{N\phi}{I}$$

Number of turns in coil *A*, $N = 1500$

Flux linking the coil *A*, $\phi = 0.04 \text{ mWb} = 0.04 \times 10^{-3}$ Wb

Current in coil *A*, $I = 4$A

Thus self-inductance of coil *A*, $L_A = \dfrac{1500 \times 0.04 \times 10^{-3}}{4} = 15 \times 10^{-3}$ H

$= 15$ mH **Ans.**

As the coils are identical, coil *B* will also have the same self-indutance.

Self-inductance of coil $B = 15$ mH

(ii) Mutual inductance between the coils *A* and *B* is given by

$$M = \frac{N_2\,\phi_2}{I} = \frac{N_2\,K\,\phi_1}{I}$$

$$K = 70\% = 0.7$$

Number of turns in coil *B*, $N_2 = 1500$

$$\phi_2 = K\,\phi_1 = 0.7 \times 0.04 = 0.028 \text{ mWb} = 0.028 \times 10^{-3} \text{ Wb}$$

Thus mutual inductance, $M = \dfrac{1500 \times 0.028 \times 10^{-3}}{4} = 10.5 \times 10^{-3}$ H

$= 10.5$ mH **Ans.**

(iii) Current in coil *A* changes from 4 to –4A in 0.02s. Thus,

$$\frac{di}{dt} = \frac{(-8)}{0.02}$$

Emf induced in coil *B* due to a change of current in coil *A*,

$$e = -M\frac{di}{dt} = -10.5 \times 10^{-3} \times \frac{(-8)}{0.02} = 4.2 \text{ V } \textbf{Ans.}$$

4.11. COEFFICIENT OF COUPLING

When two coils are placed near each other, all the flux produced by one coil does not link with the other coil, only a certain portion (say, K) of flux produced by one coil links with other coil, K being less than unity. K is called the *coefficient of coupling*.

Flux created in coil A due to current of i_1 amperes,

$$\phi_1 = \frac{N_1 i_1}{l/\mu_0 \mu_r a} \text{ wb}$$

Flux linking with coil B, $\phi_2 = K\phi_1 = \dfrac{K N_1 i_1}{l/\mu_0 \mu_r a}$ wb

Coefficient of mutual inductance, $M = \dfrac{N_2 \phi_2}{i_1} \cdot \dfrac{K N_1 N_2}{l/\mu_0 \mu_r a}$ ...(i)

Coefficient of self inductance of coil A,

$$L_1 = \frac{N_1 \phi_1}{i_1} = \frac{N_1^2}{l/\mu_0 \mu_r a} \quad ...(ii)$$

Coefficient of self inductance of coil B,

$$L_2 = \frac{N_2 \phi_2}{i_2} = \frac{N_2^2}{l/\mu_0 \mu_r a} \quad ...(iii)$$

Multiplying expressions *(ii)* and *(iii)*, we get

$$\sqrt{L_1 L_2} = \frac{N_1 N_2}{l/\mu_0 \mu_r a} \quad ...(iv)$$

Comparing expression *(i)* and *(iv)*, we get

$$M = K\sqrt{L_1 L_2}$$

or

$$K = \frac{M}{\sqrt{L_1 L_2}}$$

When the coils are tightly coupled, *i.e.*, when flux due to one coil links with other coil completely then coefficient of coupling, K is unity. If the flux due to one coil does not link with the other coil at all then the value of coefficient of coupling is zero.

Example 4.23. Two coils A and B of 600 and 1000 turns respectively are connected in series on the same magnetic circuit of reluctance 2×10^6 AT/wb. Assuming that there is no flux leakage, calculate *(a)* self-inductance of each coil; *(b)* mutual inductance between the two coils.

What would be the mutual inductance if the coefficient of coupling is 75%.

Solution :

Self inductance of coil A, $L_1 = N_1^2/S$

where $N_1 = 600$ and $S = 2 \times 10^6$ AT/wb

$\therefore$ $L_1 = (600)^2/2 \times 10^6 = 0.18$ H **Ans.**

Similarly, $L_2 = (1000)^2/2 \times 10^6 = 0.5$ H **Ans.**

Mutual inductance, $M = k\sqrt{L_1 L_2}$

when $k = 1$ $M = 1\sqrt{0.18 \times 0.5} = 0.3$ H **Ans.**

when $k = 0.75$ $M = 0.75\sqrt{0.18 \times 0.5} = 0.225$ H **Ans.**

Example 4.24. Two air-cored coils are placed closed to each other so that 80% of the flux of one coil links with the other. Each coil has mean diameter of 2 cm and a mean length of 50 cm. If there are 1800 turns of wire on one coil calculate the number of turns on the other coil to give a mutual inductance of 15 mH.

Solution : Reluctance, $S = \dfrac{l}{a\,\mu_0\,\mu_r}$

$$= \frac{0.5}{\frac{\pi}{4} \times (0.02)^2 \times 4\pi \times 10^{-7} \times 1}$$

$$= 1.2665 \times 10^9 \text{ AT/wb}$$

Now $L_1 = N_1^2/S$ and $L_2 = N_2^2/S$

$\therefore$ $\sqrt{L_1 L_2} = N_1 N_2/S$

Also $M = k\sqrt{L_1 L_2} = k N_1 N_2/S$

where $M = 15 \times 10^{-3}$ H; $N_1 = 1800$; $k = 0.8$

$\therefore$ $15 \times 10^{-3} = 0.8 \times 1800 \times N_2/1.2665 \times 10^9$

or $N_2 = \dfrac{15 \times 10^{-3} \times 1.2665 \times 10^9}{0.8 \times 1800} = 13193$ turns **Ans.**

4.12. ENERGY STORED IN A MAGNETIC FIELD

When some electrical energy is supplied to a coil, it is spent in two ways :

(*i*) A part of it is spent to meet $I^2 R$ losses which are dissipated in the form of heat and cannot be recovered.

(*ii*) The remaining part is used to create magnetic field around the coil and is stored in the magnetic field. When this field collapses, the stored energy is released by the coil and is returned to the circuit.

For instant, consider a coil having an inductance of L henry and a small resistance of R ohms connected to a d.c. supply through a switch S as shown in Fig. 4.13. When the switch is closed, the current increases in the coil gradually and attains the final steady value $I\ (= V/R)$. The increase in current is opposed by the self-induced emf produced in the coil due to this change. To overcome the opposition, some energy is supplied by the source which is stored in the magnetic field. When the switch is opened, the magnetic field collapses and the stored energy is released and returned to the circuit. This energy is dissipated in the form of heat in the coil resistance, this instant is shown in Fig. 4.13.

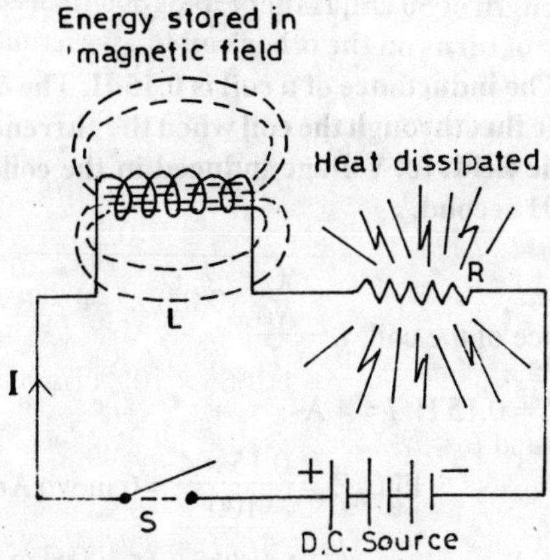

Fig. 4.13.

It can be compared with a weight when raised, some energy is supplied to it. This energy is stored in it in the form of potential energy. When the weight is released, it returns the stored energy by doing some work.

In fact, to establish magnetic field around a coil, energy is required, though no energy is needed to maintain it. This energy is stored in the magnetic field. When the field collapses, this energy is released and is supplied to the circuit.

Mathematical Expression

Let at any instant, the current flowing through the coil is i and is increasing at the rate of di/dt. The self-induced emf in the coil,

$$e = L\frac{di}{dt}$$

$$\text{Instantaneous power, } p = ei = Li\frac{di}{dt}$$

Energy stored in the magnetic field or energy supplied to the coil during a short interval of time dt,

$$dw = pdt = Li\frac{di}{dt}dt = Li\,di$$

Total energy stored in the magnetic field or supplied to the coil when current rises from 0 to I (its final value);

$$\int dw = \int_0^I Li\,di$$

or $$W = \frac{1}{2}LI^2 \text{ joules}$$

Example 4.25. The inductance of a coil is 0.15 H. The coil has 100 turns Find *(a)* total magnetic flux through the coil when the current is 4 A; *(b)* energy stored in the magnetic field; *(c)* voltage induced in the coil, when current is reduced to zero in 0.01 second.

Solution :

Inductance of the coil, $L = \frac{N\phi}{I}$

where $N = 100$ turns, $L = 0.15$ H, $I = 4$ A

$$\therefore \quad \text{Flux, } \phi = \frac{0.15 \times 4}{100} = 6 \text{ m wb } \textbf{Ans.}$$

$$\text{Energy stored} = LI^2/2 = 0.15 \times (4)^2/2 = 1.2 \text{ J } \textbf{Ans.}$$

$$\text{Voltage induced, } e = L\frac{di}{dt} = 0.15 \times \frac{4}{0.01} = 60 \text{ V } \textbf{Ans.}$$

Example 4.26. A solenoid of 1 m in length and 10 cm in diameter has 5000 turns. Calculate the energy in the magnetic field when a current of 2 A flows in the solenoid.

Solution : Inductance of the solenoid, $L = N^2 a \mu_0 \mu_r / l$

where $N = 5000$, $a = \pi d^2/4 = 25\pi \times 10^{-4}$ m^2, $l = 1$ m, $\mu_r = 1$

$$\therefore \quad L = (5000)^2 \times 25\pi \times 10^{-4} \times 4\pi \times 10^{-7} \times 1/1 = 0.2467 \text{ H}$$

Energy stored $= LI^2/2 = 0.2467 \times (2)^2/2 = 0.4934$ J **Ans.**

Example 4.27. An air core coil has 1000 turns of copper wire whose diameter is 0.5 mm. The length of the coil is 1 m and the diameter of the coil is 5 cm. The coil carries current of 5 A. Determine the following :

(a) The resistance of the coil;

(b) The inductance of the coil;

(c) The energy stored in the magnetic field.

Solution : The coil is shown in Fig. 4.14.

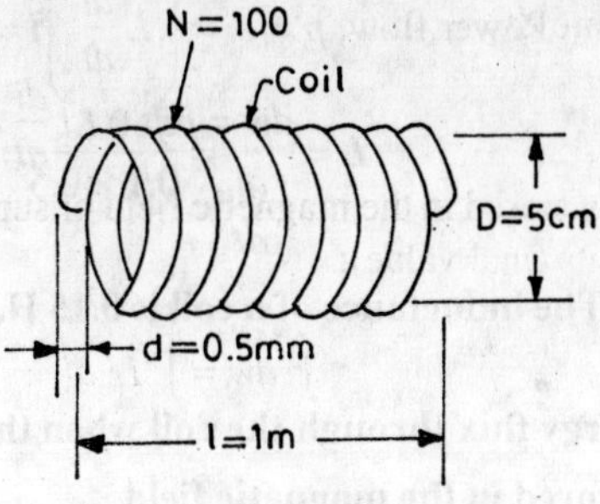

Fig. 4.14.

Length of the wire, wound in the form of a coil

$$= \pi DN$$

$$= \pi \times 5 \times 10^{-2} \times 1000 = 50\pi \text{ m}$$

Area of cross-section of wire $= \pi d^2/4 = \pi \times (0.5)^2 \times 10^{-6}/4 \text{ m}^2$

Resistivity of copper, $\rho = 1.72 \times 10^{-8}\ \Omega\text{ m}$

$\therefore$ Resistance of the coil,

$$R = \rho \frac{\text{length}}{\text{area}} = 1.72 \times 10^{-8} \times \frac{50\pi \times 4}{\pi (0.5)^2 \times 10^{-6}}$$

$$= 13.76\ \Omega \textbf{ Ans.}$$

Inductance of the coil, $L = \dfrac{N^2}{l} \times a\, \mu_0\, \mu_r$

where $N = 1000$, $l = 1$ m, $a = \pi D^2/4 = \pi \times 5 \times 5 \times 10^{-4}/4 \text{ m}^2$, $\mu_r = 1$

$$\therefore \quad L = \frac{(1000)^2}{1} \times \frac{\pi \times 25 \times 10^{-4}}{4} \times 4\pi \times 10^{-7}$$

$$= 2.467 \text{ mH } \textbf{Ans.}$$

$$\text{Energy stored} = LI^2/2 = 2.467 \times 10^{-3} \times (5)^2/2$$

$$= 30.84 \times 10^{-3} \text{ J } \textbf{Ans.}$$

Example 4.28. Find the inductance of a coil in which *(i)* current of 0.1 A yields an energy storage of 0.05 J *(ii)* a current of 0.1 A increasing at the rate of 0.5 A/s represents a power flow of 0.5 watt.

Solution :

(i) Energy stored = 0.05 J

Current through coil $I = 0.1$ A

$$\text{Inductance of coil } L = \frac{\text{Energy stored}}{1/2\ I^2} = \frac{0.05}{1/2\,(0.1)^2} = 10 \text{ H } \textbf{Ans.}$$

(ii) Power flow, $p = ei = \left(L\frac{di}{dt}\right)i = Li\frac{di}{dt}$

or $L = \frac{p}{i\frac{di}{dt}} = \frac{0.5}{0.1 \times 0.5} = 10$ H **Ans.**

Example 4.29. The inductance of a coil is 0.15 H. The coil has 100 turns. Find the following :

(i) Total energy flux through the coil when the current is 4 A.

(ii) Energy stored in the magnetic field.

(iii) Voltage induced in the coil, when current is reduced to zero in 0.01 second.

Solution : Inductance of coil. $L = 0.15$ H

Number of turns on the coil. $N = 100$

Current flowing through coil. $I = 4$ A

(i) Total magnetic flux through the coil,

$$\phi = \frac{0.15}{100} \times 4 = 6 \text{ m wb } \textbf{Ans.}$$

(ii) Energy stored in the magnetic field

$$= \frac{1}{2}LI^2 = \frac{1}{2} \times 0.15 \times 4^2 = 1.2 \text{ J } \textbf{Ans.}$$

Rate of fall of current, $\frac{di}{dt} = \frac{4}{0.01} = 400$ A/s

(iii) Voltage induced in the coil, $e = L = \frac{di}{dt} = 0.15 \times 4C0 = 60$ V **Ans.**

Example 4.30. The field winding of a d.c. electro-magnet is wound with 960 turns and has resistance of 50 Ω. The exciting voltage is 230 V and the magnetic flux linking the coil is 5 m wb. Find *(i)* self-inductance of the coil, *(ii)* the energy stored in the magnetic field.

Solution :

The number of turns on windings, $N = 960$

Exciting voltage = 230 V

Resistance of field winding = 50 Ω

Current through field winding, $I = \frac{\text{Exciting voltage}}{\text{Field winding resistance}}$

$= 230/50 = 4.6$ A

Magnetic flux produced, $\phi = 5$ m wb $= 0.005$ wb

(i) Inductance of coil, $L = \frac{N\phi}{I} = \frac{960 \times 0.005}{4.6} = 1.043$ H **Ans.**

(ii) Energy stored in the magnetic field $= \frac{1}{2} LI^2 = \frac{1}{2} \times 1.043 \times (4.6)^2$

$= 11.04$ J **Ans.**

4.13. EDDY CURRENTS AND THEIR EFFECTS

Whenever the magnetic flux linkages with a close electric circuit changes an emf is induced in the circuit and these emfs set up currents within the body of the material. These circulating currents are known as *eddy currents.* These currents circulate in large number in closed path through out the material. The current in each path is directly proportional to the induced voltage and inversely proportional to the square of current in it and owing tothe heat energy developed, the material quickly become hot.

Effects of Eddy Current

The effect of eddy current is eddy current loss. These eddy currents develop $I^2 R$ loss in the material. Eddy current loss also results in the rise of temperature of material.

Eddy currents result in loss of power with consequent heating of the material. Magnitude of power lost due to eddy currents is often a matter of considerable importance in Electrical Engineering. In electrical machines usually varying fluxes are involved.

Therefore, eddy currents induced in the iron parts causes considerable loss of power.

Eddy Current Loss

Power lost due to eddy currents is called *eddy current loss.*

The heat liberated by eddy currents usually serves no good purpose but actually represents a loss of useful energy *e.g.* in motor armature, dynamo and transformer cores.

Method of reducing this loss is to reduce the magnitude of eddy current.

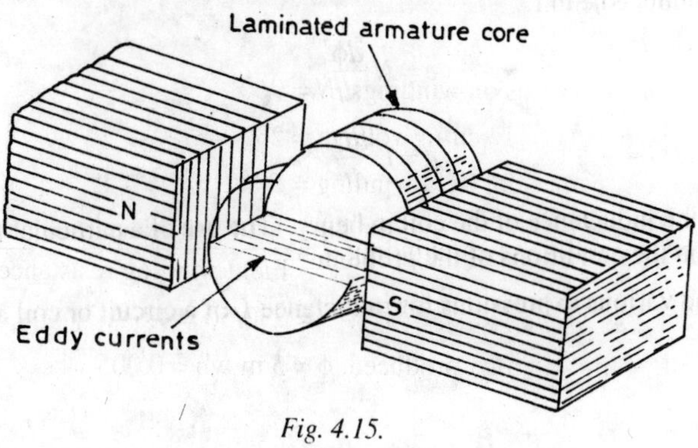

Fig. 4.15.

Firstly this can be achieved by splitting the solid block into thin sheets called laminations in planes parallel to the magnetic flux. Secondly by using high specific resistance materials, like silicon, steel. Each lamination is insulated from the other by a layer of varnish. This arrangement reduces the area of each section and hence induced emf. It also increases the resistance of eddy current paths through which the currents can pass smaller.

Both effects reduce the eddy current and hence eddy current loss.

Useful Applications of Eddy Currents

Eddy current in electrical machines are not desireable at all but there are certain cases where eddy currents are used for some useful work such as induction heating, induction type instruments such as energy meter.

SUMMARY OF FORMULAS

1. Due to electromagnetic induction, the magnitude of emf induced in a coil of N turns is given by :

$$e = \frac{N(\phi_2 - \phi_1)}{t} \text{ volts}$$

where $\phi_1 - \phi_2$ is the change of flux linking the coil in wb in time t seconds.

In differential form, the experession is $e = N\dfrac{d\phi}{dt}$

2. The magnitude of dynamically induced e.m.f. in a conductor moving at right angles to a uniform magnetic field is given by :

Dynamically induced emf, $e = B\,l\,v$ volts

where B = flux density of the field in wb/m^2

l = length of the conductor in metres

v = velocity of the conductor in m/s

If the conductor moves at an angle θ to the magnetic field, then

$$e = B\,l\,v \sin\theta \text{ volts}$$

3. The magnitude of self-induced e.m.f. in a coil of N turns is given by :

Self-induced e.m.f.

$$e = N\frac{d\phi}{dt} \quad ...(i)$$

$$e = L\frac{dI}{dt} \quad ...(ii)$$

where L is the inductance of the coil in henry. The use of a particular formula will depend upon the conditions of the problem.

4. The various expressions for inductance L of a circuit or coil are :

$$L = \frac{e}{(dI/dt)} \quad ...(i)$$

$$L = \frac{N\phi}{I} \quad ...(ii)$$

$$L = \frac{N^2 a \mu_0 \mu_r}{l} \quad ...(iii)$$

$$= \frac{N^2}{\text{reluctance } (S)}$$

Formula given in eqn. *(i)* will be used when the induced e.m.f. and rate of change of current in the coil or circuit are known. The formula of eqn. *(ii)* may be used when flux linkages and current are known. However, eqn. *(iii)* may be used when the dimensions of the magnetic circuit are known.

5. The magnitude of mutually induced e.m.f. in a coil B due to change of current in coil A is given by :

$$e_M = M \frac{dI_1}{dt} \text{ volts}$$

where M = Mutual inductance of the two coils in henry

dI_1/dt = Rate of change of current in coil A

6. The various expressions for the mutual inductance *(M)* between the two coils are :

$$M = \frac{e_M}{(dI_1/dt)} \quad ...(i)$$

$$M = \frac{N_2 \phi_{12}}{I_1} \quad ...(ii)$$

$$M = \frac{N_1 N_2 a \mu_0 \mu_r}{l} \quad ...(iii)$$

$$= \frac{N_1 N_2}{\text{reactance}(S)}$$

The use of particular formula will depend upon the conditions of the problem.

7. The co-efficient of coupling between two coils is given by :

$$\text{Co-efficient of coupling, } k = \frac{M}{\sqrt{L_1 L_2}}$$

where L_1 and L_2 are the individual inductances of the two coils and M is the mutual inductance between the coils.

8. Energy stored in a magnetic field $= \frac{1}{2} LI^2$ joules

where L and I are in henry and amperes respectively.

4.14. RISE OF CURRENT IN R-L CIRCUIT

Consider an R-L series circuit connected across a source of supply of 'V' volts as shown in Fig.4.16. Let at any instant current flowing through it be i Amp.
Resistive drop = $i\,R$ volts

$$\text{Back e.m.f. } e' = -L\frac{di}{dt}$$

Applied voltage is required to overcome resistive drop as well as back emf. Hence while current is increasing we have

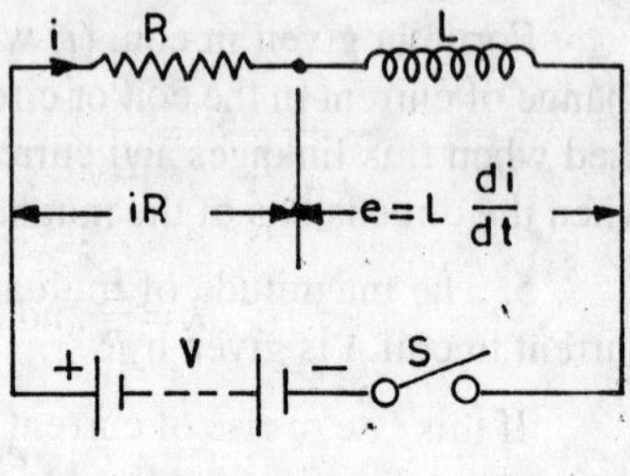

Fig. 4.16.

$$V = iR + L\frac{di}{dt}$$

or $$V - iR = L\frac{di}{dt}$$

or $$\frac{di}{V - iR} = \frac{-R}{L}\,dt$$

or $$\frac{-R\,di}{V - iR} = \frac{-R}{L}\,dt$$

Integrating both sides we get

$$\log_e (V - iR) = \frac{-R}{L}\,t + K$$

where K is a constant of integration

From initial conditions, when $t = 0, i = 0$

$\therefore$ $$K = \log_e V$$

Hence $$\log_e (V - iR) = \frac{-R}{L}\,t + \log_e V$$

or $$\log_e \frac{V - iR}{V} = \frac{-R}{L}\,t$$

or $$\frac{V - iR}{V} = e^{\frac{-R}{L}t}$$

or $$V - iR = V\,.\,e^{\frac{-R}{L}t}$$

or $$iR = V\left(1 - e^{\frac{-R}{L}t}\right)$$

or $$i = \frac{V}{R}\left(1 - e^{\frac{-R}{L}t}\right)$$

$$= I_{max}\,(1 - e^{-t/\lambda}) \qquad ...(i)$$

where $I_{max}\,(1 - e^{-t/\lambda}) = \frac{V}{R}$ the final steady value of current and $\lambda = \frac{L}{R}$, the time constant

(The time constant is also represented by English letter 'T' or Greek letter tau 'τ')

The expression $i = I_{max} = (1 - e^{-t/\lambda})$ is an exponential expression and is often referred to as *Helmholtz expression.*

Immediately after the switch is closed, the rate of rise of current *i.e.* initial rate of rise of current can be determined by differentiating expression *(i)* w.r.t. t for $t = 0$. Hence we get

$$\frac{di}{dt} = \frac{V}{R\lambda} e^{-t/\lambda}$$

$$= \frac{V}{R \times \frac{L}{R}} e^{0} = \frac{V}{L}$$

$$\because \lambda = \frac{L}{R} \text{ and } t = 0$$

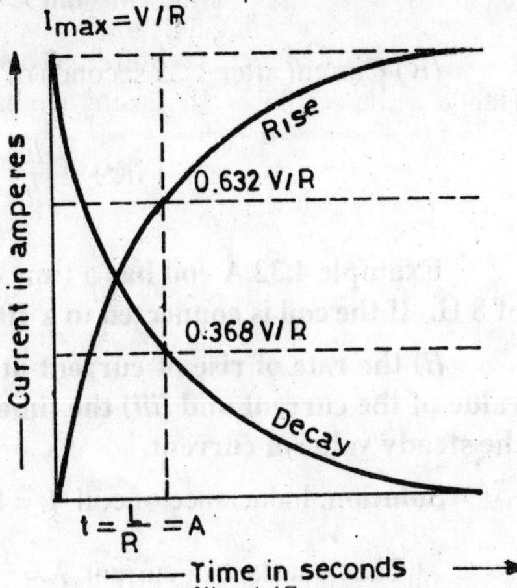

Fig. 4.17.

If this rate of rise of current were maintained then time required to reach maximum value of current $I_{max} = V/R$ would have been $\frac{V/R}{V/L} = \frac{L}{R}$ Thus time is called the *time constant* of the circuit.

Hence the *time constant,* λ may be defined as the *time required for the current to rise to its final steady value if it continuted rising at its initial rate (i.e. V/L)*

If we substitute $t = \lambda = \frac{L}{R}$ in expression *(i)*, we get

$$i = I_{max}\,(1 - e^{-1}) = 0.632\, I_{max}$$

So the *time constant* may also be defined as the time required for the current to reach 0.632 of its final steady value (*i.e.,* $0.632\, I_{max}$)

4.15. DECAY OF CURRENT IN R-L CIRCUIT

Consider an inductive coil of resistance R ohms, inductance L henrys and having a current of I amperes flowing through it. If the source of supply is removed from its circuit, the current starts decreasing and due to decrease in current, self induced emf will be set up, which will oppose the decay of current.

Let at any instant current flowing through the coil be of i amperes

$$\text{Self induced emf. } e' = -L\frac{di}{dt}$$

$$\text{Resistive drop} = iR$$

$$\text{Applied voltage } V = 0$$

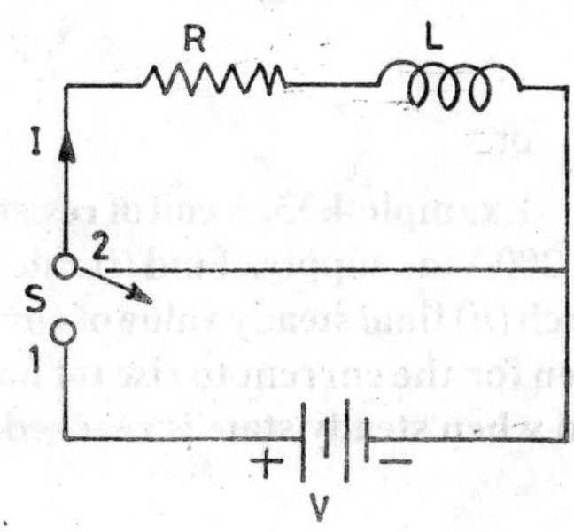

Fig. 4.18.

Since Applied voltage,

$$= \frac{24}{20} = 1.2 \text{ A/s } \textbf{Ans.}$$

(iii) Final steady state current, $I_{max} = \frac{V}{R} = \frac{24}{16} = 1.5$ A **Ans.**

$$\text{Time constant } \lambda = \frac{L}{R} = \frac{20}{16} = 1.25 \text{ seconds}$$

(iv) Current after 1.25 seconds of switching

$$i = \frac{V}{R}(1 - e^{-t/\lambda}) = 1.5\,(1 - e^{-1.25/1.25})$$

$$= 1.5 \times 0.632 = 0.948 \text{ A } \textbf{Ans.}$$

Example 4.32.A coil has a time constant of 1 second and an inductance of 8 H. If the coil is connected to a 100 volt de source, determine :

(i) the rate of rise of current at the instant of switching (ii) the steady value of the current and (iii) the time taken for the current to reach 60% of the steady value of current.

Solution. Inductance of coil, $L = 8$ H

$$\text{Time constant of current, } \lambda = \frac{L}{R} = 1 \text{ second}$$

$$\text{Resistance of circuit, } R = \frac{L}{\lambda} = \frac{8}{1} = 8\ \Omega$$

(i) The rate of rise of current at the instant of switching

$$\frac{di}{dt} = \frac{V}{L} = \frac{100}{8} = 12.5 \text{ A/s}$$

(ii) Steady value of current,

$$I_{max} = \frac{V}{R} = \frac{100}{8} = 12.5 \text{ A } \textbf{Ans.}$$

Let the time taken for the current to reach 60% of the steady value of current, *i.e.*, 0.6 × 12.5 or 7.5 A be t seconds

$$\therefore \quad i = I_{max}(1 - e^{-t/\lambda})$$

$$\therefore \quad 7.5 = 12.5\,(1 - e^{-t/1})$$

$$\text{or} \quad t = -\log_e 0.4 = 0.916 \text{ second } \textbf{Ans.}$$

Example 4.33. A coil of resistance 20 Ω and inductance 0.8 H is connected to a 200 V dc supply. Find *(i)* rate of change of current at the instant of closing switch *(ii)* final steady value of current *(iii)* time constant of the circuit *(iv)* time taken for the current to rise tot half its final value and *(v)* energy stored in the field when steady state is reached.

$$v = \text{Resistive drop} + \text{back emf}$$

$$\therefore \quad 0 = iR + L\frac{di}{dt}$$

or
$$\frac{di}{i} = \frac{-R}{L}dt$$

Integrating both sides we get

$$\log_e i = \frac{-R}{L}t + K'$$

where K' is a constant of integration

From initial conditions, *i.e.*, when $t = 0$, $i = I_{max}$

$$\therefore \quad K' = \log_e I_{max}$$

Hence
$$\log_e i = \frac{-R}{L}t + \log_e I_{max}$$

or
$$\log_e \frac{i}{I_{max}} = \frac{-R}{L}t$$

or
$$i = I_{max}\, e^{\frac{-R}{L}t} = I_{max}\, e^{-t/\lambda}$$

where $\lambda = \frac{L}{R}$, the time constant

The rise and decay of current are shown in Fig. 4.18.

Example 4.31. A choke with an inductance of 20 H and a resistance of 16 Ω is connected through a switch to a 24 V source of emf.

(i) **What is the initial current when the switch is closed?**

(ii) **What is the initial rate of change of current when the switch is closed?**

(iii) **What is the final steady state current?**

(iv) **What is the current at the instant of 1.25 s after switching?**

Solution : Resistance $R = 16\ \Omega$

Inductance, $L = 20$ H

Applied voltage, $V = 24$ volts

(i) Initial current when the switch is closed, *i.e.*, $t = 0$

$$i = I_{max}(1 - e^{-t/\lambda}) = I_{max}(1 - e^{0}) = 0 \textbf{ Ans.}$$

(ii) Rate of change of current when the switcy is closed, *i.e.*, when $t = 0$

$$\frac{di}{dt} = \frac{V}{L}$$

Solution. Resistance of coil, $R = 20\ \Omega$

Inductance of coil, $L = 0.8$ H

(i) Rate of change of current at the instant of switching

$$\frac{di}{dt} = \frac{V}{L} = \frac{200}{0.8} = 250 \text{ A/s } \textbf{Ans.}$$

(ii) Final steady value of current,

$$I_{max} = \frac{V}{R} = \frac{200}{20} = 10 \text{ A } \textbf{Ans.}$$

(iii) Time constant of circuit, $\lambda = \frac{L}{R} = \frac{0.8}{20} = 0.04$ second **Ans.**

(iv) Expression for instantaneous current while current is rising may be written as

$$i = I_{max}\,(1 - e^{-t/\lambda})$$

when $i = \frac{1}{2} I_{max}$

$$\frac{1}{2} I_{max} = I_{max}\,(1 - e^{-t/0.04})$$

or $$e^{-25t} = 0.5$$

or $$t = -0.04 \log_e 0.5 = 0.028 \text{ second } \textbf{Ans.}$$

(v) Energy stored in the field when steady state is reached

$$= \frac{1}{2} L I^2_{max} = \frac{1}{2} \times 0.8 \times (10)^2 = 40 \text{ J } \textbf{Ans.}$$

Example 4.34. The time constant of a certain inductive coil was found to be 0.0025 second. With a resistance of 80 Ω added in series, a new time constant of 0.0005 second was obtained. Determine the inductance and resistance of the coil. Determine the steady value of the current and the time taken by the current to attain 63.2% of the steady value of current when the coil alone is connected across a constant voltage of 200 volts.

Solution :

Time constant in the first case,

$$\lambda_1 = \frac{L}{R} = 0.0025 \text{ s} \qquad ...(i)$$

Time constant when a resistance of 80 Ω is added,

$$\lambda_2 = \frac{L}{R + 80} = 0.0005 \text{ s} \qquad ...(ii)$$

Dividing expression *(i)* by expression *(ii)* we get

$$\frac{R+80}{R} = \frac{0.0025}{0.0005} = 5$$

or $$R = 20\ \Omega\ \textbf{Ans.}$$

and $$L = R\,\lambda_1 = 20 \times 0.0025 = 0.05\ \text{H}\ \textbf{Ans.}$$

Steady value of current, $$I_{max} = \frac{V}{R} = \frac{200}{20} = 10\ \text{A}\ \textbf{Ans.}$$

Instantaneous current is given by the expression

$$i = \frac{V}{R}(1 - e^{-t/0.0025})$$

or $$0.632\, I = I\,(1 - e^{-t/0.0025})$$

or $$e^{-t/0.0025} = 0.368$$

or $$t = -\,0.0025 \log_e 0.368 = 0.0025 \text{ second } \textbf{Ans.}$$

Example 4.35. A coil having a resistance of 15 Ω and inductance of 10 H is connected to a constant 75 V supply. Calculate *(a)* time constant of the circuit, *(b)* final steady value of current and *(c)* the time taken for the current to reach a value of 4 A.

Solution :

(a) Time constant, $\lambda = L/R = 10/15 = 0.67$ s **Ans.**

(b) Final current, $I = V/R = 75/15 = 5$ A **Ans.**

(c) $$i = (1 - e^{-t/\lambda})$$

or $$4 = 5\,(1 - e^{-t/0.67})$$

or $$e^{-t/0.67} = 1 - (4/5) = 0.2$$

or $$e^{t/0.67} = 1/0.2 = 5$$

or $$(t/0.67) \log_e e = \log_e 5$$

$$t = 0.67 \log_e 5 = 1.078 \text{ s } \textbf{Ans.}$$

Example 4.36. The steady current flowing in an inductor is 250 mA; the current flowing 0.1 sec after connecting the supply voltage is 120 mA. Calculate *(a)* time constant of the circuit, *(b)* the time from closing the circuit at which circuit current has reached 200 mA.

Solution :

(a) $$i = I\,(1 - e^{-t/\lambda})$$

Here $i = 120$ mA; $I = 250$ mA; $t = 0.1$ sec

∴ $$120 = 250\,(1 - e^{-0.1/\lambda})$$

or $$e^{-0.1/\lambda} = 1 - (120/250) = 0.52$$

or $$e^{0.1/\lambda} = 1/0.52 = 1.923$$

or $$(0.1/\lambda) \log_e e = \log_e 1.923$$

$\therefore$ $$\text{Time constant, } \lambda = \frac{0.1}{\log_e 1.923} = 0.153 \text{ s } \textbf{Ans.}$$

(b) $$i = I(1 - e^{-t/\lambda})$$

Here $i = 200$ mA; $I = 250$ mA; $\lambda = 0.153$ sec

$$200 = 250\,(1 - e^{-t/0.153})$$

or $$e^{-t/0.153} = 1 - (200/250) = 0.2$$

$$e^{t/0.153} = 1/0.2 = 5$$

or $$(t/0.153) \log_e e = \log_e 5$$

$$t = 0.153 \log_e 5 = 0.25 \text{ s } \textbf{Ans.}$$

Example 4.37. A coil having $L = 2.4$ H and $R = 4\ \Omega$ is connected a constant 100 V supply source. How long does it take the voltage across the resistance to reach 50 V?

Solution : $$i = I(1 - e^{-t/\lambda})$$

$$= \frac{V}{R}(1 - e^{-t/\lambda})$$

or $$iR = V(1 - e^{-t/\lambda})$$

Here $iR = 50$ volts; $V = 100$ volts

$$\lambda = L/R = 2.4/4 = 0.6 \text{ s}$$

$\therefore$ $$50 = 100\,(1 - e^{-t/0.6})$$

or $$e^{-t/0.6} = 1 - (50/100) = 0.5$$

or $$e^{t/0.6} = 1/0.5 = 2$$

or $$(t/0.6) \log_e e = \log_e 2$$

$\therefore$ $$t = 0.6 \log_e 2 = 0.416 \text{ s } \textbf{Ans.}$$

Example 4.38. The time constant of a certain inductive coil was found to be 2.5 ms. With a resistance of 80 Ω added in series, a new time constant of 0.5 ms was obtained. Find the inductance and resistance of the coil.

Solution : Time constant, $\lambda = L/R$

For the first case, $L/R = 2.5$

For the secont case, $L/(R + 80) = 0.5$

$\therefore$ $$\frac{R + 80}{R} = \frac{2.5}{0.5} = 5$$

or $$R = 20\ \Omega$$

Now $$L/R = 2.5$$

$$L = 2.5\,R = 2.5 \times 20 = 50 \text{ H } \textbf{Ans.}$$

4.16. RISE OF CURRENT IN R-C CIRCUIT

Consider the capacitive circuit shown in Fig. 4.19. When switch *S* is closed, the voltage across the capacitor is zero since we started with an uncharged capacitor. The entire applied voltage *V* is dropped across *R* and charging current is maximum, *i.e.*, $I = V/R$. After having closed the switch, the charging current starts decreasing and voltage across capacitor starts increasing. The decrease of charging current and increase of voltage across the capacitor follow exponential law as shown in Fig. 4.20 and 4.21 respectively.

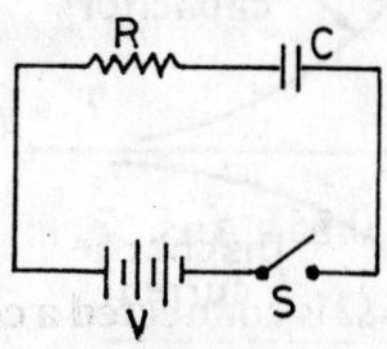

Fig. 4.19.

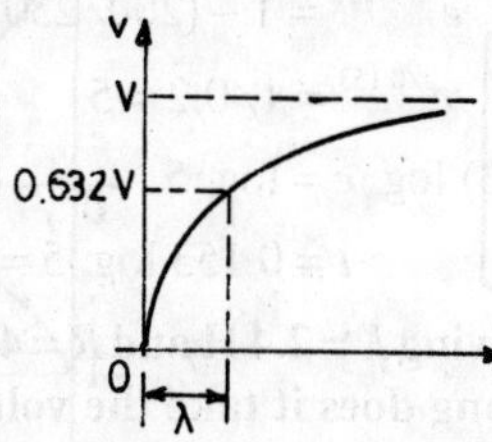

Fig. 4.20.

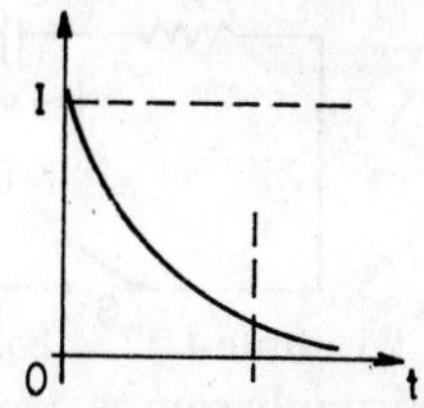

Fig. 4.21.

It can be shown mathematically that at any instant, voltage *v* across the capacitor is given by ;

$$v = V\,(1 - e^{-t/RC})$$

where V = Final steady voltage across the capacitor

t = time elapsed since switch is closed

The exponent of *e* is t/RC. The quantity *RC* is called *time constant* of the circuit. It is represented by λ.

$$\text{Time constant, } \lambda = RC \text{ seconds}$$

$$\therefore \qquad v = V\,(1 - e^{-t/\lambda}) \qquad \ldots(i)$$

Since $v = q/C$ and $V = Q/C$

$$\therefore \qquad q = Q\,(1 - e^{-t/\lambda}) \qquad \ldots(ii)$$

where q = Charge on the capacitor at any instant

Q = Final charge on the capacitor

If $t = \lambda$,

then
$$v = V\,(1 - e^{-\lambda/\lambda})$$
$$= V\,(1 - e^{-1}) = 0.632\text{ V}$$

Hence time constant may be defined as the time required for the capacitor voltage to reach 0.632 of its final steady value *V* (Fig. 4.20).

4.17. DECAY OF CURRENT IN R-C CIRCUIT

Consider a capacitor of C farad charged to p.d. of V volts and connected in series with a resistance R through a switch as shown in Fig. 4.22. When the switch is open, the voltage across capacitor is V. When switch is closed, the voltage across the capacitor decreases exponentially. The discharge current rises instantly to maximum value $I = (V/R)$ and then decays exponentially as shown in Fig. 4.23.

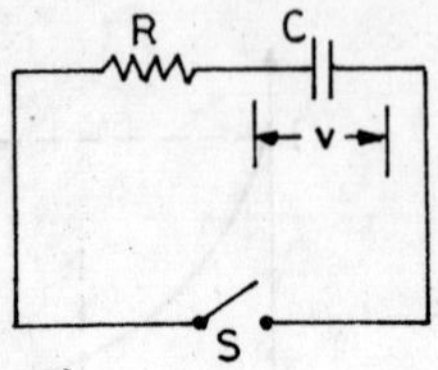

Fig. 4.22.

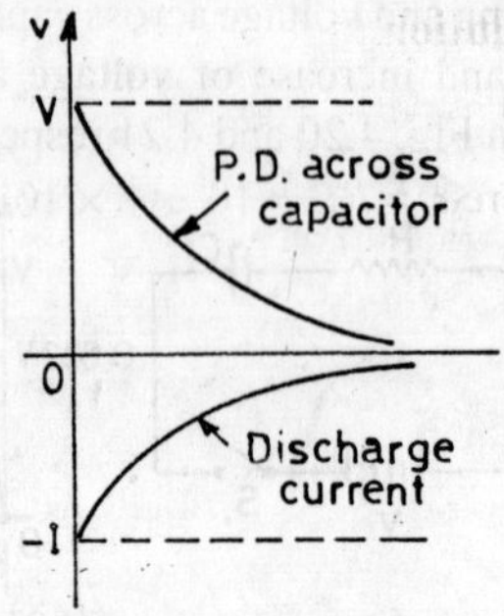

Fig. 4.23.

It can be shown mathematically that at any instant, the voltage v across the capacitor is given by :

$$v = V e^{-t/RC}$$

Again, time constant,

$$\lambda = RC \text{ seconds}$$

$$\therefore \quad v = V e^{-t/\lambda} \qquad ...(i)$$

Since $v = q/C$ and $V = Q/C$

$$\therefore \quad q = Q e^{-t/\lambda} \qquad ...(ii)$$

Note : A capacitor is almost fully charged or discharged in a time equal to 5 time constants.

Example 4.39. A 2 μ F capacitor is connected, by closing a switch, to a supply of 100 V through 1 M Ω series resistance. Calculate *(a)* time constant of the circuit, *(b)* initial charging current and *(c)* voltage across the capacitor 6 seconds after the switch has been closed.

Solution :

(a) Time constant, $\lambda = RC$

$$= (10^6) \times 2 \times 10^{-6} = 2 \text{ s} \textbf{ Ans.}$$

(b) Initial charging current, $I = \dfrac{V}{R} = \dfrac{100}{10^6} \times 10^6 = 100 \text{ μ A}$ **Ans.**

(c)

$$v = V(1 - e^{-t/\lambda})$$

Here $V = 100$ volts, $t = 6$ seconds, $\lambda = 2$ seconds

$$\therefore \quad v = 100\,(1 - e^{-6/2})$$

$$= 100\,(1 - e^{-3}) = 95.1 \text{ V } \textbf{Ans.}$$

Example 4.40. A capacitor of 8 μ F capacitance is connected to a d.c. source through a resistance of 1 M Ω. Calculate the time taken by the capacitor to receive 95 % of its final charge. How long will it take the capacitor to be fully charged?

Solution :

(a) $$q = Q\,(1 - e^{-t/\lambda})$$

Here $\lambda = RC = 10^6 \times 8 \times 10^{-6} = 8$ seconds

$$q/Q = 0.95$$

∴ $$0.95 = 1 - e^{-t/8}$$

or $$e^{-t/8} = 1 - 0.95 = 0.05$$

or $$e^{t/8} = 1/0.05 = 20$$

or $$(t/8) \log_e e = \log_e 20$$

∴ $$t = 8 \log_e 20 = 23.96 \text{ seconds } \textbf{Ans.}$$

(b) Time taken for the capacitor to be fully charged

$$= 5\,RC = 5 \times 8 = 40 \text{ seconds } \textbf{Ans.}$$

Example 4.41. A resistance *R* and a 4 μF capacitor are connected in series across a 200 V d.c. supply. Across the capacitor is connected a neon lamp that strikes at 120 V. Calculate the value of *R* to make the lamp strike after 5 seconds.

Solution : The voltage across the neon lamp has to rise to 120 V in 5 seconds.

$$v = V\,(1 - e^{-t/\lambda})$$

or $$120 = 200\,(1 - e^{-5/\lambda})$$

or $$e^{-5/\lambda} = 1 - (120/200) = 0.4$$

or $$e^{5/\lambda} = 1/0.4 = 2.5$$

or $$(5/\lambda) \log_e e = \log_e 2.5$$

∴ $$\lambda = \frac{5}{\log_e 2.5} = 5.457 \text{ seconds}$$

or $$RC = 5.457$$

∴ $$R = \frac{5.457}{4 \times 10^{-6}}$$

$$= 1.364 \times 10^6\ \Omega$$

$$= 1.364\ \text{M}\Omega$$

HIGHLIGHTS

1. **Electro-magnetic Induction :** The phenomenon by virtue of which an emf is induced in a coil when flux linking with coil changes is called electro-magnetic induction.

2. Faraday's Laws of Electro-magnetic induction

   First Law : This law states that when the flux linking with coil changes an emf is induced in it.

   Second Law : The magnitude of emf induced is directly proportional to the rate of change of flux linkages

   i.e.,
$$\text{Induced emf} = \frac{Nd\phi}{dt} = N\left(\frac{\phi_2 - \phi_1}{t}\right)$$

3. **Lenz's Law :** The direction of emf induced is such that it always opposes the cause producing it.

4. Fleming's Left Hand Rule

5. **Fleming's Right Hand Rule :** Stretch first finger, second finger and thumb of right hand mutually perpendicular to each other. If first finger represents magnetic field, thumb represents the direction of motion of conductor, then the second finger will represents the direction of induced emf.

6. **Self Induced emf :** The emf induced in a coil due to change of flux produced by it linking with its own winding is called self induced emf.
$$e = L\frac{di}{dt}$$

7. **Mutually Induced emf :** The emf induced in a coil due to change of flux produced by the other neighbouring coil linking with it is called mutually induced emf.
$$e_m = M\frac{dI_1}{dt}$$

8. **Self Inductance :** The property of coil due to which it opposes the change of current flowing through itself is called self inductance.
$$L = \frac{N^2\, a\, \mu_0\, \mu_r}{l}$$

9. **Mutual Inductance :** The property of one coil due to which it opposes the change of current in the neighbouring coil is called mutual inductance.
$$M = \frac{N_1\, N_2\, a\, \mu_0\, \mu_r}{l}$$

10. **Co-efficient of Coupling :** The fraction of magnetic flux produced by one coil linking with the other is called co-efficient of coupling between two coils.
$$K = M\,\sqrt{L_1\, L_2}$$

11. Energy stored in the magnetic field $1/2\ LI^2$ joules.

12. Rise of current on closing of circuit at any instant :

$$i = I\left(1 - e^{\frac{-R}{L}\times t}\right)$$

13. Decay of current

$$i = I\,e^{-\frac{R}{L}\times t}$$

14. Time constant $\lambda = L/R$ seconds.

SHORT QUESTIONS

1. Eddy current losses are induced in magnetic material, why ?
2. Why the time constant is measured in seconds ?
3. Why change is essential for the induced emf ?
4. Why solid piece of magnetic core is not used ?
5. Inductance disappear in the circuit when d.c. is applied, why ?

MULTIPLE QUESTIONS

1. Whenever a magnet is quickly put in an open-circuited stationery coil.

(*a*) a current is induced in it

(*b*) emf is induced in it

(*c*) Power is spent

2. The emf induced in a coil depends on

(*a*) number of its turns

(*b*) the change of flux linked with it

(*c*) the time taken to change the flux

(*d*) all of above

3. The direction of the induced emf in a coil is found by

(*a*) Faraday's Law (*b*) Lenz's Law

(*c*) Fleming's Left-hand rule

4. The energy stored in the magnetic field is

(*a*) $\frac{1}{2}LI^2$

(*b*) directly proportional to the square of current flowing

(*c*) directly proportional to the inductance of coil

(*d*) all of the above

5. To reduce eddy current loss the core of magnetic material

(*a*) the core is laminated (*b*) the core is solid one

(*c*) magnetic material has high resistivity

(*d*) both (*a*) and (*b*)

6. The emf induced is 2V when current changing at the rate of 0.5 A/s. The self inductance of the coil is

(*a*) 4 (*b*) 2

(*c*) 1 (*d*) 3

7. A 100 turns coil has an inductance of 6 mH. If the number of turns is increased to 200, other quantities remaining the same. The inductance will be

(*a*) 1.5 (*b*) 3

(*c*) 12 (*d*) 24

ANSWERS

1. (*b*)	**2.** (*d*)	**3.** (*b*)	**4.** (*d*)	**5.** (*d*)
6. (*a*)	**7.** (*b*)			

EXERCISES

1 What is meant by the term 'electro-magnetic induction'? Imagine a permanent magnet placed within a coil of few turns of wire. If the terminals of the wire are connected to a voltmeter, will it show some deflection? Justify your answer.

2. State and explain Faraday's laws of electro-magnetic induction.
3. State and explain the laws relating to electro-magnetic induction.
4. State and explain Lenz's law.
5. State Faraday's laws of electro-magnetic induction and explain.
6. Derive the expression for the dynamically induced emf. On what factor does this emf depend? Explain.
7. State and explain Fleming's right hand rule.
8. What do you understand by statically and dynamically induced emf? Write down difference between them with example.
9. Define and explain self and mutually induced emf.
10. Explain what is meant by self inductance and mutual inductance. Define the units in which each is measured.
11. What is meant by self and mutual inductances? Explain. Also give their units.
12. Explain the concept of mutual inductance. Define coefficient of coupling and derive the expresion between self inductances of two coils, mutual inductance between them and the coefficient of coupling.
13. Prove that the coefficient of mutual inductance M between two coils of self indcutances L_1 and L_2 is given by $M = \sqrt{L_1 L_2}$.
14. Obtain an expression for the energy stored in a magnetic field.

15. Derive an expression for the energy stored in an inductor.

16. Derive an expression for the voltage across resistance when R-L is suddenly connected to a dc voltage. Indicate the physical significance of 'time constant' for such a circuit.

17. Derive the equation for an R-L circuit which gives the rise and decay of currents as function of time. What is meant by the time constant of a circuit?

18. Define the time constant of a series R-L circuit. If a constant dc voltage is suddenly applied across such a circuit, find an expression for the current flow.

19. Write short notes on the following :
    - *(i)* Faraday's laws of electro-magnetic induction.
    - *(ii)* Lenz's law
    - *(iii)* Fleming's left hand rule.
    - *(iv)* Self and mutual induction.
    - *(v)* Magnetic coupling coefficient.
    - *(vi)* Energy stored in a magnetic field.

20. Define and explain the following terms :
    - *(i)* Flux linkage.
    - *(ii)* Dynamically induced emf.
    - *(iii)* Coefficient of self induction.
    - *(iv)* Coefficient of mutual induction.
    - *(v)* Coefficient of coupling between the coils.

21. Distinguish between
    - *(i)* Flux and flux linkage.
    - *(ii)* Statically induced emf and dynamically induced emf.
    - *(iii)* Self induced emf and mutually induced emf.
    - *(iv)* Self inductance and mutual inductance

22. Explain the following :
    - *(i)* In how many ways an emf can be induced?
    - *(ii)* What is indicated by the negative sign before $N\frac{d\phi}{dt}$ in expression $e = -N\frac{d\phi}{dt}$?
    - *(iii)* What is Lenz's law?
    - *(iv)* What is Fleming's right hand rule?

(v) What is the application of Fleming's right hand rule?

(vi) What is dynamically induced emf?

(vii) What is a self inductance of a coil?

(viii) What do you understand by self induced emf and mutually induced emf?

(ix) What is meant by mutual inductance?

(x) What is meant by coefficient of coupling between the coils?

23. Explain the following :

(i) In R-L circuit, prove that current at any instant is $i = \frac{V}{R}(1 - e^{-Rt/L})$.

(ii) What is the effect of self-inductance in a dc circuit?

(iii) What is meant by the 'time constant' of a series R-L circuit?

(iv) What do you mean by Eddy currents? Suggest a remedy for reducing such currents.

(v) Give a brief account of some useful applications of Eddy current effects.

PROBLEMS

1. A coil of 200 turns is linked by a flux of 8 mwb. Determine the emf induced in the coil if the flux is reversed in 0.02 seconds.

(**Ans.** 160 volts)

2. A conductor 10 cm long moves with a uniform velocity of 2 m/s at right angles to itself and to a uniform magnetic field having flux density of 1 wb/m^2. Calculate the induced emf between the ends of a wire.

(**Ans.** 0.2 V)

3. A 0.5 m long wire moves at right angles to its length at 40 m/s in uniform magnetic field of 1 wb/m^2. Calculate the emf induced in the conductor when the direction of motion is *(a)* perpendicular to field, *(b)* inclined at 30° to the direction of field.

(**Ans.** *(a)* 20 V, *(b)* 10 V)

4. A conductor of active length 30 cm and carrying a current of 100 A moves at an angle 30° with the direction of a uniform magnetic field of density of 0.4 wb/m^2. Find the emf induced in it if the velocity of the conductor is 0.5 m/s.

(**Ans.** 30 mV)

5. A coil of inductance 10 mH carries a current of 0.2 A. Find the magnitude of induced emf in the coil if the current is interrupted in 0.001 s.

(**Ans.** 2 V)

6. If a current through a coil of inductance 1 H is reduced from 10 A to –10 A in 0.05 s. Calculate the mean value of emf induced in the coil.

(**Ans.** 400 V)

7. A coil of 150 turns is linked with a flux of 0.001 wb when carrying a current of 10 A. If the current is uniformly reversed in 0.1 s, calculate the induced emf.

(**Ans.** 3 V)

8. A coil of 300 turns, wound on a core of non-magnetic material has an inductance of 10 mH. Calculate *(a)* the flux produced by a current of 5 A and *(b)* the average value of emf induced when a current of 5 A is reversed in 8 milli-seconds.

(**Ans.** *(a)* 0.1667 m wb, *(b)* 12.5 V)

9. Estimate the inductance of a solenoid of 2500 turns wound uniformly over a length of 0.5 m on a cylindrical paper tube 4 cm in diameter. The medium is air.

(**Ans.** 19.74 mH)

10. Calculate the inductance of a toroid 25 cm mean diameter and 6.25 cm^2 circular x-section wound uniformly with 1000 turns of wire. Hence calculate the emf induced when the current in it increases at the 100 A/s.

(**Ans.** 1 mH, 0.1 V)

11. Two identical coils *X* and *Y* of 1000 turns each lie in parallel planes such that 80% of the flux produced by one coil links with other. If a current of 5 A flowing in *X* produces a flux of 0.65 mwb in it, calculate mutual inductance between *X* and *Y*.

(**Ans.** 0.0Q8 H)

12. The self inductance of a coil of 5000 turns is 0.25 H. If 60% of the flux is linked with a second coil of 10000 turns, calculate mutual inductance of two coils and emf induced in the second coil when the current in first coil changes at the rate of 100 A/second. **(Ans. 0.3 H, 30 V)**

5

A. C. Theory

5.1. INTRODUCTION

We have dealt so far with cases in which the currents are steady and in one direction; this is called direct current (d.c.). The use of direct currents is limited to a few applications *e.g.*, charging of batteries, electroplating, electric traction etc. The a.c. system has offered so many advantages that at present eelctrical energy is universally generated, transmitted and used in the form of alternating current. Even when d.c. energy is necessary, it is a common practice to convert a.c. into d.c. by means of rotary converters or rectifiers.

Three principal advantages for a.c. system over the d.c. system. First, alternating voltages can be stepped up or stepped down efficiently by means of a transformer. This permits the transmission of electric power at high voltages to achieve economy. Secondly, a.c. motors (induction motors) are cheaper and simpler in construction than d.c. motors. Thirdly, the switchgear (*e.g.*, switches, circuit breakers etc.) for a.c. system is simpler than the d.c. system.

5.2. CONCEPT OF ALTERNATING VOLTAGE AND CURRENT

A voltage which changes its polarity at regular intervals of time is called an alternating voltage. When an alternating voltage is applied in a circuit, the current flows first in one direction and then in the opposite direction; the direction of current at any instant depends upon the polarity of the voltage. Fig. 5.1 shows an

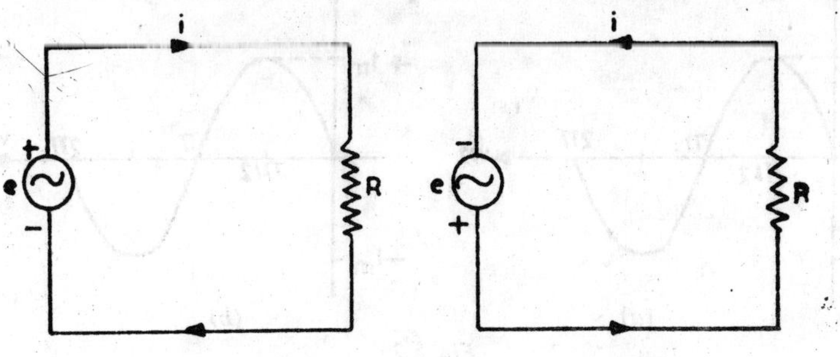

(a) (b)

Fig. 5.1.

alternating voltage source connected to a resistor *R*. In Fig. 5.1 *(a)*, the upper terminal of alternating voltage source is positive and lower terminal negative so that current flows in the current as shown in Fig. 5.1 *(a)*. After sometime (a fraction of a second), the polarities of the voltage source are reversed [See Fig. 5.1 *(b)*] so that current now flows in the opposite direction. This is called alternating current because the current flows in alternate directions in the circuit.

Sinusoidal Alternating Voltage and Current

Commercial alternators produce sinusoidal alternating voltage, *i.e.*, alternating voltage is a sine wave. A sinusoidal alternating voltage can be produced by rotating a coil with a constant angular velocity (say ω rad/sec) in a uniform magnetic field. The sinusoidal alternating voltage can be expressed by the equation:

$$e = E_m \sin \omega t$$

where e = Instantaneous value of alternating voltage

E_m = Max. value of alternating voltage

ω = angular velocity of the coil

Sinusoidal voltages always produce sinusoidal currents, unless the circuit is non-linear. Therefore a sinusoidal current can be expressed in the same way as voltage, *i.e.*, $i = i_m \sin \omega t$. Fig. 5.2 *(a)* shows the wave form of sinusoidal voltage whereas Fig. 5.2 *(b)* shows the wave form of sinusoidal current. Note that sinusoidal voltage or current not only changes direction at regular intervals but the magnitude is also changing continuously.

Wave Form : Why do we choose a sinusoidal wave rather than a "simple" curve as a square or triangular wave ? The following are the reasons for doing so:

(i) In a.c. machines and transformers, sinusoidal voltages and currents respectively produce the least iron and copper losses for a given power output. The efficiency is therefore better.

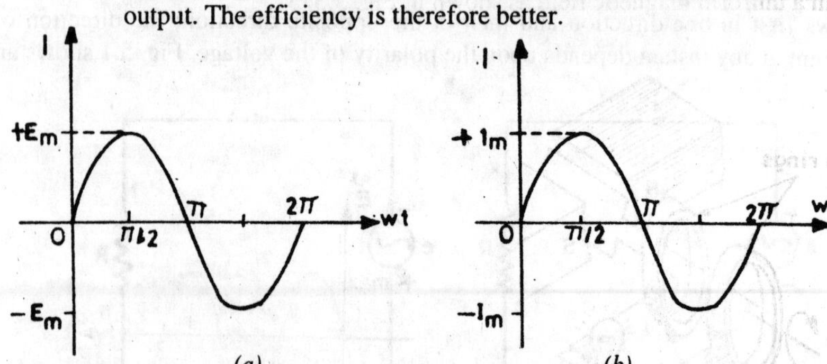

Fig. 5.2.

(ii) Sinusoidal voltages and currents produce less interference (noise) on telephone lines.

(iii) The sine wave form produces the least disturbance in the electrical circuit.

Due to above advantages, electric supply companies all over the world generate sinusoidal alternating voltages and currents. It may be noted that alternating voltage and current mean sinusoidal alternating voltage and current unless stated otherwise.

Generation of Alternating Voltage and Current

- **Principle :** When a coil or conductor is moved in magnetic field, an emf is induced in the coil or conductor.

When a conductor is moved in magnetic field, magnetic lines of force are cut. In other words flux is cut by the conductor or flux linking the conductor changes. According to Faraday's first law of electromagnetic induction an emf is induced in the conductor when magnetic flux is cut by the conductor.

The alternating voltage reverses its polarity and the current reverses its direction in the circuit.

Alternating voltage and current can be generated by :

(i) Rotating a coil in a magnetic field.

(ii) Rotating magnetic field within a stationary coil.

A C Generator : A simple AC generator consists of a rectangular coil made up of a large number of turns and mounted on a shaft which can rotate. This arrangement is called Armature. The armature is placed in a very strong magnetic field and rotated by an engine or a turbine.

To understand the principle of generation of alternating current, consider a single turn rectangular coil *ABCD* rotating in anticlockwise direction about its own axis in a uniform magnetic field as shown in Fig. 5.3.

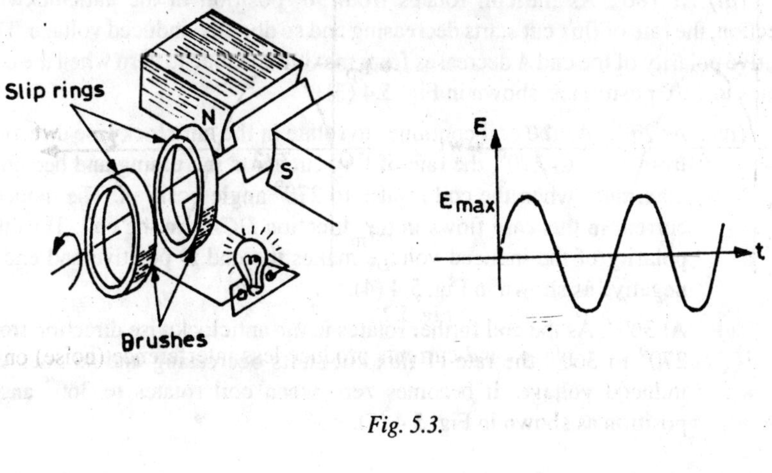

Fig. 5.3.

When coil rotates the conductors *AB* and *CD* move in the magnetic field. At different positions of the coil, the magnitudes of emf induced will be different. This is discussed below by taking five positions of the coil at 90° angle difference in one cycle of the rotation of the coil.

(i) At 0°. When the plane of the coil *ABCD* is perpendicular to the magnetic lines of force, the conductors *AB* and *CD* move parallel to the magnetic field. In this case magnetic flux is not cut, therefore, induced emf is zero as shown in Fig. 5.4 *(a)*.

(ii) At 90°. When the coil rotates in the anticlockwise direction from position (1), the rate of flux cut increases and induced emf starts increasing. It becomes maximum when the coil rotates through 90° as shown in Fig. 5.4 (2). According to Fleming's right hand rule induced current flows in the direction *ABCD* in the coil. Thus polarity of the induced voltage makes the end *A* positive and end *D* negative.

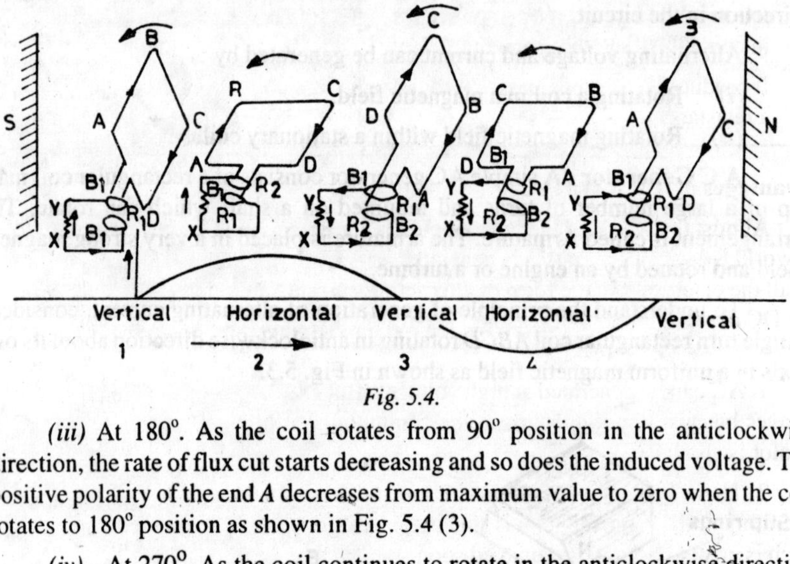

Fig. 5.4.

(iii) At 180°. As the coil rotates from 90° position in the anticlockwise direction, the rate of flux cut starts decreasing and so does the induced voltage. The positive polarity of the end *A* decreases from maximum value to zero when the coil rotates to 180° position as shown in Fig. 5.4 (3).

(iv) At 270°. As the coil continues to rotate in the anticlockwise direction from 180° to 270°, the rate of flux cut starts increasing and becomes maximum when the coil rotates to 270° angle position. The induced current in this case flows in the direction *DCBA* in the coil. Thus the polarity of the induced voltage makes the end *D* positive and end *A* negative as shown in Fig. 5.4 (4).

(v) At 360°. As the coil further rotates in the anticlockwise direction from 270° to 360°, the rate of flux cut starts decreasing and so does the induced voltage. It becomes zero when coil rotates to 360° angle position as shown in Fig. 5.4 (5).

5.3. DIFFERENCE BETWEEN AC AND DC

	Alternating Current (A.C.)	*Direct Current (D.C.)*
1.	It changes its direction 50 times per second, hence its frequency is 50 Hz.	It does not change its direction.
2.	It can be easily transferred from one place to another by varying the voltage with transformer and with less wastage of energy.	It is difficult to transfer and is more expensive.
3.	It can be changed to D.C. easily.	It is not possible.
4.	Its magnitude can be varied with choke without dissipation of energy.	Its magnitude can be varied by resistance in which it is dissipated in the form of heat.
5.	It is more dangerous if a person comes in contact with a wire carrying AC.	It is less dangerous.
6.	It produces heat when passes through a resistance.	It also produces heat.

Advantages of A.C. over D.C.

Almost the whole of the bulk amount of electrical energy used in all parts of the world for various purposes is generated as A.C. In our houses and factories we use all the electricity in the form of alternating current. Therefore, AC is preferred than DC due to a number of reasons. The following are the advantages of AC over DC.

1. AC can be generated at high voltages. But DC cannot be generated at high voltages because sparking starts at the commutator at high voltage and the commutator is spoiled.

2. High voltage AC generators are much simpler and cheaper than DC generators of the same range. It is because in AC generator there is no commutator which is costly part and is damaged soon.

3. Alternating voltages can be stepped up or stepped down with a static device called *transformer*. When voltage is stepped up current decreases to a small value. Small current produces less heat and can be transmitted through a thin conductor. Thus it is possible to transmit AC at high voltages. This reduces the size of conductor, transmission losses and increases tramsmission efficiency.

4. AC induction motors are simplest in construction, cheaper in cost and require less maintenance where as DC motors are complicated.

Derivation of Equation Sinusoidal Wave Form

Consider a rectangular coil of *N* turns rotating in a uniform magnetic field in the anticlockwise direction with an angular velocity ω radians per second as shown in Fig. 5.5 *(a)*. To start with, let the plane of coil coincide with X-axis shown in Fig. 5.5 *(b)*. In this case maximum flux (ϕ max) is cut by the coil. As the coil rotates in the anticlockwise direction, the flux is cut by it and hence an emf is induced in it. Let the coil turns through an angle θ in *t* seconds and reaches in position as shown in Fig. 5.5 *(c)*. Thus, $\theta = \omega t$.

In this position of the coil, maximum flux (ϕ max) acting vertically downwards can be resolved into two perpendicular components.

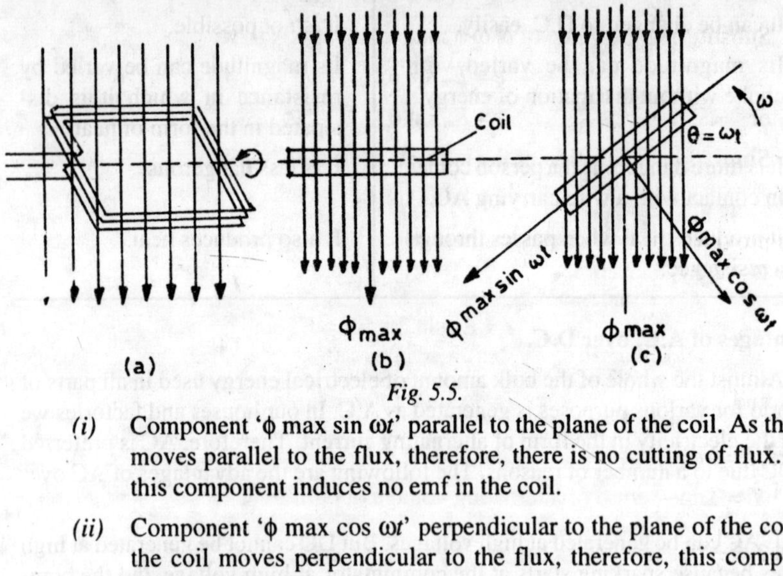

Fig. 5.5.

(i) Component 'ϕ max sin ωt' parallel to the plane of the coil. As the coil moves parallel to the flux, therefore, there is no cutting of flux. Thus this component induces no emf in the coil.

(ii) Component 'ϕ max cos ωt' perpendicular to the plane of the coil. As the coil moves perpendicular to the flux, therefore, this component induces emf in the coil.

∴ Flux linkage of the coil at this instant = No. of turns × flux cut (when coil has turned through θ)

According to Faraday's Laws of electromagnetic induction, emf induced in the coil is equal to the rate of change of flux cut by the coil. Therefore, the emf *E* induced in the coil at this instant is given by differentiating flux linkages w.r.t. time. This is the instantaneous value of emf.

$$\therefore \qquad e = -\frac{d}{dt}\,(\text{flux linkage})$$

The negative sign has been taken due to the reason that the direction of induced emf is such that it opposes the cause producing it (Lenz's Law).

$$\therefore \quad e = -\frac{d}{dt}(N\,\phi\max\cos\omega t)$$

$$\text{or} \quad e = -N\,\phi\max\frac{d}{dt}(\cos\omega t)$$

$$\text{or} \quad e = -N\,\phi\max(-\sin\omega t)\times\omega$$

$$\text{or} \quad e = N\,\omega\,\phi\max\sin\omega t \text{ volts} \qquad ...(i)$$

The value of e will be maximum (E_m) when the coil has turned through 90° ($\because$ sin 90° = 1).

$$\therefore \quad E_m = N\,\omega\,\phi\max \text{ volts} \qquad ...(ii)$$

Substituting the value of $N\,\omega\,\phi$ max in equation *(i)*, we get

$$e = E_m\sin\omega t$$

$$\text{or} \quad e = E_m\sin\theta$$

Similarly the equation of the induced alternating current is

$$i = I_m\sin\omega t$$

provided the coil circuit has been closed through a resistive load. Since the value $\omega = 2\pi f$, therefore

$$e = E_m\sin 2\pi f.t = E_m\sin\left[\frac{2\pi}{T}\right].t$$

$$\text{and} \quad i = I_m\sin 2\pi f.t = I_m\sin\left[\frac{2\pi}{T}\right].t$$

where, f = frequency

T = Time period of alternating voltage or current.

$$T = \frac{1}{f}$$

From the above equations, it is clear that the magnitude induced emf and current varies according to the sine of the angle θ or ωt. The wave shape of induced emf is as shown in Fig. 5.6.

If the equation of alternating is

$$i = I_m\sin\omega t$$

$$i = I_m\sin\theta$$

According to Faraday's laws of electromagnetic induction, the emf induced in a coil is equal to the rate of change of flux linkages of the coil. Hence the emf e at this instant is given by :

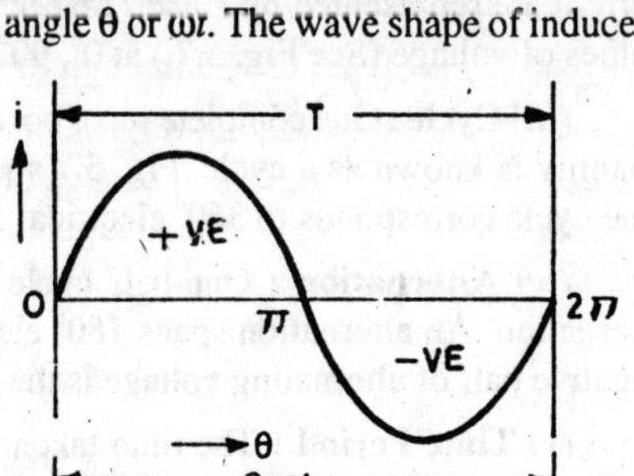

Fig. 5.6.

$$e = -\frac{d}{dt}(n\,\phi_{max}\cos\omega t)$$

$$= -n\,\phi_{max}\,\omega\,(-\sin\omega t)$$

$$= n\,\phi_{max}\,\omega\sin\omega t$$

The value of e will be maximum (call it E_m) when $\sin\omega t = 1$, *i.e.*, when the coil has turned through 90° in anticlockwise direction from the reference axis (*i.e.*, OX-axis).

$$\therefore \quad E_m = n\,\phi_{max}\,\omega$$

$$\therefore \quad e = E_m\sin\omega t \qquad \text{where } E_m = n\,\phi_{max}\,\omega$$

$$= E_m\sin\theta$$

It is clear that emf induced in the coil is sinusoidal, *i.e.*, instantaneous value of emf varies as the sine function of time angle (θ or ωt). If this alternating voltage ($e = E_m\sin\omega t$) is applied across a load, alternating current flows through the circuit which would also vary sinusoidally, *i.e.*, following a sine law. The equation of the alternating current is given by

$$i = I_m\sin\omega t$$

5.4. CONCEPT OF IMPORTANT TERMS

An alternating voltage or current changes continuously in magnitude and in direction at regular intervals of time. It rises from zero to maximum positive value, falls to zero, increases to a maximum in the revese direction and falls back to zero again (See Fig. 5.7). The important a.c. terminology is defined below:

(i) **Waveform :** The shape of the curve obtained by plotting the instantaneous values of voltage or current as ordinate against time as abcissa is called its *waveform* or *waveshape*. Fig. 5.7 shows the waveform of an alternating voltage varying sinusoidally.

(ii) **Instantaneous Value :** The value of an alternating quantity at any instant is called instantaneous value. The instantaneous values of alternating voltage and current are represented by e and i respectively. As an example, the instantaneous values of voltage (See Fig. 5.7.) at 0°, 90° and 270° are 0, $+E_m$, $-E_m$ respectively.

(iii) **Cycle :** One complete set of positive and negative values of an alternating quantity is known as a cycle. Fig. 5.7 shows one cycle of an alternating voltage. One cycle corresponds to 360° electrical or 2 π radians.

(iv) **Alternation :** One-half cycle of an alternating quantity is called an alternation. An alternation spans 180° electrical. Thus in Fig. 5.7, the positive or negative half of alternating voltage is the alternation.

(v) **Time Period :** The time taken in seconds to complete one cycle of an alternating quantity is called its time period. It is generally represented by T.

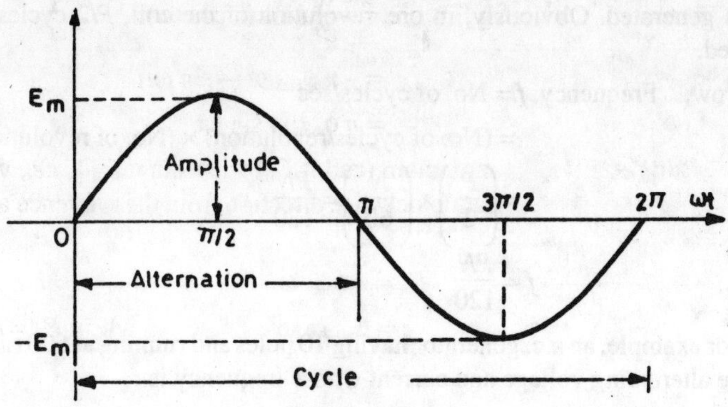

Fig. 5.7.

(vi) **Frequency :** The number of cycles made per second is called the frequency (f) of the alternating quantity. It is measured in cycles/sec (C/s) or Hertz (Hz). One Hertz is equal to 1 C/s.

The frequency of power system is low; the most common being 50 C/s or 50 Hz. It means that alternating voltage or current completes 50 cycles in one second. The 50 Hz frequency is the most popular because it gives the best results when used for operating both lights and machinery.

Important Relations

Having become familiar with a.c. terminology, we shall now establish some important relations.

(i) **Time Period and Frequency :** Consider an alternating quantity having a frequency of f C/s and time period T second.

Time taken to complete f cycles = 1 second (by definition)

Time taken to complete 1 cycle = $1/f$ second

But the time taken to complete one cycle is the time period T (by definition).

$$\therefore \quad T = \frac{1}{f} \quad \text{or} \quad f = \frac{1}{T}$$

(ii) **Angular Velocity and Frequency :** The coil is rotating with an angular velocity of ω rad/sec. In one revolution of the coil, the angle turned is 2 π radians and the voltage wave completes 1 cycle. The time taken to complete one cycle is the time period T of the alternating voltage.

$$\therefore \quad \text{Angular velocity, } \omega = \frac{\text{Angle turned}}{\text{Time taken}} = \frac{2\pi}{T}$$

$$\text{or} \quad \omega = 2\,\pi f \qquad (\because\ f = 1/T)$$

(iii) **Frequency and Speed :** Consider a coil rotating at a speed of N r.p.m. in the field of P poles. As the coil moves past a north and south pole, one complete

cycle is generated. Obviously, in one revolution of the coil, $P/2$ cycles will be generated.

Now, Frequency, f = No. of cycles/sec

= (No. of cycles/revolution) × (No. of revolutions/sec)

$$= \left(\frac{P}{2}\right) \times \left(\frac{N}{60}\right) = \frac{PN}{120}$$

$$\therefore \quad f = \frac{PN}{120}$$

For example, an a.c. generator having 10 poles and running at 600 r.p.m. will generate alternating voltage and current whose frequency is :

$$f = \frac{PN}{120} = \frac{10 \times 600}{120} = 50 \text{ Hz}$$

5.5. VALUES OF ALTERNATING VOLTAGE AND CURRENT

The voltage and current in d.c. system are constant so that there is no problem of specifying their magnitudes. Whereas, in a.c. system, the alternating voltage and current vary from instant to instant. So the question arises how to express the magnitude of alternating voltage and current. The following three ways are adopted to express the magnitude of these quantities.

(i) Peak Value

(ii) Average value or Mean value

(iii) Effective value or R.M.S. value

The r.m.s. value of an alternating quantity (voltage or current) represents the real magnitude. Whereas, the peak and average values are important in some of the engineering applications.

1. Maximum Value

The maximum value attained by an alternating quantity during one cycle is called its maximum value. This is also called peak value or crest value or amplitude. A sinusoidal alternating quantity obtains its maximum value at 90°. The maximum value of an alternating voltage and currnent is represented by E_m and I_m. The knowledge of maximum value is important in case of testing dielectric strength of insulating materials.

2. Average Value

The arithmatic average of all the instantaneous values considered of an alternating quantity (current or voltage) over one cycle is called average value.

In case of symmetrical waves the +ve half is exactly equal to –ve half, therefore, the average value over a complete cycle is zero. Since work is being done by the current in the +ve as well as in –ve half cycle, therefore, average value is determined regardless of signs. Hence, to determine average value of alternating quantities having symmetrical waves, only (+ve half) cycle is considered. The average value of alternating current or voltage can be found in two ways.

(i) **Mid Ordinate Method :** Divide the +ve half cycle into n number of equal parts as shown in Fig. 5.8. Let i_1, i_2, i_3 .... i_n be the mid-ordinates.

$$\text{Average value of current, } I_{av} = \text{mean of mid-ordinates}$$

$$= \frac{i_1 + i_2 + i_3 + ... + i_n}{n}$$

$$= \frac{\text{Area of alternation}}{\text{base}}$$

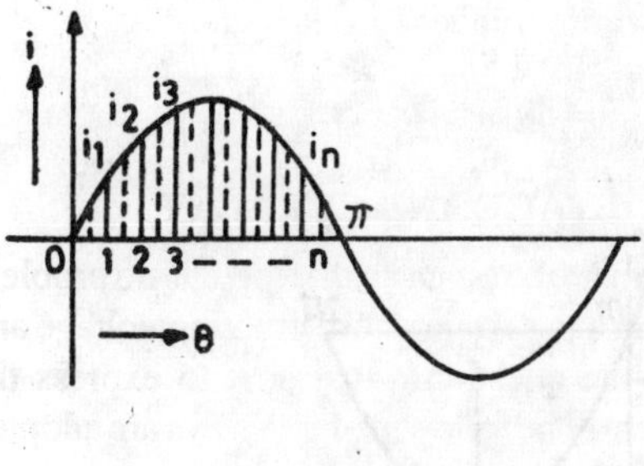

Fig. 5.8.

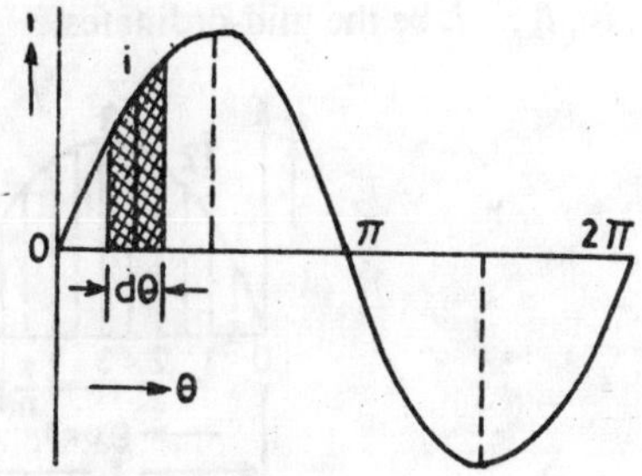

Fig. 5.9.

(ii) **Analytical Method for Average value of Sinusoidal Current :** The alternating current varying sinusoidally, as shown in Fig. 5.9, is given by the equation :

$$i = I_m \sin \theta$$

Consider an elementary strip of thickness $d\theta$ in the +ve half cycle, i be its mid-ordinates. Then

$$\text{Area of strip} = i\, d\theta$$

$$\text{Area of half cycle} = \int_0^{\pi} i\, d\theta = \int_0^{\pi} I_m \sin\theta\, d\theta$$

$$= I_m (-\cos\theta)_0^{\pi}$$

$$= I_m [-(\cos\pi - \cos 0)]$$

$$= I_m [-(-1 - 1)] = 2\, I_m$$

$$\therefore \quad \text{Average value, } I_{av} = \frac{\text{Area of alternation}}{\text{base}} = \frac{2\, I_m}{\pi}$$

$$= \frac{I_m}{\pi/2} = 0.637\, I_m$$

3. Effective or Root Mean Square Value

That steady current which when flows through a resistor of known resistance for a given time produces the same amount of heat as produced by the alternating current when flows through the same resistor for the same time is called effective or r.m.s. value of the alternating current.

(i) **Mid Ordinate Method :** Let i be the alternating current flowing through a resistor of resistance R for time t seconds which produces the same amount of heat as produced by I_{eff} (a direct current). The base of one alternation is divided into n equal parts, as shown in Fig. 5.10, so that each interval is of $\frac{t}{n}$ seconds. Let $i_1, i_2, i_3 .... i_n$ be the mid-ordinates.

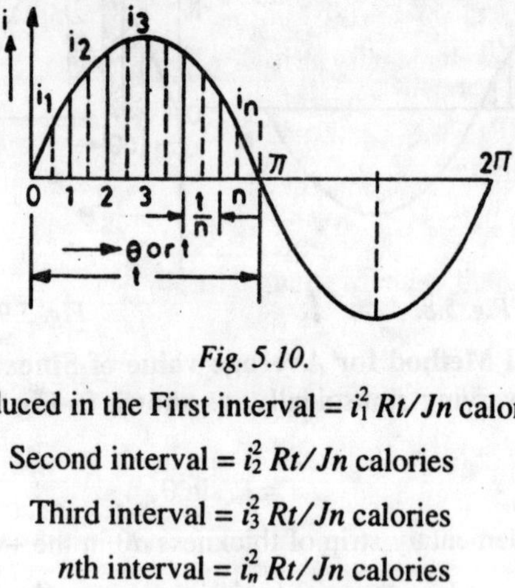

Fig. 5.10.

Then heat produced in the First interval = $i_1^2 \, Rt/Jn$ calories

Second interval = $i_2^2 \, Rt/Jn$ calories

Third interval = $i_3^2 \, Rt/Jn$ calories

nth interval = $i_n^2 \, Rt/Jn$ calories

$$\text{Total heat produced} = \frac{Rt}{J}\left(\frac{i_1^2 + i_2^2 + i_3^2 + ... + i_n^2}{n}\right) \text{calories} \quad ...(i)$$

Since I_{eff} is considered as the effective value of this current.

Then total heat produced by this current = $I_{eff}^2 \, Rt/J$ calories ...*(ii)*

Equating eqn. *(i)* and *(ii)*, we get

$$\frac{I_{eff}^2 \, Rt}{J} = \frac{Rt}{J}\left(\frac{i_1^2 + i_2^2 + i_3^2 + ... + i_n^2}{n}\right)$$

or
$$I_{eff} = \sqrt{\frac{i_1^2 + i_2^2 + i_3^2 + ... + i_n^2}{n}}$$

$$+ \sqrt{\text{mean of squares of instantaneous value}}$$

or I_{eff} = Square root of mean of squares of instantaneous values

= root-mean-square (r.m.s.) value

For example, if we say that 5A alternating current is flowing through a circuit, it means the r.m.s. value of alternating current which flows through the circuit is 5A. It transfers the same amount of energy as is transferred by 5A d.c..

The ammeters and voltmeters record the r.m.s. values of alternating currents and voltages respectively. The domestic single-phase a.c. supply is 230 V, 50 Hz. Where 230 V is the r.m.s. value of alternating voltage.

(ii) **Analytical Method :** An alternating current varying sinusoidally is given by the equation;

$$i = I_m \sin \theta$$

To determine the r.m.s. value, the squared wave of the alternating current is drawn as shown in Fig. 5.11.

Considering an elementary strip of thickness $d\theta$ in the first half cycle of the squared wave. Let i^2 be its mid- ordinate. Then

Area of strip = $i^2\, d\theta$

Area of first half cycle of squared wave

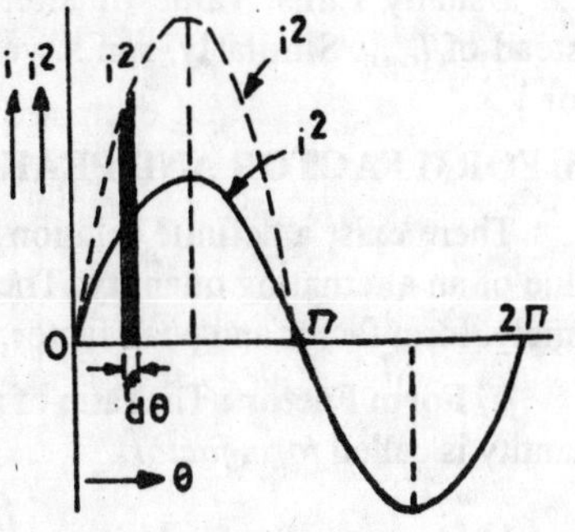

Fig. 5.11.

$$= \int_0^{\pi} i^2\, d\theta = \int_0^{\pi} (I_m \sin \theta)^2\, d\theta$$

$$= I_m^2 \int_0^{\pi} \sin^2 \theta\, d\theta$$

$$= I_m^2 \int_0^{\pi} \frac{1 - \cos 2\theta}{2}\, d\theta$$

$$\because \left(\sin^2 \theta = \frac{1 - \cos 2\theta}{2} \right)$$

$$= \frac{I_m^2}{2} \int_0^{\pi} (1 - \cos 2\theta)\, d\theta = \frac{I_m^2}{2} \left(\theta - \frac{\sin 2\theta}{2} \right)_0^{\pi}$$

$$= \frac{I_m^2}{2} \left((\pi - 0) \cdot\cdot \frac{\sin 2\pi - \sin 0}{2} \right)$$

$$= \frac{I_m^2}{2} [(\pi - 0) - (0 - 0)] = \frac{\pi I_m^2}{2}$$

Effective or r.m.s. value, $I_{r.m.s.} = \sqrt{\dfrac{\text{Area of first half of squared wave}}{\text{base}}}$

$$= \sqrt{\frac{\pi I^2 m}{2\pi}} = \frac{I_m}{\sqrt{2}}$$

or $$I_{r.m.s.} = \frac{I_m}{\sqrt{2}} = 0.707\, I_m$$

Similarly $$E_{r.m.s.} = 0.707\, E_m$$

Usually r.m.s. value of alternating current is simply represented by I instead of $I_{r.m.s.}$. Similarly, r.m.s. value of alternating voltage is represented by E or V.

5.6. FORM FACTOR AND PEAK FACTOR

There exist a definite relation among the AV value, r.m.s. value and peak value of an alternating quantity. The relationship is expressed by the two factors namely, form factor and peak factor.

(i) **Form Factor :** The ratio of r.m.s. value to average value of an alternating quantity is called *form factor.*

Mathematically, Form factor $\dfrac{I_{r.m.s.}}{I_{av}}$ or $\dfrac{E_{r.m.s.}}{E_{av}}$

$$\text{Form Factor} = \frac{I_{r.m.s.}}{I_{av}} = \frac{I_m/\sqrt{2}}{2\, I_m/\pi} = \frac{\pi\, I_m}{2\sqrt{2}\, I_m} = 1.11$$

For sinusoidal wave form factor is always 1.11.

(ii) **Peak Factor :** The ratio of maximum value to r.m.s. value of an alternating quantity is called *peak factor.*

Mathematically, Peak Factor $= \dfrac{I_m}{I_{r.m.s.}}$ or $\dfrac{E_m}{E_{r.m.s.}}$

$$\text{Peak Factor} = \frac{I_m}{I_{r.m.s.}} = \frac{I_m}{I_m/\sqrt{2}} = \sqrt{2} = 1.4142$$

Example 5.1. An alternating current is represented by $i = 70.7 \sin 520t$. Determine *(i)* the frequency, *(ii)* the current 0.0015 second after passing through zero, increasing positively, *(iii)* r.m.s. and average value.

Solution : Comparing the given equation with the standard equation :

$$i = I_m \sin \omega t$$

$$I_m = 70.7 \text{ A}$$

and $$\sin \omega t = \sin 520t$$

$\therefore$ $$2\pi f t = 520\, t$$

or $$f = 520/2\pi = 83 \text{ Hz } \textbf{Ans.}$$

$$i = 70.7 \sin 520 \times 0.0015 = 70.7 \times 0.7033 = 49.72 \text{ A } \textbf{Ans.}$$

$$I_{r.m.s.} = I_m/\sqrt{2} = 70.7/\sqrt{2} = 50 \text{ A } \textbf{Ans.}$$

$$I_{av} = 0.637\, I_m = 0.637 \times 70.7 = 45 \text{ A } \textbf{Ans.}$$

Example 5.2. A supply voltage of 230 V, 50 Hz is fed to a residential building. Write down its equation for instantaneous value.

Solution : $v = V_m \sin \omega t$

where $V_m = \sqrt{2}\, V_{r.m.s.} = \sqrt{2} \times 230 = 325.27$ V

$$\omega = 2\pi f = 2\pi \times 50 = 314.16 \text{ rad/s}$$

$\therefore$ $v = 325.27 \sin 314.16\, t$ **Ans.**

Example 5.3. An alternating current is given by the expression $i = 50 \sin 628\, t$. Determine *(i)* maximum value of current; *(ii)* r.m.s. value of current; *(iii)* frequency of current; *(iv)* value of current after $t = 0.00625$ seconds and *(v)* time taken the current to reach a value of 20 A.

Solution : Comparing the given equation with the standard equation : $i = I_m \sin \omega t$

(i) $I_m = 50$ A **Ans.**

(ii) $I_{r.m.s.} = I_m/\sqrt{2} = 50/\sqrt{2} = 35.355$ A **Ans.**

(iii) $\omega = 628$

or $2\pi f = 628$

or $f = 628/2\pi = 100$ Hz **Ans.**

(iv) $i = 50 \sin 628 \times 0.00625 = -35.3$ A **Ans.**

(v) $20 = 50 \sin 628\, t$

or $\sin 628\, t = 20/50 = 0.8$

or $628\, t = \sin^{-1} 0.8 = 0.9273$

or $t = 1.4766 \times 10^{-3}$ s **Ans.**

Example 5.4. An alternating current of frequency 60 Hz has a maximum value of 120 A. Write down the equation for its instantaneous value. Reckoning time from the instant the current is zero and is becoming positive, find : *(i)* the instantaneous value after 1/360 seconds and *(ii)* the time taken to reach 96 A for the first time.

Solution : Instantaneous value of current is given by the equation ;

$$i = I_m \sin \omega t = I_m \sin 2\pi f t$$

$$= 120 \sin 2\pi \times 60\, t = 120 \sin 377\, t \text{ } \textbf{Ans.}$$

When the reckoning time is taken from the instant the current is zero and becoming positive, equation for current is

$$i = 120 \sin 377\, t$$

(i) At : $t = 1/360$ s; $\quad i = 120 \sin 2\,\pi \times 60 \times \frac{1}{360}$

$$= 120 \sin \pi/3 = 120 \times 0.866 = 103.92 \text{ A } \textbf{Ans.}$$

(ii) Let t seconds be the time taken to reach the current to 96 A for the first time. Then

$$96 = 120 \sin 2\pi \times 60\, t$$

or $\quad \sin 120\,\pi\, t = 96/120$

or $\quad \sin 120 \times 180^\circ \times t = 0.8$

or $\quad 120 \times 180^\circ \times t = \sin^{-1} 0.8$

or $\quad 120 \times 180^\circ \times t = 53.13^\circ$

or $\quad t = \frac{53.13}{2 \times 180 \times 60} = 0.00246$ s **Ans.**

Example 5.5. A sinusoidal voltage rises from zero to one half maximum value (V_m = 100 volts) in 0.005 seconds. Find *(i)* Time period, *(ii)* frequency.

Solution : $\quad v = V_m \sin \omega t$

$$50 = 100 \sin \omega t$$

$$\sin \omega t = \frac{50}{100} = 0.5$$

$$\omega t = \sin^{-1} 0.5 = 30^\circ$$

$$\omega t = 30 \times \frac{\pi}{180} \text{ radian}$$

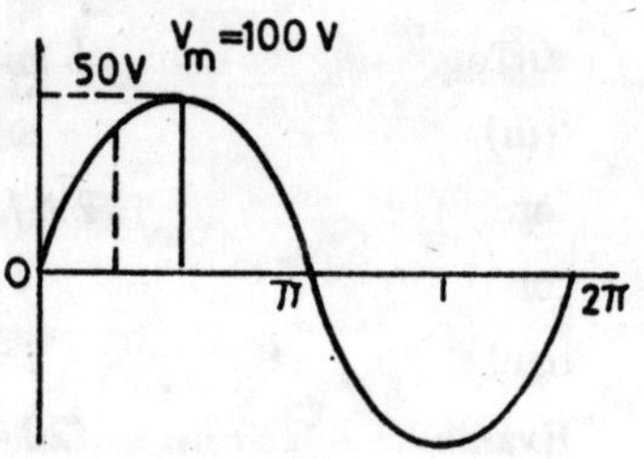

Fig. 5.12.

$$\omega = \frac{30 \times \pi}{0.005 \times 180} = 104.7 \text{ rad/sec}$$

$$\omega = 2\,\pi f$$

$$104.7 = 2 \times 3.14 \times f$$

$$f = \frac{104.7}{2 \times 3.14} = 16.67 \text{ Hz } \textbf{Ans.}$$

$$\text{Time period, } T = \frac{1}{f}$$

$$T = \frac{1}{16.67} = 0.06 \text{ sec } \textbf{Ans.}$$

Example 5.6. The equation of sine wave of voltage is given by

$$\boldsymbol{e = 100 \sin 314t}$$

Find, *(i)* Maximum value of voltage, *(ii)* Instantaneous value, when $\theta = 30^\circ$, *(iii)* Frequency.

Solution : Here, $e = 100 \sin 314\, t$

(i) Comparing with standard euqation

$$e = E_m \sin \omega t$$

We have, $E_m = 100$ is the maximum value of voltage

(ii) Instantaneous value, when $\theta = 30^\circ$

$$\therefore \quad e = 100 \sin 30 = 100 \times \frac{1}{2} = 50 \text{ volts}$$

(iii) Since $\omega = 2\pi f$

$$\therefore \quad 314 = 2 \times 3.14 \times f$$

$$\therefore \quad f = \frac{314}{2 \times 3.14} = 50 \text{ C/s}$$

Example 5.7. An alternating current of 50 Hz has a maximum value of 100 amperes. Write down the equation for its instantaneous value.

Solution : $i = I_{max} \sin \omega t$

$\therefore$ Comparing the equation with the given equation, we get

$$I_m = 100 \text{ amperes}$$

$$\text{frequency} = 50 \text{ C/s}$$

$$\therefore \quad i = 100 \sin 2 \times 3.14 \times 50 \times t$$

or

$$i = 100 \sin 314\, t$$

Example 5.8. The instantaneous value of an AC is given by the equation

$$i = 10 \sin 314\, t$$

Find, *(i)* Maximum value of current, *(ii)* frequency, *(iii)* time period, *(iv)* value of current at $t = 0.01$ sec.

Solution : The given equation is

$$i = 10 \sin 314t$$

Comparing it with the standard form

$$i = I_{max} \sin \omega t$$

(i) $\therefore$ $I_{max} = 10$ amperes

(ii) $\omega = 314$

$$\therefore \quad f = \frac{314}{2 \times 3.14}$$

or

$$f = 50 \text{ C/s}$$

(iii) Time period, $T = \frac{1}{f} = \frac{1}{50}$ 0.02 sec

(iv) When $t = 0.01$ sec

$$i = 10 \sin 3.14 \times 0.01 = 10 \sin 3.14$$

$$= 10 \sin \pi = 10 \times 0 = 0 \text{ ampere}$$

Example 5.9. An alternating current when passed through the resistance of an immersion rod placed in water for 16 minutes, it raises the temperature of water of boiling point. When a direct current of 4 A was passed through the same resistance under identical conditions, it took 25 minutes so just boil the water. Find the peak value of alternating current varying sinusoidally.

Solution : Let I be the r.m.s. value of the alternating current.

Heat produced by alternating current $= I^2 \frac{Rt}{J}$

$$= I^2 R \times 16 \times 60/J \text{ calories} \qquad ...(i)$$

Heat produced by direct current $= I_{d.c.}^2 \frac{Rt}{J}$

$$= 4 \times 4 \times R \times 25 \times 60/J \text{ calories} \qquad ...(ii)$$

Equating eqn. *(i)* and *(ii)*, we get,

$$I^2 R \times 16 \times 60/J = 4 \times 4 \times R \times 25 \times 60/J$$

or $$I^2 = 4 \times 4 \times 24/16 = 25$$

or $$I = \sqrt{25} = 5 \text{ A Ans.}$$

Example 5.10. Calculate the average value, r.m.s. value, form factor and peak factor of a periodic current wave having values for equal time intervals changing suddenly from one value to next : 0, 30, 45, 70, 90, 70, 45, 30, 0, –30, –45, –70 etc. amperes. What would be the average and the r.m.s. value of a sine wave having the same peak value ?

Solution : The periodic wave form of the alternating current is shown in Fig. 5.13.

Average value of current, $I_{av} = \frac{i_1 + i_2 + ... + i_n}{n}$ (where $n = 8$)

$$= \frac{0 + 30 + 45 + 70 + 90 + 70 + 45 + 30}{8}$$

$$= 47.5 \text{ A Ans.}$$

R.M.S. value of current, $I_{r.m.s.} = \sqrt{\frac{i_1^2 + i_2^2 + ... + i_n^2}{n}}$

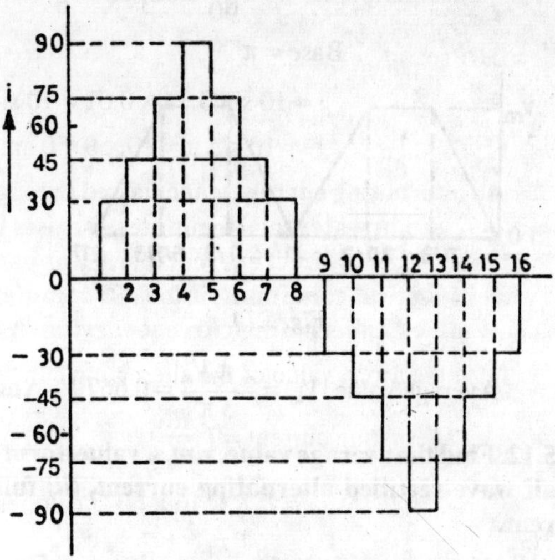

Fig. 5.13.

$$= \sqrt{\frac{0 + (30)^2 + (45)^2 + (70)^2 + (90)^2 + (70)^2 + (45)^2 + (30)^2}{8}}$$

$$= \sqrt{\frac{23750}{8}} = 54.486 \text{ A } \textbf{Ans.}$$

$$\text{Form Factor, } \frac{I_{r.m.s.}}{I_{av}} = \frac{54.486}{47.5} = 1.147 \textbf{ Ans.}$$

$$\text{Peak Factor} = \frac{I_m}{I_{r.m.s.}} = \frac{90}{54.486} = 1.652 \textbf{ Ans.}$$

Average value of a sinusoidal a.c. having peak value of 90 A;

$$I_{av} = 0.637\, I_m$$

$$= 0.637 \times 90 = 57.33 \text{ A } \textbf{Ans.}$$

$$\text{R.M.S. value, } I_{r.m.s.} = \frac{I_m}{\sqrt{2}} = \frac{90}{\sqrt{2}} = 63.64 \text{ A } \textbf{Ans.}$$

Example 5.11. Find the average value of the periodic function shown in Fig. 5.14.

Solution : $\text{Average Value, } V_{av} = \frac{\text{Area of alternation}}{\text{Base}}$

$$\text{Area of alternation} = \frac{1}{2} V_m \times \frac{\pi}{3} + V_m \left(\frac{2\pi}{3} - \frac{\pi}{3} \right) + \frac{1}{2} V_m \times \frac{\pi}{3}$$

$$= \frac{\pi V_m}{6} + \frac{\pi V_m}{3} + \frac{\pi V_m}{6} = \frac{2\pi V_m}{3}$$

Base = π

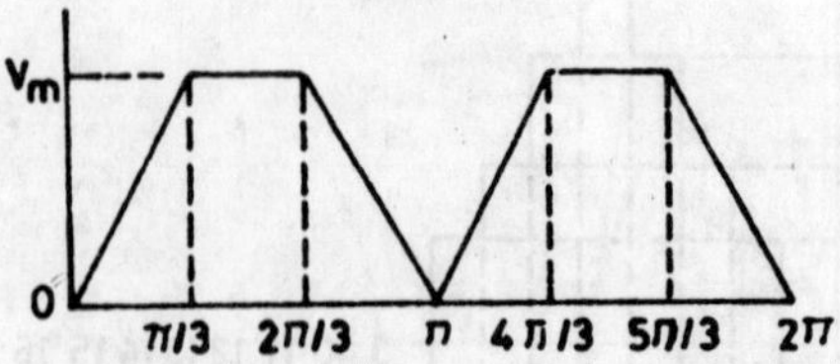

Fig. 5.14.

$$\text{Average value, } V_{av} = \frac{2\pi V_m}{3\pi} = 0.667\ V_m \textbf{ Ans.}$$

Example 5.12. Find the average value, r.m.s. value, form factor and peak factor for *(i)* half wave rectified alternating current, *(ii)* full wave rectified alternating current.

Solution : A half wave rectified a.c. wave is shown in Fig. 5.15.

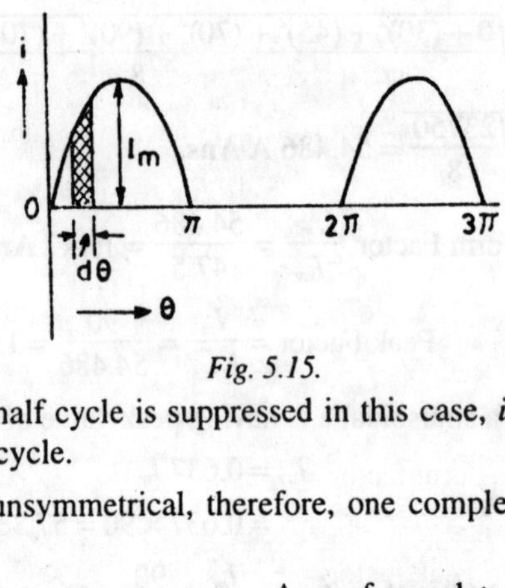

Fig. 5.15.

The negative half cycle is suppressed in this case, *i.e.*, current flows only during positive half cycle.

The wave is unsymmetrical, therefore, one complete cycle is to be considered.

(i) $$\text{Average value, } I_{av} = \frac{\text{Area of complete cycle}}{\text{base}}$$

Area of complete cycle = Area of half cycle

$$= \int_0^{\pi} i\, d\theta = \int_0^{\pi} I_m \sin\theta\, d\theta$$

$$= I_m \left[-\cos\theta \right]_0^{\pi} = 2\, I_m$$

Base = 2 π

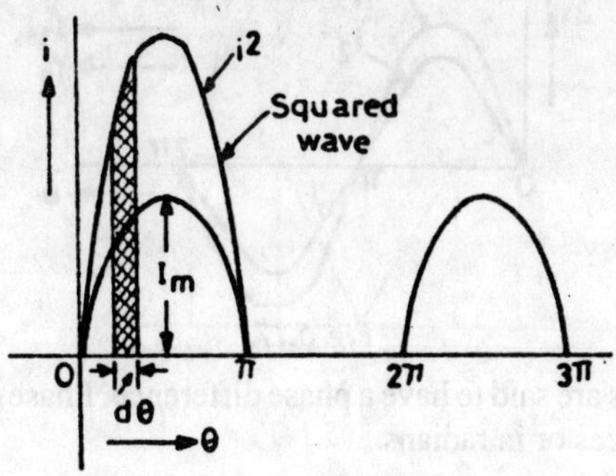

Fig. 5.16.

$$\therefore \quad I_{av} = \frac{2\, I_m}{2\,\pi} = \frac{I_m}{\pi} = 0.3183\, I_m \textbf{ Ans.}$$

(ii) The squared wave of a half wave rectified wave is shown in Fig. 5.16.

$$\text{r.m.s. value, } I_{r.m.s.} = \sqrt{\frac{\text{Area of squared wave over a complete cycle}}{\text{base}}}$$

Area of squared wave over a complete cycle = Area of squared wave in half cycle

(as negative half cycle is suppressed)

$$= I_m^2 \int_0^{\pi} \left(\frac{1 - \cos 2\theta}{2} \right) d\theta = \frac{\pi\, I_m^2}{2}$$

Base = $2\,\pi$

$$\therefore \quad I_{r.m.s.} = \sqrt{\frac{\pi\, I_m^2}{2 \times 2\pi}} = \frac{I_m}{2} \textbf{ Ans.}$$

(iii) $\text{Form factor} = \frac{I_{r.m.s.}}{I_{av}} = \frac{I_m \times \pi}{2 \times I_m} = \frac{\pi}{2} = 1.57$ **Ans.**

(iv) $\text{Peak factor} = \frac{I_m}{I_{r.m.s.}} = \frac{I_m}{I_m/2} = 2$ **Ans.**

5.7. PHASE AND PHASE DIFFERENCE

Two alternating quantities (either two voltages or two currents or one voltage and one current) are said to be in phase with each other if they reach their maximum and zero values at the same time. In Fig. 5.17 currents I_1 and I_2 are in phase since they start simultaneously and also reach their maximum and zero values simultaneously.

When two alternating quantities are considered simultaneously, the frequency being the same, they may not pass through a particular point at the same time.

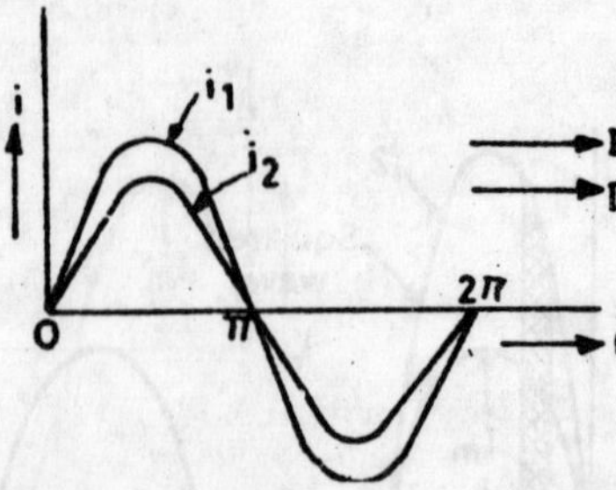

Fig. 5.17.

These two quantities are said to have a phase difference. Phase difference is always given either in degrees or in radians.

In Fig. 5.18 current I_1 represented by vector OA leads the current represented by vector OB by ϕ or current I_2 lags behind the current I_1 by ϕ (since vectors are assumed to rotate in the anti- clock wise direction).

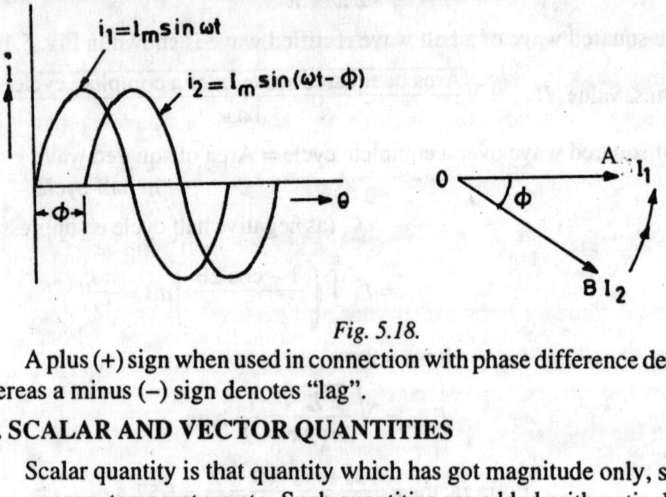

Fig. 5.18.

A plus (+) sign when used in connection with phase difference denotes "lead" whereas a minus (–) sign denotes "lag".

5.8. SCALAR AND VECTOR QUANTITIES

Scalar quantity is that quantity which has got magnitude only, such as time, mass, energy, temperature etc. Such quantities are added arithmatically such as 2 kgms plus 4 kgms is equal to 6 kgms.

Vector quantity is that quantity which has magnitude as well as direction, *e.g.*, force, acceleration, current, voltage etc. These are graphically represented by straight lines called vectors. The length of line represents the magnitude of the quantity and arrow head placed at one end gives direction in which the quantity acts.

Example 5.13. An alternating current of 50 Hz has maximum value of 20 Amps. Represents the alternating current *(i)* by equation, *(ii)* by graph *(iii)* by vector diagram.

Solution : *(i)* $i = I_{max} \sin \omega t$ ...*(i)*

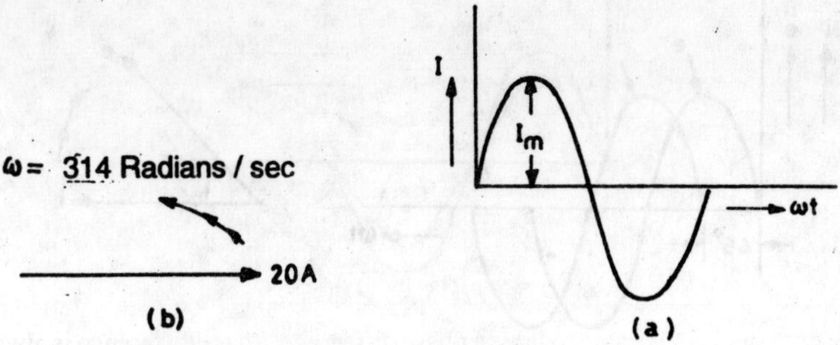

Fig. 5.19.

Here I_{max} = 20 amperes, $\omega = 2\pi f = 2\pi \times 50 = 314$ radians/sec

Substituting the values of I_{max} and ω in equation *(i)*, we get

$$\therefore \quad i = 20 \sin 314\, t$$

(ii) Representation by graph is shown in Fig. 5.19 *(a)*.

(iii) Representation by vector diagram is shown in Fig. 5.19 *(b)*.

Example 5.14. An alternating voltage and current are represented by the following equations :

$$i = 15 \sin 314\, t$$

$$e = 200 \sin (314\, t + 45°)$$

Find

(*i*) Phase difference between current and voltage,

(*ii*) Does the current lead or lag behind the voltage?

(*iii*) Show the current and voltage graphically and vectorially.

(*iv*) Find the frequency.

Solution : *(i)*

$$i = 15 \sin 314\, t$$

$$e = 200 \sin (314\, t + 45°)$$

The phase difference between current and voltage = 45°

(ii) Current lags behind the voltage by 45°.

(iii) See Fig. 5.20.

(iv) $\omega = 314$

or $2\pi f = 314$

$$\therefore \quad \text{Frequency, } f = \frac{314}{2\pi} = 50 \text{ Hz.}$$

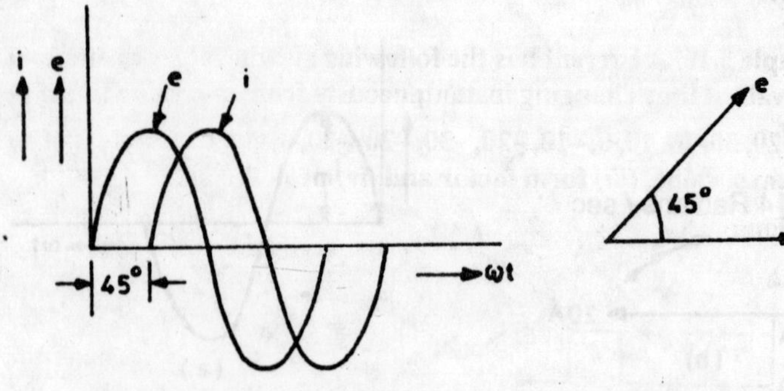

Phase Diagram *Vector Diagram*

Fig. 5.20.

Example 5.15. Two alternating currents I_1 and I_2 of frequency 50 Hz have maximum values of 20 amperes and 30 amperes respectively. Current I_1 leads the current I_2 by 60°.

(i) Write down equations for the current I_1 and I_2.

(ii) Show the currents graphically and vectorially.

Solution : Maximum value of $I_1 = 20$ amperes.

Maximum value of $I_2 = 30$ amperes

Phase angle, $\phi = 60°$

Frequency, $f = 50$ Hz

$\therefore$ $\omega = 2\,\pi f = 2\,\pi \times 50 = 314$ radians/sec

Now $i = I_{max} \sin \omega t$

If current I_1 is taken as a reference vector then the equations for I_1 and I_2 are

$i_1 = 20 \sin 314\, t$ and $i_2 = 30 \sin (314\, t - 60°)$

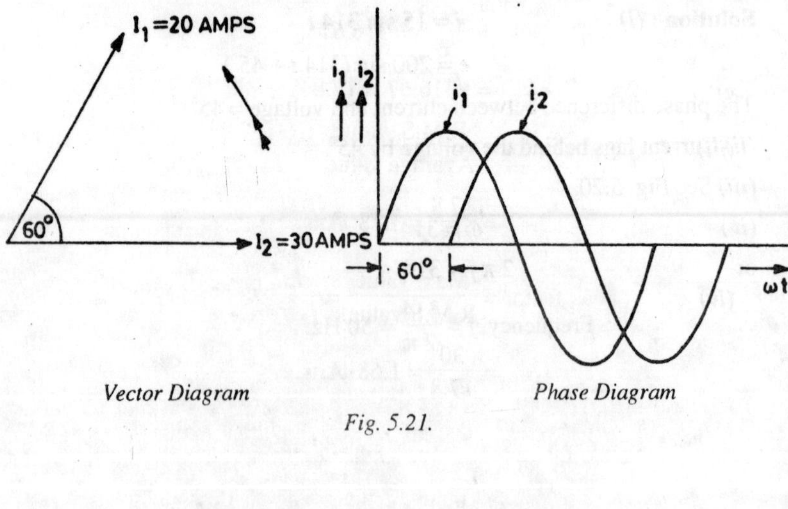

Vector Diagram *Phase Diagram*

Fig. 5.21.

Example 5.16. A current has the following steady values in amperes for equal intervals of time changing instantaneously from one value to the next.

0, 10, 20, 30, 20, 10, 0, –10, –20, –30, –20, –10, 0, etc. Calculate *(i)* average value, *(ii)* r.m.s. value, *(iii)* form factor and *(iv)* peak factor.

Solution :

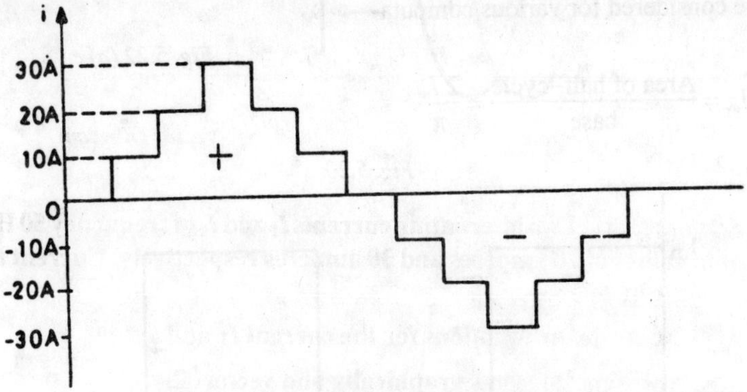

Fig. 5.22.

(i)
$$I_{av} = \frac{i_1 + i_2 + i_3 + i_4 + i_5 + i_6}{6}$$

$$= \frac{0 + 10 + 20 + 30 + 20 + 10}{6} = \frac{90}{6} = 15 \text{ A}$$

(ii)
$$I^2_{r.m.s.} = \frac{i_1^2 + i_2^2 + i_3^2 + i_4^2 + i_5^2 + i_6^2}{6}$$

$$= \frac{0^2 + 10^2 + 20^2 + 30^2 + 20^2 + 10^2}{6}$$

$$= \frac{1900}{6} = 316.67$$

or
$$I_{r.m.s.} = \sqrt{316.67} = 17.8 \text{ A}$$

(iii)
$$\text{Form factor} = \frac{\text{R.M.S. value}}{\text{Average value}} = \frac{I_{R.M.S.}}{I_{av}}$$

$$= \frac{17.8}{15} = 1.18 \textbf{ Ans.}$$

(iv)
$$\text{Peak factor} = \frac{\text{Max. value}}{\text{R.M.S.value}} = \frac{I_m}{I_{R.M.S.}}$$

$$= \frac{30}{17.8} = 1.68 \textbf{ Ans.}$$

Full-Wave rectified A.C.

Fig. 5.22 *(a)* shows full-wave rectified a.c. in which both half- cycles appear in the output, *i.e.,* current flows in the same direction for both half-cycles. Since the wave is symmetrical, half-cycle may be considered for various computations.

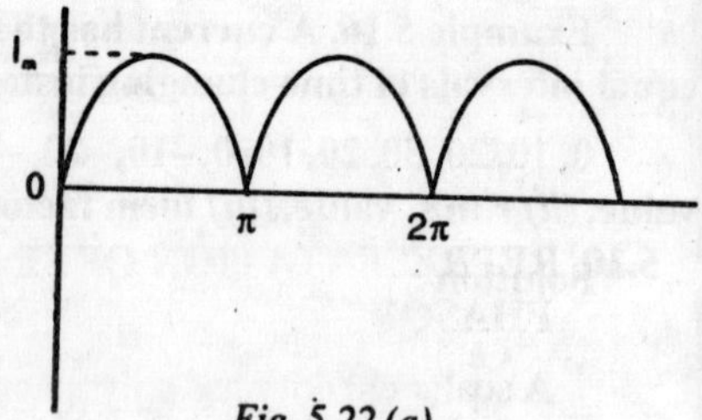

Fig. 5.22 *(a)*

$$I_{av} = \frac{\text{Area of half–cycle}}{\text{base}} = \frac{2\,I_m}{\pi}$$

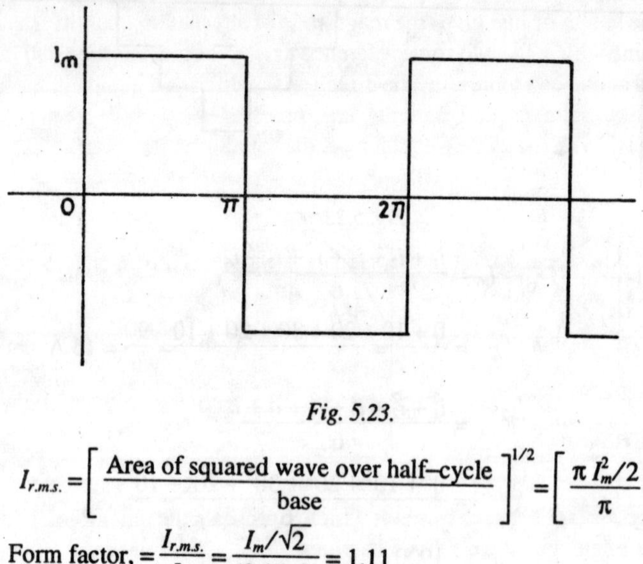

Fig. 5.23.

$$I_{r.m.s.} = \left[\frac{\text{Area of squared wave over half–cycle}}{\text{base}}\right]^{1/2} = \left[\frac{\pi\, I_m^2/2}{\pi}\right]^{1/2} = \frac{I_m}{\sqrt{2}}$$

$$\text{Form factor,} = \frac{I_{r.m.s.}}{I_{av}} = \frac{I_m/\sqrt{2}}{(2/\pi)\,I_m} = 1.11$$

$$\text{Peak factor,} = \frac{I_m}{I_{r.m.s.}} = \frac{I_m}{I_m/\sqrt{2}} = 1.414$$

5.9. R.M.S VALUE OF SQUARE WAVE

An alternating symmetrical square wave is shown in Fig. 5.23.

Instantaneous value of a square wave is given by

$$i = I_m \text{ for } \theta = 0 \text{ to } \pi = -\,I_m \text{ for } \theta = \pi \text{ to } 2\pi$$

$$I_{r.m.s.}^2 = \frac{1}{\pi}\int_0^{\pi} i^2\, d\theta = \frac{1}{\pi}\int_0^{\pi} I_m\, d\theta$$

$$= \frac{I_m^2}{\pi} \int_0^{\pi} 1 \,.\, d\theta = \frac{I_m^2}{\pi} \left| \theta \right|_0^{\pi} = \frac{I_m^2}{\pi} (\pi - 0)$$

$$\therefore \qquad I_{r.m.s.} = I_m^2$$

5.10. REPRESENTATION OF SINUSOIDAL QUANTITIES BY PHASORS

A scalar quantity has only magnitude and no direction such as 10 metres, 15 seconds and 10 rupees etc.

A vector quantity is that which has both magnitude and direction. Such quantities are known only when particulars of their magnitude, direction or the sense in which they act, are given. They are represented by straight lines called vectors. The length of line gives the magnitude of alternating quantity, the inclination of the line with respect to some reference axis gives the direction and the arrow head placed at one end represents the direction in which that quantity acts. The AC quantities like voltages and currents are represented by such vectors rotating anti-clockwise with the same frequency as that of alternating quantity.

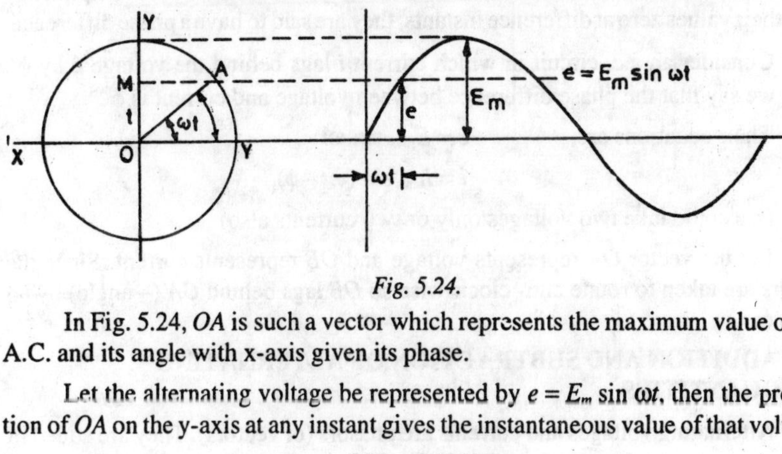

Fig. 5.24.

In Fig. 5.24, *OA* is such a vector which reprsents the maximum value of the A.C. and its angle with x-axis given its phase.

Let the alternating voltage be represented by $e = E_m \sin \omega t$, then the projection of *OA* on the y-axis at any instant gives the instantaneous value of that voltage.

$$OM = OA \sin \omega t$$

$$\therefore \qquad e = OA \sin \omega t$$

$$\text{or} \qquad e = E_m \sin \omega t$$

Here it should be noted that a line *OA* can be made to represent alternating voltage or current, if it satisfies the following conditions.

(i) Its length should be equal to the maximum value of sinusoidal alternating voltage to a suitable scale.

(ii) It should be in the horizontal position at the same instant when the alternating quantity is zero and increasing positively.

(iii) Its angular velocity (ω) should be such that it completes one revolution in the same time as taken by the alternating quantity to complete one cycle.

Phase : The phase of an alternating quantity (current or voltage) at any particular instant is defined as the fractional part of a cycle through which the quantity has advanced from selected origin.

In practice, we are more concerned with the phase difference between the two alternating quantities.

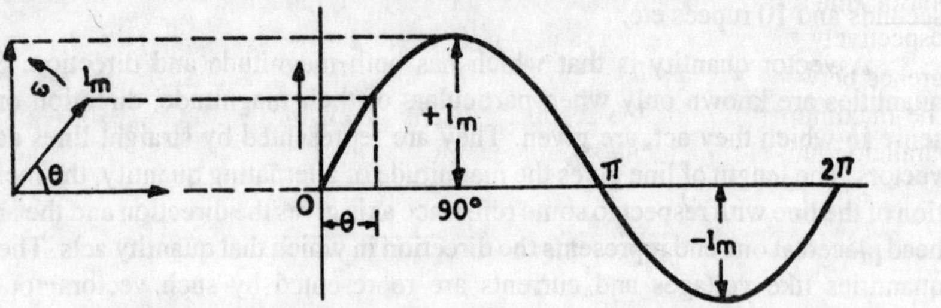

Fig. 5.25.

Phase Difference : When two alternating quantities of the same frequency have their values zero at difference instants, they are said to have a phase difference.

Consider an a.c. circuit in which current i lags behind the voltage e by ϕ^o. Then we say that the phase difference between voltage and current is ϕ^o.

Their equations are, $e = E_{max} \sin \omega t$

$i = I_{max} \sin (\omega t - \phi)$

(we could take two voltages only or two currents also)

Let the vector OA represents voltage and OB represents current. Since the vectors are taken to rotate anti- clock-wise so OB lags behind OA (– angle) by an angle ϕ.

5.11. ADDITION AND SUBTRACTION OF ALTERNATING QUANTITIES

Alternating voltages and currents are phasors (or vectors). They are added in the same manner as forces are added. Only phasors of the same kind are added, *i.e.*, either all the voltages are added or all the currents are added, but voltages and currents are never added to each other.

The following methods are used :

1. Parallelogram Method
2. Method of Components.

1. Parallelogram Method

This method is used for adding two phasors (vectors) at a time. If the two quantities are represented in magnitude and direc-

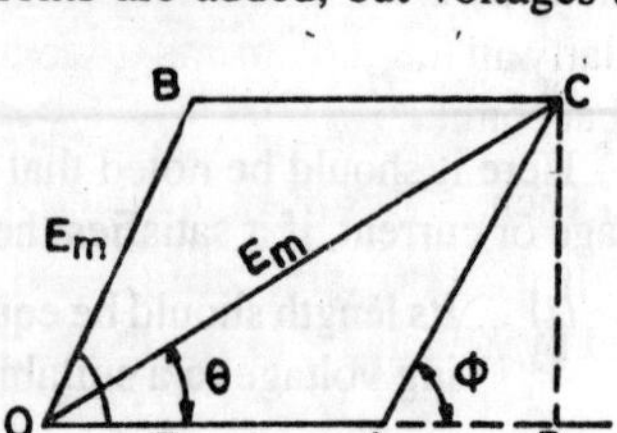

Fig. 5.26.

tion by the two adjacent sides of the parallelogram, then the diagonal of the parallelogram gives the maximum value of the resultant.

Let us suppose that we are to add two phasors as shown in the Fig. 5.26.

$$e_1 = E_{1m} \sin \omega t$$

$$e_2 = E_{2m} \sin (\omega t + \phi)$$

The maximum values of the two voltages E_{1m} and E_{2m} are represented in phasor diagram as the two sides of the parelleloram, *i.e.*, sides OA and OB respectively of the parallelogram $OACB$ as shown in Fig. 5.26. The phase difference between the two is ϕ^o. Voltage E_{1m} leads the voltage E_{1m} by an angle ϕ^o. The maximum value of the resultant E_{2m} is given by the diagonal OC of the parallelogram. In right angled triangle ODC.

$$OC = \sqrt{OD^2 + DC}$$

or
$$OC = \sqrt{(OA + OD)^2 + (DC)^2}$$

∴
$$E_{2m} = \sqrt{(E_{1m} + E_{2m} \cos \phi)^2 + (E_{2m} \sin \phi)^2}$$

or
$$= \sqrt{E_{1m}^2 + E_{2m}^2 [\cos^2 \phi + \sin^2 \phi] + 2 E_{1m} . E_{2m} \cos \phi}$$

$$= \sqrt{E_{1m}^2 + E_{2m}^2 + 2 E_{1m} . E_{2m} \cos \phi}$$

Here,
$$\text{phase angle } \theta = \tan^{-1} \frac{CD}{OD}$$

∴
$$\theta = \tan^{-1} \left[\frac{E_{2m} \sin \phi}{E_m + E_{1m} \cos \phi} \right]$$

∴
$$\text{Phase angle, } \theta = \tan^{-1} \frac{E_{1m} \sin \phi}{E_{1m} + E_{2m} \cos \phi}$$

The instantaneous value of the resultant voltage is given by $e_r = E_{2m} \sin (\omega t + \theta)$.

This method is not convenient when the number of phasors is more than two.

2. Method of Components

This method is very convenient for adding two or more phasors. Each phasor is resolved into its horizontal and vertical components. Then all the horizontal components are added up algebraically to give resultant horizontal component. Similarly all the vertical components are added algebraically to obtain the resultant vertical component.

Then, $\text{Resultant} = \sqrt{(\text{Horizontal component})^2 + (\text{Vertical component})^2}$

If the horizontal component, vertical component and Resultant are represented by X, Y and R respectively, then

$$R = \sqrt{X^2 + Y^2}$$

The phase angle of resultant, $\tan\phi = \dfrac{Y}{X}$

$$\therefore \qquad \phi = \tan^{-1}\frac{Y}{X}$$

The instantaneous value of resultant voltage is given by

$$e_r = E_{2m}\sin(\omega t + \phi)$$

Example 5.17. Two currents I_1 and I_2 are represented by $I_1 = 25\sin\omega t$ and $I_2 = \sin(\omega t + \pi/4)$. Find the resultant current in the form of $I_p = I_{max}\sin(\omega t + \phi)$.

Solution : Let vector OA represents current I_1 and vector OB represents current I_2 leading the current I_1 by $\pi/4$ or 45°. (shown in Fig. 5.27.) Then the resultant I_2 can be determined by using the formula.

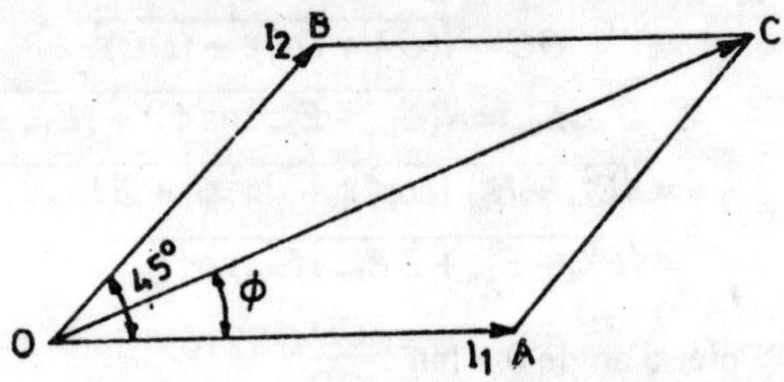

Fig. 5.27.

$$I_2 = \sqrt{I_1^2 + I_2^2 + 2\,I_1\,.\,I_2\,.\cos 45^\circ}$$

$$= \sqrt{(25)^2 + (10)^2 + 2\,.\,25\,.\,10\,.\cos 45^\circ}$$

$$= \sqrt{625 + 100 + 500 \times 0.707}$$

$$\therefore \qquad I_2 = 32.84 \text{ (Max. value)}$$

$$\tan\phi = \frac{I_{2m}\,.\sin\phi}{I_{1m} + I_{2m}\cos\phi}$$

$$= \frac{10\sin 45^\circ}{25 + 10\cos 45^\circ}$$

$$\text{or} \qquad \tan\phi = \frac{10 \times 0.707}{25 + 10 \times 0.707} = \frac{7.07}{32.07}$$

$$\therefore \qquad \tan\phi = 0.2204$$

$$\phi = \tan^{-1} 0.2204 = 12.4^\circ$$

$$= 32.84\sin(\omega t + 12.4^\circ) \textbf{ Ans.}$$

Example 5.18. Add the following currents $I_1 = 7\sin\omega t$ and $I_2 = 10\sin(\omega t + \pi/3)$.

Solution : Let vector OA gives the maximum value of current I_1 and vector OB gives the maximu value of current I_2 leading the current I_1 by 60°. OC gives the diagonal of the parallelogram $OACB$ and also the resultant current I_3. Then using the formula.,

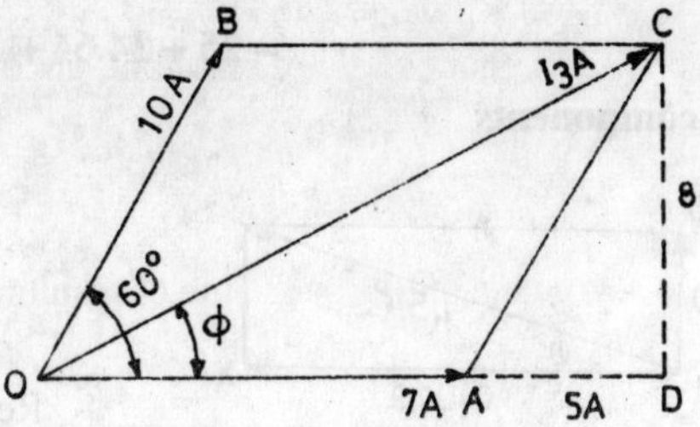

Fig. 5.28.

$$I_2 = \sqrt{I_1^2 + I_2^2 + 2 \,.\, I_1 \,.\, I_2 \,.\, \cos 60}$$

$$= \sqrt{(7)^2 + (10)^2 + 2 \,.\, 7 \,.\, 10 \,.\, \cos 60}$$

$$= \sqrt{49 + 100 + 140 \times 0.5}$$

or $$I_2 = \sqrt{219} = 14.8$$

$$\tan \phi = \frac{I_{2m} \sin \phi}{I_{1m} + I_{2m} \cos \phi} = \frac{10 \times \sin 60}{7 + 1 \times \cos 60^\circ}$$

$$= \frac{10 \times 0.866}{7 + 0 \times 0.5} = \frac{8.66}{12} = 0.7216$$

or $$\phi = \tan^{-1} 0.7216 = 35.8^\circ$$

$\therefore$ $$I_r = I_3 \sin (\omega t + \phi)$$

$$= 14.8 \sin (\omega t + 35.8^\circ) \textbf{ Ans.}$$

Example 5.19. Three impedences in series have voltage drops across them which can be represented by

$$\boldsymbol{e_1 = 25 \sin \omega t}$$

$$\boldsymbol{e_2 = 40 \sin (\omega t + \pi/6)}$$

$$\boldsymbol{e_3 = 20 \sin (\omega t - \pi/4)}$$

Sketch the phasor diagram to represent the three voltages and their resultant. Express the resultant voltage in the same form as the three individual voltages expressed above.

Solution : Phasors representing the maximum values of the three voltages are shown in Fig. 5.29.

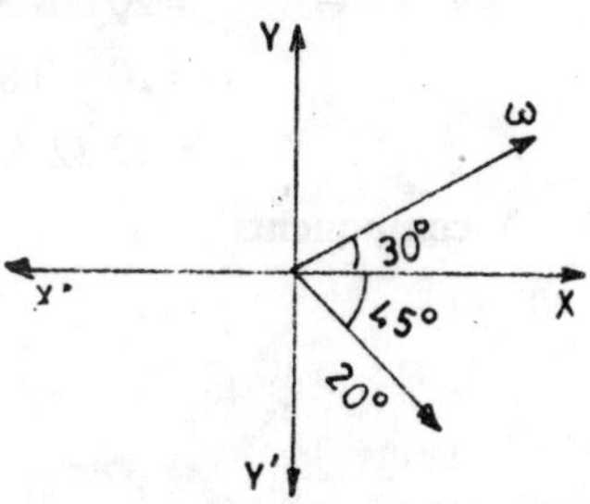

Fig. 5.29.

Resolving the phasors into horizontal and vertical components, we get

X-components

$= 25 \cos 0^\circ + 40 \cos 30^\circ + 20 \cos 45^\circ$

$= 25 \times 1 + 40 \times 0.866 + 20 \times 0.707$

$$= 25 + 34.64 + 14.14 = 73.78 \text{ V}$$

Y-components

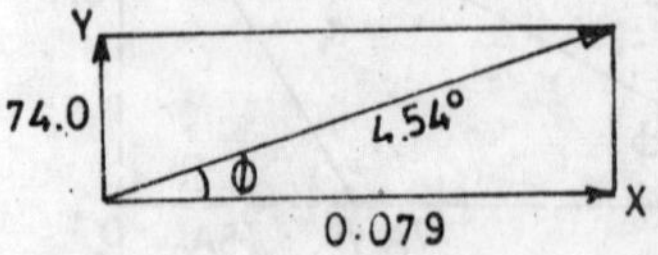

Fig. 5.30.

$$= 25 \sin 0 + 40 \sin 30^\circ - 20 \sin 45^\circ$$
$$= 0 + 40 \times 0.5 - 20 \times 0.707$$
$$= 20^\circ - 14.14 = 5.86 \text{ V}$$

Resultant, $R = \sqrt{X^2 + Y_2}$

$$= \sqrt{(73.78)^2 + (5.86)^2}$$
$$= \sqrt{5443.48 + 34.33}$$
$$= 74.0 \text{ volts}$$

$$\tan \phi = \frac{Y}{X} = \frac{5.86}{73.78} = 0.079$$

$\therefore$ $\phi = \tan^{-1} 0.079 = 4.54^\circ$

$\therefore$ $e_r = 74.0 \sin (\omega t + 4.54^\circ)$

Example 5.20. Two currents I_1 and I_2 are given by the expressions

$$i_1 = 20 \sin (314t + \pi/6)$$
$$i_2 = 10 \sin (314t - \pi/3)$$

Find $(i_1 - i_2)$ and express the answer in the same form as individual currents. Find also the r.m.s. value and frequency of the resultant current.

Solution : The two currents are represented by phasors as shown in Fig. 5.31. Since current i_2 is to be subtracted from i_1, therefore, reverse the phasor I_{2m}. Now find the vector sum of I_{1m} and reversed vector of I_{2m}.

Resolving into their horizontal and vertical components

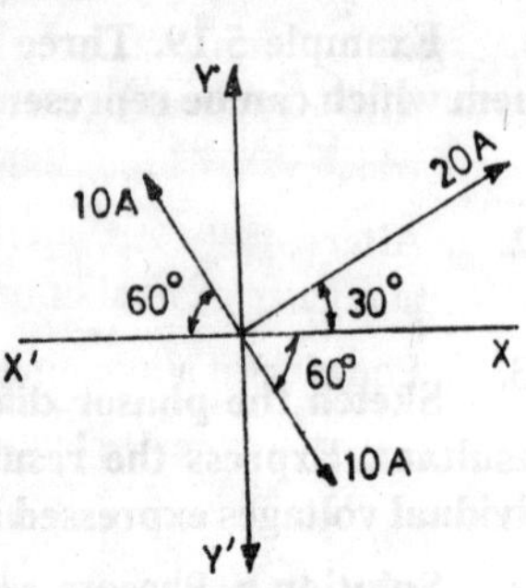

Fig. 5.31.

X-components $= I_1 \cos \phi_1 - I_2 \cos \phi_2$

$$= 20 \cos 30^\circ - 10 \cos 60^\circ$$
$$= 20 \times 0.866 - 10 \times 0.5$$
$$= 12.32 \text{ A}$$

Y-components $= I_1 \sin \phi_1 + I_2 \sin \phi_2$

$$= 20 \sin 30^\circ + 10 \sin 60^\circ$$
$$= 20 \times 0.5 + 10 \times 0.866$$
$$= 18.66 \text{ amperes}$$

$\therefore$ Maximum value of resultant current

$$I_r = \sqrt{X^2 + Y^2}$$

$$= \sqrt{(12.32)^2 + (18.66)^2}$$

$$= 22.36 \text{ amperes}$$

Let ϕ be the angle which the resultant current makes with the x-axis.

$$\therefore \quad \tan \phi = \frac{Y}{X} = \frac{18.66}{12.32} = 1.5146$$

$$\text{or} \quad \phi = \tan^{-1} 1.5146 = 56.56^\circ$$

Hence the expression for the resultant current is

$$I_r = 22.36 \sin (314t + 56.56^\circ) \textbf{ Ans.}$$

$$\text{r.m.s. value of current} = \frac{I_{max}}{\sqrt{2}}$$

$$= \frac{22.36}{\sqrt{2}} = 15.81 \text{ A } \textbf{Ans.}$$

$$\text{Since,} \quad \omega = 2\pi f$$

$$\therefore \quad 314 = 2 \times 3.14 \times f$$

$$\text{or} \quad f = \frac{314}{2 \times 3.14} = 50$$

$\therefore$ Frequency, $f = 50$ Hz **Ans.**

HIGHLIGHTS

1. **Alternating Quantity :** An alternating quantity (i.e., voltage or current) which varies according to sine of angle is known as alternating quantity.
2. **Alternating Current :** The current which changes its polarity and magnitude at regular interval of time is called alternating current.
3. Equation of alternating voltage and alternating current.

   Alternating voltage $e = E_m \sin \omega t$

   Alternating current $i = I_m \sin \omega t$
4. **Wave Form :** It is a graph plotted between alternating quantity and time.
5. **Cycle :** When alternating quantity goes through a complete set of positive and negative value is called cycle.
6. **Alternation :** One half cycle (+ve of –ve) is called one alternation.
7. **Instantaneous Value :** The value of alternating quantity at any instant is called instantaneous value. It is represented by small letter *e* or *i* respectively.
8. **Time Period :** The time taken in seconds by an alternating quantity to complete one cycle is called time period and denoted by *T*.

9. **Frequency :** The number of cycle made per second by an alternating quantity is called frequency. It is measured in Hertz (Hz) and is denoted by *f*. In India the supply frequency is 50 Hz.

10. **Maximum value or Amplitude :** It is the maximum value attained by an alternating quantity during one cycle is called its amplitude or peak value or maximum value.

11. **Average Value :** The arthmatic average of all the instantaneous value of voltage or current over one cycle is called average value

$$e_{av} = \frac{e_1 + e_2 + ... + e_n}{n}$$

For sinusoidal wave $e_{av} = \frac{2E_m}{\pi} = 0.637\ E_m$

12. **RMS value :** It is also called effective value. It is defined as the steady value of current which when flows through a given circuit for a given time produces the same amount of heat as produced by alternating current when flowing through the same circuit for same time is called r.m.s. value. It is generally represented by capital letter such *V* for r.m.s. value of voltage and *I* for r.m.s. value of current.

$$I_{r.m.s.} = \sqrt{\frac{i_1^2 + i_2^2 + i_3^3 + ... + i_n^2}{n}}$$

For sinusoidal $I_{r.m.s.} = \frac{I_m}{\sqrt{2}} = 0.707\ I_m$

13. **Peak Factor :** It is the ratio of maximum value of alternating quantity to the r.m.s. value.

$$K_P = \frac{E_m}{E_{r.m.s.}}$$

Its value is 1.414.

14. **Form Factor :** It is the ratio of r.m.s. value to average value of an alternating quantity is called form factor. It is denoted by K_f. Its value for sinusoidal quantity is 1.11.

SHORT QUESTIONS

1. Why alternating voltage are added vertorially ?
2. Why r.m.s. value is called effective value ?
3. Why the value of form factor is 1.11 for sinusoidal quantity ?
4. What factors are responsible for the induced emf ?
5. What are the characteristics of the sine a.c. wave form.

MULTIPLE CHOICE QUESTIONS

1. The equation of instantaneous value of emf is

(a) $e = E_{rms} \sin \omega t$ (b) $e = E_{max}$

(c) $e = E_{max} \sin \omega t$

2. The frequency of an alternating voltage is given by

(a) $f = \frac{w}{2\pi}$ (b) $f = \frac{W}{2\,n \sin \theta}$

(c) $f = \frac{W}{2\,\pi f}$

3. One cycle of a sinusoidal wave is equal to

(a) $\frac{\pi}{2}$ radians (b) π radians

(c) $2\,\pi$ radians

4. The rms of sinusoidal alternating voltage is given by the relation

(a) $E_{rms} = \frac{E_{max}}{2}$ (b) $E_{rms} = \frac{E_{max}}{\sqrt{2}}$

(c) $E_{rms} = \sqrt{2}\, E_{max}$ (d) $E_{rms} = \frac{E_{max}}{\sqrt{2}}$

5. For a sinusoidal wave the value of peak factor is

(a) 1.212 (b) 1.11

(c) 1.414 (d) 1.155

6. A sine-wave has a frequency of 50 Hz. Its angular velocity is ...... rad/sec.

(a) $50/\pi$ (b) $50/2\,\pi$

(c) $50\,\pi$ (d) $100\,\pi$

5. If the frequency of supply is 50 Hz, the time period of one cycle will be

(a) 0.02 sec (b) 0.1667 sec

(c) 16.67 sec

ANSWERS

1. (c) 2. (a) 3. (c) 4. (a) 5. (c)
6. (d) 7. (a)

TRUE-FALSE TYPE QUESTIONS

1. The value of instantaneous emf will be maximum when $\omega t = 0$. (False)

2. One cycle equals 2π radians. (True)

3. One cycle is also called an alternation. (False)

4. Peak factor is defined as the ratio of maximum value and rms value. (True)

EXERCISE

1. How an alternating emf is generated? What is its wave shape ?
2. Derive an expression for the instantaneous value of alternating current varying sinusoidally.
3. In an a.c. system, define the following :

   (i) Amplitide, *(ii)* Cycle, *(iii)* Time period, *(iv)* Frequency,

   (v) phase difference.
4. Define the following:

   (i) Instantaneous value, *(ii)* Angular velcoity, *(iii)* r.m.s. value,

   (iv) Average value.
5. Define the form factor and peak factor.
6. Prove that for a sine wave

$$\text{r.m.s. value} = \frac{\text{Max. value}}{\sqrt{2}}$$

7. How are the two alternating quantities added?
8. What do you understand by phase and phase difference in case of alternating quantities.
9. Derive an expression for average and r.m.s. value of a sinusoidally varying quantity.
10. Explain how sinusoidal quantities can be represented by vectors.

PROBLEMS

1. An alternating quantity is represented by $e = 50 \sin 157\, t$. Find

   (i) Maximum value of voltage

   (ii) R.M.S. value

   (iii) Frequency

   (**Ans.** 50 V, 35.35 V, 25 Hz)
2. An alternating current of frequency 60 Hz has an r.m.s. value of 84.853 A.

   (i) Write down the equation for its instantaneous value.

   (ii) Find the instantaneous value of current 0.002083 sec after passing through zero and becoming positive.

   (iii) Time taken to reack 96 A, for the first time after passing through positive maximum value.

   (**Ans.** *(i)* 120 sin 377 *t*, *(ii)* 84.85 A, *(iii)* 1.7 ms)
3. An alternating voltage is represented by :

$$e = 141.4 \sin 377\, t$$

Find *(i)* the maximum value, *(ii)* frequency, *(iii)* time period and *(iv)* the instantaneous value of voltage when t is 3 ms.

(Ans. *(i)* 141.4 V, *(ii)* 60 Hz, *(iii)* 16.67 ms, *(iv)* 127.8 V)

4. An alternating current of frequency 50 Hz has a maximum value of $200\sqrt{2}$ A. Reckning the time from the instant the current is zero and becoming positive, find the time taken by the current to reach a value of 141.4 A for a first and second time.

(Ans. 1.67 ms; 21.67 ms

5. For a full wave rectified alternating current determine :

(i) Average value, *(ii)* r.m.s. value, *(iii)* Form factor, *(iv)* Peak factor.

(Ans. 0.637 I_m, 0.707 I_m, 1.11, 1.4142)

6. Calculate the r.m.s. value of a periodic voltage having the following values for equal intervals changing from one value to the next, 0, 10, 30, 60, 30, 10, 0, –10, –30, –60, –30, –10 etc.

(Ans. 30.55 V)

7. Two currents taken by two parallel circuits are 12 A in phase with the applied voltage and 20 A lagging 30° behind the applied voltage respectively. Determine the current taken by the combined circuit and its phase with respect to the applied voltage.

(Ans. 31 A, 1

8. Three voltages represented by $e_1 = 20 \sin(\omega t + \pi/3)$, e_2 - $\cos(\omega t + \pi/6)$ and $e_3 = 40 \sin(\omega t - \pi/6)$ act together in a circuit. Fi expression for the resultant voltage.

[Ans. 37.7 sin (ωt + 38.1)]

9. The following four currents are fed into a common conductor

$$i_1 = 50 \sin \omega t$$
$$i_2 = 25 \sin(\omega t + \pi/3)$$
$$i_3 = 40 \cos \omega t$$
$$i_4 = 30 \sin(\omega t - \pi/2)$$

Find the expression for the resultant in the form $i = I_m \sin(\omega t \pm \phi)^2$.

(Ans. 70.05 sin (ωt + 26.2°))

10. Two currents i_1 and i_2 are given by the expression $i_1 = 15 \sin(\omega t + \pi/3)$ and $i_2 = 5 \sin(\omega t - \pi/3)$. Find $(i_1 - i_2)$ and express the answer in the same form as individual currents.

(Ans. 18.028 sin (ωt + 73.90))

5.12. PHYSICAL CONCEPT OF ALTERNATING VOLTAGE APPLIED TO PURE RESISTANCE

When an alternating voltage is applied across pure resistance, then current flow in one direction for the first half-cycle of the supply and then flow in the opposite direction during the next half-cycle, thus constituting alternating current in the circuit.

Consider a circuit containing a pure resistance of $R\ \Omega$ connected across an alternating voltage source. Let the alternating voltage be given by the equation ;

$$v = V_m \sin \omega t \quad ...(i)$$

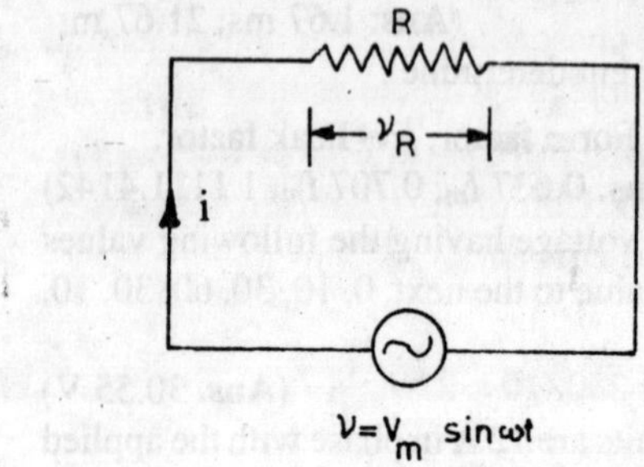

Fig. 5.32.

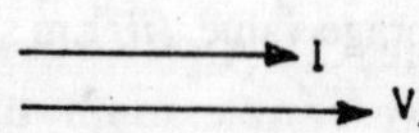

Fig. 5.33.

As a result of this voltage, an alternating current *i* will flow in the circuit. The applied voltage has to overcome the drop in the resistance only, *i.e.*,

$$v = iR$$

or $$i = \frac{v}{R}$$

Substituting the value of *v*, we get

$$i = \frac{V_m}{R} \sin \omega t \quad ...(ii)$$

The value of *i* will be maximum when $\sin \omega t = 1$

$\therefore$ $$I_m = V_m / R$$

$\therefore$ eqn. *(ii)* becomes, $$i = I_m \sin \omega t \quad ...(iii)$$

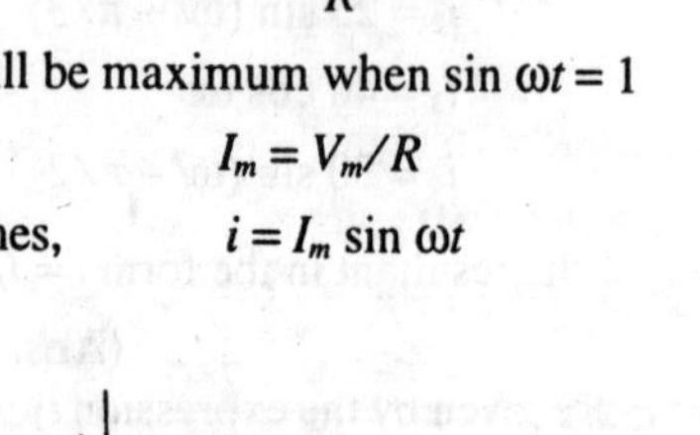

Fig. 5.34.

$$\text{In terms of r.m.s. values, } \frac{V_m}{\sqrt{2}} = \frac{I_m}{\sqrt{2}} \times R$$

or $$V_R = IR$$

(i) **Phase Angle :** It is clear from eqns. *(i)* and *(iii)* that the applied voltage and the circuit current are in phase with each other, *i.e.*, they pass through their zero values at the same instant and attain their positive and negative peaks at the same instant. This is also indicated by the phasor diagram shown in Fig. 5.33. Note that r.m.s. values have been used in drawing the phasor diagram. The wave diagram shown in Fig. 5.34 also shows that currrent is in phase with the applied voltage.

Example 5.21. An a.c. circuit consists of a pure resistance of 10 Ω and is connected across an a.c. supply of 230 V, 50 Hz. Calculate *(i)* current, *(ii)* power consumed and *(iii)* equations for voltage and current.

Solution :

(i) Current, $I = V/R = 230/10 = 23$ A

(ii) Power, $P = VI = 230 \times 23 = 5290$ W

(iii) Now,

$$V_m = \sqrt{2}\ V = \sqrt{2} \times 230 = 325.27 \text{ volts}$$

$$I_m = \sqrt{2}\ I = \sqrt{2} \times 23 = 32.52 \text{ A}$$

$$\omega = 2\pi f = 2\pi \times 50 = 314$$

∴ Equations of voltage and current are :

$$e = 325.27 \sin 314\,t \text{ volts}$$

$$i = 32.52 \sin 314\,t \text{ Amps}$$

Example 5.22. An a.c. circuit consists of a pure resistance of 25 ohms and is connected across an a.c. supply of 250 V, 50 Hz. Calculate, *(i)* current, *(ii)* power consumed and *(iii)* write down the equations of voltage and current.

Solution :

Resistance of the circuit, $R = 25$ ohms

Applied voltage, $V = 250$ volts

Supply frequency, $f = 50$ Hz

(i) Current in the circuit $I = \frac{V}{R} = \frac{250}{25} = 10$ A **Ans.**

(ii) Power consumed $P = V \cdot I = 250 \times 10 = 2500$ watts **Ans.**

(iii) Max. value of voltage $V_m = \sqrt{2} \times V_{r.m.s.}$

$$= 1.4142 \times 250$$

$$= 353.5 \text{ volts}$$

∴ $$v = 353.5 \sin 314\,t$$ **Ans.**

Maximum value of current $= I_m = \sqrt{2} \times I_{r.m.s.}$

$$= 1.4142 \times 10 = 14.142 \text{ Amp.}$$

$\therefore$ $$i = 14.142 \sin 314\, t \text{ Ans.}$$

5.13. ALTERNATING VOLTAGE APPLIED TO PURE INDUCTANCE

When an alternating current flows through a pure inductive coil, a back emf ($L\, di/dt$) is induced due to the inductance of the coil. This back emf at every instant opposes the change in current through the coil. Since there is no ohmic drop, the applied voltage has to overcome the back emf only.

$\therefore$ Applied alternating voltage = Back emf

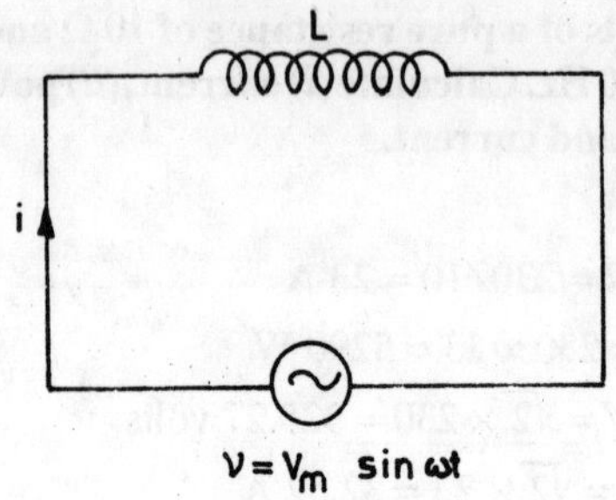

Fig. 5.35.

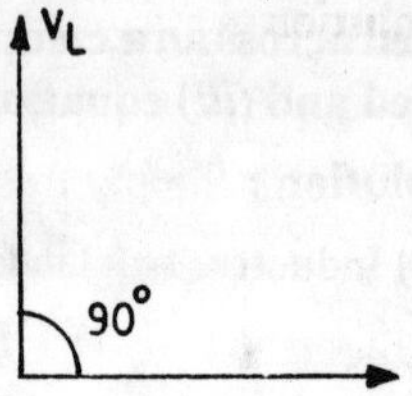

Fig. 5.36.

Consider an alternating voltage applied to a pure inductance of L henry as shown in Fig. 5.35. Let the equation of the applied alternating voltage be :

$$v = V_m \sin \omega t \qquad ...(i)$$

Clearly, $$V_m \sin \omega t = L \frac{di}{dt}$$

or $$di = \frac{V_m}{L} \sin \omega t \, dt$$

Integrating both sides, we get

$$i = \frac{V_m}{L} \int \sin \omega t \, dt$$

$$= \frac{V_m}{\omega L} (-\cos \omega t)$$

or $$i = \frac{V_m}{\omega L} \sin (\omega t - \pi/2) \qquad ...(ii)$$

The value of i will be maximum (*i.e.*, I_m) when $\sin (\omega t - \pi/2)$ is unity.

$\therefore$ $$I_m = V_m/\omega L$$

Substituting the value of $V_m/\omega L = I_m$ in eqn. *(ii)*, we get

$$i = I_m \sin (\omega t - \pi/2) \qquad ...(iii)$$

(i) **Phase Angle :** It is clear from eqns. *(i)* and *(iii)* that current lags behind the voltage by $\pi/2$ radians or 90°. Hence in a pure inductance, current lags the voltage by 90°. This is also indicated by the phasor diagram shown in Fig. 5.36. Note that r.m.s. values have been used in drawing the phasor diagram. The wave diagram also shows that current lags the voltage by 90°.

Example 5.23. A 0.014 H choke coil with negligible resistance is connected to a 220 V, 50 Hz supply. Determine *(i)* inductive reactance of the coil, *(ii)* current flowing through the coil, *(iii)* power consumed by the coil. Write down the expressions for voltage and current and give the wave and phasor diagrams.

Solution :

$$\text{Applied voltage, } V_{r.m.s.} = 220 \text{ V}$$

$$\text{Supply frequency, } f = 50 \text{ Hz}$$

(i) Inductive reactance of coil, $X_L = 2\,\pi f L$

$$= 2\,\pi \times 50 \times 0.014 = 4.4\ \Omega \textbf{ Ans.}$$

(ii) Current flowing through the coil, $I = \dfrac{V}{X_L} = \dfrac{220}{4.4} = 50$ A **Ans.**

(iii) Power consumed by the coil, P = zero **Ans.**

Expression for the voltage, $v = V_{max} \sin 2\,\pi f t$

$$= 220\sqrt{2} \sin 2\,\pi \times 50\, t = 311 \sin 314\, t \textbf{ Ans.}$$

Expression for the current,

$$i = I_{max} \sin\left(2\,\pi f t - \frac{\pi}{2}\right)$$

$$= 50\sqrt{2} \sin\left(314\, t - \frac{\pi}{2}\right) \textbf{Ans.}$$

Wave and phasor diagramas are similar to those shown in Fig. 5.36 *(a)* and Fig. 5.36 *(b)* respectively.

5.14. ALTERNATING VOLTAGE APPLIED TO PURE CAPACITANCE

When a capacitor is connected to an alternating voltage the capacitor is charged in one direction and then in other direction as the voltage reverses.

Let an alternating voltage represented by $v = V_{max} \sin \omega t$ be applied across a capacitor of capacitance C farads.

Change on the Capacitor at any instant

$$q = C\, V_{max} \sin \omega t$$

Since the capacitor current is equal to the rate of change of charge, therefore, the capacitor current may be obtained by differentiating the above equation

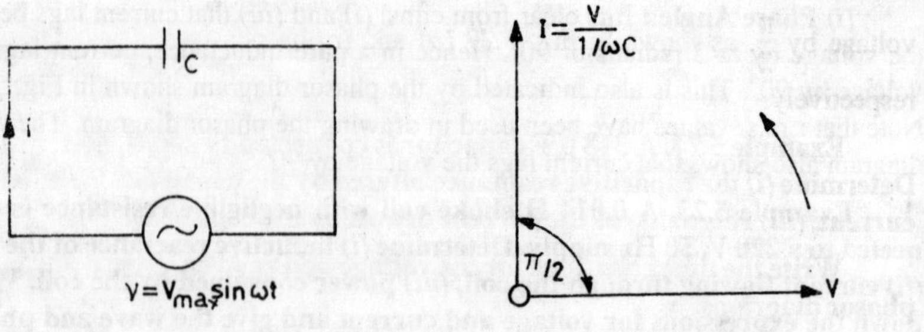

(a) *(b)*

Fig. 5.37.

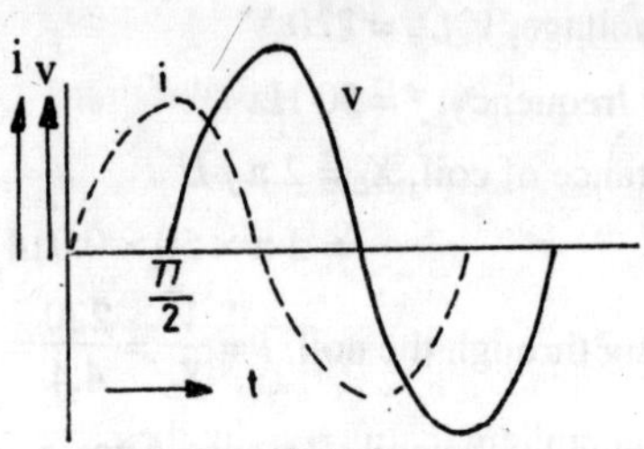

Wave diagram

Fig. 5.37. (c)

$$i = \frac{dq}{dt} = \frac{d}{dt}\,[C\,V_{\max}\;\sin\omega t]$$

$$= \omega\,C\,V_{\max}\cos\omega t$$

$$= \frac{V_{\max}}{1/\omega\,C}\sin\left(\omega t + \frac{\pi}{2}\right)$$

Current is maximum (I_m) when $\sin\left(\omega t + \frac{\pi}{2}\right) = 1$

$$\therefore \qquad I_{\max} = \frac{V_{\max}}{1/\omega\,C}$$

Substituting $\frac{V_{\max}}{1/\omega\,C} = I_{\max}$ in the above expression for instantaneous current, we get

$$i = I_{\max}\sin\left(\omega t + \frac{\pi}{2}\right)$$

From the expressions of instantaneous applied voltage and instantaneous current flowing through capacitance, it is observed that the current leads the applied voltage by $\frac{\pi}{2}$. as shown in Fig. 5.37 *(b)* and *(c)* by phasor and wave diagrams respectively.

Example 5.24. A 318 μ F capacitor is connected to 200 V, 50 Hz supply. Determine *(i)* the capacitive reactance offered by the capacitor, *(ii)* maximum current, *(iii)* rms value of the current drawn by the capacitor.

Write the time equations for the voltage and current. Give the wave and phasor diagrams.

Solution :

Applied voltage, $V_{r.m.s.} = 200$ V

Supply frequency, $f = 50$ Hz

$$\omega = 2\pi f = 314 \text{ radians}$$

(i) Capacitance reactance, $X_C = \frac{1}{\omega C}$

$$= \frac{1}{314 \times 318 \times 10^{-6}} = 10\ \Omega \textbf{ Ans.}$$

(ii) Maximum value of current drawn by the capacitor,

$$I_{max} = \frac{V_{max}}{X_C} = \frac{\sqrt{2} \times 200}{10} = 28.28 \text{ A } \textbf{Ans.}$$

(iii) RMS value of current drawn by the capacitor,

$$I_{r.m.s.} = \frac{V_{r.m.s.}}{X_C} = \frac{200}{10} = 20 \text{ A } \textbf{Ans.}$$

Expression for the voltage $V = V_{max} \sin \omega t$ *Assuming that at time* $t = 0$, $v = 0$ *and is increasing in positive direction*

$$= 200\sqrt{2} \sin 314\, t \textbf{ Ans.}$$

Expression for current, $i = I_{max} \sin\left(\omega t + \frac{\pi}{2}\right)$

$$= 28.28 \sin\left(314t + \frac{\pi}{2}\right) \textbf{Ans.}$$

Phasor and wave diagram are similar to those shown in Fig. 5.37 *(b)* and 5.37 *(c)* respectively.

5.15. INDUCTIVE REACTANCE

Inductance not only causes the current to lag behind the voltage but it also limit the magnitude of current in the circuit. We know that

$$I_m = \frac{V_m}{\omega L}$$

$$\frac{V_m}{I_m} = \omega L$$

where ωL is the opposition offered by the inductance to the flow of current and is called the inductive reactance X_L of the coil and measured in ohms.

$$\therefore \quad I_m = \frac{V_m}{X_L}$$

$$\frac{I_m}{\sqrt{2}} = \frac{V_m/\sqrt{2}}{X_L}$$

$$I = \frac{V}{X_L}$$

where inductive reactance $X_L = \omega L = 2\,\pi f L$

X_L in ohm if L is in henry and f in Hertz.

Also inductive reactance is directly proportional to frequency. The inductive reactance (X_L) is considered as positive.

5.16. CAPACITIVE REACTANCE

Capacitance not only causes the voltage to lag behind current but it also limits the magnitude of current.

We know that $\quad I_m = \omega L\, V_m$

$$\frac{V_m}{I_m} = \frac{1}{\omega L}$$

If V and I are the r.m.s. values, then

$$\frac{V_m/\sqrt{2}}{I_m/\sqrt{2}} = \frac{1}{\omega L}$$

$$\frac{V}{I} = \frac{1}{\omega L}$$

Clearly, the opposition offered by the capacitance to current flow is $\frac{1}{\omega L}$. This quantity $\frac{1}{\omega L}$ is called the capacitive reactance X_L of the capacitor. It has same dimention as resistance and is, therefore, measured in ohm.

$$\therefore \quad I = \frac{V}{X_L}$$

where X_L is the capacitive reactance.

$$X_L = \frac{1}{\omega L} = \frac{1}{2\,\pi f C}$$

X_L will be in ohm, if C is in farad and f in Hz.

Also the capacitive reactance is inversly proportional to the frequency. The capacitive reactance (X_L) is considered as negative.

5.17. ALTERNATING VOLTAGE APPLIED TO RESISTANCE AND INDUCTANCE IN SERIES

A circuit that contains a pure resistance R ohms connected in series with a coil having pure inductance of L henry is known as R—L series circuit. This is the general case we met across in actual practice.

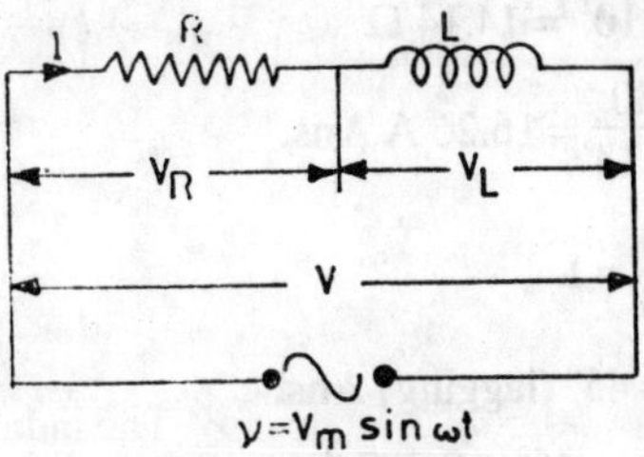

Fig. 5.38.

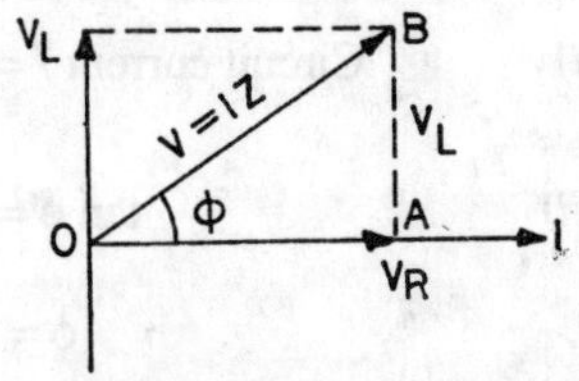

Fig. 5.39.

An R—L series circuit and its phasor diagram is shown in Fig. 5.38 and 5.39 respectively. To draw the phasor diagram, current I is taken as the reference vector. Voltage drop in resistance V_R (= IR) is taken in phase with current vector, whereas, voltage drop in inductive reactance V_L (= IX_L) is taken 90° ahead of the current vector (since current lags behind the voltage by 90° in pure inductive circuit). The vector sum of these two voltages (drops) is equal to the applied voltage V.

Now $$V_R = IR$$

and $$V_L = IX_L \qquad \text{(where } X_L = 2\pi f L)$$

In right angle triangle OAB

$$V = \sqrt{(V_R)^2 + (V_L)^2} = \sqrt{(IR)^2 + (IX_L)^2} = I\sqrt{R^2 + X_L^2}$$

or $$I = \frac{V}{\sqrt{R^2 + X_L^2}} = \frac{V}{Z}$$

where $Z = \sqrt{R^2 + X_L^2}$ is the total opposition offered to the flow of alternating current by an R—L series circuit and is called *impedance* of the circuit. It is measured in ohms.

Phase Angle

From the phasor diagram in Fig. 5.39, it is clear that current in this circuit lags behind the applied voltage by an angle ϕ, called *phase angle*.

From phasor diagram; $$\tan \phi = \frac{V_L}{V_R} = \frac{IX_L}{IR} = \frac{X_L}{R}$$

or $$\phi = \tan^{-1} X_L/R$$

Example 5.25. A coil having a resistance of 10 ohms and inductance of 31.8 mH is connected to 230 V, 50 Hz supply. Calculate : *(i)* the circuit current, *(ii)* phase angle, *(iii)* power factor and *(iv)* power consumed.

Solution : Resistance of coil, $R = 10$ ohms

Inductive Reactance, $X_L = 2\,\pi f \,.\, L$

or $X_L = 2 \times 3.14 \times 50 \times 31.8 \times 10^{-3} = 10\,\Omega$

$\therefore$ Coil Impedence, $Z = \sqrt{10^2 + 10^2} = 14.14\,\Omega$

(i) Circuit current $I = \dfrac{V}{Z} = \dfrac{230}{14.14} = 16.26$ A **Ans.**

(ii) $\tan\phi = \dfrac{X_L}{R} = \dfrac{10}{10} = 1$

$\therefore$ $\phi = \tan^{-1} 1 = 45°$ (lagging) **Ans.**

(iii) Power factor $= \cos\phi = \cos 45° = 0.707$ **Ans.**

(iv) Power consumed, $P = V \,.\, I \,.\, \cos\phi$

$= 230 \times 16.26 \times 0.707$

$= 2644$ Watts **Ans.**

Example 5.26. A coil having a resistance of 10 ohms and inductive reactance of 25 ohms is connected across a 220 volt, 50 Hz supply. Find *(i)* Angle of phase difference, *(ii)* Impedence of the coil, *(iii)* Current, *(iv)* Angle of phase difference, *(v)* Power factor, *(vi)* True power and *(vii)* Sketch the vector diagram.

Solution :

Resistance of coil, $R = 10\,\Omega$

Inductive reactance of coil, $X_L = 25\,\Omega$

Supply voltage, $V = 220$ Volts

Supply frequency, $f = 50$ Hz

(i) Since, $X_L = \omega L$

$\therefore$ $L = \dfrac{X_L}{\omega} = \dfrac{X_L}{2\,\pi f}$

$\therefore$ $L = \dfrac{25}{2 \times 3.14 \times 50} = 79.58$ mH **Ans.**

(ii) Impedence, $Z = \sqrt{R^2 + X_L^2}$

$= \sqrt{(10)^2 + (25)^2} = 26.925\,\Omega$ **Ans.**

(iii) Current, $I = \dfrac{V}{Z} = \dfrac{220}{26.925} = 8.17$ A **Ans.**

(iv) $$\tan\phi = \frac{X_L}{R} = \frac{25}{10} = 2.5$$

$$\therefore \quad \phi = \tan^{-1} 2.5 = 68.198° \textbf{ Ans.}$$

(v) $$\text{Power factor} = \frac{R}{Z} = \frac{10}{26.925} = 0.3714 \text{ (lagging) } \textbf{Ans.}$$

(vi) $$\text{True Power} = V \,.\, I \cos\phi$$

$$= 220 \times 8.17 \times 0.3714 = 667.3 \text{ Watts } \textbf{Ans.}$$

(vii) Vector diagram :

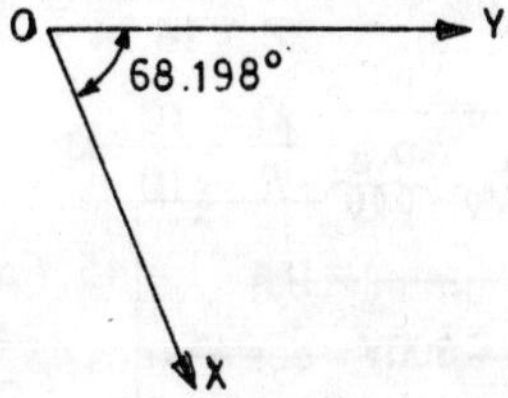

Fig. 5.40.

Example 5.27. A coil when connected across a 100 V d.c. supply dissipates 800 watts of power. When connected across a 100 V a.c. supply of frequency 50 Hz, it dissipates 400 watts. Calculate the value of resistance and inductance of the coil.

Solution : (For d.c. supply, because frequency is 0, therefore, inductive reactance $X_L = 2\pi f \,.\, L = 0$).

Hence for d.c. supply, we have to take into account the resistance only. But for a.c. supply, both resistance and inductance of the coil offer opposition to the flow of current.

For d.c. supply :

$$\text{Resistance of the coil, } R = \frac{V^2}{P} = \frac{100 \times 100}{800} = 12.5\ \Omega \textbf{ Ans.}$$

For a.c. supply :

$$\text{Power consumed } P = V \,.\, I \cos\phi$$

$$\therefore \quad P = V \,.\, \frac{V}{Z} \,.\, \frac{R}{Z} = \frac{V^2 \,.\, R}{Z^2}$$

$$\therefore \text{ Impedence of the coil, } Z = \sqrt{\frac{V^2 \,.\, R}{P}}$$

$$= \sqrt{\frac{100^2 \times 12.5}{400}} = 17.67\ \Omega$$

$$\text{Reactance of the coil, } X_L = \sqrt{Z^2 - R^2}$$

$$= \sqrt{(17.67)^2 - (12.5)^2} = 12.48\ \Omega$$

$$\therefore \qquad L = \frac{X_L}{2\pi f} = \frac{12.48}{2\pi \times 50} = 0.039 \text{ henry } \textbf{Ans.}$$

Example 5.28. A circuit consists of a pure resistance and a coil connected in series. Power dissipated in the resistance and in the coil are 1000 watts and 250 watts respectively. Voltage drop across the resistance and the coil are 200 V and 300 V respectively. Determine the reactance of the coil and the supply voltage.

Solution : The pure resistance *R* and coil have been shown connected in series in Fig. 5.41 *(a)*. The phasor diagram showing different voltages is shown in Fig. 5.41 *(b)*.

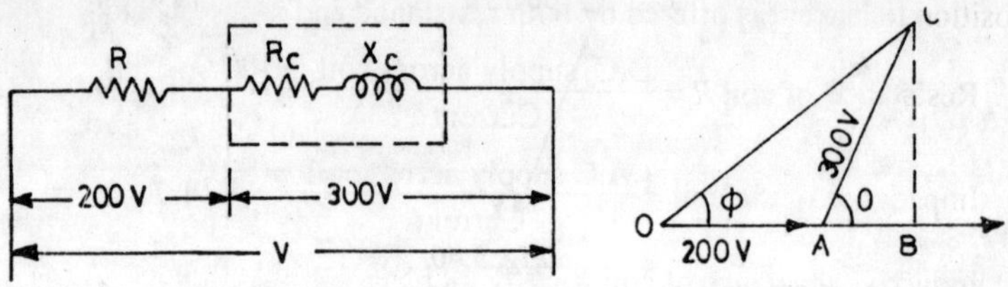

(a) *(b)*

Fig. 5.41

Power consumed in resistance = 1000 W

Power consumed in coil = 250 W

Voltage across resistance *R* = 200 V

Voltage across the coil V_C = 300 V

$$\therefore \qquad R = \frac{V^2}{P} = \frac{200 \times 200}{1000} = 40\ \Omega$$

Voltage across pure resistance = 200 V

$$\therefore \qquad \text{Current in resistance, } I = \frac{200}{40} = 5 \text{ amperes}$$

Power consumed in coil, $P = I^2 . R_C$

$$\therefore \qquad 250 = (5)^2 \times R_C$$

$$\therefore \qquad R_C = \frac{250}{25} = 10\ \Omega$$

$$\text{Coil impedence, } Z_{coil} = \frac{300}{5} = 60\ \Omega$$

Inductive reactance of the coil $= \sqrt{60^2 - 10^2} = 59.16\ \Omega$ **Ans.**

Total impedence of the circuit, $Z = \sqrt{(R + R_C)^2 + (X_L)^2}$

$$= \sqrt{(40 + 10)^2 + (59.16)^2} = 77.46 \text{ ohms}$$

Total voltage across the circuit $V = IZ$

$$V = 5 \times 77.46 = 387.3 \text{ volts } \textbf{Ans.}$$

Example 5.29. A coil of wire when connected to a D.C. supply of 100 V takes a current of 10 Amps. When the coil is connected to an A.C supply of 100 V at a frequency of 50 Hz, the current is 5 Amps. Explain the difference and calculate the inductance of the coil.

Solution : The coil possesses both resistance and inductance when D.C. supply is applied across the coil, the opposition to current is only due to resistance R of the coil as X_L of DC is zero (For D.C. supply frequency is zero hence $X_L = 2\pi fL = 0$ and hence offeres no opposition). On the other hand for A.C. supply, opposition to current is offered by both resistance and inductance of the coil.

$$\text{Resistance of coil } R = \frac{\text{D.C. supply across coil}}{\text{Current}} = \frac{100}{10} = 10\ \Omega$$

$$\text{Impedance of the coil} = \frac{\text{A.C supply across coil}}{\text{Current}} = \frac{100}{5} = 20\ \Omega$$

$$\text{Inductive reactance of coil } X_L = \sqrt{Z^2 - R^2} = \sqrt{(20)^2 - 10^2} = 17.32\ \Omega$$

$$\text{Inductance of coil } L = \frac{X_L}{2\pi f} = \frac{17.32}{2\pi \times 50} = 0.0552 \text{ H } \textbf{Ans.}$$

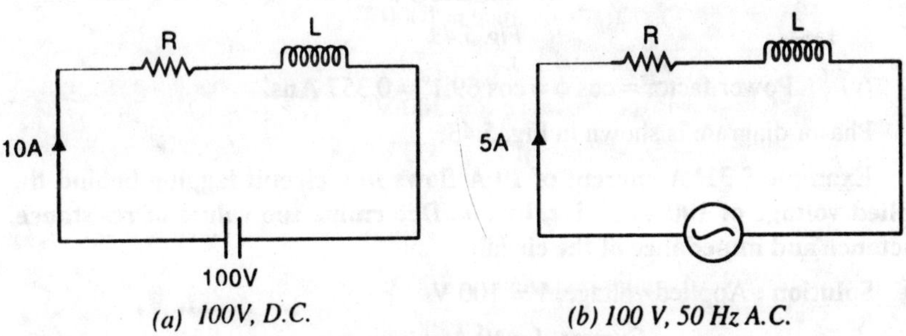

(a) 100V, D.C. *(b) 100 V, 50 Hz A.C.*

Fig. 5.42

Example 5.30. A coil having a resistance of 12 Ω and an inductance of 0.1 H is connected across a 100 V, 50 Hz supply. Calculate *(i)* the reactance and the impedance of the coil, *(ii)* the current, *(iii)* the phase difference between the current and the applied voltage and *(iv)* the power factor. Draw also the phasor diagram; showing voltages and current.

Solution : Supply voltage $V = 100$ V

Resistance of coil, $R = 12\ \Omega$

Inductance of coil, $L = 0.1$ H

Supply frequency, $f = 50$ Hz

(i) Coil reactance, $X_L = 2\pi \times 50 \times 0.1 = 31.416\ \Omega$ **Ans.**

(ii) Coil impedance, $Z = \sqrt{R^2 + X_L^2} = \sqrt{(12)^2 + (31.416)^2} = 33.63\ \Omega$ **Ans.**

(iii) Coil current, $I = \frac{V}{Z} = \frac{100}{33.63} = 2.974$ A **Ans.**

(iv) Phase difference between coil current and applied voltage,

$$\phi = \tan^{-1}\frac{X_L}{R} = \tan^{-1}\frac{31.416}{12} = 69.1° \textbf{ Ans.}$$

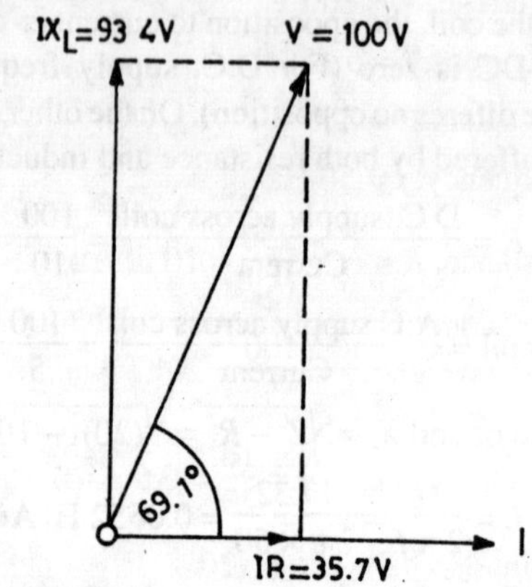

Fig. 5.43.

(v) Power factor $= \cos\phi = \cos 69.1° = 0.357$ **Ans.**

Phasor diagram is shown in Fig. 5.43.

Example 5.31. A current of 10 A flows in a circuit lagging behind the applied voltage of 100 V, 50 Hz, by 30°. Determine the values of resistance, reactance and impedance of the circuit.

Solution : Applied voltage, $V = 100$ V

Current, $I = 10$ A

Circuit impedance, $Z = \frac{V}{I} = \frac{100}{10} = 10\ \Omega$ **Ans.**

Circuit phase angle, $\phi = 30°$ (lag)

Circuit resistance, $R = Z\cos\phi = 10\cos 30° = 10\frac{\sqrt{3}}{2} = 8.66\ \Omega$

Circuit reactance, $X_L = Z\sin\phi = 10\sin 30° = 10 \times \frac{1}{2} = 5\ \Omega$ **Ans.**

Note : Since phase angle is of lag, therefore reactance will be inductive.

Example 5.32. A sinusoidal alternating supply has an r.m.s. value of 100 V and a frequency of 50 Hz and is connected to a series circuit having both

resistance and inductance. The current taken from the supply has an r.m.s. value of 5 A and power taken is 250 W. Calculate *(i)* the resistance of the circuit, *(ii)* the impedance of the circuit, *(iii)* the power factor of the circuit, *(iv)* the inductance, *(v)* the peak value of the current, *(vi)* the time taken for one complete cycle of supply and *(vii)* the apparent power taken by the circuit.

Assuming supply voltage as reference phasor, write expression for instantaneous current as a function of time.

Solution : Supply voltage, $V = 100$ V

Supply current, $I = 5$ A

Power taken, $P = 250$ W

Supply frequency, $f = 50$ Hz

(i) Circuit resistance, $R = \frac{P}{I^2} = \frac{250}{(5)^2} = 10\ \Omega$ **Ans.** $\quad \because \ P = I^2 R$

(ii) Circuit impedance, $Z = \frac{V}{I} = \frac{100}{5} = 20\ \Omega$ **Ans.**

(iii) Circuit power factor, $\cos\phi = \frac{R}{Z} = \frac{10}{20} = 0.5$ **Ans.**

Reactance of the coil, $X_L = \sqrt{Z^2 - R^2} = \sqrt{20^2 - 10^2} = 17.32\ \Omega$ **Ans.**

(iv) Circuit inductance, $L = \frac{X_L}{2\pi f} = \frac{17.32}{2\pi \times 50} = 0.055$ H

(v) Peak value of current $I_{max} = \sqrt{2}\ I_{r.m.s.} = \sqrt{2} \times 5 = 7.07$ A **Ans.**

(vi) Time taken for one complete supply cycle, $T = \frac{1}{f} = \frac{1}{50} = 0.02$ s **Ans.**

(vii) Apparent power taken by circuit

$= V \times I = 100 \times 5 = 500$ volt-amperes **Ans.**

(viii) Instantaneous current,

$$i = I_{max} \sin(\omega t - \phi) = 7.07 \sin\left(2\pi \times 50\, t - \frac{\pi}{3}\right) \textbf{ Ans.}$$

$\because \ \phi = \cos^{-1} 0.5 = 60^\circ$ or $\pi/3$ radians

Example 5.33. The power dissipated is 450 watts when a voltage of 240 V at 50 Hz is applied to a coil and the current drawn is 10 A. When the same voltage is applied to another coil the current drawn is 15 amperes and the power dissipated is 960 watts. Calculate *(i)* The inductive reactance of each coil, *(ii)* The current and power taken when both the coils are connected in series across the said voltage.

Solution : Applied voltage, $V = 240$ volts

Current drawn by first coil, $I_1 = 10$ amperes

Impedance of first coil, $Z_1 = \frac{V}{I_1} = \frac{240}{10} = 24\ \Omega$

Resistance of first coil, $R_1 = \frac{P_1}{I_1^2} = \frac{450}{10^2} = 4.5\ \Omega$ $\quad \because\ P_1 = I_1^2 R_1$

Inductive reactance of first coil,

$$X_{L_1} = \sqrt{Z_1^2 - R_1^2} = \sqrt{24^2 - 4.5^2} = 23.57\ \Omega \textbf{ Ans.}$$

Current drawn by second coil, $I_2 = 15$ amperes

Impedance of second coil, $Z_2 = \frac{V}{I_2} = \frac{240}{15} = 16\ \Omega$

Resistance of second coil, $R_2 = \frac{P_2}{I_2^2} = \frac{960}{15^2} = 4.267\ \Omega$

Inductive reactance of second coil,

$$X_{L_2} = \sqrt{Z_2^2 - R_2^2} = \sqrt{16^2 - 4.267^2} = 15.42\ \Omega \textbf{ Ans.}$$

(ii) When both coils are connected in series

Resistance of combination, $R = R_1 + R_2 = 4.5 + 4.267 = 8.767\ \Omega$

Reactance of combination, $X_L = X_{L_1} + X_{L_2} = 23.57 + 15.42 = 38.99\ \Omega$

Impedance of combination, $Z = \sqrt{R^2 + X_L^2} = \sqrt{8.767^2 + 38.99^2} = 40\ \Omega$

Current drawn from 240 V supply, $I = \frac{V}{z} = \frac{240}{40} = 6$ A **Ans.**

Power taken from 240 V supply, $P = I^2 R = 6^2 \times 8.767 = 315.6$ watts **Ans.**

Example 5.34. Define power factor in an a.c. circuit. A 0.8 pf (lagging) load draws 100 A from a 250 V, 1 — ϕ a.c. supply mains. Find *(i)* the value of the true power and apparent power, *(ii)* the value of circuit components.

Solution : Supply voltage, $V = 250$ V

Supply current, $I = 100$ A

Power factor, $\cos\phi = 0.8$ (lagging)

(i) True value of power,

$$P = VI\cos\phi = 250 \times 100 \times 0.8 = 20000 \text{ W or } 20 \text{ Kw } \textbf{Ans.}$$

$$\text{Apparent power} = VI\ \frac{250 \times 100}{1000} = 25 \text{ KVA } \textbf{Ans.}$$

(ii) Circuit impedance, $Z = \frac{V}{I} = \frac{250}{100} = 2.5\ \Omega$

Circuit resistance, $R = Z\cos\phi = 2.5 \times 0.8 = 2\ \Omega$ **Ans.**

Circuit reactance, $X = \sqrt{Z^2 - R^2} = \sqrt{2.5^2 - 2^2} = 1.5\ \Omega$ (inductive) **Ans.**

$\because$ power factor is lagging.

Example 5.36. A 120 V, 60 W lamp is to be operated on 220 V, 50 Hz supply mains. Calculate what value of *(a)* non-inductive resistance, *(b)* pure inductance would be required in order that lamp is run on correct voltage? Which method is preferable and why?

Solution : Compared with the resistance of the filament of a lamp its inductance is negligible. As such a lamp is considered to be non-inductive resistance.

Rated power of lamp, $P = 60$ W

Rated voltage of lamp, $V = 120$ V

Current drawn by the lamp when operated on rated voltage

$$I = \frac{P}{V} = \frac{60}{120} = 0.5 \text{ A}$$

(i) Let a non-inductive resistance of R ohms be connected in series with the lamp, as shown in Fig. 5.44 *(a)*.

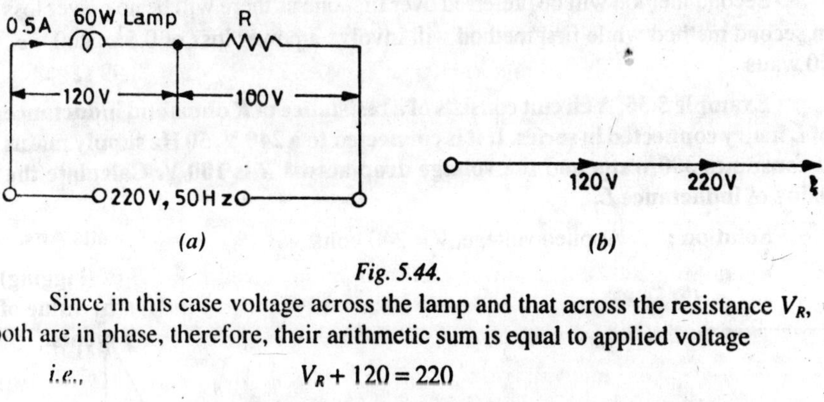

Fig. 5.44.

Since in this case voltage across the lamp and that across the resistance V_R, both are in phase, therefore, their arithmetic sum is equal to applied voltage

i.e., $$V_R + 120 = 220$$

or $$V_R = 220 - 120 = 100 \text{ V}$$

$$\text{Resistance, } R = \frac{V_R}{I} = \frac{100}{0.5} = 200\ \Omega \textbf{ Ans.}$$

(ii) Let pure inductance of L henry be connected in series with the lamp, as shown in Fig. 5.45 *(a)*.

Since in this case voltage drop across the inductance leads the current by 90° and $\pi/2$ radians while that across the lamp will be phase with it, therefore, from phasor diagram shown in Fig. 5.49 *(b)*, we have

$$120^2 + V_L^2 = 220^2$$

or $$V_L = \sqrt{220^2 - 120^2} = 184.4 \text{ volts}$$

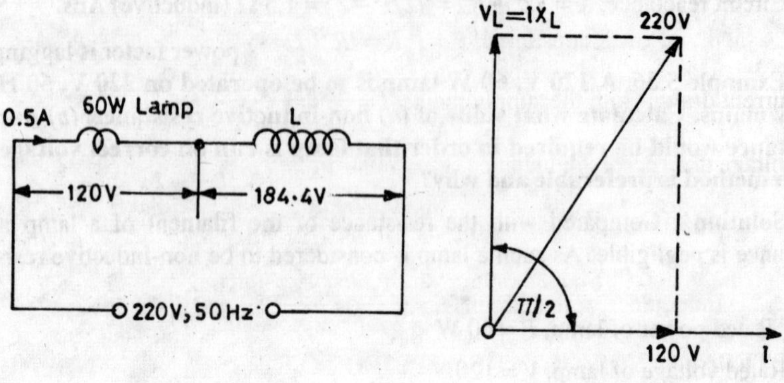

(a) (b)

Fig. 5.45.

or $$2 \pi f L I = 184.4 \text{ V}$$

or $$L = \frac{184.4}{2 \pi \times 50 \times 0.5} = 1.174 \, H \textbf{ Ans.}$$

Second method will be preferred over first one as there will be no power loss in second method while first method will involve a power loss of $0.5^2 \times 200$, *i.e.*, 50 watts.

Example 5.36. A circuit consists of a resistance of *R* ohms and inductance of *L* henry connected in series. It it is connected to a 240 V, 50 Hz supply mains it consumes 300 watts and the voltage drop across *R* is 100 V. Calculate the value of inductance *L*.

Solution : Applied voltage, $V = 240$ volts

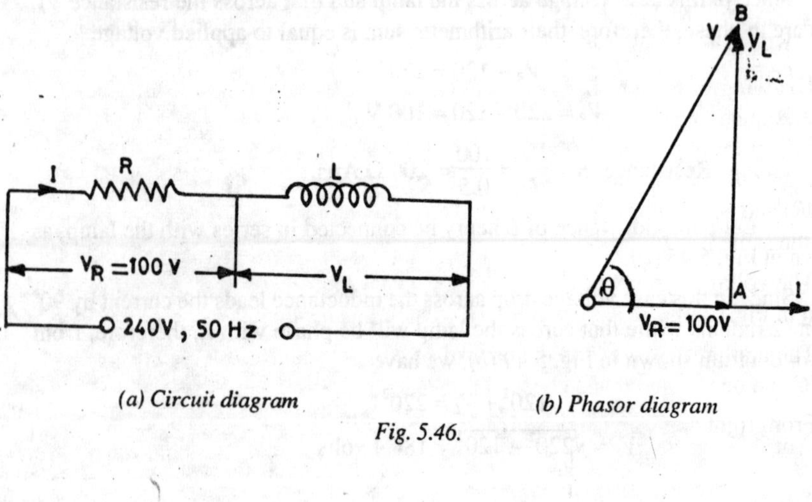

(a) Circuit diagram *(b) Phasor diagram*

Fig. 5.46.

Voltage drop across the resistor, $V_R = IR = 100$ V

Power consumed, $P = I^2 R = IR \times I = V_R I$

Current drawn from the supply mains, $I = \dfrac{P}{V_R} = \dfrac{300}{100} = 3$ A

Voltage drop across inductance, $V_L = \sqrt{V^2 - V_R^2} = \sqrt{240^2 - 100^2} = 218.2$ V

Reactance of the circuit, $X_L = \dfrac{V_L}{I} = \dfrac{218.2}{3} = 72.7\ \Omega$

Inductance of the circuit, $L = \dfrac{X_L}{2\pi f} = \dfrac{72.7}{2\pi \times 50} = 0.2315$ H **Ans.**

5.18. ALTERNATING VOLTAGE APPLIED TO RESISTANCE AND CAPACITANCE IN SERIES

The Fig. 5.47 *(a)* shows a simple circuit containing pure resistance *R* ohms and a pure capacitance in series.

A voltage of r.m.s. value *V* is applied across the combination. Let the current flowing through the circuit is *I* amperes.

Let V_R be the voltage across *R* and V_C is the voltage across capacitance *C*. The circuit can be shown by drawing the phasor diagram. (Fig. 5.47 *(b)*).

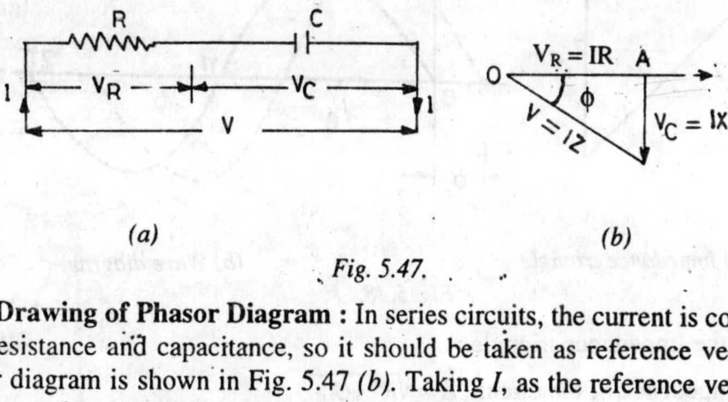

Fig. 5.47.

Drawing of Phasor Diagram : In series circuits, the current is common to both resistance and capacitance, so it should be taken as reference vector. The phasor diagram is shown in Fig. 5.47 *(b)*. Taking *I*, as the reference vector, it is shown as *OB* drawn along x-axis. The voltage across pure resistance *R* is ($V_R = IR$) and acts in phase with the current as shown in Fig. 5.47 *(b)*. It is taken as $\vec{OA}$ in the same line of current.

The voltage across pure capacitance is ($V_C = I \times X_C$) lags the current by 90° shown in Fig.) and is represented by $\vec{AC}$ to the same scale.

The applied voltage V is the sum of phasors *OA* and *AC*. This is given by $\vec{OC}$. *OC* can be measured and checked as per scale.

From right angled triangle *OAC* :

$$OC^2 = OA^2 + AC^2$$

or $$V^2 = (V_R)^2 + (V_C)^2$$

$$V^2 = (IR)^2 + (IX_C)^2$$

$$V = I\sqrt{R^2 + X_C^2}$$

∴ $$V = I \cdot Z$$

or $$I = \frac{V}{Z}$$

and $$Z = \frac{V}{I}$$

Here $\sqrt{R^2 + X_C^2}$ is known as Impedence of the circuit. It is denoted by Z and is measured in ohms.

The Impedence Triangle : Since the current is common to all V_R, V_C and V, so it may be omitted and impedence triangle constructed with only R, X_C and Z as shown in Fig. 5.48 *(a)*.

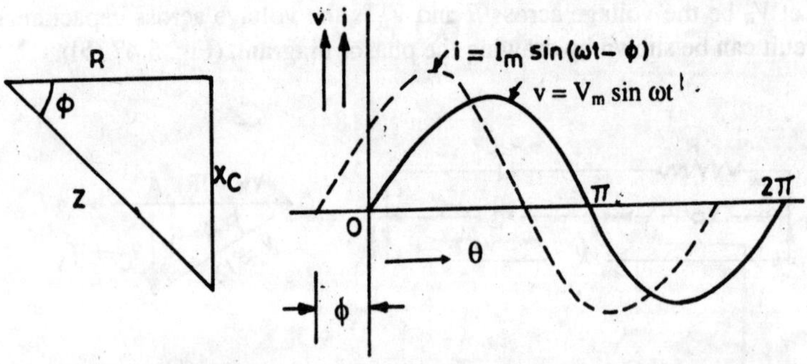

(a) Impedance triangle *(b) Wave diagram*

Fig. 5.48.

From the Impedence triangle :

Impedence of the circuit, $Z = \sqrt{R^2 + X_C^2}$

where $X_C = \dfrac{1}{\omega C} = \dfrac{1}{2\pi fc}$ is the capacitive resistance.

$\tan\phi = \dfrac{X_C}{R}$, $\phi = \tan^{-1}\dfrac{X_C}{R}$ is known as the phase angle. The current leads the voltage by this angle ϕ.

Example 5.37. A resistance of 10 ohms and a capacitor of 100 μF (micro farad) capacitance are connected in series across a 250 volts, 50 Hz supply. Calculate *(i)* Capacitive reactance, *(ii)* Impedence, *(iii)* Current flowing, *(iv)*

Angle of phase difference, *(v)* Power factor, *(vi)* True power and *(vii)* Phasor diagram.

Solution : Value of resistance, $R = 10$ ohms

Value of capacitance, $C = 100\ \mu F = 100 \times 10^{-6}$ F

Supply voltage, $V = 250$ volts

Supply frequency, $f = 50$ Hz

(i) Capacitive reactance, $X_C = \dfrac{1}{\omega C} = \dfrac{10^6}{2 \times 3.14 \times 50 \times 100} = 31.84\ \Omega$

(ii) Impedence, $Z = \sqrt{R^2 + X_C^2}$

$= \sqrt{(10)^2 + (31.84)^2} = 33.37$ ohms **Ans.**

(iii) Current flowing, $I = \dfrac{V}{Z} = \dfrac{250}{33.37} = 7.489$ A **Ans.**

(iv) Phase angle $\phi = \tan^{-1} \dfrac{X_C}{R}$

$\therefore$ $\tan \phi = \dfrac{31.84}{10} = 3.184$

$\therefore$ $\phi = \tan^{-1} 3.184 = 72.5°$

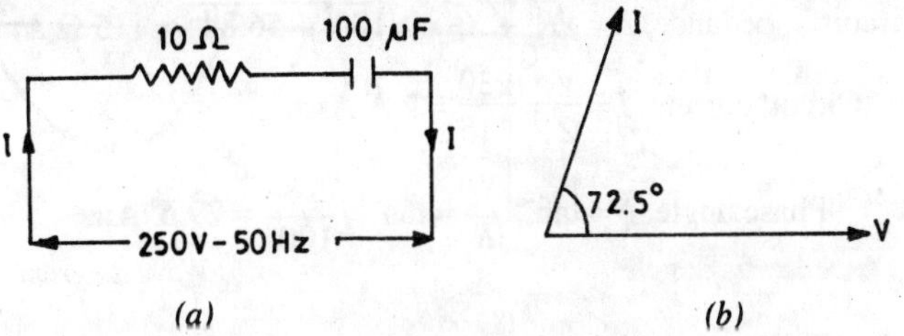

Fig. 5.49.

(v) Power factor, $\cos \phi = \cos 72.5 = 0.3$ (leading) **Ans.**

(vi) True Power $= V \,.\, I \,.\, \cos \phi$

$= 250 \times 7.489 \times 0.3 = 561.5$ Watts **Ans.**

(vii) The phasor diagram is shown in Fig. 5.49 *(b)*.

Example 5.38. A voltage of 125 V at 60 Hz is applied across a non-inductive resistor connected in series with a condenser. The current in the circuit is 2.2 A. The power loss in the resistor is 96.8 watts and that in the condenser is negligible. Calculate the resistance and capacitance.

Solution : Applied voltage, $V = 125$ volts

Circuit current, $I = 2.2$ amperes

$$\text{Circuit impedance, } Z = \frac{V}{I} = \frac{125}{2.2} = 56.82\ \Omega$$

$$\text{Power loss in the resistor} = I^2 R = 96.8 \text{ watts}$$

$$\text{Resistance of the circuit, } R = \frac{I^2 R}{I^2} = \frac{96.8}{2.2^2} = 20\ \Omega \textbf{ Ans.}$$

Since power loss in the condenser is negligible, therefore, resistance of the capacitor is zero.

Capacitive reactance of the circuit, $X_C = \sqrt{Z^2 - R^2} = \sqrt{56.82^2 - 20^2} = 53.2\ \Omega$

Capacitance of the circuit,

$$C = \frac{1}{\omega X_C} = \frac{1}{2\pi f X_C} = \frac{1}{2\pi \times 60 \times 53.2} = 50\ \mu\text{ F } \textbf{Ans.}$$

Example 5.39. A resistor of 100 Ω is connected in series with a 56 μ F capacitor to a supply at 230 V, 50 Hz. Find *(i)* the impedance, *(ii)* the current, *(iii)* the phase angle, *(iv)* the pf, *(v)* the voltage across the resistor and across the capacitor. Draw the phasor diagram.

Solution : Resistance of the circuit, $R = 100\ \Omega$

$$\text{Capacitive reactance of the circuit, } X_C = \frac{1}{2\pi f C} = \frac{1}{2\pi \times 50 \times 56} = 56.84\ \Omega$$

(i) Circuit impedance, $Z = \sqrt{R^2 + X_C^2} = \sqrt{100^2 + 56.84^2} = 115\ \Omega$ **Ans.**

(ii) Circuit current, $I = \frac{V}{Z} = \frac{230}{115} = 2$ A **Ans.**

(iii) Phase angle, $\phi = \tan^{-1}\frac{X_C}{R} = \tan^{-1}\frac{56.84}{100} = 29.6°$ **Ans.**

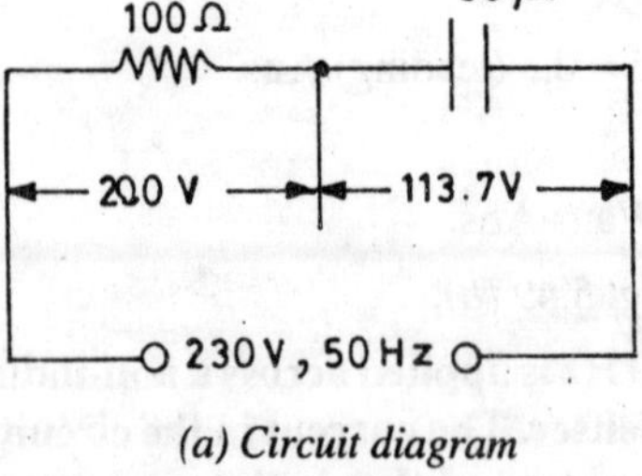

(a) Circuit diagram

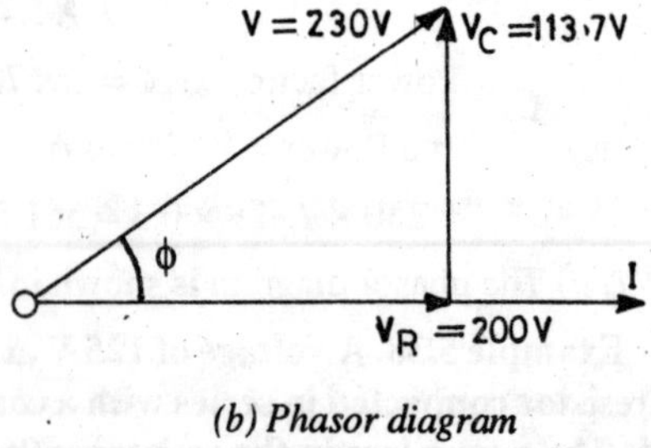

(b) Phasor diagram

Fig. 5.50.

(iv) Power factor, $\cos\phi = \cos 29.6° = 0.87$ **(lead) Ans.**

(v) Voltage across resistor, $V_R = IR = 2 \times 100 = 200$ V **Ans.**

(vi) Voltage across capacitor, $V_C = IX_C = 2 \times 56.84 = 113.7$ V **Ans.**

Example 5.40. A capacitor of capacitance 79.5 μF is connected in series with a non-inductive resistance of 30 Ω across 100 V, 50 Hz supply. Find *(i)* impedance, *(ii)* current, *(iii)* phase angle and *(iv)* equation for the instantaneous value of current.

Solution : *(i)* Capacitive reactance, $X_C = \dfrac{1}{2\pi fC}$

$$= \frac{10^6}{2\pi \times 50 \times 79.5} = 40\ \Omega$$

Circuit impedance, $Z = \sqrt{R^2 + X_C^2} = \sqrt{30^2 + 40^2} = 50$ *OEMGA* **Ans.**

(ii) Circuit current, $I = \dfrac{V}{Z} = \dfrac{100}{50} = 2$ A **Ans.**

(iii) $\tan\phi = \dfrac{X_C}{R} = \dfrac{40}{30} = 1.33$

∴ Phase angle, $\phi = \tan^{-1} 1.33 = 53°$ Lead **Ans.**

(iv) $I_m = 2 \times \sqrt{2} = 2.838$ A

$\omega = 2\pi f = 2\pi \times 50 = 314$

∴ $i = 2.828 \sin(314t + 53°)$ **Ans.**

Example 5.41. A capacitor and resistor are connected in series across a 120 V, 50 Hz supply. The circuit draws a current of 1.144 A. If power loss in the circuit is 130.8 W, find the values of resistance and capacitance.

Solution : Power loss occurs in resistance only as capacitor consumes no power.

$$P = I^2 R$$

∴ Resistance, $R = \dfrac{P}{I^2} = \dfrac{130.8}{(1.144)^2} = 100\ \Omega$ **Ans.**

Circuit impedance, $Z = \dfrac{V}{I} = \dfrac{120}{1.144} = 104.9\ \Omega$

Capacitive reactance, $X_C = \sqrt{Z^2 - R^2} = \sqrt{(104.9)^2 - (100)^2} = 31.7\ \Omega$

∴ Capacitance, $C = \dfrac{1}{2\pi f X_C} = \dfrac{1}{2\pi \times 50 \times 31.7} = 100 \times 10^{-6}$ F **Ans.**

Example 5.42. A 10 Ω resistor and 400 μF capacitor are connected in series to a 60 V sinusoidal supply. The circuit current is 5A. Calculate the supply frequency and phase angle between the current and voltage.

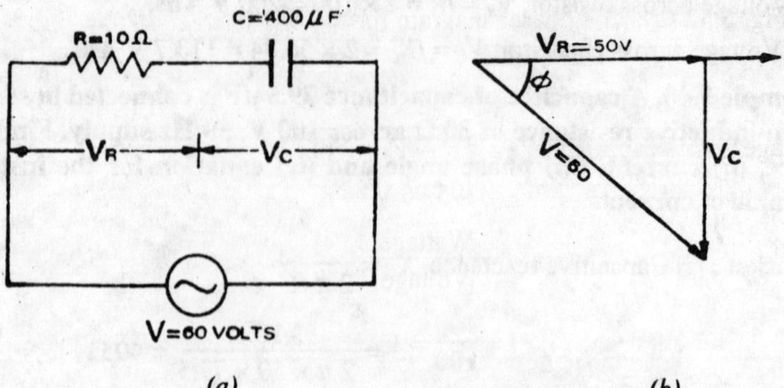

Fig. 5.51.

Solution : Voltage across R, $V_R = IR = 5 \times 10 = 50$ V

Voltage across C, $V_C = \sqrt{V^2 - V_R^2}$ [See phasor diagram]

$$= \sqrt{60^2 - 50^2} = 33.17 \text{ V}$$

Reactance of capacitor, $X_C = \dfrac{V_C}{I} = \dfrac{33.17}{5} = 6.634\ \Omega$

Now, $$X_C = \frac{1}{2\pi fC}$$

Supply frequency, $f = \dfrac{1}{2\pi C X_C} = \dfrac{10^6}{2\pi \times 400 \times 6.634} = 60$ Hz **Ans.**

$$\tan\phi = \frac{V_C}{V_R} = \frac{33.17}{60} = 0.6634$$

$\therefore$ Phase angle, $\phi = \tan^{-1} 0.6634 = 33.6^\circ$ lead **Ans.**

Example 5.43. Calculate the capacitance of a condenser to be connected in series with a 100 V, 100 W lamp to enable it to be used on a 220 V, 50 Hz supply. Also calculate power factor.

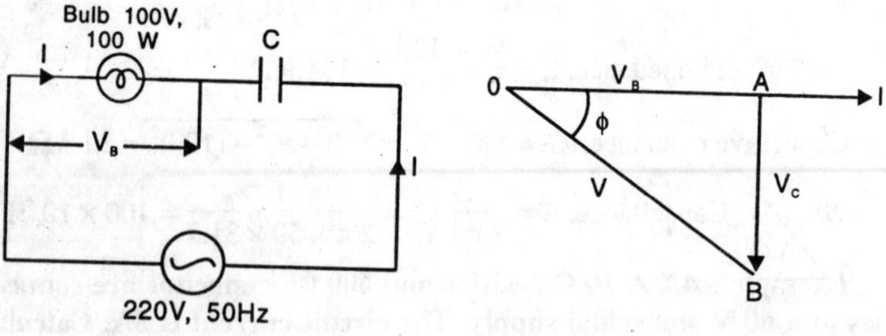

(a) Circuit diagram *(b) Phasor diagram*

Fig. 5.52.

Solution : Let C be the pure capacitance connected in series with the bulb as shown in Fig. 5.52 *(a)*. The phasor diagram for the circuit as shown in Fig. 5.52 *(b)*.

From phasor diagram

Voltage across capacitor, $V_C = \sqrt{V^2 - V_B^2} = \sqrt{(220)^2 - (100)^2}$

$$V_C = 195.96 \text{ V}$$

$$\text{Current } I = \frac{\text{Wattage}}{\text{Voltage}}$$

$$I = \frac{100}{100} = 1 \text{ A}$$

$$X_C = \frac{V_C}{I} = \frac{195.96}{1} = 195.96\ \Omega$$

$$X_C = \frac{1}{2\pi fC} = 195.96\ \Omega$$

or Capacitance of capacitor, $C = \dfrac{1}{2\pi \times 50 \times 195.96} = 16.24\ \mu\text{F}$ **Ans.**

For power factor, $R = \dfrac{V^2}{P} = \dfrac{(100)^2}{100} = 100\ \Omega$

$$Z = \frac{V}{I} = \frac{220}{1} = 200\ \Omega$$

Power factor, $\dfrac{R}{Z} = \dfrac{100}{200} = 0.5$ **Ans.**

5.19. IMPEDANCE TRIANGLE AND PHASE ANGLE

Impedance Triangle

The simplified phasor diagram of R—L series circuit is shown in Fig. 5.53 *(a)*. When each side of this phasor diagram is divided by a common factor I, we

(a) *(b) Impedance triangle*

Fig. 5.53.

get another right angled triangle as shown in Fig. 5.53 *(b)*, whose sides represent R, X_L and Z. Such triangle is known as impedance triangle.

Thus a right angle triangle whose base represents circuit resistance, perpendicular represents reactance and hypotaneous represents circuit impedance is called an impedance triangle.

The concept of impedance triangle is useful since it enable us to calculate :

(i) The impedance of the circuit $Z = \sqrt{R^2 + X^2}$

(ii) The power factor of the circuit $\cos \phi = R/Z$

(iii) Phase angle, $\phi = \tan^{-1} X/R$

(iv) Wheather current leads or lags the voltage.

Therefore, it is always helpful to draw the impedance triangle while analysing an a.c. circuit.

5.20. PHASE AND PHASE ANGLE

By phase of an alternating current is meant the fraction of the time period of that alternating current that has elapsed since the current last passed through the zero position of reference.

The phase angle of any quantity means the angle of the vector representing the quantity makes with the reference line to be taken at zero degrees or radian. It is represented by ϕ.

Phase angle ϕ is positive if the practical circuit is having resistance and inductance or inductive circuit. On the other hand phase angle is negative if the circuit is capacitive.

HIGHLIGHTS

1. **A.C. Circuit :** The path for the flow of alternating current is called a.c. circuit.
2. **Pure resistive circuit :** Circuit contains pure resistance.

   $v = V_m \sin \omega t$; $i = I_m \sin \omega t$; $I_m = V_m/R$; $P = VI$; $p.f. = 1$;

   Current is in phase with voltage vector.
3. **Pure inductive circuit :** Circuit contains pure inductance.

   $v = V_m \sin \omega t$; $i = I_m \sin (\omega t - \pi/2)$; $I_m = V_m/X_L$; $X_L = 2\pi f L$;

   $P = 0$; $p.f. = 0$ lag;

   Current lags behind voltage vector of 90° (*i.e.*, $\pi/2$).
4. **Pure capacitive circuit :** Circuit contains pure capacitance.

   $v = V_m \sin \omega t$; $i = I_m \sin (\omega t + \pi/2)$; $I_m = V_m/X_C$; $X_C = 1/2\pi f C$;

   $P = 0$; $p.f. = 0$ lead ;

   Current leads the voltage vector by 90° (*i.e.*, $\pi/2$).

5. **R-L series circuit :** Circuit contains resistance and inductance in series. It is generally called inductive circuit.

$v = V_m \sin \omega t$; $i = I_m \sin (\omega t - \phi)$; $\phi = \tan^{-1} X_L/R$; $I_m = V_m/Z$;

$Z = \sqrt{R^2 + X_L^2}$; $P = VI \cos \phi = I^2 R$; $p.f. = \cos \phi = R/Z$ lag ;

6. **True power :** It is the base or horizontal component of the power triangle which represents the actual or true power consumed in an a.c. circuit.

$P = VI \sin \phi$; Units of true power are watts (W), kilowatts (kw)or megawatts (MW)

7. **Reactive power :** It is the perpendicular or vertical component of the power triangle which represents reactive power of the circuit. This is the power which reacts in the circuit, in fact, what so ever power is supplied by the source in quarter cycle the same is fed back in the next quarter cycle.

$P_r = VI \sin \phi$; units of reactive power are volt-ampere-reactive (VAR), KVAR or MVAR.

8. Apparent Power : It is the hypotenous of the power triangle. It is the power which looks to be consumed in the circuit but actually it is not so.

$P_a = VI$; units of reactive power are volt-ampere (VA), kVA or MVA.

9. **Power factor and its importance :** p.f. = $\cos \phi = R/Z$ = true power/apparent power.

Ill-effects of poor p.f. *(i)* greater conductor size, *(ii)* poor efficiency, *(iii)* poor voltage regulation and *(iv)* larger kVA rating of equipment.

10. **R-C series circuit :** Circuit contains resistance and capacitance in series. It is generally called capacitive circuit.

$v = V_m \sin \omega t$; $i = I_m \sin (\omega t + \phi)$; $\phi = \tan^{-1} X_C/R$; $I_m = V_m/Z$;

$Z = \sqrt{R^2 + X_C^2}$; $P = VI \cos \phi = I^2 R$; $p.f. = \cos \phi = R/Z$ leading.

SHORT QUESTIONS

1. The power consumed in a pure resistance is not zero, why ?
2. Why power consumed in a pure inductive circuit is zero ?
3. Why pure capacitance circuit consumed zero power ?
4. Why the value of phase angle is less than 90° ?
5. Why inductive reactance increases with increase of frequency ?
6. Why capacitive reactance decreases with increase of frequency ?

5.21. COMPLEX NUMBERS

The solution of A.C. circuits becomes easier if the alternating electrical quantities such as voltage, current and impedance are represented by phasors. The solution of A.C. series and parallel circuits involve addition, substraction, multiplication and division of such phasors.

5.22. VARIOUS FORM

A phasor quantity can be represented into number of ways which enable different mathematical operations to be performed. The following are various methods in which the phase quantity is represented.

(i) Rectangular form or symbolic J-method

(ii) Trigonometric form

(iii) Polar form

(iv) Exponential from

1. RECTANGULAR FORM

Fig. 5.54 shows the axis of X and the axis of Y. As we know, towards right is +X and left is –X. Similarly towards top is +Y and bottom is –ve Y. We further know that thus the plane has been divided into Four Quardrants I, II, III and IV.

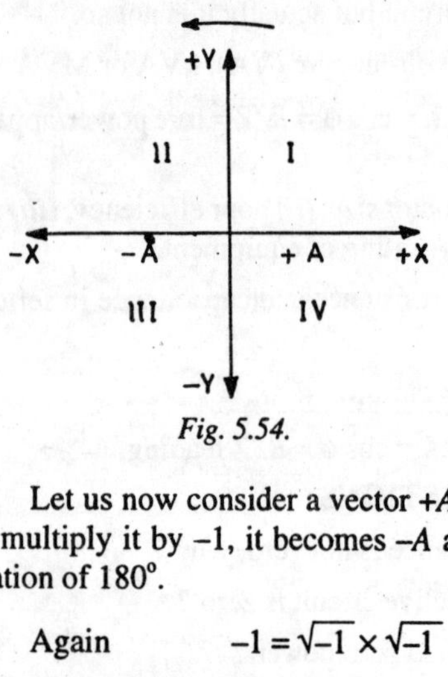

Fig. 5.54.

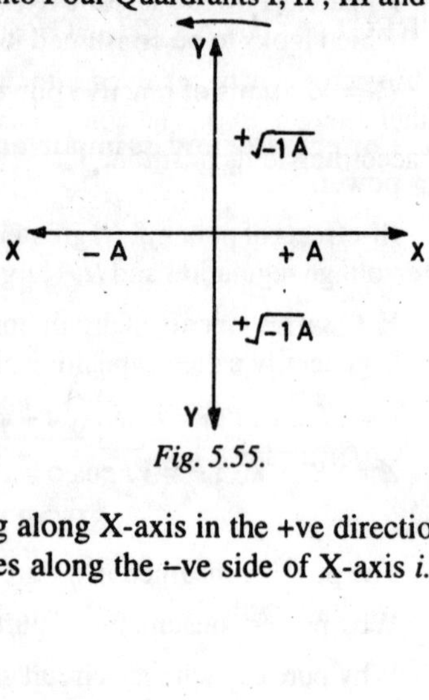

Fig. 5.55.

Let us now consider a vector +*A* lying along X-axis in the +ve direction. If we multiply it by –1, it becomes –*A* and lies along the –ve side of X-axis *i.e.,* a rotation of 180°.

Again $\quad -1 = \sqrt{-1} \times \sqrt{-1}$

Thus it means that if we operate the vector +*A* by $\sqrt{-1}$, it will rotate through 90° in the +ve or anticlockwise direction. It will now lie along +ve direction of Y-axis. Again if we operate the vector by $\sqrt{-1}$, it will rotate through further 90° and lie along –ve direction of X-axis (See Fig. 5.55).

Further operation of vector –A by $\sqrt{-1}$ will make it lie along –ve directon of Y-axis as $\sqrt{-1} \times (-A)$. Operating it further by $\sqrt{-1}$ will result the vector to lie along +X-axis as $\sqrt{-1} \times \sqrt{-1} \times -A = +A$.

Thus we have completed one rotation.

$\sqrt{-1}$ in mathematics is an imaginary quantity and is name as i (iota). However, here we shall denote it by I. The vector +A will lie along +ve direction of X-axis, +JA (*i.e.,* $\sqrt{-1}$ A) lies along +ve direction of Y-axis, –A lies along –ve direction of X-axis and $-JA(\sqrt{-1} - A)$ lies along –ve direction of Y-axis. (Fig 5.56)

The X-axis is known as axis of reals and the Y-axis as axis of imaginaries.

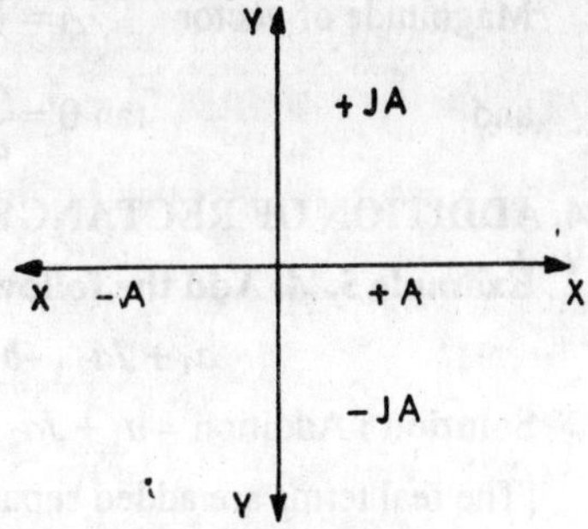

Fig. 5.56.

Saying the above conclusion in the other way, the operation vector with +ve J will cause it to Advance by 90° and with –ve J to Retard it by the same angle.

The plane represented by co-ordinate axis is known as complex plane.

5.23. RECTANGULAR PHASORS

A vector may be resolved into two components, one lying along X- axis and the other along Y-axis. The component lying along Y-axis will be denoted by +J or –J according to its position.

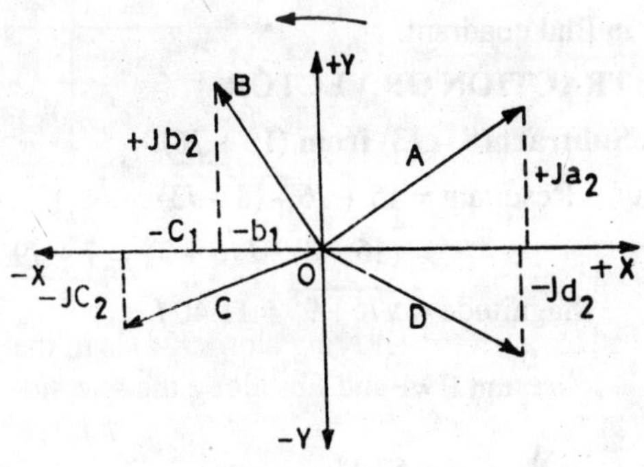

Fig. 5.57.

Any vector A in the first quadrant when resolved into two components may be represented by $+ a_2 + Ja_2$.

Similarly vector B, C and D in IInd, IIIrd and IVth quardrants are represented by

$$\text{Vector } B \text{ as } - b_1 + jb_2$$

$$\vec{C} \text{ as } - c_1 - jc_2$$

$$\vec{D} \text{ as } + d_1 - jd_2$$

It is again reminded at $J^2 = -1$

Magnitude of vector $A = |A| = \sqrt{a_1^2 + a_2^2}$

and $\tan\theta = \dfrac{a_2}{a_1}$

5.24. ADDITION OF RECTANGULAR PHASORS

Example 5.44. Add the following vectors :

$$a_1 + Ja_2\,,\ -b_1 + jb_2\,,\ +c_1 - jc_2$$

Solution : Addition $= a_1 + Ja_2 - b_1 + jb_2 + c_1 - Jc_2$

[The real terms are added separately and the imaginary terms separately].

Example 5.45. Add the following vectors:

$$6 + J9\ ;\ -7 - J3\ ;\ -8 + J4$$

Solution : Addition $= 6 + J9 - 7 - J3 - 8 + J4$

Addition $= (6 - 7 - 8) + J\,(9 - 3 + 4) = (9 + J\,10)$ is the resultant vector.

$$\text{Magnitude of the vector} = \sqrt{(-9)^2 + 10^2} = 13.4536$$

It makes an angle θ with +ve direction of X-axis such that

$$\tan\theta = +\frac{10}{9} = +1.111$$

$$\therefore \qquad \theta = +48^\circ$$

The vector lies in IInd quadrant.

5.25. PHASOR SUBTRACTION OR VECTOR

Example 5.46. Subtract (8 – J3) from (15 + J6).

Solution : Resultant $= 15 + J6 - (8 - J3)$

$$= (15 - 8) + J\,(6 + 3) = 7 + J9$$

$$\text{Magnitude} = \sqrt{7^2 + 9^2} = 11.40\,I$$

$$\tan\theta = \frac{9}{7}$$

$$\therefore \qquad \theta = 52.1^\circ$$

The vector lies in 1st quadrant.

5.26. MULTIPLICATION OF RECTANGULAR PHASORS OR VECTOR

Example 5.47 Determine the product of

$$(6 + J4) \times (9 - J3)$$

Solution : $(6 + J4) \times (9 - J3) = 6 \times 9 - J\,(6 \times 3) + J9 \times 4 - J^2\, 3 \times 4$

$= 54 - J18 + J36 + 12$ [Because $J^2 = -1$]

$= 66 + J18$

$$\text{Magnitude} = \sqrt{66^2 + 18^2} = 68.41$$

$$\tan\theta = \frac{18}{66} = 0.2727$$

$$\therefore \qquad \theta = 15^\circ - 15'$$

The vector lies in 1st quadrant.

5.27. RECIPROCAL OF PHASORS

Example 5.48. Find $\frac{1}{8 - J6}$

Solution : $\frac{1}{8 - J6}$ can be solved by Rationalising the vector, *i.e.*, multiplying and dividing it by $8 + J6$.

$$= \frac{1}{8 - J6} \times \frac{8 + J6}{8 + J6} = \frac{8 + J6}{64 + 36} \qquad [\because (a - b)(a + b) = a^2 - b^2]$$

$$= \frac{8 + J6}{100} = 0.08 + J\,0.06$$

The magnitude and the angle can be determined in the usual way.

5.28. DIVISION OF PHASORS

Example 5.49. Divide $(8 - J\,10)$ by $(6 + J\,4)$

Solution : $\frac{8 - J\,10}{6 + J4}$

Rationalising the denominator, *i.e.*, multiplying by Numerator and Denominator by $6 - J4$, we get

$$= \frac{8 - J\,10}{6 + J4} \times \frac{(6 - J4)}{(6 - J4)} = \frac{48 - J32 - J60 + J^2\,40}{36 + 16}$$

$$= \frac{8 - J9^2}{52} = 0.154 - J\,1.77\,. \qquad [\because J^2\,40 = -40]$$

$$\text{Magnitude} = \sqrt{0.154^2 + (-1.77)^2} = 1.78$$

$$\tan\theta = \frac{1.77}{0.154} = 11.493 \qquad \therefore \ \theta = 85^\circ$$

The phasor lies in the 4th quadrant.

EXERCISES

1. Multiply :
   (*i*) $(6 + J8) \times (-3 + J9)$
   (*ii*) $(9 - J9) \times 3 + J7$
   (*iii*) $(-8 - J4) \times (4 + J3)$

(iv) $(2 - J6) \times (-3 + J5)$

2. Divide :

(i) $(16 - J4)$ by $(3 + J1)$

(ii) $(0.8 + J\,0.3)$ by $(-1 + J\,0.2)$

5.29. SIGNFICANCE OF OPERATOR J

The letter j used in above expressions is symbol of an operation, just as symbols X, +, – etc. It is used to indicate the anti-clockwise rotation of a vector by 90°. It is assigned a numerical value of $\sqrt{-1}$. The double operation of J on a vector rotates it anti-clockwise through an angle of 180° and hence reverses its sense.

$$j \,.\, J = j^2 = (\sqrt{-1}\,)^2 = -1$$

When a vector E is operated by j, we get a new vector displaced by 90° in the anti-clockwise direction. If the vector E is operated by j twice we get $j^2 E$ or $-E$.

If operator j is applied thrice to a vector E we get

$$J^3 E = J \,.\, J^2 E = -j\,E$$

If operator j is applied to a vector E four times we get

$$j^4 E = -j\,(-jE) = -j^2 E = E$$

The above facts may be summarised as

$j = 90°$ Anti-clockwise rotation $= \sqrt{-1}$

$j^2 = 180°$ Anti-clockwise rotation $= \sqrt{(-1)^2} = -1$

$j^3 = 270°$ Anti-clockwise rotation $= \sqrt{(-1)^3} = -\sqrt{-1} = -j$

$j^4 = 360°$ Anti-clockwise rotation $= \sqrt{(-1)^4} = 1$

Conjugate Complex Numbers

Two complex numbers are said to be conjugate, if they differ only in algebraic sign of their imaginary parts.

For example the complex number $(a - jb)$ is conjugate of $a + jb$

5.30. TRIGONOMETRICAL FORM OF PHASOR REPRESENTATION

As may be seen from Fig 5.58, X-component of $V = V \cos \theta$ and Y-component $= V \sin \theta$. Hence phasor V can be expressed in the trigonometrical form as

$$V = V \cos \theta + j\,V \sin \theta = V\,(\cos \theta + j \sin \theta)$$

This is equivalent to the symbolic or rectangular form $V = (a + jb)$ because $a = V \cos \theta$ and $b = V \sin \theta$.

In general $V = V\,(cos\ \theta \pm j \sin \theta)$

The above expression represents a phasor of numerical value V and having phase angle of $\pm\theta$ with the reference axis, positive angle being measured in counter- clockwise direction from the reference axis and negative angle in clockwise direction from the reference axis.

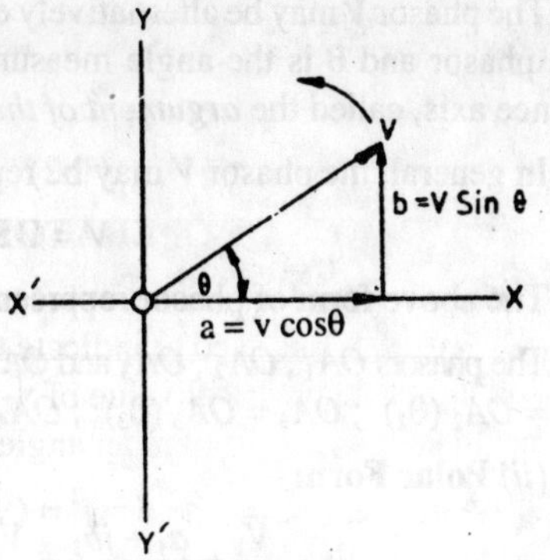

Fig. 5.58.

The phasors OA_1, OA_2, OA_3 and OA_4 may be represented thus:

$$OA_1 = OA_1 (\cos\theta_1 + j\sin\theta_1)$$
$$OA_2 = OA_2 (\cos\theta_2 + j\sin\theta_2)$$
$$OA_3 = OA_3 (\cos\theta_3 - j\sin\theta_3)$$
$$OA_4 = OA_4 (\cos\theta_4 - j\sin\theta_4)$$

5.31. PRODUCT OF TRIGONOMETRIC OF PHASORS

$$V_1 = V_1 (\cos\theta_1 + j\sin\theta_2)$$
$$V_2 = V_2 (\cos\theta_2 - j\sin\theta_2)$$

Their product

$$V_1 V_2 = V_1 [(\cos\theta_1 + j\sin\phi_2)] [(V_2 \cos\phi_2 - j\sin\theta_2)]$$

$$= V_1 V_2 (\cos\theta_1 \cos\theta_2 - j\cos\theta_1 \sin\theta_2 + j\sin\theta_1\cos\theta_2 - j^2 \sin\theta_1 \sin\theta_2)$$

$$= V_1 V_2 [(\cos\theta_1 \cos\theta_2 + \sin\theta_1 \sin\theta_2)] + [j (\sin\theta_1 \cos\theta_2 - \sin\theta_2 \cos\theta_1)]$$

Division of Trignometric Phasors

$$V_1 = V_1 (\cos\theta_1 + j\sin\theta_1)$$
$$V_2 = V_2 (\cos\theta_2 - j\sin\theta_2)$$

or

$$\frac{V_1}{V_2} = \frac{V_1 (\cos\theta_1 + j\sin\theta_1)}{V_2 (\cos\theta_2 - j\sin\theta_2)}$$

$$= \frac{V_1 (\cos\theta_1 + j\sin\theta_1)}{V_2 (\cos\theta_2 - j\sin\theta_2)} \times \frac{\cos\theta_2 + j\sin\theta_2}{\cos\theta_2 + j\sin\theta_2}$$

$$= \frac{\frac{V_1}{V_2} [\cos\theta_1 \cos\theta_2 - \sin\theta_1 \sin\theta_2 + j (\sin\theta_1 \cos\theta_2 + \sin\theta_2 + \cos\theta_1)]}{\cos^2\theta_2 + \sin^2\theta_2}$$

$$= \frac{V_1}{V_2} [\cos(\theta_1 + \theta_2) + j\sin(\theta_1 + \theta_2)]$$

5.32. POLAR FORM OF PHASOR REPRESENTATION

The phasor V may be alternatively expressed as $V\ \theta$ where V is the magnitude of the phasor and θ is the angle measured in counter -clockwise direction from reference axis, called the *argument of the phasor.*

In general, the phasor V may be represented as

$$V = V \pm \theta$$

The above form of phasor representation is known as polar form.

The phasors OA_1, OA_2, OA_3 and OA_4 may be represented thus $OA_1 = OA_1\ (\theta_1)$; $OA_2 = OA_2\ (\theta_2)$; $OA_3 = OA_3\ (\theta_3)$; $OA_4 = OA_4\ (\theta_4)$.

***(ii)* Polar Form**

$$\frac{V_1}{V_2} = \frac{a_1 + jb_1}{a_2 - jb_2} = \frac{V_1 (\cos\theta_1 + j\sin\theta_1)}{V_2(\cos\theta_2 - j\sin\theta_2)}$$

$$= \frac{V_1 (\theta_1)}{V_2(-\theta_2)}$$

The quotient is $\dfrac{V_1}{V_2} = \dfrac{V_1 (\cos\theta_1 + j\sin\theta_1)}{V_2(\cos\theta_2 - j\sin\theta_2)}$

$$= \frac{V_1 (\cos\theta_1 + j\sin\theta_1)}{V_2(\cos\theta_2 - j\sin\theta_2)} \times \frac{(\cos\theta_2 + j\sin\theta_2)}{(\cos\theta_2 + j\sin\theta_2)}$$

$$= \frac{V_1}{V_1}\left[\frac{\cos\theta_1\cos\theta_2 - \sin\theta_1\sin\theta_2 + j(\sin\theta_1\cos\theta_2 + \sin\theta_2\cos\theta_1)}{\cos^2\theta_2 + \sin^2\theta_2}\right]$$

$$= \frac{V_1}{V_2}[\cos(\theta_1 + \theta_2) + j\sin(\theta_1 + \theta_2)]$$

We thus see that, when we divide one phasor by another phasor, the quotient is a phasor having a modulus equal to the quotient of the modulus of the two phasors and an argument equal to the algebraic difference of the arguments of the two phasors.

In polar form we have $\dfrac{V_1}{V_2} = \dfrac{V_1}{V_2}[\theta_1 - (-\theta_2)]\ ! = \dfrac{V_1}{V_2}(\theta_1 + \theta_2)$

Example 5.50. Determine the product of phasors $30\angle 45^\circ$ and $7.5\angle -30^\circ$.

Solution : $30\angle 45^\circ \times 7.5\angle -30^\circ$

$$= 30 \times 7.5 \angle 45^\circ + (-30^\circ) = 225\angle 15^\circ \text{ Ans.}$$

Example 5.51. Divide phasor $30\angle 45^\circ$ by phasor $7.5\angle -30^\circ$.

Solution : $\dfrac{30\angle 45^\circ}{7.5\angle -30^\circ} = \dfrac{30}{7.5}\angle 45^\circ - (-30^\circ) = 4\angle 75^\circ$ **Ans.**

Example 5.52. Three phasors are $A = 10 + j\,10$; $B = 50 \angle 120°$ and $C = 8 - j\,6$. Evaluate $\dfrac{AB}{C}$ and express it in rectangular and polar forms.

Solution : Phasor $A = (10 + j\,10)$ have modulus of $\sqrt{10^2 + 10^2}$, *i.e.*, $10\sqrt{2}$ and argument of $\tan^{-1}\dfrac{10}{10}$, *i.e.*, 45° or $\dfrac{\pi}{4}$ radians.

So phasor A in polar form will be $10\sqrt{2} \angle 45°$.

Phasor B in polar form is $50 \angle 120°$.

Phasor $C = (8 - j\,6)$ and have modulus of $\sqrt{8^2 + (-6)^2}$, *i.e.*, 10 and argument of $\tan^{-1}\dfrac{-6}{8}$ or $-36.87°$.

So $\qquad C = 10 \angle -36.87°$

Hence
$$\frac{AB}{C} = \frac{10\sqrt{2} \angle 45° \times 50 \angle 120°}{10 \angle -36.87°}$$
$$= \frac{10\sqrt{2} \times 50}{10} = \angle 45° + 120° - (-36.87°)$$
$$= 50\sqrt{2} \angle 201.87° \text{ **Ans.**}$$

In rectangular form $\dfrac{AB}{C}$ will be $= 50\sqrt{2}\cos 201.87° + j\,50\sqrt{2}\sin 201.87°$
$$= -65.622 - j\,26.34 \text{ **Ans.**}$$

Example 5.53. A resistance of 4Ω is connected in series with an inductive reactance of 5 Ω. Draw the impedance phasor for this circuit and express the impedance in the symbolic form and polar form.

Solution : Circuit resistance, $R = 4\ \Omega$

Circuit resistance, $X = 5\ \Omega$ (inductive)

The impedance phasor of the circuit is shown in Fig 5.59.

Magnitude of impedance, $Z = \sqrt{4^2 + 5^2} = 6.4\ \Omega$

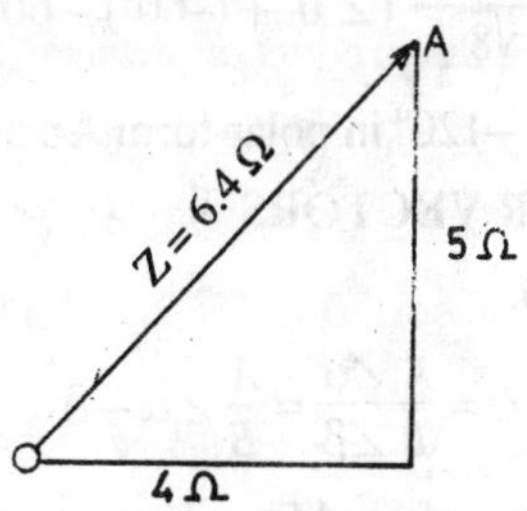

Fig. 5.59.

$$\theta = \tan^{-1}\frac{5}{4} = 51.34^\circ \text{ or } 0.285 \text{ radians}$$

$\therefore$ Impedance, $Z = (4 + j\,5)\ \Omega$ in symbolic form **Ans.**

or $Z = 6.4 \angle 51.34^\circ\ \Omega$ in polar form **Ans.**

Example 5.55. $V = 200 \angle 36.9^\circ$ volts ; $Z = 20 \angle -53.1$ ohms, find I.

Solution : Circuit voltage, $V = 200 \angle 36.9^\circ$ volts

Circuit impedance, $Z = 20 \angle -53.1$ ohms

$$\text{Circuit current, } I = \frac{V}{Z} = \frac{200 \angle 36.9^\circ}{20 \angle -53.1} = \frac{200}{20} \angle 36.9^\circ - (-53.1)$$

$$= \angle\, 10\, 90^\circ \text{ A } \textbf{Ans.}$$

Example 5.55. Given: $I_1 = 10\sqrt{2} \angle 0^\circ$, $I_2 = 10\sqrt{2} + j\,10\sqrt{6}$ and $I_3 = 15 \angle -60^\circ$. Find in polar forms *(i)* $I_1 + I_2 + I_3$ and $\dfrac{I_1 \times I_3}{I_2}$.

Solution : $I_1 = 10\sqrt{2} \angle 0^\circ = 10\sqrt{2}\,(\cos 0^\circ + j \sin 0^\circ) = 10\sqrt{2} + j\,0$

$$I_2 = 10\sqrt{2} + j\,10\sqrt{6} = \sqrt{(10\sqrt{2})^2 + (10\sqrt{6})^2}\left[\angle \tan^{-1}\frac{10\sqrt{6}}{10\sqrt{2}}\right]$$

$$= 10\sqrt{8} \angle 60^\circ$$

(i) $I_1 + I_2 + I_3 = (10\sqrt{2} + j\,0) + (10\sqrt{2} + j\,10\sqrt{6}) + (7.5 - j\,7.5\sqrt{3})$

$= 35.784 + j\,11.505$ in rectangular form **Ans.**

$$= \sqrt{(35.784)^2 + (11.505)^2}\left[\angle \tan^{-1}\frac{11.505}{35.784}\right]$$

$= 37.588 \angle 17.82^\circ$ in polar form **Ans.**

(ii)
$$\frac{I_1 \times I_3}{I_2} = \frac{10\sqrt{2} \angle 0^\circ \times 15 \angle -60^\circ}{10\sqrt{8} \angle 60^\circ}$$

$$= \frac{10\sqrt{2} \times 15}{10\sqrt{8}}\,(\angle 0^\circ + (-60^\circ) - 60^\circ)$$

$= 7.5 \angle -120^\circ$ in polar form **Ans.**

5.33. DIVISION OF POLAR VECTORS

Divide $A\angle\alpha$ with $B\angle\beta$.

$$\therefore \quad C = \frac{A\angle\alpha}{B\angle\beta} = \frac{A}{B}\angle\alpha - \beta$$

N.B. An angle may be transferred from Denominator to Numerator and vice versa by reversing its sign.

Example 5.56. Multiply 20 ∠36° and 25 ∠74°.

Solution :

$$20 \angle 36^\circ \times 25 \angle 74^\circ$$
$$= 20 \angle 36^\circ \times 25 \angle -74^\circ$$
$$= 20 \times 25 \angle 36^\circ - 74^\circ$$
$$= 500 \angle -38^\circ = 500 \angle 38^\circ$$

Example 5.57. Divide 50 ∠ 35° by 5 ∠ 45°.

Solution : $$\frac{50 \angle 35}{5 \angle 45} = \frac{50}{5} \angle 35^\circ - 45^\circ$$
$$= 10 \angle -10^\circ = 10 \angle 10^\circ$$

Example 5.58. Divide 35 ∠ 13° by 7 ∠ 30°.

Solution : $$\frac{35 \angle 130^\circ}{7 \angle 30^\circ} = \frac{35 \angle 130^\circ}{7 \angle -30^\circ}$$
$$= 5 \angle 130^\circ + 30^\circ$$
$$= 5 \angle 160^\circ$$

Example 5.59. Express the following in the polar form :

(i) $60 + J\,80$

(ii) $-3 + J\,5$

(iii) $6 - J\,15$

Solution :

(i) Absolute value $= \sqrt{60^2 + 80^2} = \sqrt{3600 + 6400}$
$$= \sqrt{10000} = 100$$
$$\tan \theta = \frac{80}{60} = 1.333$$
$$\therefore \quad \theta = 53^\circ — 08'$$

∴ Vector is 100 ∠ 53° — 08′

(ii) Absolute value $= \sqrt{3^2 + 5^2} = \sqrt{34} = 5.831$
$$\tan \theta = \frac{5}{3} = 1.667$$
$$\therefore \quad \theta = 59^\circ — 02'$$

Angle with X-axis = 18° – (59° — 02′) = 120 – 58′

or Vector is 5.831 ∠ 120° — 58′

(iii) (6 – J 15)

$$\text{Modulus} = \sqrt{6^2 + 15^2} = \sqrt{36 + 225} = \sqrt{261} = 16.16$$

$$\tan\theta = -\frac{15}{6} = -2.5$$

$$\therefore \quad \theta = -68^\circ — 12' = \angle 68^\circ — 12'$$

Vector is $16.16 \angle 68^\circ — 12'$.

HIGHLIGHTS

1. j is an operator which indicates an operation of rotating the phasor through 90° in clockwise direction,

$$j = \sqrt{-1}\ ; j^2 = -1$$

$$j^3 = -1\ ; j^4 = 1$$

2. The conjugate of $2 + j\,3$ is $2 - j\,3$. The conjugate of $-4 + j\,3$ is $-4 - j\,3$. In the polar form, conjugate of $5 \angle 30^\circ$ is $5 \angle -30^\circ$.

3. In the polar form,

$$A^n = A^n \angle n\,\theta \qquad ...(i)$$

$$\sqrt[n]{A} = \sqrt[n]{A} \angle \frac{\theta}{n} \qquad ...(ii)$$

4. Applying complex algebra to R-L series a.c. circuit,

(i) Rectangular form $I = I + j\,0$ ...Reference phasor

$$Z = R + j\,X_L$$

$$\therefore \quad V = IZ = (I + j\,0)\,(R + j\,X_L)$$

The solution of V can be obtained in the standard form.

This will enable us to quickly determine the magnitude and the phase angle of V.

Taking voltage as the reference phasor, we have,

$$V = V + j\,0 \qquad \text{...Reference phasor}$$

$$\therefore \quad I = \frac{V}{Z} = \frac{V + j\,0}{R + j\,X_L}$$

The magnitude and phase angle of I can be readily determined.

(ii) Polar form

$$I = I \angle 0^\circ \qquad \text{...Reference phasor}$$

$$Z = Z \angle \phi^\circ \quad \text{where } Z = \sqrt{R^2 + X_L^2}$$

and

$$\phi = \tan^{-1}\frac{X_L}{R}$$

$$\therefore \quad V = IZ = I \angle 0^\circ \times Z \angle \phi^\circ = IZ \angle \phi^\circ$$

If voltage is taken as reference phasor, then

$$V = V \angle 0^{\circ} \text{ ...Reference phasor}$$

$$\therefore \qquad I = \frac{V}{Z} = \frac{V \angle 0^{\circ}}{Z \angle \phi^{\circ}} = \frac{V}{Z} \angle -\phi^{\circ}$$

5. Applying complex algebra to R-C series a.c. circuit,

(i) Rectangular form

$$I = I + j\,0 \qquad \text{...Rectangular phasor}$$

$$Z = R - j\,X_C$$

$$\therefore \qquad V = IZ = (I + j\,0)\,(R - j\,X_C)$$

The magnitude and phase angle of V can be calculated.

Taking voltage as the reference phasor,

$$V = V + j\,0 \text{ ...Reference phasor}$$

$$\therefore \qquad I = \frac{V}{Z} = \frac{V - j\,0}{R - j\,X_C}$$

The solution of I can be obtained in the standard form $a \pm j\,b$. This will enable us to quickly determine the magnitude and phase angle of I.

(ii) Polar form

$$I = I \angle 0^{\circ} \qquad \text{...Reference phasor}$$

$$Z = Z \angle -\phi^{\circ} \quad \text{where } Z = \sqrt{R^2 + X_C^2}$$

and $$\phi = \tan^{-1} \frac{X_C}{R}$$

$$V = IZ = I \angle 0^{\circ} \times Z \angle -\phi^{\circ} = IZ\,\phi - \phi^{\circ}$$

6. Applying complex algebra to R-L-C series a.c. circuit,

(i) Rectangular form

$$I = I + j\,0 \qquad \text{...Reference phasor}$$

$$Z = R + j\,(X_L - X_C)$$

$$\therefore \qquad V = IZ = [(I + j\,0)]\,[R + j\,(X_L - X_C)]$$

The circuit will be inductive or capacitive depending upon whether X_L or X_C is greater in magnitude.

If voltage is taken as the reference phasor, then,

$$V = V + j\,0 \text{ ...Reference phasor}$$

$$\therefore \qquad I = \frac{V}{Z} = \frac{V + j\,0}{R + j\,(X_L - X_C)}$$

(ii) Polar form

$$I = I\angle 0^\circ \quad \text{...Reference phasor}$$

$$Z = Z\angle \pm\phi^\circ \text{ where } Z = \sqrt{R^2 + (X_L - X_C)^2}$$

and $$\tan\phi = \frac{X_L - X_C}{R}$$

$\therefore$ $$V = IZ = I\angle 0^\circ \times Z\angle \pm\phi^\circ = IZ\angle \pm\phi^\circ$$

If voltage is taken as the reference phasor, then,

$$V = V\angle 0^\circ \quad \text{...Reference phasor}$$

$$I = \frac{V}{Z} = \frac{V\angle 0^\circ}{Z\angle \pm\phi^\circ} = \frac{V}{Z}\angle \pm\phi^\circ$$

7. Apparent power in complex form

= Phasor voltage × Conjugate of phasor current

SHORT QUESTIONS

1. Why polar method is used ?
2. Why symbolic J-method is preffered for solving a.c. circuits ?
3. A.c. quantities are added vectorially, why ?

MULTIPLE CHOICE QUESTIONS

1. The operator J has a value of

(a) +1 *(b)* –1
(c) $\sqrt{-1}$ *(d)* $\sqrt{1}$

2. The conjugate of complex impedance, $Z = R\,j\,X$ is

(a) $-R - jx$ *(b)* $-R + jx$
(c) $R - jx$ *(d)* $jx - R$

3. The symbol J represents anticlockwise rotation of a vector by

(a) 180° *(b)* 360°
(c) 270° *(d)* 90°

4. The vector $J^4 E$ is same as vector

(a) JE *(b)* $J^2 E$
(c) E *(d)* $J^3 E$

ANSWERS

1. *(c)* 2. *(c)* 3. *(d)* 4. *(c)*

EXERCISES

1. Explain *j*-notation as applicable to electric quantities.
2. Explain the method of representing a phasor by symbolic or *j* notation.
3. Explain how alternating quantities may by represented by complex numbers?
4. Is impedance a phasor quantity, if yes then how it is expressed in phasor or complex form?
5. Explain the conjugate method of determining power in ac circuits.

PROBLEMS

1. Give the current expressed in the polar form $I = 10 \angle 53.1^\circ$ amperes in symbolic and trigonometrical forms.

   (Ans. $(6 + J\,8)A$; $10\,(\cos 53.1^\circ + j\,53.1^\circ)\,A$**)**
2. Divide phasor $(5 + J\,6)$ by phasor $(4 - J\,3)$.

   (Ans. $0.08 + j\,1.56$**)**
3. Add phasors $30 \angle 45^\circ$ and $7.5 \angle -30$.

   (Ans. $27.7 + j\,17.46$**)**
4. Determine the current flowing through a circuit having impedance of $(5 + j\,12)\,\Omega$ connected across 230 V, 50 Hz supply.

   (Ans. $17.69 \angle -67.38^\circ$**)**
5. A coil of resistance 20 Ω and inductance 0.1 H is connected in series with 50 μF capacitor across a 100 V, 50 Hz supply. Estimate *(i)* current in the circuit *(ii)* power *(iii)* power factor *(iv)* voltage across the capacitor.

   (Ans. *(i)* **2.636A,** *(ii)* **138.9 W,** *(iii)* **0.5271 leading,** *(iv)* **167.8V)**
6. A resistance of 50 Ω, inductance 100 mH and a capacitor of 100 μF are connected in series across 200 V, 50 Hz supply. Determine the impedance, current, phase angle, pf, active and reactive power.

   [Ans. $50 \angle -0.5^\circ\,\Omega$; $4 \angle 0.5^\circ$ A ; **0.5 (lead), 0.99997 (lead); 799.97 W ; 7 VARs (capacitive)]**
7. A coil of resistance 10 Ω and inductance 0.1 H is connected in series with a capacitor of 150 μF across 200 V, 50 Hz supply. Determine the impedance, circuit current, active and reactive power.

   (Ans. $14.28 \angle 45.57^\circ$; $14 \angle -45.57^\circ$ A ; **1960 W ; 1960 VARs (inductive))**
8. Two impedances $(10 + j\,5)$ ohms and $(8 + j\,6)$ ohms are joined in parallel and connected across a voltage of $(200 + j\,0)$ volts. Calcutate *(i)* the total current drawn from the supply; *(ii)* the phase angle of total current with the supply voltage; *(iii)* the branch currents; and *(iv)* the total power.

   (Ans. *(i)* **37.74 A;** *(ii)* **32° (lagging);** *(iii)* **17.89 A;** *(iv)* **6.4 kw)**
9. Two impedances $Z_1 = (8 + j\,6)$ ohms and $Z_2 = (6 - j\,8)$ ohms are connected in parallel across 200 V ac mains. Calculate the *(i)* current in each branch; and *(ii)* total current and power factor of the combination, stating whether it is lagging or leading.

(**Ans.** *(i)* $20\angle -36.9^\circ$ A; $20\angle -53.1^\circ$ A, *(ii)* $28.28\angle 8.13^\circ$ A, 0.99 (leading))

10. Two impedances $Z_1 = (6 - j\,8)$ ohms and $Z_2 = (16 + j\,12)$ ohms are in parallel. If the total current of the combination is $(20 + j\,10)$ amperes, find the complex or power taken by each impedance. Draw and explain the complete phasor diagram.

(**Ans.** $(2400 + j\,3200)$, $(1600 - j\,1200)$)

11. Two impedances $Z_1 = (8 + j\,6)$ ohms and $Z_2 = (3 - j\,4)$ ohms are in parallel. If the total current of this combination is 25 A, find the power taken by each impedance. Draw and explain the complete phasor diagram.

(**Ans.** 1000 watts, 1,500 watts)

5.34. SOLUTION AND PHASOR DIAGRAMS FOR PRACTICAL R-L-C SERIES CIRCUITS

A circuit which contains a pure resistance *R*, a pure inductance *L* and pure capacitance *C* connected in series is called R-L-C series circuit. The Fig. 5.60 *(a)* shows a R-L-C series circuit.

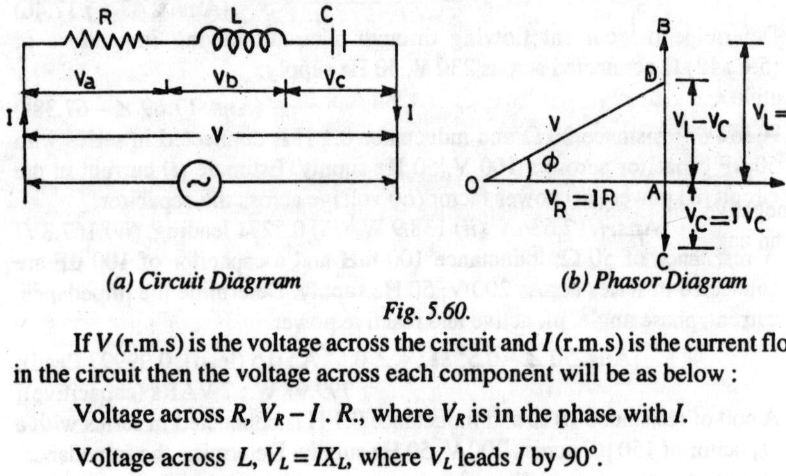

(a) Circuit Diagrram *(b) Phasor Diagram*

Fig. 5.60.

If *V* (r.m.s) is the voltage across the circuit and *I* (r.m.s) is the current flowing in the circuit then the voltage across each component will be as below :

Voltage across *R*, $V_R - I\,.\,R.$, where V_R is in the phase with *I*.

Voltage across *L*, $V_L = IX_L$, where V_L leads *I* by 90°.

Voltage across *C*, $V_C = IX_C$ where V_C lags *I* by 90°.

The phasor diagram (shown in Fig. 5.60 *(b)*. *OA* represents V_R. *AB* represents V_L and *AC* represents V_C. It is clear from the phasor diagram, that the circuit can either be effectively inductive or capacitive, depending upon the voltage drop (*i.e.*, when $V_L > V_C$) it is inductive and on the other hand (*i.e.*, when $V_C > V_L$) it is capacitive for the case considered, $V_L > V_C$, so that net voltage drop across $L - C$ combination is $(V_L - V_C)$ and is represented by AD. Therefore, the applied voltage V is the phasor sum of V_R and $(V_L - V_C)$.

$$\therefore \quad V = \sqrt{V_R^2 + (V_L - V_C)^2}$$

$$= \sqrt{(I.R.)^2 + (IX_L - IX_C)^2}$$

$$= I\sqrt{R^2 + (X_L - X_C)^2}$$

$$V = I \cdot Z$$

where $Z = \sqrt{R^2 + (X_L - X_C)^2}$, and offers opposition to the flow of current and is called impedance of the circuit.

Phase Angle : From phasor diagram

$$\text{Circuit power factor, } \cos\phi = \frac{R}{Z} = \frac{R}{\sqrt{R^2 + (X_L - X_C)^2}}$$

$$\tan\phi = \frac{V_L - V_C}{V_R} = \frac{(X_L - X_C)}{R}$$

Since X_L , X_C and R and known, therefore, phase angle ϕ of the circuit can be determined.

Three Cases of R-L-C Series Circuit : In the previous article, we have seen that the impedance of an R-L-C series circuit is given by :

$$Z = \sqrt{R^2 + (X_L - X_C)^2}$$

Case I : When $(X_L - X_C)$ is positive *i.e.*, if $X_L > X_C$ then phase angle ϕ is positive and the circuit will be inductive. In such cases, current lags in the voltage by an angle ϕ.

Case II : When $(X_L - X_C)$ is negative *i.e.*, $(X_L < X_C)$, then the phase angle ϕ is negative and the circuit is capacitive. In such circuits current leads the voltage by an angle ϕ.

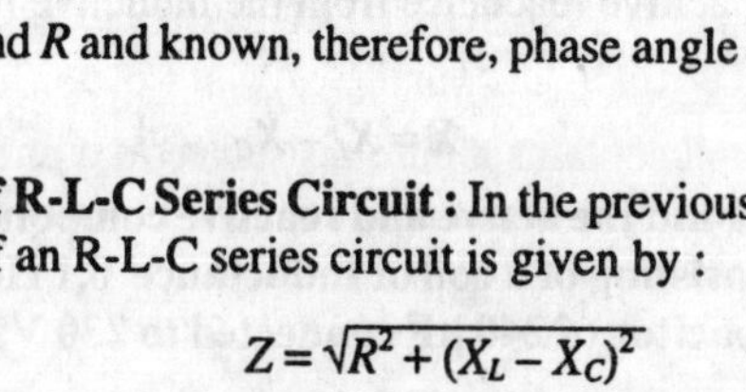

Fig. 5.61.

Case III : When $(X_L - X_C)$ is zero (*i.e.*, $X_L = X_C$) then the phase angle $\phi = 0°$ and the circuit is pure resistive. In such cases, the current is in phase with the voltage. The power factor of such circuits is unity. If the equation for the applied voltage is $v = V_{max} \sin \omega t$

Then the equations of the currents can be :

(i) $i = I_{max} \sin(\omega t - \phi)$ ... case I; $X_L > X_C$

(ii) $i = I_{max} \sin(\omega t + \phi)$ ... case II; $X_C > X_L$

(iii) $i = I_{max} \sin \omega t$ ... case III; $X_L = X_C$

5.35. REACTANCE

Inductive reactance, X_L is directly proportional to frequency being equal to ωL or $2\pi f L$ and capacitive reactance, X_C is inversely proportional to frequency being equal to $\frac{1}{\omega C}$ or $\frac{1}{2\pi f C}$.

Inductive reactance causes the current to lag behind the applied voltage, while the capacitive reactance causes the current to lead the voltage. So when inductance and capacitance are connected in series, their effects neutralize each other and their combined effect is then their difference. The combined effect of inductive reactance and capacitive reactance is called the reactance and is found by subtracting the capacitive reactance from the inductive reactance or according to equation

$$X = X_L - X_C$$

Example 5.60. Find the active and reactive components of current taken by a series circuit consisting of a coil of Inductance 0.1 Henry and a resistance of 10 ohms and a capacitor of 240 μF connected to 230 V, 50 Hz supply mains.

Solution : Value of resistance, $R = 10$ ohms.

Value of inductance, $L = 0.1$ Henry

Value of capacitance, $C = 240\ \mu F = 240 \times 10^{-5}$ F

Supply voltage, $V = 230$ volts.

Supply frequency, $f = 50$ Hz

Inductive reactance, $X_L = L = 2\pi f L$

$$= 2\pi \,.\, 50 \times 0.1 = 31.4 \text{ ohms}$$

$$\text{Capacitive reactance, } X_C = \frac{1}{\omega C} = \frac{1}{2\pi f C}$$

$$= \frac{10^6}{2\pi \,.\, 50 \,.\, 240} = 13.26 \text{ ohms}$$

Resultant reactance, $X_L - X_C = 31.4 - 13.26 = 18.14\ \Omega$

$$\text{Impedance, } Z = \sqrt{R^2 + (X_L - X_C)^2}$$

$$= \sqrt{10^2 + 18.14^2} = 20.71\ \Omega$$

$$\therefore \quad \text{Current, } I = \frac{V}{Z} = \frac{230}{20.71} = 11.105 \text{ A}$$

$$\text{Power factor, } \cos\phi = \frac{R}{Z} = \frac{10}{20.71} = 0.4828$$

$$\sin\phi = \left(\frac{X_L - X_C}{Z}\right) = \frac{18.14}{20.71} = 0.8759$$

Active component of current $= I \cos \phi$

$$= 11.105 \times 0.4828$$

$$= 5.36 \text{ A } \textbf{Ans.}$$

Reactive component of current, $= I \sin \phi$

$$= 11.105 \times 0.8759$$

$$= 9.72 \text{ A } \textbf{Ans.}$$

Example 5.61. Two impedances Z_1 and Z_2 when connected separately across a 230 V, 50 Hz supply, consume powers of 100 W, 60 W at power factors of 0.5 lagging and 0.6 leading respectively. If the two impedances are now connected together in series across the same supply. Calcuate *(i)* current *(ii)* power absorbed *(iii)* circuit power factor.

Solution : Let Z_1 and Z_2 be the two impedances. Since the power factor of Z_1 is lagging and of the Z_2 leading, therefore, Z_1 consists of an R-L circuit whereas Z_2 consists of an R-C circuit.

Case No. 1. Let Z_1 consists of R-L series circuit.

Voltage across, $Z_1 = 230$ V

Power consumed, $P_1 = 100$ W

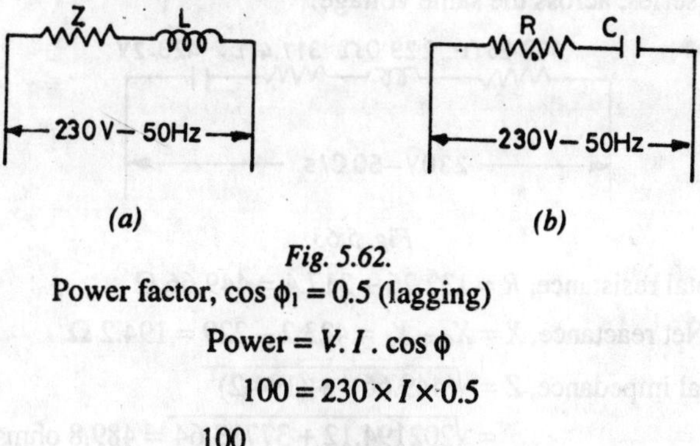

(a) *(b)*

Fig. 5.62.

Power factor, $\cos \phi_1 = 0.5$ (lagging)

Since $\quad$ Power $= V . I . \cos \phi$

or $\quad 100 = 230 \times I \times 0.5$

$\therefore \quad I = \dfrac{100}{230 \times 0.5} = 0.8695$ A

Also $\quad P = I^2 R$

$\therefore \quad R = \dfrac{P}{I^2} = \dfrac{100}{0.8695^2} = 132.26\ \Omega$

$$\cos \phi = \frac{R}{Z}$$

$\therefore \quad Z = \dfrac{R}{\cos \phi} = \dfrac{132.26}{0.5} = 264.52\ \Omega$

Inductive reactance, $X_L = \sqrt{Z^2 - R^2} = \sqrt{(264.52)^2 - (132.26)^2} = 229.0\ \Omega$

Case No. 2. Let Z_2 consists of R-C circuit [Fig. 5.62 *(b)*].

Voltage across, $Z_2 = 230$ V

Power consumed, $P_2 = 60$ W

Power factor, $\cos\phi_2 = 0.6$ (lagging)

$$\text{Power} = V \cdot I \cdot \cos\phi$$

or $$60 = 230 \times I \times 0.6$$

$\therefore$ $$I = \frac{60}{230 \times 0.6} = 0.4347 \text{ A}$$

$$\text{Impedence, } Z_2 = \frac{230}{0.4347} = 529\ \Omega$$

$$\text{Power factor, } \cos\phi = \frac{R}{Z}$$

$\therefore$ $$R = Z\cos\phi = 529 \times 0.6 = 317.4\ \Omega$$

Capacitive reactance, $X_C = \sqrt{Z^2 - R^2}$ or $X_c = \sqrt{529^2 - 317.4^2} = 423.2\ \Omega$

The circuit can be re-arranged in series when the two impedances are connected in series, across the same voltage.

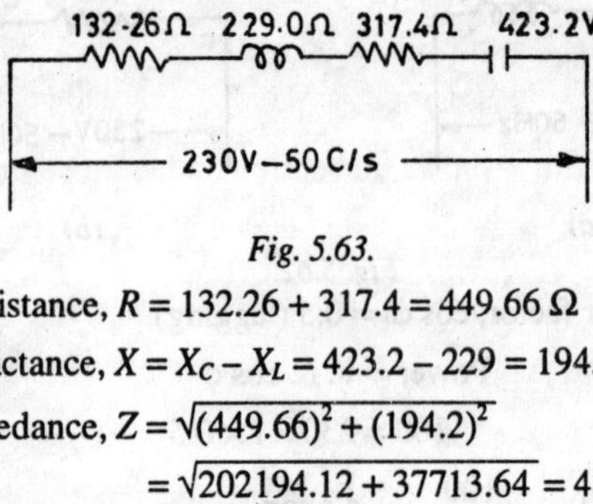

Fig. 5.63.

Total resistance, $R = 132.26 + 317.4 = 449.66\ \Omega$

Net reactance, $X = X_C - X_L = 423.2 - 229 = 194.2\ \Omega$

Total impedance, $Z = \sqrt{(449.66)^2 + (194.2)^2}$

$= \sqrt{202194.12 + 37713.64} = 489.8$ ohms

$$\text{Current, } I = \frac{V}{Z} = \frac{230}{489.8} = 0.47 \text{ A } \textbf{Ans.}$$

Power absorbed, $I^2 R = (0.47)^2 \times 449.66 = 99$ Watts **Ans.**

Circuit power factor, $\cos\phi = \frac{R}{Z} = \frac{449.66}{489.8} = 0.918$ **Ans.**

Example 5.62. A 230 V, 50 Hz supply is applied to a coil of 0.06 H inductance and 2.5 Ω resistance connected in series with a 6.8 μ F capacitor. Calculate *(i)* impedance, *(ii)* current, *(iii)* phase angle between current and voltage, *(iv)* power factor and *(v)* power consumed.

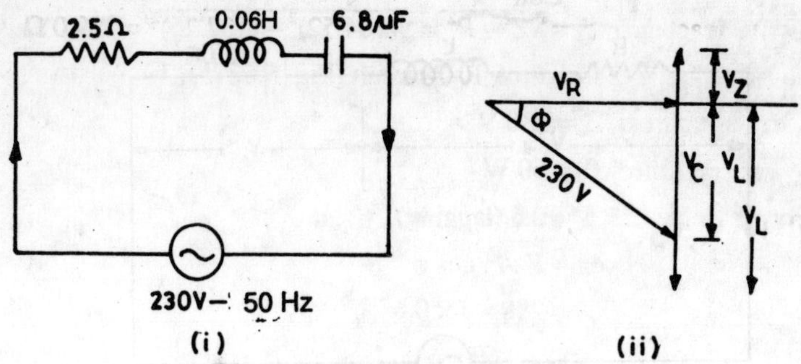

Fig 5.64.

Solution : $X_L = 2\pi f L = 2\pi \times 50 \times 0.06 = 18.85\ \Omega$

$$X_C = \frac{1}{2\pi f C} = \frac{10^6}{2\pi \times 50 \times 6.8} = 468\ \Omega$$

(i) Circuit impedance, $Z = \sqrt{R^2\ (X_L - X_C)^2}$

$$= \sqrt{(2.5)^2 + (18.85 - 468)^2} = 449.2\ \Omega$$

(ii) Circuit current, $I = V/Z = 230/449.2 = 0.512$ A

(iii) $\tan\phi = \dfrac{X_L - X_C}{R} = \dfrac{18.85 - 468}{2.5} = -179.66$

$\therefore$ Phase angle, $\phi = \tan^{-1} -179.66 = -89.7^\circ$

$= 89.7^\circ$ lead

The negative sign with ϕ shows that current is leading the voltage [See the phasor diagram in Fig . 5.64 *(ii)*.

(iv) Power factor, $\cos\phi = \dfrac{R}{Z} = \dfrac{2.5}{449.2} = 0.00557$ lead

(v) Power consumed, $P = VI \cos\phi$

$$= 230 \times 0.512 \times 0.00557 = 0.656\ \text{W}$$

Example 5.63. A coil of p.f. 0.8 is connected in series with a 110 μF capacitor. The supply frequency is 50Hz. The p.d. across the coil is found to be equal to the p.d. across the capacitor. Calculate the resistance and inductance of the coil.

Solution :

Reactance of capacitor, $X_C = \dfrac{1}{2\pi f C} = \dfrac{10^6}{2\pi \times 50 \times 110} = 29\ \Omega$

Now, $I Z_{coil} = I X_C$

$\therefore$ $Z_{coil} = X_C = 29\ \Omega$

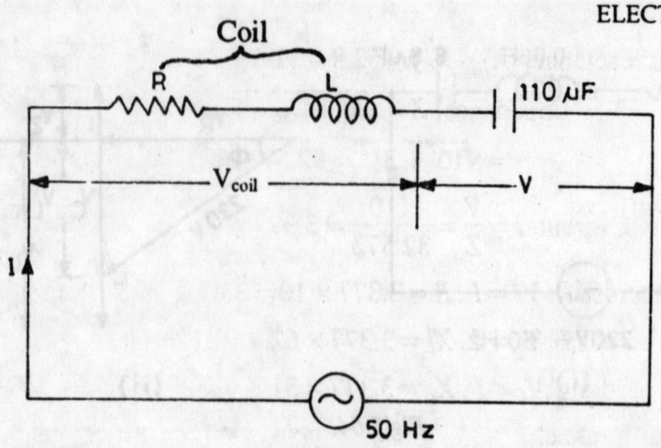

Fig. 5.65.

For the coil, $\cos \phi = R/Z_{coil}$

$\therefore$ $R = Z_{coil} \cos \phi = 29 \times 0.8 = 23.2\ \Omega$

Reactance of coil, $X_L = Z_{coil} \sin \phi = 29 \times 0.6 = 17.4\ \Omega$

$\therefore$ Inductance of coil, $L = \dfrac{X_L}{2\pi f} = \dfrac{17.4}{2\pi \times 50} = 0.055$ H

Example 5.64. An ac circuit having a resistance of 10 ohms, an inductance of 0.2 H and a capacitance of 100 μF in series, is connected across a single phase 110 volts, 50 Hz supply. Calculate the following for this circuit.

(i) **Resultant reactance**

(ii) **Impedance**

(iii) **Current**

(iv) **Voltage across R, L and C**

(v) **Phase difference between supply voltage and current**

(vi) **Draw the phasor diagram.**

Solution : Value of resistance, $R = 10$ ohms

Value of inductance, $L = 0.2$ Henry

Value of capacitance, $C = 100 \times 10^{-6}$ Farads

Supply frequency, $f = 50$ Hz

Supply voltage, $V = 110$ V

Inductive reactance, $X_L = \omega L = 2\pi \times 50 \times 0.2 = 62.8$ ohms

Capacitive reactance, $X_C = \dfrac{1}{\omega C}$

$$= \frac{10^6}{2\pi \times 50 \times 100} = 31.8 \text{ ohms}$$

(i) Resultant reactance, $(X_L - X_C) = 62.8 - 31.8 = 31$ ohms **Ans.**

(ii) Impedance, $Z = \sqrt{R^2 + (X_L - X_C)^2}$

$$= \sqrt{10^2 + 31^2} = 32.573 \text{ ohms } \textbf{Ans.}$$

(iii) Current, $I = \dfrac{V}{Z} = \dfrac{110}{32.573} = 3.377$ A **Ans.**

(iv) Voltage across R, $V_R = I \cdot R = 3.377 \times 10 = 33.77$ volts

Voltage across L, $V_L = I \cdot X_L = 3.377 \times 62.8 = 212$ volts

Voltage across C, $V_C = I \cdot X_C = 3.377 \times 31.8 = 107.388$ volts **Ans.**

(v) Phase difference,

$$\phi = \tan^{-1} \frac{X_L - X_C}{R}$$

$$= \tan^{-1} \frac{31}{10} = 72.12° \text{ (lagging)}$$

(vi) Phasor diagram:

$OA = 33.77$ volts

$AB = 212$ volts

$AC = 107.388$ volts

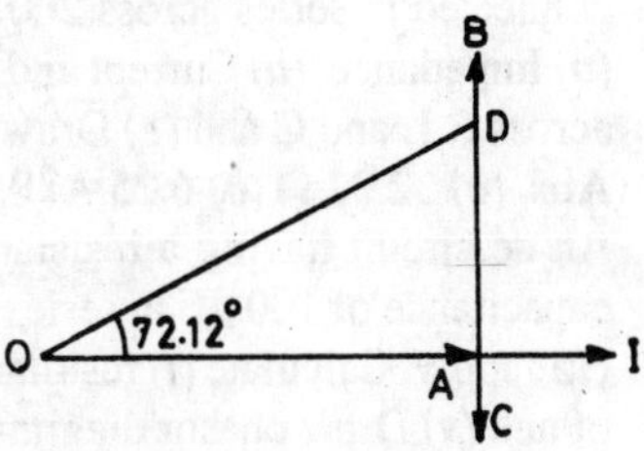

Fig. 5.66.

HIGHLIGHTS

1. In R-L-C series a.c. circuit,

$$\text{Impedence, } Z = \sqrt{R^2 + (X_L - X_C)^2}$$

$$\text{Circuit current, } I = V/Z$$

$$\text{Phase angle, } \phi = \tan^{-1} \frac{X_L - X_C}{R}$$

$$\text{Power factor, } \cos\phi = R/Z$$

$$\text{Power consumed, } P = VI \cos\phi = I^2 R$$

The circuit will be inductive if $X_L - X_C$ is positive, *i.e.*, circuit current will lag behind the applied voltage by ϕ. However, if $X_L - X_C$ is negative, the circuit will be capacitive, *i.e.*, the circuit current will lead the applied voltage by ϕ.

PROBLEMS

1. A coil of resistance 10 Ω and inductance 0.1 H connected in series with a condenser of capacitance 150 μF across a 200 V, 50 Hz supply. Determine the following : *(i)* Inpedance, *(ii)* Current, *(iii)* Power fact, *(iv)* Voltage across the coil and *(v)* Voltage across condenser.

(**Ans.** *(i)* 14.28 N, *(ii)* 14.0 A, *(iii)* 0.7 lag, *(iv)* 461.57 V, *(v)* 299.0 V)

2. A resistance of 20 Ω, an inductance of 0.2 H and a capacitance of 100 μF are connected in series across 220 V, 50 Hz mains. Determine the following : *(a)* impedance, *(b)* current, *(c)* voltage across R,L and C, *(d)* power in watts and VA and *(e)* pf and angle of lag.

(**Ans.** *(a)* 36.9 Ω *(b)* 5.96 A *(c)* 119.3 V, 374.7 V, 189.8 V *(d)* 710 W; 1311 VA, *(e)* 0.54 (lag); 57.2°)

3. A series circuit having $R = 100\ \Omega$, $L = 120$ mH and $C = 28.27$ μF is fed from a 100V, 50Hz supply. Calculate the current, active power, impedance, rms value of voltage across resistance, inductance and capacitance.

(**Ans.** 0.8 A, 64 W, 125 Ω, 80 V, 30.16 V, 90.08 V)

4. A resistance of 8 Ω, inductance of 200 mH and capacitor of 100 μF are connected in series across 200 V, 50 Hz supply. Determine the following : *(a)* Impedance, *(b)* Current and power factor, *(c)* Power in watts, *(d)* Voltage across R,L and C and *(e)* Draw voltage triangle.

(**Ans.** *(a)* 32.02 Ω *(b)* 6.25 A; 0.25 *(c)* 312.5 W, *(d)* 50 V, 392.7 V, 198.9 V)

5. An ac circuit having a resistance of 10 Ω , an inductance of 0.2 H and a capacitance of 100 μF in series is connected across a single phase 110 V, 50 Hz supply. Calculate *(i)* resultant reactance, *(ii)* impedance, *(iii)* current, *(iv)* pf and *(v)* Draw phasor diagram.

(**Ans.** *(i)* 31 Ω inductive *(ii)* 32.6 Ω *(iii)* 3.38 A *(iv)* 0.307 (lag))

6. A series circuit contains an inductor of 10 Ω resistance and 6 Ω reactance and a condenser of 25 Ω capacitive reactance and negligible resistance. If the voltage is 240 V, what current flows through the circuit ?

(**Ans.** 11.178 A leading applied voltage by 62.24°)

EXERCISE

1. Explain what is meant by impedance of a circuit and determine its value for a circuit possessing resistance *R*, inductance *L* and capacitance *C*.
2. What do you understand by the term impedance ?
3. Derive an expression for the impedance of a circuit containing resistance, inductance and capacitance.

5.35.1. A.C. Parallel CirCuits

The a.c. circuits in which number of branches are connected in such a manner so that voltage across each branch is the same but current flowing through them is different, are called *a.c. parallel circuits.*

The parallel circuits are used more frequently in a.c. system because of the following reasons :

(i) Almost all the electrical appliances (or devices) of different ratings are operated at the same supply voltage and are connected in parallel.

(*ii*) Each device is required to be operated independently (with a switch) without disturbing the operation of other devices. Hence, connected in parallel.

5.35.2. Methods of Solving Parallel A.C. Circuits

In parallel circuits, number of branches are connected in parallel. Each branch, generally, contains number of components like resistance, inductance and capacitance forming series circuits. Therefore, each branch is analysed separately as a series circuit and then the effects of separate branches are combined together. While carrying out circuit calculations, the magnitudes and phase angles of voltages and currents are taken into account. The following methods may be applied for solving a.c. parallel circuits.

(*i*) Phasor (or vector) method.

(*ii*) Admittance method.

(*iii*) Symbolic method or *J*-method

Here we will discuss vector method.

The method to be applied for the solution, depends upon the conditions of the problem. However, in general, the method which yields quick results is applied.

5.36. PHASOR (OR VECTOR) METHOD

To solve parallel a.c. circuits by this method, proceed as follows:

Step I : Draw the circuit as per the given problem as shown in Fig. 5.67 *(a)*. (Here for illustration, we have considered two branches connected in parallel. One branch contains resistance and inductance in series, where as second branch contains resistance and capacitance in series. The supply voltage is *V* volt.

Step II : Find the impedance of each branch of the circuit separately, *i.e.*,

$$Z_1 = \sqrt{R_1^2 + X_{L_1}^2} \text{ where } X_{L_1} = 2\pi f L_1$$

$$Z_2 = \sqrt{R_2^2 + X_{C_2}^2} \text{ where } X_{C_2} = 2\pi f C_2$$

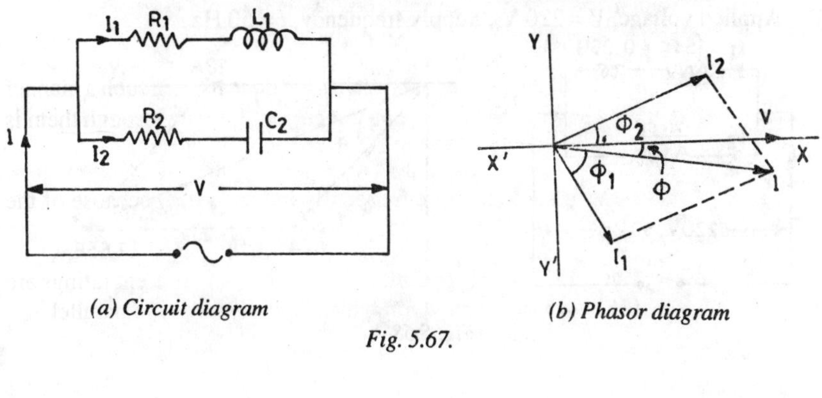

(a) Circuit diagram *(b) Phasor diagram*

Fig. 5.67.

Step III : Determine the magnitude of current and phase angle with the voltage in each branch.

$$I_1 = \frac{V}{Z_1}\ ;\ \phi_1 = \tan^{-1}\frac{X_{L_1}}{R_1} \text{ (lagging) [for inductive branch]}$$

$$I_2 = \frac{V}{Z_2}\ ;\ \phi_2 = \tan^{-1}\frac{X_{C_2}}{R_2} \text{ (leading) [for capacitive branch]}$$

Step IV : Draw the phasor diagram taking voltage as the reference phasor. Represent the branch currents on it, as shown in Fig. 5.67 *(b)*.

Step V : Find the phasor sum of branch currents by the method of components.

$$I_{XX} = I_1 \cos\phi_1 + I_2 \cos\phi_2$$

$$I_{YY} = -I_1 \sin\phi + I_2 \sin\phi_2 \text{ (negative)}$$

$$I = \sqrt{(I_{XX})^2 + (I_{YY})^2}$$

Step VI : Find the phase angle ϕ between the total current I and circuit voltage V.

$$\phi = \tan^{-1}\frac{I_{YY}}{I_{XX}} \text{ lagging (Since, } I_{YY} \text{ is negative)}$$

Power factor of the circuit = $\cos\phi$ (lagging)

or $\quad$ Power factor $= \dfrac{I_{XX}}{I}$ (lagging)

Example 5.65. A coil resistance 15 ohm and inductance 0.05 H is connected in parallel with a non-inductive resistance of 20 ohm. Find *(i)* current in each branch of the circuit, *(ii)* total current supplied, *(iii)* phase angle and p.f. of combination when a voltage of 200 volt at 50 Hz is applied *(iv)* power consumed in the circuit.

Solution : The circuit is shown in Fig. 5.68.

Given data :

Applied voltage, V = 220 V ; Supply frequency, f = 50 Hz

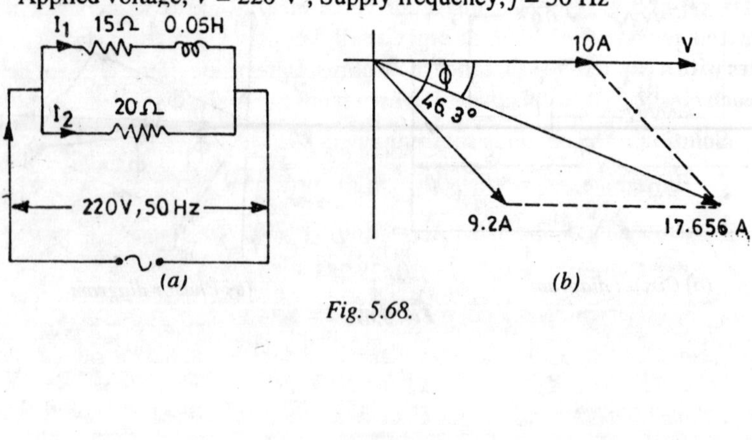

Fig. 5.68.

Branch I

$$R_1 = 15\ \Omega\ ;\ X_{L_1} = 2\pi f L_1 = 2\pi \times 50 \times 0.05 = 15.7\ \Omega$$

$$\text{Impedance, } Z_1 = \sqrt{R_1^2 + X_L^2} = \sqrt{(15)^2 + (15.7)^2} = 21.72 \text{ ohm}$$

$$\text{Current in the coil, } I_1 = \frac{V}{Z_1} = \frac{200}{21.72} = 9.2 \text{ A } \textbf{Ans.}$$

$$\text{Phase angle, } \phi_1 = \tan^{-1}\frac{X_{L_1}}{R} = \tan^{-1}\frac{15.7}{15} = 46.3^\circ \text{ lagging}$$

Branch II

$$\text{Resistance, } R_2 = 20 \text{ ohm}$$

$$\text{Branch current, } I_2 = \frac{V}{R_2} = \frac{200}{20} = 10 \text{ A } \textbf{Ans.}$$

$$\text{Phase angle, } \phi_2 = 0\ (I_2 \text{ is in phase with } V)$$

The two currents are shown vectorially in Fig. 5.68 *(b)*. Resolving the currents horizontally and vertically,

$$I_{XX} = I_2 + I_1 \cos \phi_1$$
$$= 10 + 9.2 \cos 46.3^\circ = 10 + 9.2 \times 0.6909 = 16.356 \text{ A}$$
$$I_{YY} = 0 - I_1 \sin \phi_1$$
$$= 0 - 9.2 \sin 46.3^\circ = -9.2 \times 0.723 = -6.65 \text{ A}$$

$$\text{Total current supplied, } I = \sqrt{I_{XX}^2 + I_{YY}^2}$$
$$= \sqrt{(16.356)^2 - (6.65)^2} = 17.656 \text{ A } \textbf{Ans.}$$

$$\text{Phase angle, } \phi = \tan^{-1}\frac{I_{YY}}{I_{XX}} = \tan^{-1}\left(\frac{-6.65}{16.356}\right) = -22.126^\circ \textbf{ Ans.}$$

$$\text{Power factor of the circuit, } \cos\phi = \cos(-22.126^\circ) = 0.9264 \text{ (lagging) } \textbf{Ans.}$$

$$\text{Power, } P = VI\cos\phi = 200 \times 17.656 \times 0.9264 = 3271.3 \text{ W } \textbf{Ans.}$$

Example 5.66. The circuits A and B are connected in parallel to a 230 V, 50 Hz supply. Circuit A consists of resistance 20 ohms in series with an inductive reactance of 20 ohms and circuit B consists of resistance 40 ohms in series with a capacitive reactance of 20 ohms. Determine *(i)* the current drawn by each circuit ; *(ii)* total current drawn from the mains.

Solution : The circuit is shown in Fig. 5.69.

$$\text{Impedance, } Z_1 = \sqrt{R_1^2 + X_{L_1}^2} = \sqrt{(20)^2 + (20)^2} = 28.284\ \Omega$$

$$\text{Impedance, } Z_2 = \sqrt{R_2^2 + X_{C_2}^2} = \sqrt{(40)^2 + (20)^2} = 44.72\ \Omega$$

$$\text{Current, } I_1 = V/Z_1 = 230/28.284 = 8.13 \text{ A } \textbf{Ans.}$$

$$\text{Current, } I_2 = V/Z_2 = 230/44.72 = 5.143 \text{ A } \textbf{Ans.}$$

Phase angle, $\phi_1 = \tan^{-1} X_{L_1}/R_1 = \tan^{-1} 20/20 = 45^\circ$ lag

Phase angle, $\phi_2 = \tan^{-1} X_{C_2}/R_2 = \tan^{-1} 20/40 = 26.56^\circ$

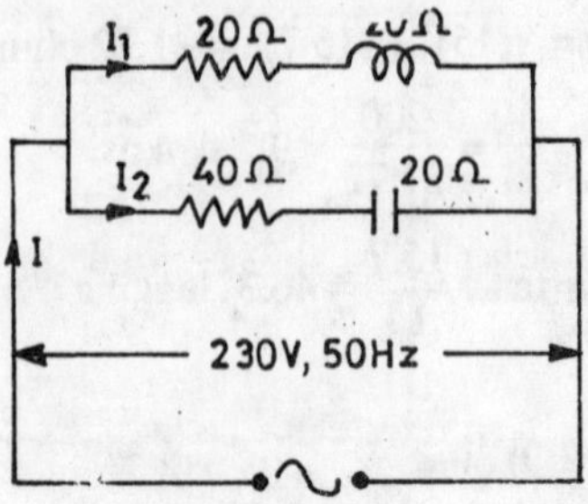

Fig. 5.69.

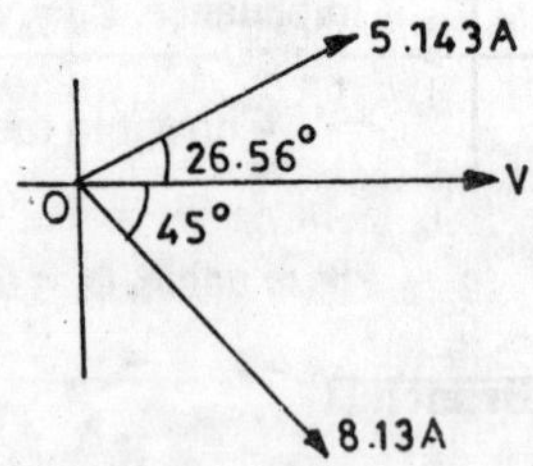

Fig. 5.70.

The phasor diagram representing voltage as reference phasor is shown in Fig. 5.70.

Resolving the currents horizontally and vertically, we get,

$$I_{XX} = I_1 \cos \phi_1 + I_2 \cos \phi_2$$

$$= 8.13 \cos 45^\circ + 5.143 \times \cos 26.56^\circ = 10.35 \text{ A}$$

$$I_{XX} = -I_1 \sin \phi_1 + I_2 \sin \phi_2$$

$$= -8.13 \times \sin 45^\circ + 5.143 \sin 26.56^\circ = -3.45 \text{ A}$$

Current drawn from the mains,

$$I = \sqrt{I_{XX}^2 + I_{YY}^2} = \sqrt{(10.35)^2 + (3.45)^2} = 10.91 \text{ A } \textbf{Ans.}$$

Example 5.67. A resistance of 10 Ω, an inductive reactance of 8 Ω and capacitance reactance of 15 Ω are connected in parallel across 120 V, 50 Hz. Determine :

(i) Current in each circuit.

(ii) The total current.

(iii) The phase difference between total current and applied voltage.

Show the respective current vectorially. What will be the power factor and power expended ?

Solution : I_R, current through, $R = \dfrac{V}{I} = \dfrac{120}{10} = 12$ A (In phase with V) **Ans.**

I_L, current through $L = \dfrac{V}{X_L} = \dfrac{120}{8} = 15$ A Lagging behing V by $\dfrac{\pi}{2}$ **Ans.**

I_C, current through $C = \dfrac{V}{X_C} = \dfrac{120}{15} = 8$ A (Leading the voltage by $\pi/2$) **Ans.**

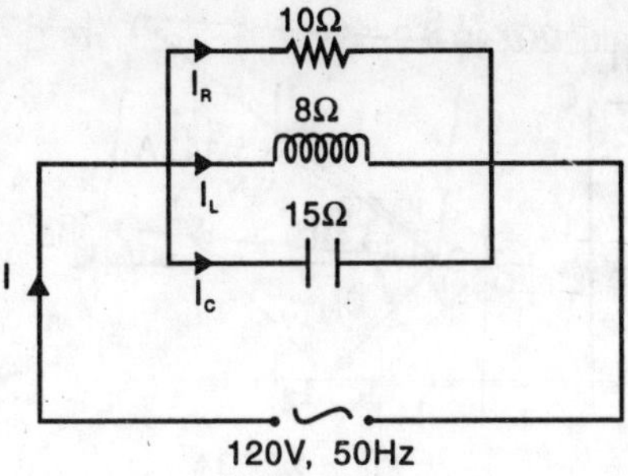

Fig. 5.71. Circuit diagram

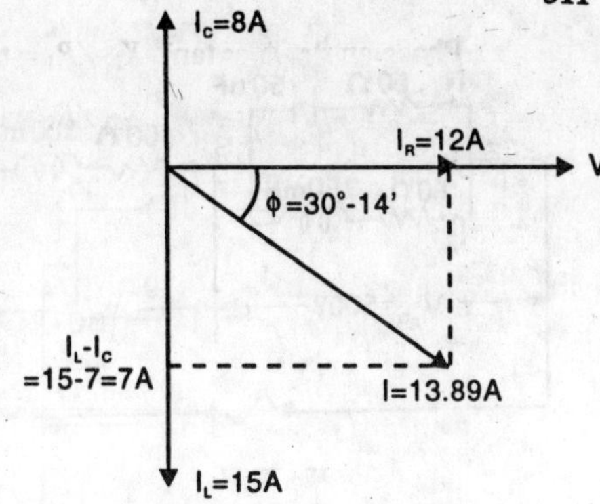

Fig. 5.72. Vector diagram

Voltage is taken as reference vector. Resultant current is the vector sum of branch currents.

$$I_L - I_C = 15 - 8 = 7 \text{ A}$$

Total current,

$$I = \sqrt{I_R^2 + (I_L - I_C)^2} = \sqrt{(12)^2 + (15-8)^2}$$
$$= \sqrt{(2)^2 + (7)^2} = 13.89 \text{ A } \textbf{Ans.}$$

$$\text{Power factor, } \cos\phi = \frac{R}{Z} = \frac{12}{13.89} = 0.864$$

$$\text{Phase angle, } \phi = \cos^{-1} 0.864 = 30^\circ\ 14' \textbf{ Ans.}$$

Power expended, $P = VI \cos\phi = 120 \times 13.89 \times 0.864 = 1440$ W **Ans.**

Example 5.68. An a.c. circuit includes two sections *AB* and *BC* in series. The section AB consists of two branches in parallel. The first of these is formed of resistance of 60 ohms in series with a capacitor of 50 μF, while the second consists of resistance of 60 ohm having an inductance of 250 mH. The section *BC* consists of a resistance of 100 ohm having an inductance of 300 mH. The frequency of the current is 50 Hz. The voltage across section *AB* is 500 V. What is the voltage across the section *BC*.

Solution : The circuit for the given problem is shown in Fig. 5.74.

Considering section *AB* ;

Branch I

$$\text{Capacitive reactance, } X_C = \frac{1}{2\pi f C} = \frac{2}{2\pi \times 50 \times 50 \times 10^{-6}} = 63.66 \text{ ohm}$$

$$\text{Impedance, } Z_1 = \sqrt{R_1^2 + X_{C_1}^2} = \sqrt{(60)^2 + (63.66)^2} = 87.48 \text{ ohm}$$

$$\text{Current, } I_1 = \frac{V_{AB}}{Z_1} = \frac{500}{87.48} = 5.72 \text{ A}$$

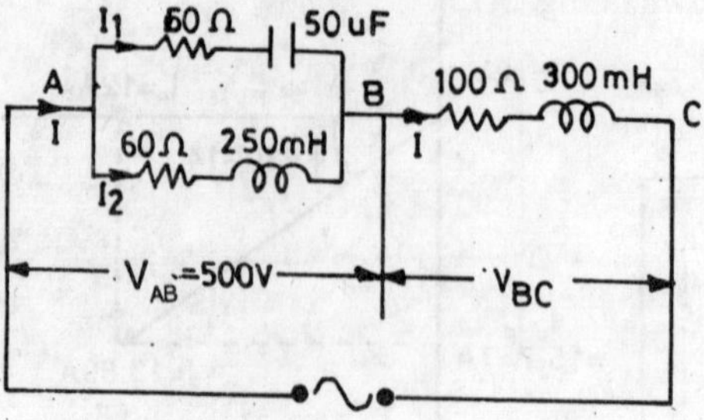

Fig. 5.73.

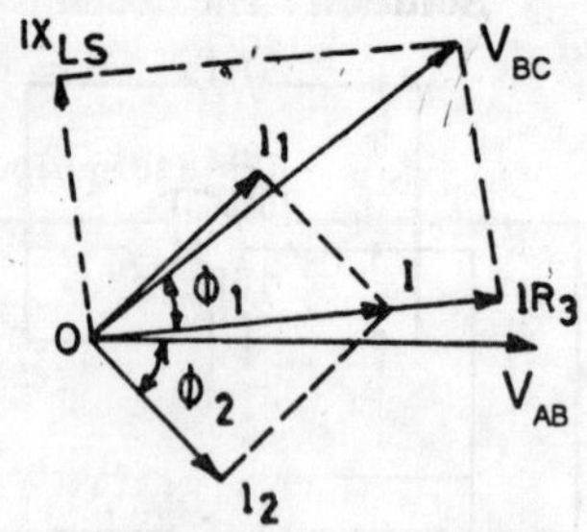

Fig. 5.74.

$$\text{Phase angle, } \phi_1 = \tan^{-1}\frac{X_{C_1}}{R_1} = \tan^{-1}\frac{63.66}{60} = 46.7^\circ \text{ (leading)}$$

Branch II

Inductive reactance, $X_{L_2} = 2\pi f L = 2\pi \times 50 \times 250 \times 10^{-3} = 78.54$ ohm

$$\text{Impedance, } Z_2 = \sqrt{R_2^2 + X_{L_2}^2} = \sqrt{(60)^2 + (78.54)^2} = 98.84 \text{ ohm}$$

$$\text{Current, } I_2 = \frac{V_{AB}}{Z_2} = \frac{500}{98.84} = 5.06 \text{ A}$$

$$\text{Phase angle, } \phi_2 = \tan^{-1}\frac{X_{L_2}}{R_2} = \tan^{-1}\frac{78.54}{60} = 52.62^\circ \text{ (lagging)}$$

The two currents I_1 and I_2 are shown vectorially in Fig. 5.74.

Resolving the currents horizontally and vertically;

$$I_{XX} = I_1 \cos\phi_1 + I_2 \cos\phi_2$$

$$= 5.72 \times \cos 46.7^\circ + 5.06 \times \cos 52.62^\circ = 6.9948 \text{ A}$$

$$I_{YY} = I_1 \sin\phi_1 - I_2 \sin\phi_2$$

$$= 5.72 \times \sin 46.7^\circ - 5.06 \times \sin 52.62^\circ = 0.142 \text{ A}$$

$$\text{Total current, } I = \sqrt{I_{XX}^2 + I_{YY}^2} = \sqrt{(6.9948)^2 + (0.142)^2} = 6.996 \text{ A}$$

Considering section BC ;

Inductive reactance, $X_{L_3} = 2\pi f L_3 = 2\pi \times 50 \times 300 \times 10^{-3} = 94.25$ ohm

$$\text{Impedance, } Z_3 = \sqrt{R_3^2 + X_{L_3}^2} = \sqrt{(100)^2 + (94.25)^2} = 137.41 \text{ ohm}$$

Current, $I = 6.996$ A

Voltage across BC, $V_{BC} = I\,Z_3 = 6.996 \times 137.41 = 961.35$ V **Ans.**

Example 5.69. A single phase motor takes 5 A current at 230 V, 50 Hz supply at a p.f. 0.707 lagging. It is required to improve the p.f. of the motor to 0.9 by connecting a capacitor in parallel with it. Determine the capacitance of capacitor.

Solution : The circuit is shown in Fig. 5.75 .

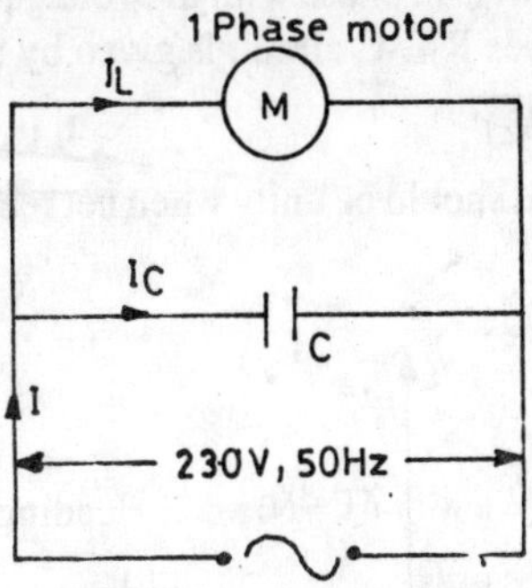

Fig. 5.75.

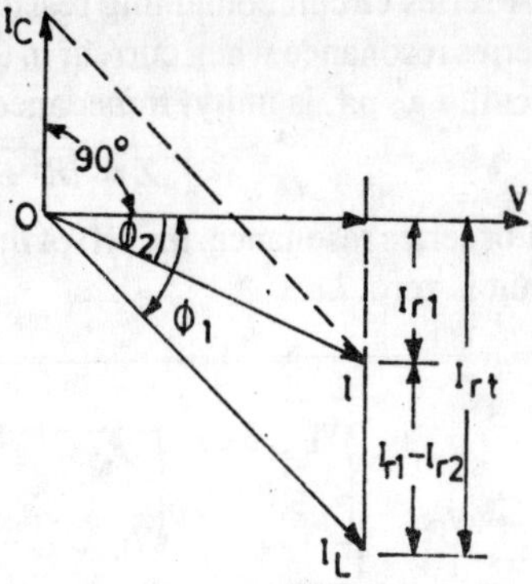

Fig. 5.76.

Active component of current, $I_a = I_m \cos \phi_1 = 5 \times 0.707 = 3.535$ A

Since load on the motor remains the same the active component of current drawn by the motor remains the same.

Now $\qquad \cos \phi_1 = 0.707$; $\tan \phi_1 = \tan \cos^{-1} 0.707 = 1$

Reactive component of current at p.f. 0.707;

$$I_{r_1} = I_a \tan \phi_1 = 3.535 \times 1 = 3.535 \text{ A}$$

When p.f. = $\cos \phi_2 = 0.9$;

$$\tan \phi_2 = \tan \cos^{-1} 0.9 = 0.4843$$

Reactive component of current at p.f. 0.9 ;

$$I_{r_2} = I_a \tan \phi_2 = 3.535 \times 0.4843 = 1.712 \text{ A}$$

From phasor diagram shown in Fig. 5.77 ;

Now, $\qquad I_C = V/X_C = V/(1/2\pi f C) = 2\pi f C V$

or $\qquad C = I_C/2\pi f V = 1.823/2\pi \times 50 \times 230 = 25.23$ μF **Ans.**

5.37. INTRODUCTION TO SERIES AND PARALLEL RESONANCE AND ITS CONDITIONS

An a.c. circuit containing reactive elements (*L* and *C*) is said to be in resonance when the circuit power factor is unity.

When applied voltage and circuit current in an a.c. circuit are in having phase angle zero or p.f. unity the circuit is said to be in *electrical resonance.* If this condition exists in a series a.c. circuit, it is called *series resonance.* On the other hand, if this condition exists in a parallel a.c. circuit, it is called *parallel resonance.* The frequency at which resonance occurs is called *resonant frequency* (fr).

5.38. SERIES RESONANCE

A series circuit containing resistance, inductance and capacitance is said to have series resonance when current in the circuit is in phase with the voltage across the circuit, *i.e.*, p.f. is unity. Impedance of series R-L-C circuit is given by :

$$Z = \sqrt{R^2 + (X_L - X_C)^2}$$

For series resonance, the p.f. of the circuit should be unity when net reactance of circuit is zero, *i.e.*,

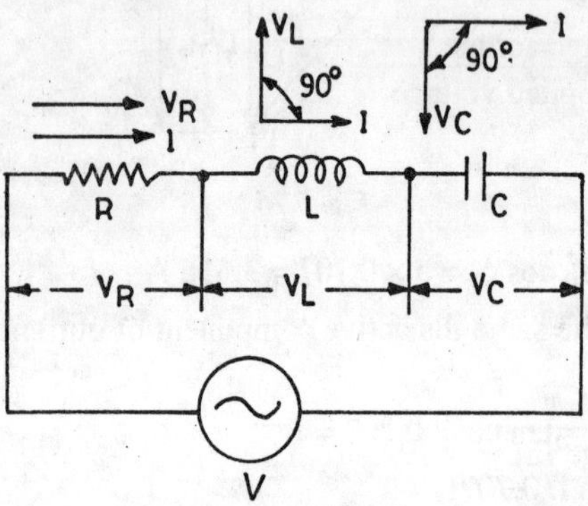

Fig. 5.77.

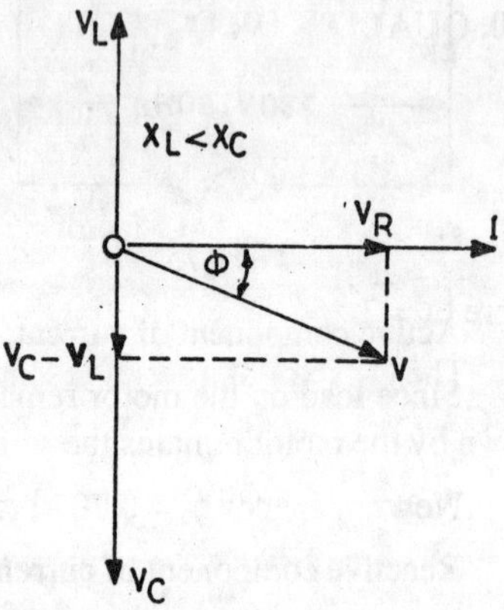

Fig. 5.78.

At series resonance, $X_L = X_C$

or $2\pi f_r L = \dfrac{1}{2\pi f_r C}$

Resonant frequencty, $f_r = \dfrac{1}{2\pi \sqrt{LC}}$

If L and C are measured in henry and farad respectively, then f will be in Hz.

5.39. CONDITIONS OF SERIES RESONANCE

When series resonance occurs, the effect on the circuit is the same as though neither inductance nor capacitance is present. The current under this condition is dedpedent solely on the resistance of the circuit and voltage across it.

(i) The impedance of the circuit is minimum and is equal to the resistance of the circuit *i.e.*,

$$Z = R$$

(ii) The current in the circuit is maximum as it is limited by the resistance of the circuit alone *i.e.*,

$$I = V/Z = V/R$$

As the current is at its maximum value, the power circuit will also be at its maximum value.

(iii) The power factor is unity.

(iv) Since at series resonance the curent flowing in the circuit is very large, the voltage drops across L and C are also very large. In fact, these drops are much greater than the applied voltage. However, voltage drop across L, C combination as a whole will be zero because these drops are equal in magnitude but 180° out of phase with each other.

5.40. QUALITY OR Q-FACTOR OF SERIES RESONANT CIRCUIT

$$\text{Q-factor} = \frac{\text{Voltage across } L \text{ or } C}{\text{Applied voltage}} = \frac{I_r X_L}{I_r R} = \frac{X_L}{R}$$

$$\therefore \quad \text{Q-factor} = \frac{\omega_r L}{R} \qquad ...(i)$$

where $\omega_r = 2\pi f_r$

The Q-factor of a series resonant circuit can also be expressed in terms of L and C.

We know, $$f_r = \frac{1}{2\pi\sqrt{LC}}$$

or $$2\pi f_r = \frac{1}{\sqrt{LC}}$$

or $$\omega_r = \frac{1}{\sqrt{LC}}$$

Substituting the value of ω_r in eqn. *(i)* above, we have,

$$\text{Q-factor} = \frac{1}{R}\sqrt{L/C}$$

The value of Q-factor depends entirely upon the design of coil. With a well designed coil. the Q-factor or quality factor can be 200 or more.

5.41. BANDWIDTH OF A SERIES CIRCUIT

The bandwidth of a series circuit is defined *as the range of frequencies over which circuit current is equal to or greater than 70.7% of maxmium current.*

The current in R-L-C series circuit changes with frequency. Referring to the resonance curve (*i.e.*, current/frequency curve) in Fig. 5.80, it is clear that for any frequency lying between f_1 and f_2 the circuit current is equal to or greater than 70.7% of maximum current (*i.e.*, $I_r = V/R$). Therefore, $f_1 - f_2$ is the band-width of the circuit.

Note that f_1 and f_2 are the limiting frequencies at which current is exactly equal to 70.7% of maximum value. The frequecny f_1 (*i.e.*, on the lower side) is

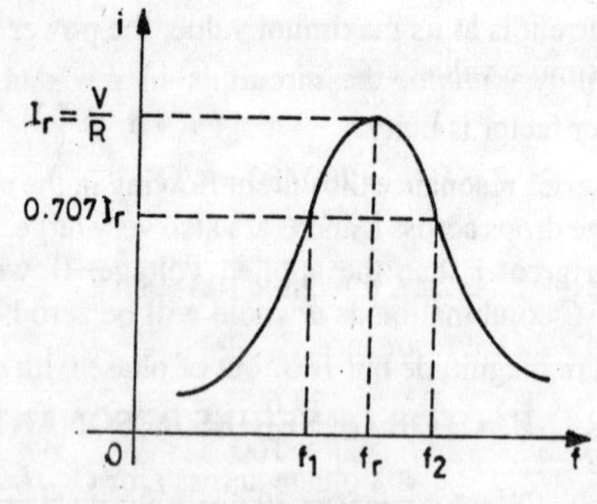

Fig. 5.79.

called *lower cut-off frequency* and the frequency f_2 (*i.e.*, on the higher side) is called *upper cut-off frequecny*. The band-width of a series circuit means that the circuit will offer very low impedance to this frequency range. It can be proved mathematically that the following relation exists between Q-factor and band width of a series circuit.

$$\text{Q-factor} = \frac{f_r}{BW}$$

Example 5.70. A coil of resistance 100 Ω and inductance 100 mH is connected in series with a 100 pF capacitor. The circuit is connected to a 10 V variable frequency source. Calculate *(i)* the resonant frequency *(ii)* current at resonance *(iii)* voltage across L and C at resonance and *(iv)* Q-factor of the circuit.

Solution :

(i) Resonant frequency, $f_r = \frac{1}{2\pi\sqrt{LC}} = \frac{1}{2\pi\sqrt{10^{-4} \times 10^{-10}}}$

$= 1.59 \times 10^6$ Hz

(ii) Current at resonance, $I_r = V/R = 10/100 = 0.1$ A

(iii) At resonance, $X_L = 2\pi f_r L = 2\pi \times 1.59 \times 10^6 \times 10^{-4} = 1000\ \Omega$

At resonance, $V_L = I_r X_L = 0.1 \times 1000 = 100$ V

At resonance, $V_C = I_r \times X_C = 0.1 \times 1000 = 100$ V

$$\text{Q-factor} = \frac{1}{R}\sqrt{L/C} = \frac{1}{100}\sqrt{10^{-4}/10^{-10}} = 10$$

Example 5.71. A choking coil is connected in series with a 20 μF capacitor. With a constant supply voltage of 200 V, it is found that the circuit takes its maximum current of 50 A when the supply frequency is 100 Hz. Calculate *(i)* resistance and inductance of the choking coil and *(ii)* voltage across the capacitor. What is the Q-factor of the coil?

Solution :

Since current is maximum, the circuit is in resonance *i.e.*, $I_r = 50$ A, $f_r = 100$ Hz.

(i) $$R = V/I_r = 200/50 = 4\ \Omega$$

$$X_C = \frac{1}{2\pi f_r C} = \frac{10^6}{2\pi \times 100 \times 20} = 79.6\ \Omega$$

Now $X_C = X_L = 79.2\ \Omega$ at resonance

$$\therefore \quad L = \frac{X_L}{2\pi f_r} = \frac{79.6}{2\pi \times 100} = 0.127\ \text{H}$$

(ii) $$V_C = I_r X_C = 50 \times 79.6 = 3980\ \text{V}$$

$$\text{Q-factor} = V_C/V = 3980/200 = 19.9$$

Example 5.72. A resistance, a capacitor and a variable inductor are connected in series across a 200 V, 50 Hz, supply. The maximum current which can be obtained by varying the inductance is 314 mA and the voltage across capacitance is then 300 V. Calculate the capacitance of a capacitor and values of resistance and inductance.

Solution : Voltage across the circuit, $V = 200$ V

Supply frequency, $f = 50$ Hz

Maximum value of current, $I_m = 314$ mA

Voltage across capacitance $V_C = 300$ V

At resonance, the current is maximum and equal to $\frac{V}{R}$.

$$I_m = \frac{V}{R}$$

or $$314 \times 10^{-2} = \frac{200}{R}$$

$$\therefore \quad R = \frac{200}{314 \times 10^{-2}} = 636.9 \text{ ohms } \textbf{Ans.}$$

Also, at resonance, the voltage drop across capacitance is equal to voltage drop across inductance.

Voltage drop across capacitance, $V_L = IX_C$

or $$300 = 314 \times 10^{-3}\, X_C$$

$$\therefore \quad X_C = \frac{300}{314 \times 10^{-3}}$$

$$X_C = \frac{1}{\omega C}$$

$$\therefore \quad C = \frac{1}{\omega X C} = \frac{314 \times 10^{-3} \times 10^{6}}{2\pi \times 50 \times 300} = 3.33 \ \mu\text{F} \textbf{ Ans.}$$

Voltage drop across inductance, $V_L = V_C$

$$\therefore \quad 300 = I \,.\, X_L$$

or $$X_L = \frac{300}{3.14 \times 10^{-3}}$$

$$\therefore \quad L = \frac{X_L}{\omega} = \frac{300}{314 \times 10^{-3} \times 314} = 3.04 \text{ Henry } \textbf{Ans.}$$

Example 5.73. A resistance of 30 Ω, an inductance of 4 H and capacitance of 25 μF are connected in series across 230V supply. Calculate *(i)* the frequency at which current will be maximum *(ii)* the current at this frequency *(iii)* the potential difference across the inductance.

Solution : Value of resistance, $R = 30\ \Omega$

Value of inductance, $L = 4$ H

Value of capacitance, $C = 25\ \mu$F

Supply voltage, $V = 230$ V

The current will be maximum, when resonance occurs and the value of that frequency $f_r = \dfrac{1}{2\pi\sqrt{LC}}$

$$\therefore \quad f_r = \frac{10^3}{2\pi \,.\, \sqrt{4 \times 25}} = \frac{10^3}{20\pi} = 15.9 \text{ Hz } \textbf{Ans.}$$

The value of max. current, $I_m = \dfrac{V}{R} = \dfrac{230}{30} = 7.67$ A **Ans.**

The potential difference across inductance

$$V_L = I \,.\, X_L = I \,.\, \omega L$$

$$= 7.67 \times 2\pi \times 15.9 \times 4$$

$$= 3063.46 \text{ volts } \textbf{Ans.}$$

Example 5.74. A choking coil of resistance 5 ohms and inductance 0.6 H is in series with a capacitance of 10 μF. If a voltage of 230 V is applied and the frequency is adjusted to resonance, find the voltage across the coil and capacitor.

Solution : Inductance, $L = 0.6$ H

Capacitance, $C = 10\ \mu$F

$$\text{Resonant frequency, } f_r = \frac{1}{2\pi\sqrt{LC}} = \frac{1}{2 \times 3.14\sqrt{0.6 \times 10 \times 10^{-6}}} = 65 \text{ Hz}$$

$$\text{Current at resonance, } = \frac{230}{5} = 46 \text{ A}$$

$$\text{Inductive reactance, } X_L = \omega L = 2\pi f L$$

$$= 2 \times 3.14 \times 65 \times 0.6 = 245\ \Omega$$

$$Z_L = \sqrt{R^2 + X_L^2}$$

$$= \sqrt{(5)^2 + (245)^2} = 245\ \Omega$$

Voltage across coil $= 46 \times 245 = 11270$ volts

$$\text{Capacitive reactance, } X_C = \frac{1}{2\pi f C} = \frac{10^6}{2 \times 3.14 \times 65 \times 10} = 245\ \Omega$$

Voltage across capacitor, $V_C = 245 \times 46 = 11270$ volts

5.42. PARALLEL RESONANCE

An a.c. circuit containing an inductor and capacitor in parallel is said to be in parallel resonance when the circuit current is in phase with the applied voltate circuit p.f. is unity.. Consider an inductor of L henry having some resistance R ohms connected in parallel with a capacitor of capacitance C farads across a supply voltage of V volts as shown in Fig. 5.80. The phasor diagram of the circuit is shown in Fig. 5.81.

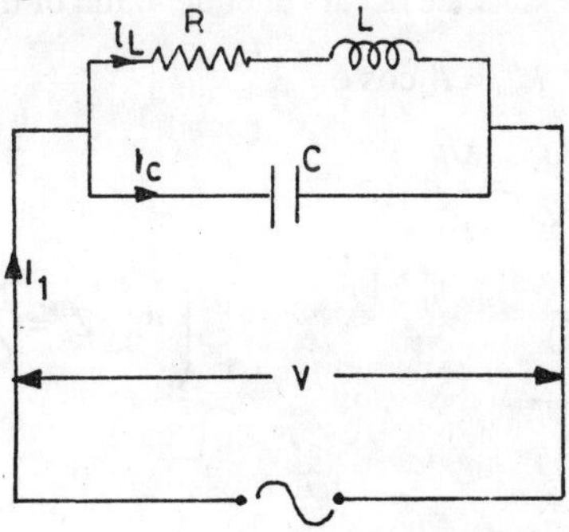

Fig. 5.80.

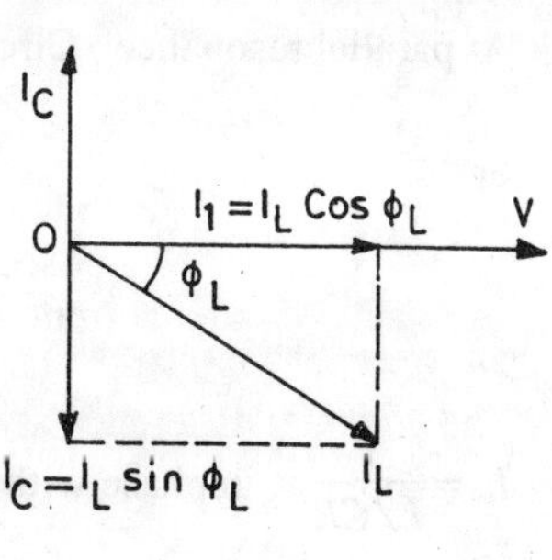

Fig. 5.81.

The circuit current I_r will only be in phase with the supply voltage when

$$I_C = I_L \sin \phi_L$$

Resonant Frequency

The value of X_L and X_C can be changed by changing the supply frequency. When frequency increases, the value of X_L increases. This decreases the magnitude of current I_m which also lags behind the voltage V. On the other hand, the value of X_C decreases and the value of I_C increases. At resonance **frequency,** $I_C = I_L \sin \phi_L$ and resonance occurs.

$\therefore$ At parallel resonance, $I_C = I_L \sin \phi_L$

where, $I_L = V/Z_L$; $\sin \phi_L = X_L/Z_L$ and $I_C = V/X_C$

$$\therefore \quad \frac{V}{X_C} = \frac{V}{Z_L} \times \frac{X_L}{Z_L}$$

$$\text{or} \quad X_L X_C = Z_L^2$$

$$\text{or} \quad \frac{\omega L}{\omega C} = Z_L^2 = (R^2 + X_L^2) \qquad \ldots(i)$$

$$\text{or} \quad \frac{L}{C} = R^2 + (2\pi f_r L)^2$$

$$\text{or} \quad 2\pi f_r L = \sqrt{\frac{L}{C} - R^2}$$

$$\text{or} \quad f_r = \frac{1}{2\pi L}\sqrt{\frac{L}{C} - R^2} = \frac{1}{2\pi}\sqrt{\frac{1}{LC} - \frac{R^2}{L^2}} \qquad \ldots(ii)$$

If R is very small as compared to L, then

$$\text{Resonant frequency, } f_r = \frac{1}{2\pi\sqrt{LC}}$$

Conditions of Parallel Resonance

At parallel resonance ; Circuit current, $I_{min} = I_L \cos \phi$

$$\text{or} \quad \frac{V}{Z_r} = \frac{V}{Z_L} \times \frac{R}{Z_L} = \frac{VR}{Z_L^2}$$

$$= \frac{VR}{L/C} \qquad \left(\because \; Z_L^2 = \frac{L}{C}\right)$$

$I_m = \dfrac{V}{L/CR}$ is in phase with V

$$\therefore \quad \text{Circuit impedance, } Z_r = \frac{V}{I_{min}} = \frac{V}{\dfrac{V}{L/CR}} = \frac{L}{CR}$$

It shows that

(*i*) Circuit impedance, Z_r (= L/CR) is a pure resistive because there is no frequency term present. If the value of L, R and C is in henry, ohm and farads then the value of Z_r is in ohms.

(*ii*) The value of Z_r is very high because the ratio L/C is very large at parallel resonance.

(iii) The value of circuit current I_r (= V/Z_r) is very small because the value of Z_r is very high.

(iv) The current flowing through the capacitor and coil is much greater than the line current because the impedance of each branch is quite low than circuit impedance Z_r.

Since a parallel resonant circuit can draw a very small current and power from the mains, it is often regarded as **rejector circuit.**

Resonance Curve : A current-frequency curve for a typical parallel resonant circuit is shown in Fig. 5.82. The value of line current I (= V/Z) is minimum at resonance.

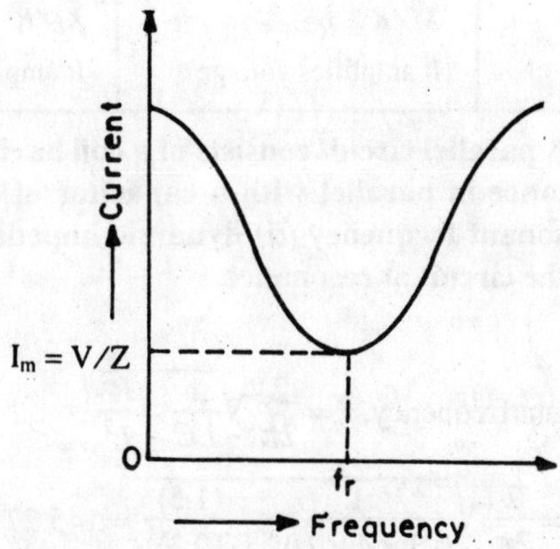

Fig. 5.82.

5.43. Q-FACTOR OF A PARALLEL RESONANT CIRCUIT

We have seen that at parallel resonance, the current circulating between the two branches is many times greater than the line current drawn from the mains. This current amplification produced by the resonance is called the *Q-factor of the parallel resonant circuit.*

$$\therefore \quad \text{Q-factor} = \frac{\text{Current circulating between } L \text{ and } C}{\text{Circuit current}} = \frac{I_C}{I_{min}}$$

Now, $I_C = V/X_C = 2\pi f_r CV$

and $I_{min} = V/(L/CR)$

$$\therefore \quad \text{Q-factor} = \frac{2\pi f_r CV}{V} \times \frac{L}{CR} = \frac{2\pi f_r L}{R} \quad \text{........same as for series circuit}$$

or $$= \frac{2\pi L}{R} \cdot \frac{1}{2\pi\sqrt{LC}} \qquad \left(\because f_r = \frac{1}{2\pi\sqrt{LC}} \text{ neglecting } R \right)$$

or $$= \frac{1}{R}\frac{\sqrt{L}}{C}$$

5.44. COMPARISON OF SERIES AND PARALLEL RESONANT CIRCUITS

The comparison of series and parallel resonant circuits is given below:

S.No.	*Particulars*	*Series Circuit*	*Parallel Circuit*
1.	Impedance	Minimum *i.e.*, $Z_r = R$	Maximum *i.e.*, $Z_r = L/CR$
2.	Current	Maximum *i.e.*, $I_{max} = V/R$	Minimum *i.e.*, $I_{min} = V/Z_r$
3.	Resonant frequency	$f_r = \frac{1}{2\pi\sqrt{LC}}$	$f_r = \frac{1}{2\pi}\sqrt{\frac{1}{LC} - \frac{R^2}{L^2}}$
4.	Power factor	Unity	Unity
5.	Q-factor	X_L/R	X_L/R
6.	Amplification	It amplifies voltage	It amplifies the current

Example 5.75. A parallel circuit consists of a coil having 15 Ω resistance and 300 m H inductance in parallel with a capacitor of capacitance 4 μF. Determine *(i)* the resonant frequency *(ii)* dynamic impedance of the circuit and *(iii)* Q-factor of the circuit at resonance.

Solution :

(i) Resonant frequency, $$f_r = \frac{2}{2\pi}\sqrt{\frac{1}{LC} - \frac{R^2}{L^2}}$$

or $$f_r = \frac{2}{2\pi}\sqrt{\frac{1}{0.3 \times 4 \times 10^{-6}} - \frac{(1.5)^2}{(0.3)^2}} = 145.27 \text{ Hz } \textbf{Ans.}$$

(ii) Dynamic impedance, $Z_r = L/CR = 0.3/4 \times 10^{-6} \times 15 = 5000\ \Omega$ **Ans.**

(iii) $$\text{Q-factor} = \frac{2\pi f_r L}{R} = \frac{2\pi \times 145.27 \times 0.3}{15} = 18.255 \textbf{ Ans.}$$

Example 5.76. A coil of 15 Ω resistance and 0.3 H inductance is connected in parallel with a variable capacitor across a 230 V, 50 Hz supply. Calculate *(i)* the capacitance of the capacitor for resonance *(ii)* the effective impedance of the circuit and *(iii)* the current supplied from the mains.

Solution : The circuit is shown in Fig. 5.83.

At resonance ; $$f_r = \frac{1}{2\pi}\left(\frac{1}{LC} - \frac{R^2}{L^2}\right)^{1/2}$$

or $$\frac{1}{LC} = 4\pi^2 f_r^2 + \frac{R^2}{L^2}$$

$$\frac{1}{0.3 \times C} = 4\pi^2 (50)^2 + \frac{(15)^2}{(0.3)^2}$$

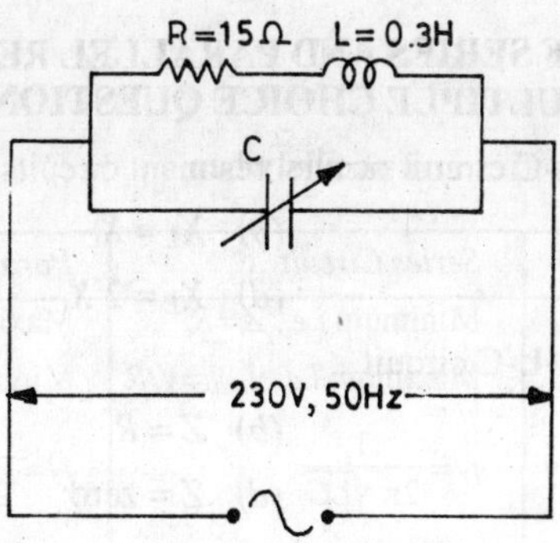

Fig. 5.83.

or $$C = 32.94\ \mu\text{F}\ \textbf{Ans.}$$

Effective impedance of the circuit,

$$Z_r = L/CR = 0.3/32.94 \times 10^{-6} \times 15 = 607.18\ \Omega\ \textbf{Ans.}$$

Line current, $I_r = V/Z_r = 230/607.18 = 0.3788$ A **Ans.**

HIGHLIGHTS

1. In a series a.c. circuit, resonant frequency f_r is given by :

$$f_r = \frac{1}{2\pi\sqrt{LC}}$$

Impedance at resonance, $Z = R\sqrt{R^2 + (X_L - X_C)^2}$

Current at resonance, $I = V/Z = V/R$

$$\text{Q-factor} = \frac{2\pi f_r L}{R}$$

Also $$\text{Q-factor} = \frac{1}{R}\sqrt{\frac{L}{C}}$$

2. In a parallel a.c. circuit, the resonant frequency f_r is given by ;

$$f_r = \frac{1}{2\pi}\sqrt{\frac{1}{LC} - \frac{R^2}{L^2}}$$

Impedance at resonance, $Z_r = L/CR$

Current at resonance, $I_r = V/Z_r$

$$\text{Q-factor} = \frac{2\pi f_r L}{R}$$

Also $$\text{Q-factor} = \frac{1}{R}\sqrt{\frac{L}{C}}$$

MULTIPLE CHOICE QUESTIONS

1. Resonance in R-L-C circuit occurs when
   (a) $R = X_L - X_C$ (b) $X_L = X_C$
   (c) $X < R$ (d) $X_L = 2\,X_C$
2. At resonance in R-L-C circuit
   (a) $Z = \sqrt{R^2\,X^2}$ (b) $Z = R$
   (c) $Z = X_L - X_R$ (d) $Z =$ zero
3. At resonance in R-L-C circuit, resonant frequency is given by
   (a) $f_r = \dfrac{1}{LC}$ (b) $f_r = \dfrac{1}{\sqrt{LC}}$
   (c) $f_r = \sqrt{\dfrac{2\pi}{LC}}$ (d) $f_r = \dfrac{1}{2\pi}\sqrt{\dfrac{1}{LC}}$
4. At resonance in series circuit, current is
   (a) minimum (b) maximum
   (c) in phase with voltage (d) (b) and (c)
5. Q-factor R-L-C circuit is
   (a) $\dfrac{1}{R}\sqrt{\dfrac{L}{C}}$ (b) $R\sqrt{\dfrac{L}{C}}$
   (c) $\dfrac{1}{L}\sqrt{\dfrac{R}{C}}$ (d) $\dfrac{1}{R}\sqrt{\dfrac{C}{L}}$

ANSWERS

1. (b) 2. (b) 3. (d) 4. (d) 5. (a)

EXERCISE

1. Write the expressions of currents in different branches and the total curren of a circuit consisting of an inductance, a capacitance and a non-inductiv resistance connected in parallel. Draw also the phasor diagram.
2. What do you understand by resonance in R-L-C circuit.
3. Derive an expression for resonance frequency in R-L-C series circuits.
4. What do you understand by band width and selectivity of R-L-C serie resonant circuit.
5. Define Q-factor of a R-L-C series circuit.
6. Describe the conditions for parallel resonant circuit.
7. Explain the difference between series resonance and parallel resonance.

PROBLEMS

1. A circuit containing a resistance of 40 Ω in parallel with an inductive reactance of 30 Ω is connected across 240-V a.c. mains. Calculate *(i)* the current in each branch and the total current drawn from the mains *(ii)* the power factor of each branch and the power factor of the whole circuit *(iii)* the power drawn by each branch and the total power drawn from the mains.

   (Ans. *(i)* 6 A, 8A, 10A, *(ii)* 1.0, zero, 0.6 (lag), *(iii)* 1.440 watts, zero; 1,440 watts**)**

2. An inductance of 0.6 H, a resistor of 100 Ω and a 30 μF capacitor are connected in parallel across a 230 V, 50 Hz supply. Calculate the line current, circuit phase angle, power dissipated, total circuit impedance and parameters of equivalent series circuit.

   (Ans. 2.49 A, 22.39^o (lead), 529 W, 92.53 Ω ; 80.51 Ω, 35.25 Ω**)**

3. A resistance of 100 Ω, an inductance, of 0.1 H and a capacitance of 150 μF are connected in parallel across 200 V, 50 Hz ac supply. Determine the branch currents, total current and the total power taken from the supply.

   (Ans. 2 A, 6.366 A, 9.42 A, 3.655 A; 400 watts**)**

4. A series combination of *R* and *C* is in parallel with a 20 Ω resistor across 50 Hz source results in a total current of 7 A, a current through the 20 Ω resistor of 5 A and a current in the *R*-*C* branch of 3 A. Determine *R* and *C*.

   (Ans. R = 16.667 Ω; C = 28.8675 Ω**)**

5. An inductor of 0.5 H inductance and 90 Ω resistance is connected in parallel with a 2 μF capacitor. A voltage of 230 V at 50 Hz is maintained across the circuit. Determine the total power taken from the source.

   (Ans. 1476 watts**)**

6. Two parallel circuits comprise respectively of *(i)* a coil of resistance 20 Ω and inductance 0.07 H and *(ii)* a capacitor of capacitance 60 μF in series with resistance of 50 Ω.. Calculate the current in each branch and the current taken from the supply and power factor when connected to 220 V, 50Hz supply.

   (Ans. 7.4 A; 3 A; 7.8A; 0.907 (lag)**)**

7. A coil having resistance of 100 Ω and inductance of 3 m H is connected in series with a capacitor 30 μF across 230 V a.c. supply. Determine the supply frequency to make the circuit resonant. Also determine the current and power factor of the circuit at this frequency.

   (Ans. 530.52 Hz; 2.3 A;1.0**)**

8. An R-L-C series circuit consists of a resistance of 1000 Ω, an inductance of 100 mH and a capacitance of 10 μF. If the voltage of 100 V is applied across the combination, find *(i)* the resonance frequency *(ii)* Q-factor of the circuit and *(iii)* the half-power points.

   (Ans. *(i)* $10^6/2\pi$ Hz, *(ii)* 100, *(iii)* 158.4 KHz and 160 KHz**)**

9. A circuit comprises an inductance of 26 μH associated with resistance of 0.1 Ω in parallel with a capacitance of 0.16 μF. Find the ratio of the circulating current to the supply current at resonant frequency.

(**Ans.** 127.5)

10. An inductive coil when connected across 500 V, 50 Hz supply draws 1.0 A at 0.8 p f. What capacitance is connected in parallel with the coil to make the power factor of the combination unity ? What will be the current *(i)* drawn from the supply; and *(ii)* in each branch with this new capacitor.

(**Ans.** 3.82 μF; *(i)* 0.8 A, *(ii)* 1.9 A, 0.6 A)

11. An a.c. voltage of 200 V and variable frequency is applied to a coil of 20 Ω resistance and 0.3 H inductance connected in parallel with a capacitance of 200 μF. Determine the parallel resonant frequency and the input current at this frequency.

(**Ans.** 20.5 Hz; 2.667 A)

12. An inductive circuit of resistance 2 Ω and inductance 0.01 H is connected to 250-V, 50Hz supply. What capacitance placed in parallel will produce resonance ? Find the total current taken from the supply and the current in the branch circuits.

(**Ans.** 721 μF; 36.05 A; 125 A; 67.13 A ∠ – 57.52° A; 56.63 ∠ 90° A)

5.45. POWER IN PURE RESISTANCE

In an a.c. circuit, electric power consumed at any instant is the product of voltage and current at that instant, *i.e.*,

$$\text{Instantaneous power} = v \cdot i$$

$$= V_m \sin \omega t \,.\, I_{max} \sin \omega t$$

$$= V_m \,.\, I_m \,.\, \sin^2 \omega t$$

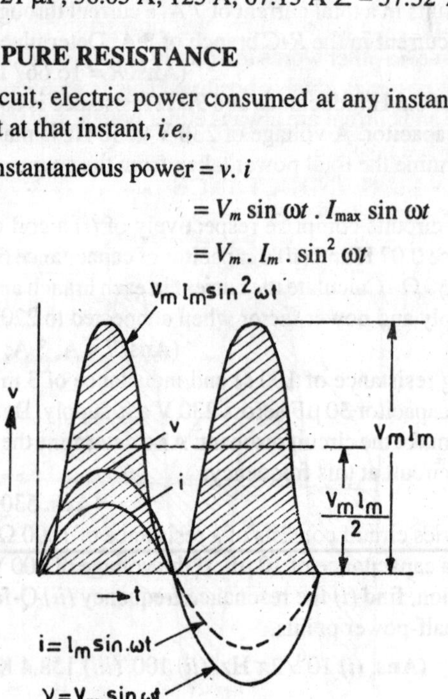

Fig. 5.84.

$$= \frac{V_m}{\sqrt{2}} \cdot \frac{I_m}{\sqrt{2}} \cdot 2 \sin^2 \omega t$$

$$= \frac{V_m}{\sqrt{2}} \cdot \frac{I_m}{\sqrt{2}} \cdot (1 - \cos 2\omega t)$$

$$= \frac{V_m I_m}{\sqrt{2}} - \frac{V_m I_m}{2} \cos 2\omega t$$

Power consists of two parts, constant: part $\left(\frac{V_m I_m}{2}\right)$ and a fluctuating part $\left(\frac{V_m I_m}{2}\right) \cos 2\omega t$. For a complete cycle, the average value of $\frac{V_m I_m}{2} \cos 2\omega t$ is zero.

Hence, power for whole cycle is $\frac{V_m I_m}{2}$

$$\therefore \qquad P = \frac{V_m}{\sqrt{2}} \times \frac{I_m}{\sqrt{2}} = V \times I \text{ watts}$$

where V is r.m.s. value of applied voltage, I is r.m.s value of the current.

Hence Power, $P = V \times I$ watts.

It is seen from the power curve that no part of the power cycle becomes negative at any time. In other words in a purely resistive circuit power is never zero. So it is clear that power is always positive. This is so because the instantaneous values of voltage and current are always either both positive or negative and hence the product is always positive.

5.46. POWER IN PURE INDUCTANCE

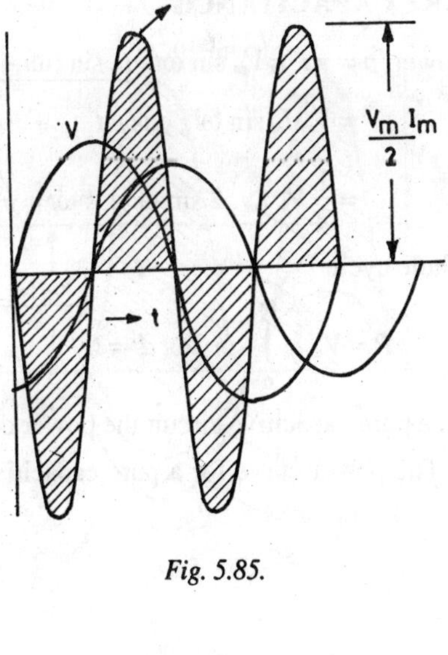

Fig. 5.85.

Instantaneous power, $p = e \cdot i = V_m \sin \omega t \cdot I_m \cdot \sin(\omega t - \pi/2)$

$$= -V_m I_m \sin \omega t \cdot \cos \omega t$$

$$= -\frac{V_m I_m}{2} 2 \sin \omega t \cos \omega t \qquad [\sin 2\theta = 2 \sin\theta \cos\theta]$$

$$= -\frac{V_m I_m}{2} \cdot \sin 2\omega t$$

$\therefore$ Average power, P = Average of p over one cycle

$$= \frac{1}{2\pi} \int_0^{2\pi} -\frac{V_m I_m}{2} \cdot \sin 2\omega t \, d(\omega t) = 0$$

Hence power consumed in a pure inductive circuit is zero.

Phase Angle : From equations *(i)* and *(iii)* it is found that current flowing through a pure inductive circuit lags behind the applied voltage by $\pi/2$ radians as shown in Fig. 5.85.

Hence in an a.c. circuit containing pure inductance, current lags behind the applied voltage by 90°.

Power Curve : From the power curve diagram it is clear that the average power in a half cycle is zero, as the negative and positive loop area under power curve is the same. Fig. 5.85 shows the power curve for a pure inductive circuit. During the first quarter cycle (90°) the power supplied by the source that is stored in the magnetic field set up around the coil. In the next quarter cycle (*i.e.,* 90°), the magnetic field collapses and the power stored in the field is returned to the source. This process is repeated in each and every half cycle. Hence no power is consumed in any pure inductive circuit.

5.47. POWER IN PURE CAPACITANCE

Instantaneous power, $p = v\, i = V_m \sin \omega t \cdot I_m \sin(\omega t + \pi/2)$

$$= V_m I_m \sin \omega t \cdot \cos \omega t$$

$$= \frac{1}{2} V_m I_m \cdot 2 \sin \omega t \cos \omega t = \frac{V_m I_m}{2} 2 \sin \omega t$$

Power for the whole cycle

$$P = V_m I_m \int_0^{2\pi} \sin 2\omega t \, dt = 0$$

So we see that in a pure capacitive circuit the power consumed is zero.

Power Curve : The power curve for a pure capacitive circuit is shown in Fig. 5.85 *(a)*.

It is clear from the curve, that average power in half cycle is zero. Since the positive and negative loop area under power curve is same.

In the first quarter, the power (energy supplied by the source that is stored in the electric field set up between the capacitor plates. In the next quarter, the electric field collapses and the power (Energy) stored in the electric field is returned to the source. This process is repeated in each and every half cycle. Hence no power is consumed by the circuit.

Phase Angle : From equation *(i)* and *(iii)*, it is clear that current passing in any pure capacitive circuit leads the applied voltage by 90° as shown in Fig. 5.85 *(a)*.

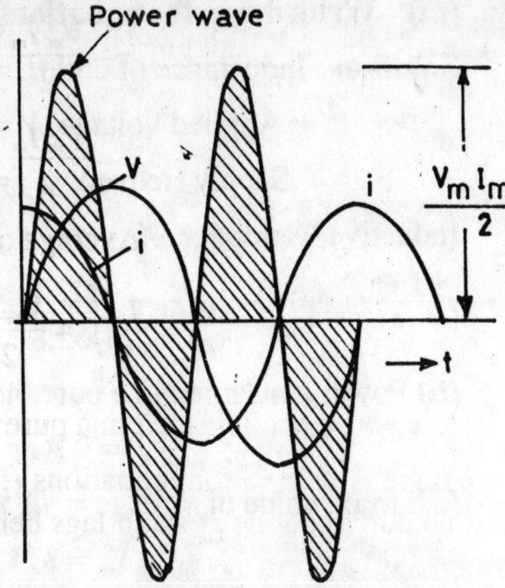

Fig. 5.85. (a)

Example 5.77. An a.c. circuit consists of a pure resistance of 25 ohms and is connected across an a.c. supply of 250 V, 50Hz. Calculate

(i) Current

(ii) Power consumed

(iii) Write down the equations of voltage and current.

Solution : Resistance of the circuit, $R = 25$ ohms.

Applied voltage, $V = 250$ volts

Supply frequency, $f = 50$Hz.

(i) Current in the circuit, $I = \frac{V}{R} = \frac{250}{25} = 10$ **A Ans.**

(ii) Power consumed, $P = V.\,I = 250 \times 10 = 2500$ watts. **Ans.**

(iii) Max. value of voltage, $V_m = \sqrt{2} \times V_{r.m.s.}$

$= 1.4142 \times 250 = 353.5$ volts

$\therefore \quad v = 353.5 \sin 314\, t$ **Ans.**

Maximum value of current $I_m = \sqrt{2} \times I_{r.m.s.}$

$= 1.414 \times 10 = 14.142$ Amp.

$\therefore \quad i = 14.142 \sin 314\, t$ **Ans.**

Example 5.78. An inductive coil has negligible resistance and inductance of 0.1 Henry. It is connected across 220 volts, 100 Hz Supply. Find

(i) Current

(ii) **Power**

(iii) **Write down the equations of voltage and current.**

Solution : Inductance of coil, $L = 0.1$ Henry

Applied voltage, $V = 220$ volts

Supply frequency, $f = 100$ Hz

Inductive Reactance, $X_L = \omega L = 2 \times 3.14 \times 100 \times 0.1 = 62.8$ ohms

(i) Current, $I = \frac{220}{62.8} = 3.5$ A **Ans.**

(ii) Power consumed in a pure inductive circuit is zero.

$\therefore$ $P = 0$ **Ans.**

(iii) Max. value of voltage $= \sqrt{2} \times 220$

$V_m = 311$ volts

$\therefore$ $v = 311 \sin 62.8\, t$

(iv) Max. value of current $= \sqrt{2} \times 3.5 = 4.95$ A

$\therefore$ $i = 4.95 \sin (62.8t - \pi/2)$ **Ans.**

Example 5.79. A capacitor of capacitance 100 μF is connected across a 250 volt, 50 Hz supply. Calculate ;

(i) **Current**

(ii) **Power**

(iii) **Write down the equations of voltage and current.**

Solution : Capacitance of the circuit, $C = 100\ \mu$F

Voltage applied, $V = 250$ volts

Frequency, $f = 50$ Hz

Capacitive reactance, $X_C = \frac{1}{\omega C} = \frac{10^{-6}}{2 \times 3.14 \times 50 \times 100} = 31.76$ ohms

(i) $I = \frac{V}{X_C} = \frac{250}{31.76} = 7.852$ A **Ans.**

(ii) Power consumed in any pure capacitive circuit is zero.

$\therefore$ $P = 0$ **Ans.**

(iii) Max. value of voltage, $V_m = \sqrt{2} \times 250 = 353.5$ volts

$\therefore$ $v = 353.5 \sin 314\, t$

Max. value of current, $I_m = \sqrt{2} \times 7.852 = 11.1$ A

$\therefore$ $i = 11.1 \sin (314t + \pi/2)$ **Ans.**

5.48. POWER IN R-L SERIES CIRCUIT

$$\text{Instantaneous Power, } P = ei$$

$$= [V_m \sin \omega t] [I_m \sin (\omega t - \phi)]$$

$$= V_m I_m \sin \omega t \sin (\omega t - \phi)$$

$$= \frac{V_m I_m}{2} [2 \sin \omega t \sin (\omega t - \phi)]$$

$$= \frac{V_m I_m}{2} [\cos \phi - \cos (2 \omega t - \phi)]$$

Thus power consists of two parts :

(i) Constant part $\frac{V_m I_m}{2} \cos \phi$, whose average value

(ii) Sinusoidally varying part $\frac{V_m I_m}{2} \cos (2\omega t - \phi)$ whose average value over a complete cycle is zero.

$$\therefore \quad \text{Average Power, } P = \frac{V_m I_m}{2} \cos \phi = V_m \sqrt{2} \times \frac{I_m}{\sqrt{2}} \cos \phi$$

$$= VI \cos \phi$$

i.e., Power = rms value of voltage × rms value of current × cosine of angle between E and I

The term $\cos \phi$ is called power factor of the circuit.

From vector diagram, $\cos \phi = \frac{IR}{IZ} = \frac{R}{Z}$

Alternatively : Power is consumed in resistance only as inductance does not consume any power.

$$\therefore \quad \text{Power Consumed} = I^2 R = I (IR)$$

$$= \frac{V}{Z} (IR) = VI \cos \phi \; (\because \; R/Z = \cos \phi)$$

Power Curve : The power curve for R-L series circuit is shown in Fig. 5.87. The curve indicates that greater part is positive and smaller part is negative so that net power over the cycle is positive.

During the time interval a to b, applied voltage and current are in the opposite direction so that power is negative. Under such condition, the inductance L returns power to the circuit.

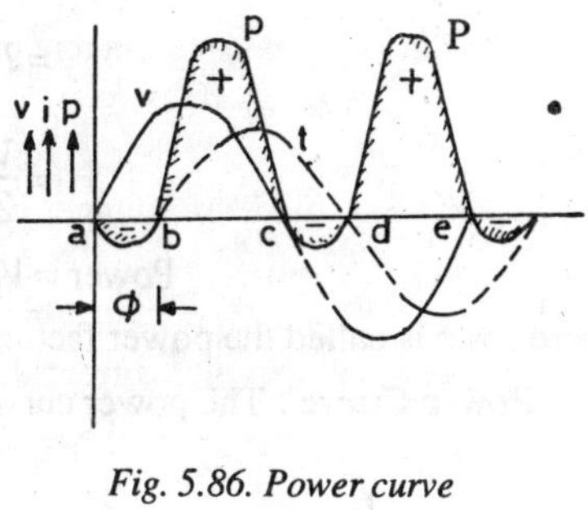

Fig. 5.86. Power curve

During the interval between *b* and *c*, the applied voltage and current are in the same direction so that power is positive and, therefore, power is put in the circuit. Similarly, during interval between *c* and *d*, inductance *L* returns power to circuit while during interval between *d* and *e*, power is put into the circuit. It should be noted that power absorbed by resistance R is converted into heat and is not returned.

Power in R-C Series Circuit

Power : If the applied voltage is expressed as :

$$e = V_m \sin \omega t$$

then $$i = I_m \sin(\omega t + \pi/2)$$

$$\text{Average Power, } P = \text{Average of } ei$$
$$= VI \cos \phi$$

Alternatively :

$$\text{Power consumed, } P = \text{Power in } R + \text{Power in } C$$
$$= I^2 R + 0 = (IR)\, I$$
$$= IR \frac{V}{Z}$$
$$= VI \cos \phi$$

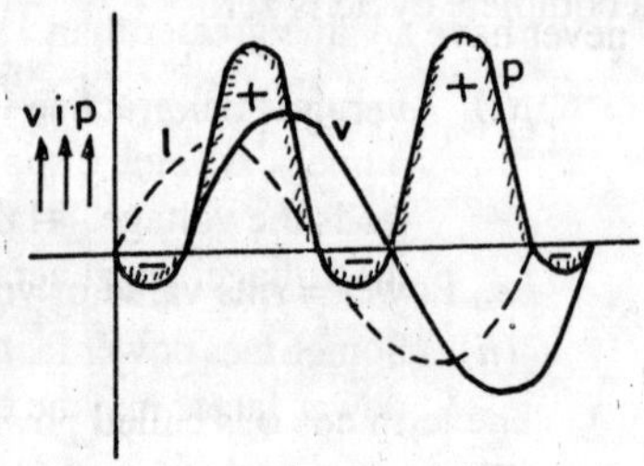

Fig. 5.87.

Power Curves : The power curve for R-C series circuit is shown in Fig. 5.88. The curve indicates that greater part is positive and smaller part is negative so that net power is positive.

5.49. POWER IN R-L-C SERIES CIRCUIT

Consider the R-L-C series circuit, the instantaneous value of applied voltage is given by the equation

$$v = V_m \sin \omega t$$
$$i = I_m \sin(\omega t - \phi)$$
$$\text{Power} = I^2 R$$
$$= I \,.\, I R \qquad \left(\because \; I = \frac{V}{Z}\right)$$
$$= \frac{V}{Z} I \,.\, R \qquad \left(\frac{R}{Z} = \cos \phi\right)$$
$$\text{Power} = VI \cos \phi$$

where cos ϕ is called the power factor of the circuit.

Power Curve : The power curve is same as in case of R-L series circuit.

5.50. POWER FACTOR

The power factor (*i.e.*, cos ϕ) of a circuit can be defined in one of the following ways:

(i) Power factor = cos ϕ = cosine of angle between V and I

(ii) Power factor $= \dfrac{R}{Z} = \dfrac{\text{Resistance}}{\text{Impedance}}$.

(iii) Power factor $= \dfrac{VI\cos\phi}{VI} = \dfrac{\text{True power}}{\text{Apparent power}}$

For example, in a resistor, the current and voltage are in phase *i.e.*, $\phi = 0^\circ$. Therefore, power factor of a pure resistive circuit is $\cos 0^\circ = 1$. Similarly, phase difference between voltage and current in a pure inductance or capacitance is 90°. Hence power factor of pure L or C is zero. This is the reason that power consumed by pure L or C is zero. For a circuit having R, L and C in varying proportions, the value of power factor will lie between 0 to 1. It may be noted that power factor can never have a value greater than 1.

(i) It is a usual practice to attach the word 'lagging' or leading with the numerical value of power factor to signify whether the current lags or leads the voltage. Thus if a circuit has a p.f. of 0.5 and the current lags the voltage, we generally write p.f. as 0.5 lagging.

(ii) Sometimes power factor is expressed as a percentage. Thus 0.8 lagging power factor may be expressed as 80%

The power factor for an inductive circuit is always lagging where as it is always leading for a capacitive circuit.

Disadvantages of Low Power Factor : As power consumed in any single phase a.c. circuit, $P = VI\cos\phi$.

or $$I = \frac{P}{V\cos\phi} \qquad ...(i)$$

But power consumed in three phase a.c. circuit is

$$P = \sqrt{3}\,.\,V_L\,.\,I_L\cos\phi$$

or $$I_L = \frac{P}{\sqrt{3}\,V_L\,.\cos\phi} \qquad ...(ii)$$

From equations *(i)* and *(ii)* we see that for a fixed value of power and voltage, the load currents are inversly proportional to the power factor. Lower the power factor, higher is the current, and vice versa. The higher current results in the following disadvantages.

1. Greater Conductor Size : To transmit or distribute a fixed amount of power at a fixed voltage at low power factor, the conductor will have to carry large amount of current which will necessitate a large conductor size.

2. Large Copper Losses : Due to more currents carried by the conductors, $I^2 R$ losses are increased which results in poor efficiency of transmission or distribution.

3. Poor Voltage Regulation : Due to low power factor and increased currents, the voltage in transformers, alternators and transmission distribution lines drops more which again results in poor voltage regulation.

4. Large kVA Rating of the Equipment : The output power of electrical machines e.g. alternators and transformers is always rated in apparent power *i.e.*, kVA

Because $$kVA = \frac{kW}{\cos \phi}$$

Thus, we see that the rating of the equipment is inversly proportional to the power factor. At low power factor, the value of kVa will increase. Therefore, at low power factor, equipment of higher kVA has to be installed which will increase the size and cost of the equipment.

5. Terminal Voltage : It causes fall in terminal voltage.

Causes of Low Power Factor : The following are the main causes of low power factor.

1. Transformers and chokes are the inductive devices which cause the current to lag behind the voltage *i.e.*,, they operate at a lagging power factor. However, at normal loads, the magnetising current does not effect too much, but at low loads or light loads, the magnetising current effects too much and lowers the power factor.

2. The motors which we use in our industry are mostly induction motors *i.e.*, having inductive loads, which have low power factor. Normally the power factor of induction motors vary from 0.8 to 0.85 when operating at full load and 0.3 to 0.5 when running lightly *i.e.*, at small loads.

3. Arc lamps, heating furnaces (Induction and arc furnaces also operate at a low power factor (lagging).

Power Factor Improvement : The low power factor is generally due to the inductive loads in nature and also they take the current at lagging power factor. In order to improve the power factor, a static capacitor is connected in parallel with the load as shown in the Fig. 5.88 *(a)*. The static capacitor draws a current which leads the voltage and thus neutralises the lagging wattless component partly or completely.

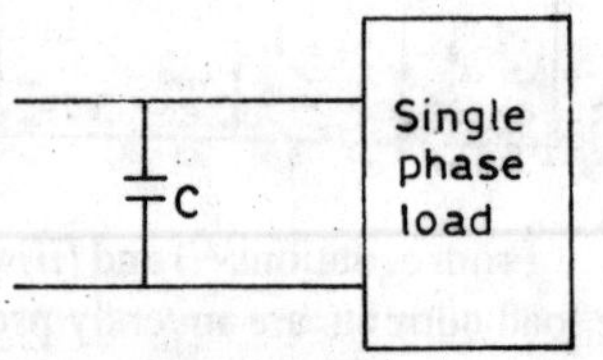

Fig. 5.88 (a)

For three phase load or 3 phase induction motors, the capacitors can be connected in star or delta as shown in the Fig. 5.88 *(b)* and 5.88 *(c)* respectively.

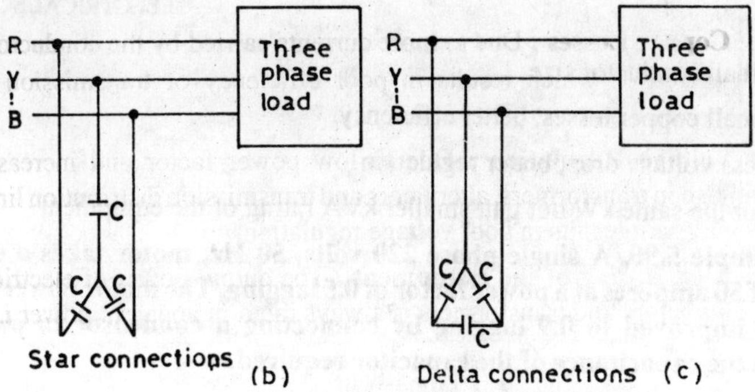

(b) *(c)*

Fig. 5.88.

Generally, in practice the capacitors are connected in delta because in delta connections, the size of the capacitors will be one third of the size in star connections. There is not much power lost in the capacitors and also they need not much maintenance and are also available in small sizes.

Consider an inductive load as shown in the circuit diagram. The load current I lags by an angle ϕ_1.

A capacitor C is connected in parallel with the load. The capacitor current I_C leads the voltage by 90°. The resultant current I_1 is the vector sum of the two currents I and I_C and its angle of lag is now ϕ_2, which is less than ϕ_1. The value of $\cos \phi_2$ is more than $\cos \phi_1$. Hence the power factor of the load is improved from $\cos \phi_1$ to $\cos \phi_2$.

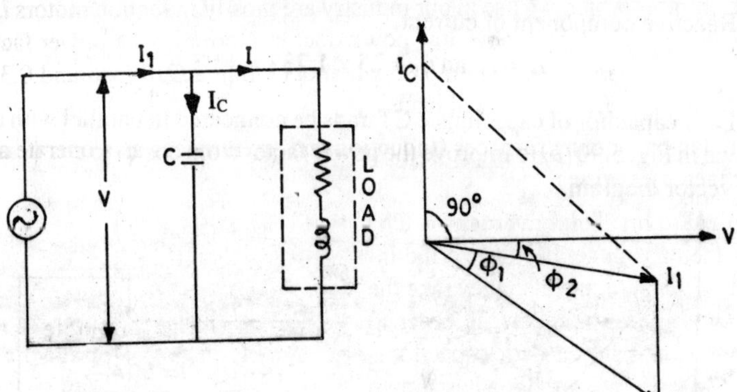

(a) *(b)*

Fig. 5.89.

Advantages of Improved Power Factor : When the power factor of the system is improved, then the current flowing through whole of the system decreases. Due to decreased current, the following are the advantages.

1. Small conductor size.
2. Small copper losses, better efficiency.
3. Less voltage drop, better regulation.
4. For the same kW out put, smaller kVA rating of the equipment.

Example 5.90. A single phase 220 volts, 50 Hz, motor takes a supply current of 50 amperes at a power factor of 0.5 lagging. The motor power factor has been improved to 0.9 lagging by connecting a condensor in parallel. Calculate the capacitance of the capacitor required.

Solution : Applied voltage, $V = 220$ volts

Current taken from supply, $I = 50$ A

Supply frequency, $f = 50$ Hz

Power factor of the motor $\cos\phi_1 = 0.5$ (lagging)

$\therefore \qquad \phi_1 = \cos^{-1} 0.5 = 60^\circ$

Active component of current, $I_a = I \cos\phi = 50 \times 0.5 = 25$ A

Since load on the motor is constant.

The active component of current remains the same.

$$\tan\phi_1 = 1.732$$

i.e., $\qquad \tan 60^\circ = 1.732$

Reactive component of current

$$I_{r_1} = I_a \tan\phi_1 = 25 \times 1.732 = 43.3 \text{ A}$$

Let a capacitor of capacitance C farads be connected in parallel with the load as shown in Fig. 5.90 *(b)* to improve the power factor from $\cos\phi_1$ to $\cos\phi_2$ as shown in the vector diagram.

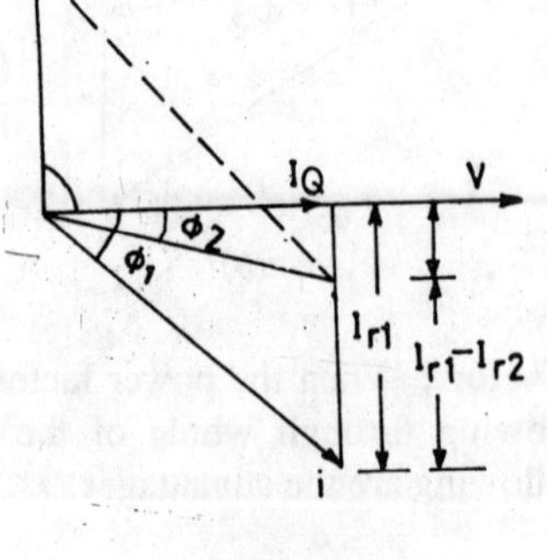

Fig. 5.90. (a)

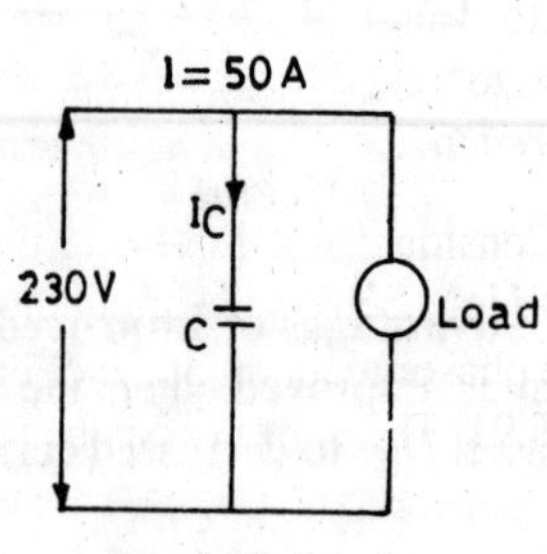

Fig. 5.90. (b)

Now $\cos \phi_2 = 0.9$

$\therefore \qquad \tan \phi_2 = \tan \cos^{-1} 0.9 = 0.4843$

Reactive component I_{r_2} of the current $= I_a \tan \phi. = 25 \times 0.4843 = 12.1$ A

$\therefore$ Current through capacitor $I_C = (I_{r_1} - I_{r_2})$

$$I_C = (43.3 - 12.1) = 31.2 \text{ A}$$

Now $\qquad I_C = \frac{V}{X_C} = 2\pi f C V$

$$\therefore \qquad C = \frac{I_C}{2\pi f V} = \frac{31.2 \times 10^6}{2 \times 3.14 \times 50 \times 220} = 451 \ \mu\text{F}$$

5.51. PRACTICAL IMPORTANCE OF POWER FACTOR

The apparent power drawn by a circuit has two components viz. *(i)* true power and *(ii)* reactive power. True power component should be as large as possible because it is this component which does the useful work in the circuit. This is possible only if the reactive power component is small. As seen from the power triangle the smaller the phase angle ϕ (*i.e.*, greater the p.f. $\cos \phi$), the smaller is the reactive power component. Thus when $\phi = 0^\circ$ (*i.e.*, $\cos \phi = 1$), the reactive power component is zero and the true power is equal to the apparent power. That means the whole of apparent power drawn by the circuit is being utilised by it. Thus power factor of a circuit is a measure of its effectiveness in utilising the apparent power drawn by it. The greater the power factor of a circuit, the greater is its ability to utilise the apparent power. Thus 0.5 p.f. (*i.e.*, 50% p.f.) of a circuit means that it will utilise only 50% of the apparent power whereas 0.8 p.f. would mean 80% utilisation of apparent power. For this reason, we wish that the power factor of the circuit to be as near to 1 as possible.

5.52. ACTIVE AND REACTIVE CURRENT AND THEIR SIGNIFICANCE

Active Current : The current component which is in phase with circuit voltage (*i.e.*, $I \cos \phi$) and contributes to active or true power of the circuit is called *active component or wattful component or in-phase component of current.*

Reactive Current : The current component which is in quadrature (or 90° out of phase) to circuit voltage (*i.e.*, $I \sin \phi$) and contributes to reactive power of the circuit is called *reactive component* of current.

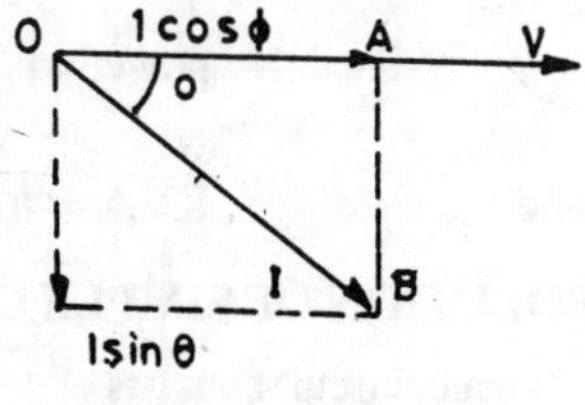

Fig. 5.91. Active and Reactive components

Consider an inductive circuit in which current I lags behind the applied voltage V by ϕ°. The phasor diagram of the circuit is shown in Fig 5.91. The current I can be resolved into two rectangular components *viz.* *(i)* $I \cos \phi$ in phase with V and *(ii)* $I \sin \phi$; 90° out phase with V.

$$\therefore \quad \text{True power, } P = V \times I \cos \phi = VI \cos \phi \text{ Watts or kW}$$

5.53. APPARENT POWER, TRUE POWER AND REACTIVE POWER

(i) **Apparent Power :** The product of rms values of current and voltage, VI is called the *apparent power* and is measured in volt-amperes or kilo-volt amperes (kva).

(ii) **True Power :** The product of voltage and current and cosine angle between them is called true power. The *true power* in an a.c. circuit is obtained by multiplying the apparent power by the power factor and is expressed in watts or kilo-watts (kw). It is also called real power.

(iii) **Reactive Power :** The product of apparent power, VI and the sine of the angle between voltage and current, $VI \sin \phi$ is called the *reactive power*. This is also known as *wattless power* and is expressed in reactive volt-amperes or kilo-volt amperes reactive (kvar).

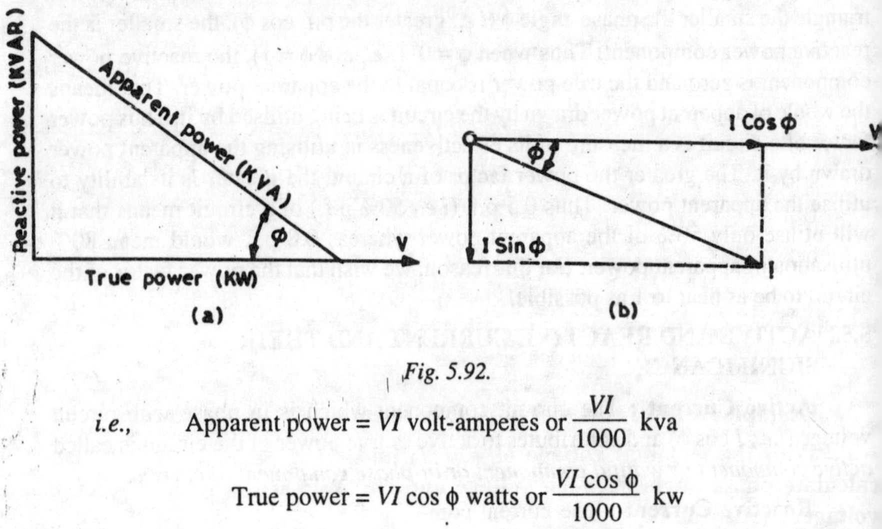

Fig. 5.92.

i.e., $$\text{Apparent power} = VI \text{ volt-amperes or } \frac{VI}{1000} \text{ kva}$$

$$\text{True power} = VI \cos \phi \text{ watts or } \frac{VI \cos \phi}{1000} \text{ kw}$$

$$\text{Reactive power} = VI \sin \phi \text{ var or } \frac{VI \sin \phi}{1000} \text{ kvar}$$

and $$\text{KVA} = \sqrt{(kw)^2 + (kvar)^2}$$

5.54. SOLUTION OF SIMPLE CIRCUIT USING J-NOTATION

If an a.c. circuit consists of several branches the vector diagram may become too difficult/complicated to give an easy solution. In order to overcome this difficulty J-method is employed.

In J-method, a vector is represented by complex number or complex quantity.

Applications of Complex Number to A.C. Circuits

(a) R-L series circuit

Let V be the applied a.c. voltage and I be the current

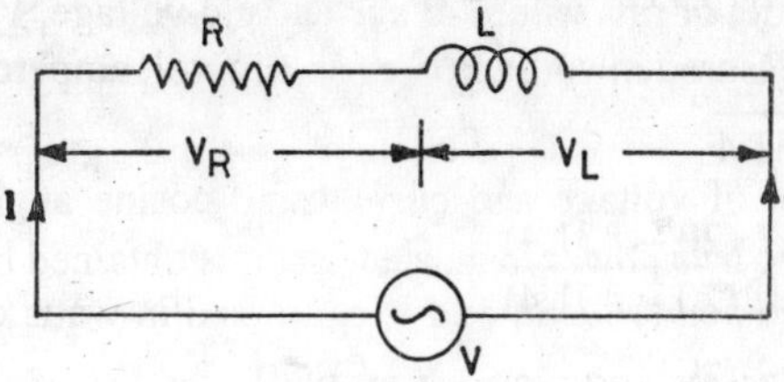

Fig. 5.93.

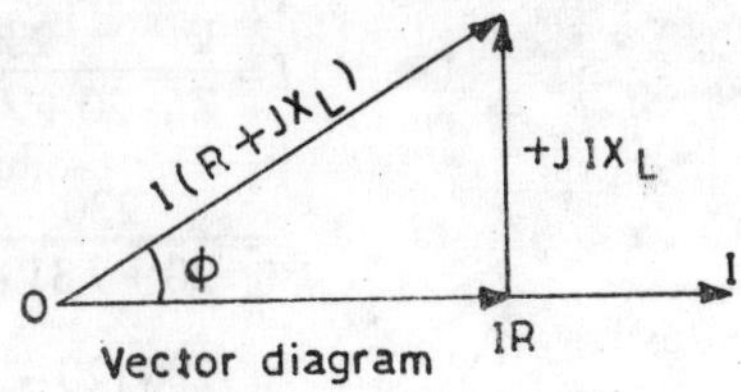

Fig. 5.94.

Draw V.D. taking current as reference vector

$$V_R = IR \text{ volts}$$

$$V_L = IX_L \text{ volts}$$

V can be represented in complex form

$$V = V_R + J\,V_L$$

$$V = IR + J\,IX_L$$

$$V = I\,(R + J\,X_L)$$

But $\quad V = IZ$

or $\quad Z = R + J\,X_L$

$$\text{Magnitude of } Z = \sqrt{R^2 + X_L^2}$$

$$\tan\phi = \frac{X_L}{R}$$

Example 5.81. A resistance of 20 Ω in series with an inductance of 0.1H. With the terminal voltage of 230 V, find the value of current. Also calculate phase angle between voltage and current relative to the applied voltage. Use J-Method.

Solution :

$$R = 20\ \Omega$$

$$X_L = 2\pi f L = 2\pi \times 50 \times 0.1 = 31.4\ \Omega$$

$$\text{Impedance, } Z = R_1 + JX_L = 20 + J\,31.4$$

Taking applied voltage V as reference vector.

$$V = 230 + J\,0$$

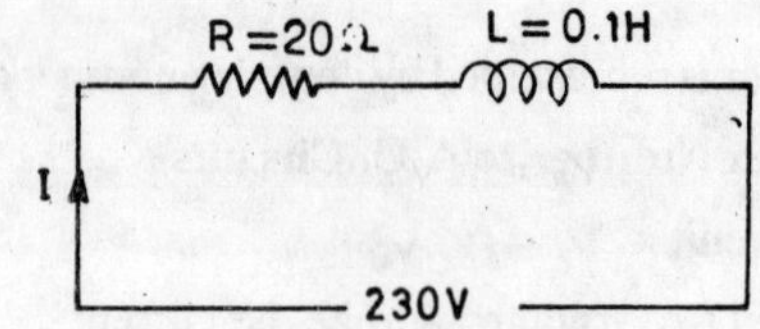

Fig. 5.95.

$$I = \frac{V}{Z} = \frac{230}{20 + J\,31.4}$$

$$= \frac{230}{20 + J\,31.4} \times \frac{(20 - J\,31.4)}{(20 - J\,31.4)}$$

$$= \frac{230\,(20 - J\,31.4)}{(20)^2 + (31.4)^2} = \frac{230\,(20 - J\,31.4)}{400 + 985}$$

$$= \frac{230 \times 20 - J\,230 \times 31.4}{1385} = \frac{4600}{1385} - J\frac{7222}{1385}$$

$$I = 3.32 - J\,5.22$$

Magnitude of $I = \sqrt{(3.32)^2 + (5.22)^2} = 6.18$ A

$$\tan\phi = \frac{5.22}{3.32} = 1.621$$

$\therefore$ $\phi = \tan^{-1} 1.621 = 58°\,21'$ **Ans.**

(b) R-C Series Circuit

Fig. 5.96 shows circuit having resistance R and capacitive reactance X_C connected in series.

Let V be the applied voltage, a.c. voltage and I be the current

Draw the vector diagram taking current as reference vector

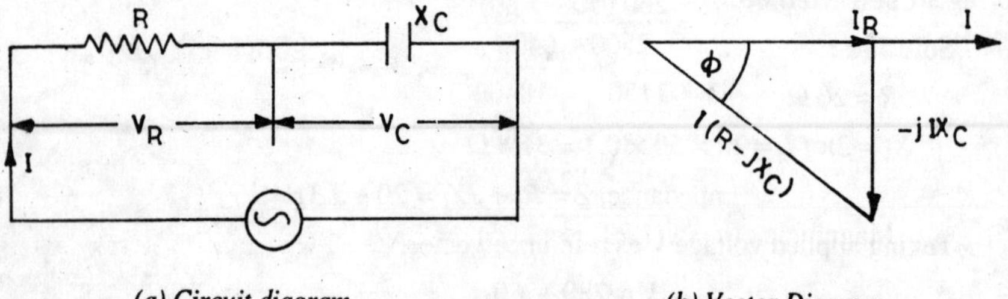

(a) Circuit diagram *(b) Vector Diagram*

Fig. 5.96.

$$V_R = IR \text{ volts}$$

$$V_L = IX_C \text{ volts}$$

$$\therefore \quad V = V_R + J\,V_C$$

$$V = IR + J\,IX_L$$

$$V = I\,(R + J\,X_C)$$

But $$V = IZ$$

$$Z = R + J\,X_L$$

$$\text{Magnitude of } Z = \sqrt{R^2 + X_C^2}$$

$$\tan\phi = \frac{X_C}{R}$$

Exercise 5.82. A resistance of 50 Ω in series with capacitance of 40 μF is connected across 230 V suppy. Calculate the value of current. Also calculate phase angle between current and voltage. Use J-method.

Solution :

$$R = 50\ \Omega$$

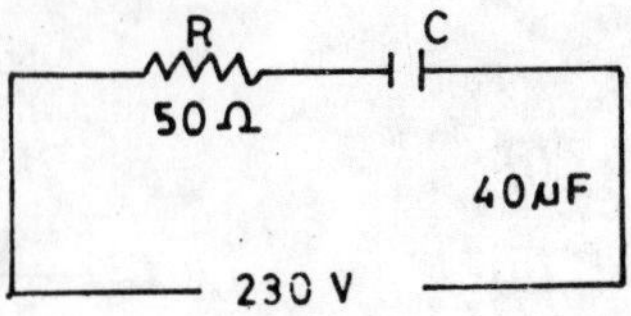

Fig. 5.97.

$$X_C = \frac{1}{2\pi \times 50 \times \frac{40}{10^6}} = 80\ \Omega$$

$$Z = R - J\,X_C \qquad \text{–ve being } X_C$$

$$V = 230 + J\,0$$

$$I = \frac{V}{Z} = \frac{230}{50 - J\,80}$$

$$= \frac{230}{50 - J\,80} \times \frac{50 + J\,80}{50 + J\,80}$$

$$= \frac{230\,(50 + J\,80)}{2500 + 6400}$$

$$= \frac{1150}{8900} + J\,\frac{18400}{8900}$$

$$I = 1.29 + J\,2.06$$

$$\text{Magnitude of } I = \sqrt{(1.29)^2 + (2.06)^2} = 2.3 \text{ A}$$

$$\phi = \tan^{-1}\frac{2.06}{1.29} = 1.59 = 57^\circ\,57' \textbf{ Ans.}$$

Example 5.83. A coil of resistance 10 ohms and Inductive reactance 0.1 H is connected in series with a condenser of capacitance 150 μF across a 200 volt 50 Hz supply. Calculate

(i) Complex equation for impedance

(ii) Complex equation for current

(iii) Absolute value of current

(iv) Power factor

(v) Power consumed by the circuit.

Solution : Fig 5.98.

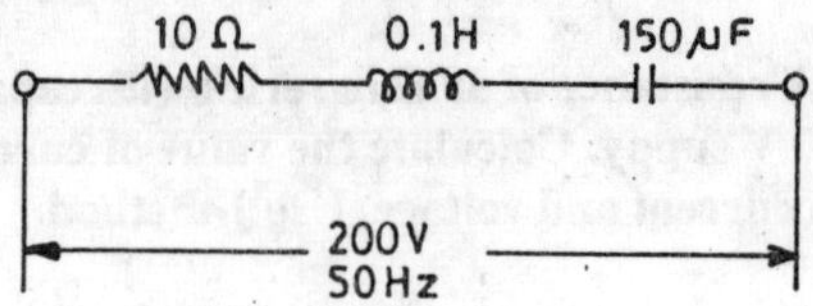

Fig. 5.98.

(i) $X_L = 2\pi f L = 31.416\ \Omega$

(ii) $X_C = \dfrac{1}{2\pi f C} = 21.22\ \Omega$

(i) $Z = R + J X_L - J X_C$

$= 10 + J\,31.416 - J\,21.22$

$Z = 10 + J\,10.196$ ohms is the complex equation for impedance.

(ii) $I = \dfrac{V}{Z} = \dfrac{200}{10 + J\,10.196}$

$= \dfrac{200\,(10 - J\,10.196)}{(10 + J\,10.196)\,(10 - J\,10.196)}$

$= \dfrac{2000 - J\,2039.2}{100 + 104}$

$= \dfrac{2000 - J\,2039}{204} = 9.8 - J\,9.997$

(iii) $|\,I\,| = \sqrt{9.8^2 + 9.997^2} = 14$ Amps.

(iv) $\tan\phi = \dfrac{9.997}{9.8} = 1.02$

or $\phi = 45° — 34'$

Power factor $= \cos\phi = \cos 45° — 34' = 0.7$ **Ans.**

(v) Power = Sum of (Product of Real quantities and imaginary quantities of V and I with sign)

$$V = 200\,(1 + J\,0) = 200 + J\,0$$

$$I = 9.8 - J\,9.997$$

$$\text{Power} = 200 \times 9.8 + (-9.997 \times 0) = 1960 \text{ watts}$$

Example 5.84. Two circuits of impedances $2 + J\,4$ and $3 + J\,4$ ohms are connected in parallel and A.C. voltage of 100 V is applied across the parallel combination. Calculate the magitude of current as well as p.f. of each circuit and magnitude of total current for the parallel combination and circuits p.f.

Solution :

$$V = 100 + J\,0$$

$$I_1 = \frac{V}{Z_1} = \frac{100 + J\,0}{2 + J\,4}$$

$$= \frac{100 + J\,0}{2 + J\,4} \times \frac{2 - J\,4}{2 - J\,4} = \frac{200 - J\,400}{20}$$

$$I_1 = 10 - J\,20$$

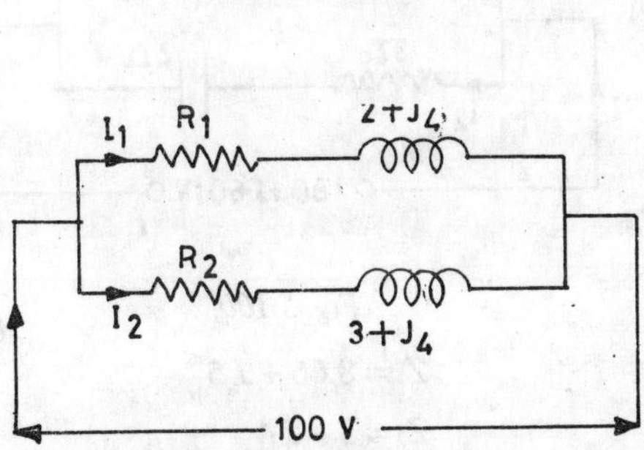

Fig. 5.99.

$$\text{Magnitude of current } I_1 = \sqrt{(\text{Real comp})^2 + (\text{Imaginary comp})^2}$$

$$= \sqrt{(10)^2 + (-20)^2} = 22.36 \text{ A}$$

$$\text{p.f.} = \frac{\text{Real component of current}}{\text{Magnitude of current}} = \frac{100}{22.36} = 0.4472 \text{ lagging}$$

$$\text{Similarly, } I_2 = \frac{100 + J\,0}{3 + J\,4} \times \frac{3 - J\,4}{3 - J\,4} = \frac{300 - J\,400}{25} = 12 - J\,16$$

$$\text{Magnitude of current } I_2 = \sqrt{(12)^2 + (16)^2} = \sqrt{44 + 256} = 20 \text{ A}$$

$$\text{p.f.} = \frac{12}{20} = 0.6 \text{ lagging}$$

$$\text{Total current} = I_1 + I_2 = (10 - J\,20) + (12 - J\,16)$$

$$I = 22 - J\,36$$

$$\text{Magnitude of } I = \sqrt{(22)^2 + (-36)^2} = \sqrt{484 + 1296} = 42.2 \text{ A}$$

$$\text{p.f. of combination} = \frac{12}{42.2} = 0.521 \text{ lagging } \textbf{Ans.}$$

Example 5.85. An Impedance (8.66 + *J* 5) ohm is connected in parallel with another impedance of (3 – J 4) ohm across a supply (80 + *J* 60) volts, calculate

(i) **Current in each branch**

(ii) **Total current taken from supply**

(iii) **p.f. of the whole circuit.**

Solution :

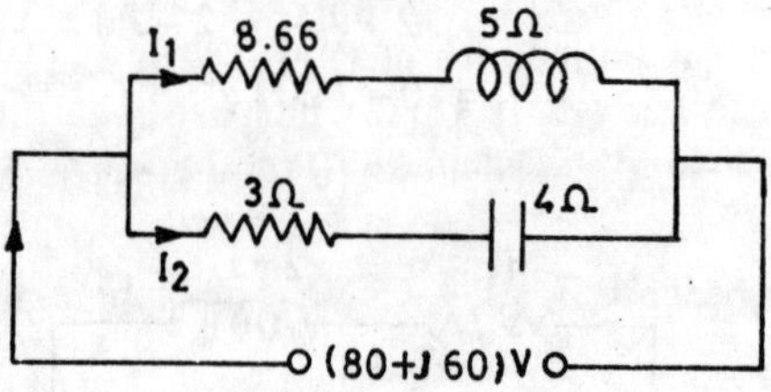

Fig. 5.100.

$$Z_1 = 8.66 + J\,5$$

$$Z_2 = 3 - J\,4$$

$$E = 80 + J\,60$$

$$I_1 = \frac{E}{Z_1} = \frac{80 + J\,60}{8.66 + J\,5} \text{ (Rationalising)}$$

$$= \frac{80 + J\,60}{8.66 + J\,5} \times \frac{8.66 - J\,5}{8.66 - J\,5} = \frac{892.8 + J\,519.6 - J\,400 + 300}{(8.66)^2 + (5)^2}$$

$$= \frac{9.928 + J\,119.6}{100} = 9.928 + J\,1.196$$

$$\text{Magnitude of current } I_1 = \sqrt{(9.928)^2 + (1.96)^2} = 10 \text{ A } \textbf{Ans.}$$

Current in second impedance

$$I_2 = \frac{E}{Z_2} = \frac{80 + J\,60}{3 - J\,4}$$

$$= \frac{80 + J\,60}{3 - J\,4} \times \frac{3 + J\,4}{3 + J\,4}$$

$$= \frac{240 + J\,180 + J\,320 - 240}{(3)^2 + (4)^2} = \frac{0 + J\,500}{25}$$

$$I_2 = 0 + J\,20$$

$$\text{Magnitude of } I_2 = \sqrt{(0)^2 + (20)^2} = 20 \text{ A}$$

$$\text{Total current, } I = I_1 + I_2 = (9.928 + J\,1.196) + (0 + J\,20)$$

$$= 9.928 + J\,21.196$$

$$\text{Magnitude of } I = \sqrt{(9.928)^2 - (21.196)^2} = 23.4 \text{ A}$$

$$\tan\phi = \frac{\text{Imginary component}}{\text{Real component}} = \frac{21.196}{9.928} = 2.1349$$

$$\phi = \tan^{-1} 2.1349 = 64° \, 54'$$

$$p.f. = \frac{\text{Real component}}{\text{Magnitude of current}} = \frac{9.928}{23.4} = 0.884 \text{ leading}$$

Example 5.86. An impedance of (2+*J*5) Ω is connected in series impedance of (10 + *J* 4) Ω and (12 – *J* 18) Ω which are in parallel of whole circuit, is connected to 200 V supply. Calculate *(i)* total current *(ii)* p.f. of the whole circuit.

Solution : Impedance $Z_1 = 2 + J\,5$

$$Z_2 = 10 + J\,4$$

$$Z_3 = 12 - J\,18$$

Equivalent Impedance of parallel circuit

$$\frac{1}{Z_p} = \frac{1}{Z_2} + \frac{1}{Z_3} \qquad J^2 = -1$$

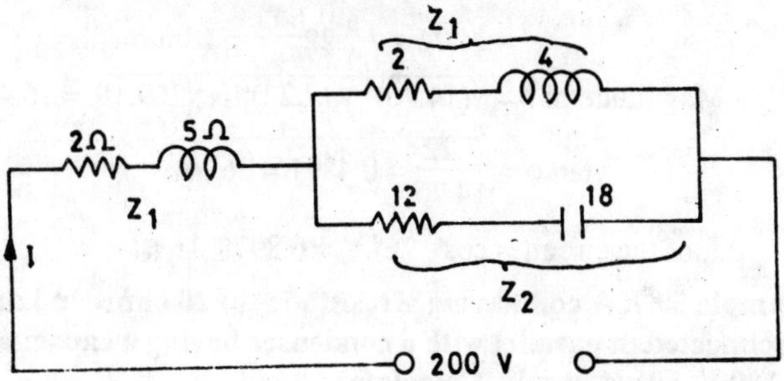

Fig. 5.101.

$$\therefore \quad \frac{1}{Z_p} = \frac{1}{10+J\,4} + \frac{1}{12-J\,18}$$

$$= \frac{(12-J\,18)+(10+J\,4)}{(10+J\,4)\,(12-J\,18)}$$

$$= \frac{22-J\,14}{120-J\,180+48+72}$$

$$\frac{1}{Z_p} = \frac{22-J\,14}{192-J\,132}$$

$$Z_p = \frac{192-J\,132}{192-J\,14} \text{ (Rationalising)}$$

$$= \frac{192-J\,132}{22-J\,14} \times \frac{22+J\,14}{22+J\,14}$$

$$= \frac{(192-J\,132)\,(22+J\,14)}{(22)^2+(14)^2}$$

$$= \frac{4224-J\,2904+J\,2688+1848}{484+196}$$

$$Z_p = \frac{6072-J\,216}{680}$$

$$= \frac{6072}{680} - J\,\frac{216}{280} = 8.93 - J\,0.32$$

Total circuit Impedance $Z = Z_1 + Z_p$

$$= (2+J\,5)+(8.93-J\,0.32) = 10.93+J\,4.68$$

$$I = \frac{E}{Z} = \frac{200}{10.93+J\,4.68}$$

$$= \frac{200}{10.93+J\,4.68} \times \frac{10.93-J\,4.68}{10.93-J\,4.68}$$

$$= 14.76 - J\,7.22$$

Magnitude of $I = \sqrt{(14.76)^2+(7.22)^2} = \sqrt{269.19} = 16.4$ A

$$\tan\phi = \frac{7.22}{14.76} = 0.4891 = 26^\circ\,4'$$

p.f. of the circuit $= \cos^{-1} 26^\circ\,4' = 0.8978$ **Ans.**

Example 5.87. A coil having a resistance of 20 ohms and an inductance 0.1 H is connected in parallel with a condenser having a capacitance of 50 μF across a 100 V, 50 c/s supply. Calculate

(a) **The current in each branch**

(b) **The resultant current**

(*c*) **The power factor of the whole Circuit**

(*d*) **Total Power**

(*e*) **Sketch the vector diagram.**

Solution : Fig. 5.102.

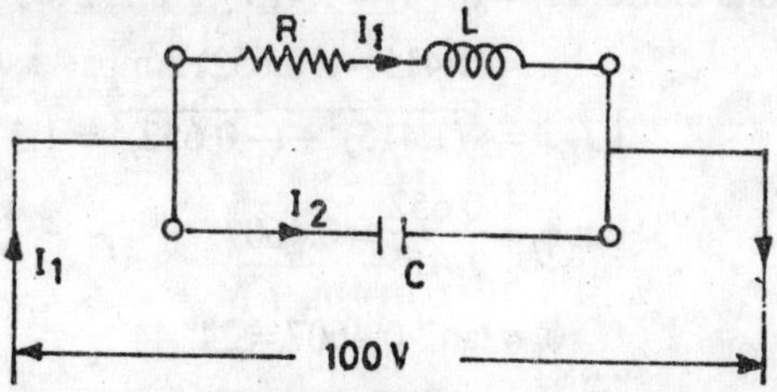

Fig. 5.102.

Branch 1 : $R_1 = 20\ \Omega$

$$X_{L_1} = 2\pi f L_1 = 31.416\ \Omega$$

$$Z_1 = R_1 \times J X_{L_1} = 20 + J\,31.416$$

$$I_1 = \frac{V}{Z_1} = \frac{100}{20 + J\,31.416}$$

$$= \frac{100\,(20 - J\,31.416)}{(20 + J\,31.416)\,(20 - J\,31.416)}$$

or $$I_1 = \frac{2000 - J\,3141.6}{400 + 1014}$$

$$= \frac{2000}{1414} - J\frac{3141.6}{1414} = 1.415 - J\,2.222$$

∴ $$|\,I_1\,| = \sqrt{(1.415)^2 + (2.222)^2}$$

$$= \sqrt{2 + 4.937} = \sqrt{6.937} = 2.633$$

$$\tan \phi_1 = \frac{2.222}{1.415} = 1.57$$

∴ $$\phi_1 = 57^\circ - 30'\ \text{lag}$$

Branch 2 : $$X_C = \frac{1}{2\pi f C} = \frac{1}{2 \times 3.1416 \times 50 \times 50} = 63.66\ \Omega$$

$$R_2 = 0$$

∴ $$Z_2 = 0 - J\,63.66$$

$$I_2 = \frac{V}{Z_2} = \frac{100}{-J\,63.66} \times \frac{J\,63.66}{J\,63.66}$$

$$= \frac{J\,6366}{4053} = +J\,1.57 \text{ Amps}$$

$$\therefore \quad |I_2| = \sqrt{(1.57)^2} = 1.57 \text{ Amps}$$

$$\phi_2 = 90^\circ \text{ lead}$$

(b) Total Current $I_T = I_1 + I_2 = 1.415 - J\,2.222 + J\,1.57$

$$= (1.415 - J\,0.652) \text{ Amps}$$

$$\therefore \quad |I_T| = \sqrt{(1.415)^2 + (-0.652)^2} = 1.57 \text{ Amps}$$

(c) $$\tan \phi_t = \frac{0.652}{1.415} = 0.4607$$

$$\therefore \quad \phi_t = \tan^{-1} 0.4607 = 24^\circ\, 44'$$

$$\therefore \quad \cos \phi_t = p.f. = \cos 24^\circ\, 44' = 0.9083 \text{ Lag}$$

(d) Total Power = $(100 \times 1.415) + (-0.652 \times 0) = 141.5$ watts

(e) Vector Diagram : See Fig. 5.103.

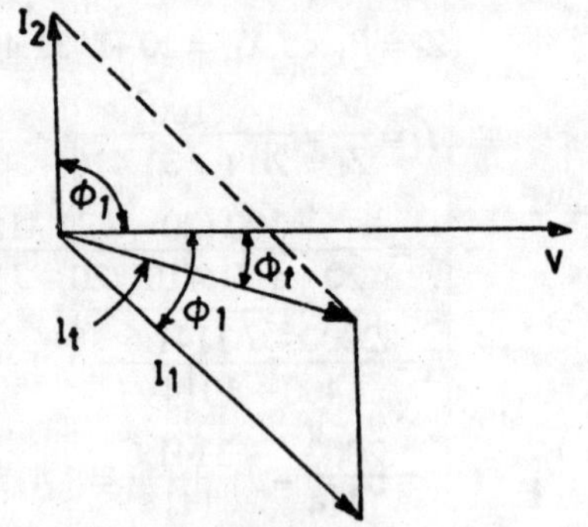

Fig 5.103.

Z is taken as Ref. vector. Current I_1 is lagging by 57° 30′

Current I_2 is leading by 90.

Resultant Current

I_t lags behind by 24° 44′.

5.55. USE OF ADMITTANCE, CONDUCTANCE AND SUSCEPTANCE

(a) ADMITTANCE METHOD

1. Admittance (Denoted by *Y*) : The reciprocal of impedance of a circuit is called admittance of the circuit. It is represented by 'Y'.

$$\therefore \quad Y = \frac{1}{Z}$$

or $$Y = \frac{1}{V/I}$$

or $$\frac{I}{V} = \frac{\text{Current}}{\text{Voltage}}$$

The unit of admittance is MHO (ohm spelled back words) and its symbol is ***.

Just as impedance Z of circuit has two components resistance R and reactance X.

Similarly admittance Y also has two components known as conductance G and susceptance B.

2. Conductance : The reciprocal of resistance is called conductance. It is denoted by G. The unit of conductance is mho.

3. Susceptance : The reciprocal of reactance. It is denoted by capital letter B. Its units is mho.

For Inductive Circuit

Fig. 5.104 shows the impedance and admittance triangle for inductive circuit.

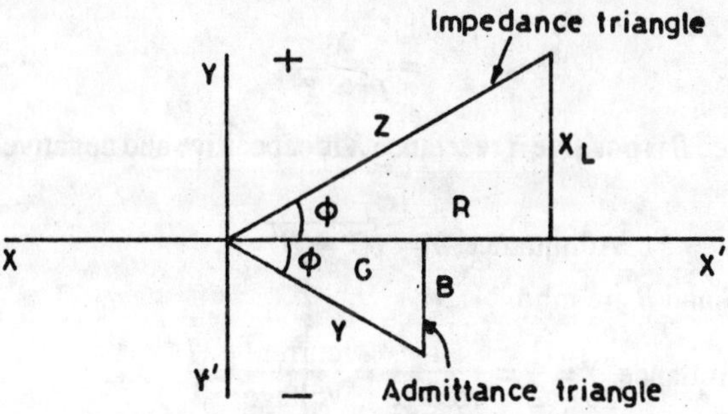

Fig. 5.104. Inductive Circuit.

It is clear that susceptance B is –ve (being below X-axis). Hence inductive susceptance is –ve.

For Capacitive Circuit

It is clear that susceptance is +ve being above X-axis. Hence capacitive susceptance is +ve.

Relations : The value of G and B in term of resistance R and reactance X can be easily found out. Referring to admittance and impedance Δ.

Conductance $$G = Y \cos \phi$$

$$G = \frac{1}{Z} \cdot \frac{R}{Z} \qquad \cos \phi = \frac{R}{Z} \text{ and } Y = \frac{1}{Z}$$

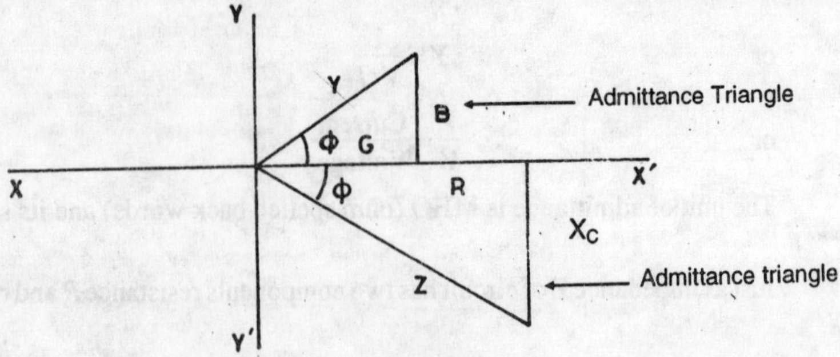

Fig. 5.105. Capacitive circuit

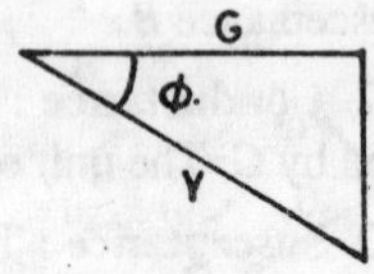

Fig 5.106. Admittance triangle

$$G = \frac{R}{Z^2} = \frac{R}{R^2 + X^2}$$

Conductance G is always positive.

Similarly Suceptance $B = Y \sin \phi$

$$B = \frac{1}{Z} \cdot \frac{X}{Z} = \frac{X}{Z^2}$$

$$= \frac{X}{R^2 + X^2}$$

Susceptance B is positive if reactance X is capacitive and negative if reactance is inductive.

$$\text{Admittance, } Y = \sqrt{G^2 + B^2}$$

Units of G and B are mho.

$$\text{Admittance, } Y = \frac{1}{Z} = \frac{1}{\frac{\text{voltage}}{\text{current}}} = \frac{\text{current}}{\text{voltage}} = \frac{I}{V}$$

$$I = VY \text{ Amps} \qquad \left(\because \; I = \frac{V}{Z} = V\frac{1}{Z} = VY\right)$$

5.56. APPLICATION OF ADMITTANCE METHOD

Consider two parallel a.c. circuit.

Consisting two branchs impedances Z_1 (R_1 , C_1) and Z_2 (R_2 , L_2) V volts (rms) is applied across parallel circuit.

Branch I

$$\text{Impedance, } Z_1 = \sqrt{R_1^2 + X_{C_1}^2}$$

$$\phi_1 = \tan^{-1} \frac{X_c}{R}$$

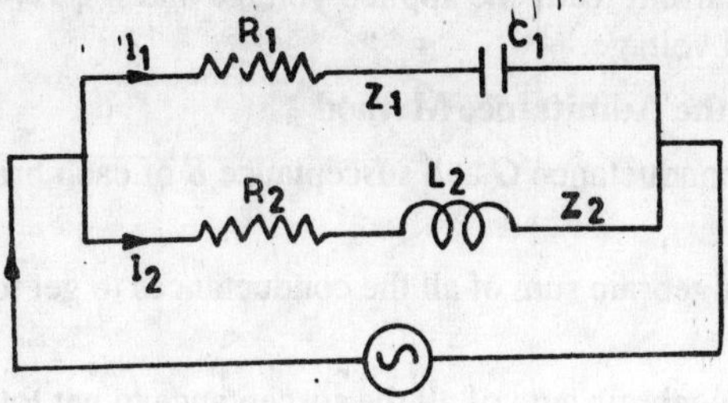

Fig. 5.107.

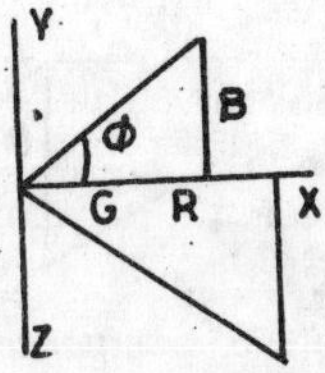

Fig. 5.108.

$$\text{Conductance, } G_1 = \frac{R_1}{Z_1^2} = \frac{R_1}{R_1^2 + X_{C_1}^2}$$

$$\text{Susceptance, } B_1 = \frac{X_{C_1}}{Z_1^2} = \frac{X_{C_1}}{R_1^2 + X_{C_1}^2}$$

+ve being capacitive

Branch II

$$\text{Impedance, } Z_2 = \sqrt{R_2^2 + X_{L_2}^2}\ \Omega$$

$$\text{Conductance, } G_2 = \frac{R_2}{Z_2^2} = \frac{R_2}{R_2^2 + X_{L_2}^2}$$

$$\phi_1 = \tan^{-1}\frac{X_L}{R}$$

$$\text{Susceptance, } B_2 = \frac{X_{L_2}}{Z_2^2} = -\frac{X_{L_2}}{R^2 + X_L^2} \quad \text{–ve being inductive}$$

$$\text{Total Conductance, } G = G_1 + G_2$$

$$\text{Total Susceptance of circuit, } B = B_1 + (-B_2) = B_1 - B_2$$

$$\text{Admittance, } Y = \sqrt{G^2 + B^2}$$

$$\text{Line current, } I = VY$$

$$\text{Phase angle, } \phi = \tan^{-1}\frac{B}{G} \qquad \tan\phi = \frac{B}{G}$$

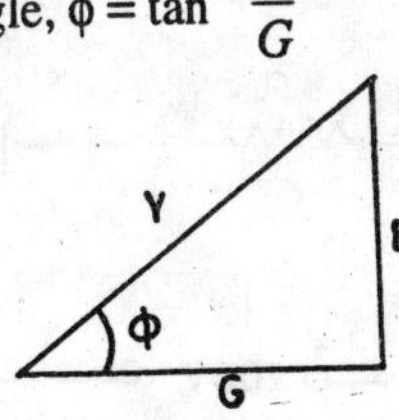

Fig. 5.109.

If ϕ is +ve current leads the applied voltage and if ϕ is –ve the current lags behind the applied voltage.

Various Steps of the Admittance Method

1. Find the conductance G and susceptance B of each branch of the parallel circuit.

2. Find the algebraic sum of all the conductances to get total conductance G of the circuit.

3. Find the algebraic sum of all the susceptance to get total susceptance B.

4. Find admittance Y of the circuit from the relation

$$Y = \sqrt{G^2 + B^2}$$

5. The total line current $I = VY$.

6. Phase angle $\phi = \tan^{-1} \dfrac{B}{G}$

7. p.f. $\cos \phi = \dfrac{G}{Y}$

Example 5.88. A circuit having a resistance of 6 Ω and reactance of 8 ohms is connected across 200 c/s mains is parallel with another circuit having a resistance of 8 ohms and a capacitance reactance of 6 ohms. Find :

(i) **Magnitude of current and Power factor for each circuit.**

(ii) **Magnitude of total current and power factor.**

Use Admittance method.

Solution : Applied voltage $V = 200$ volts

Branch 1

Resistance, R = 6Ω

Reactance, $X_{L_1} = 8\ \Omega$

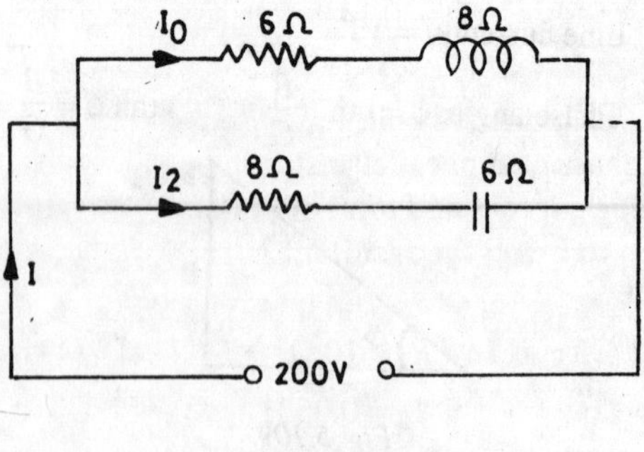

Fig. 5.110.

$$Z_1^2 = R_1^2 + X_{L_1}^2 = 6^2 + 8^2 = 100$$

$$\text{Conductance, } G = \frac{R_1}{Z_1^2} = \frac{6}{100} = 0.06 \text{ Mho}$$

$$\text{Susceptance, } B_1 = \frac{X_{L_1}}{Z_1^2} = \frac{8}{100} (\ 0.08 \text{ Mho}) \quad \text{(being inductives)}$$

$$\text{Admittance of the first branch, } Y_1 = \sqrt{G^2 + B^2} = \sqrt{(0.06)^2 + (0.08)^2} = 0.1 \text{ Mho}$$

$$\text{Current, } I = V\,Y_1 = 200 \times 0.1 = 20 \text{ A}$$

$$\text{p.f., } \cos\phi_1 = \frac{R_1}{Z_1} = \frac{6}{10} = 0.6 \text{ lagging } \textbf{Ans.}$$

Branch 2

$$R_2 = 8\ \Omega$$

$$X_{C_2} = 6\ \Omega$$

$$Z_2^2 = R_2^2 + X_{C_2}^2 = 8^2 + 6^2 + 100$$

$$\text{Conductance, } G_2 = \frac{R_2}{Z_2^2} = \frac{8}{100} = 0.08 \text{ Mho}$$

$$\text{Susceptance, } B_2 = \frac{X_{C_2}}{Z_2^2} = \frac{6}{100} = 0.06 \text{ Mho (+ve)}$$

$$\text{Admittance of second branch, } Y_2 = \sqrt{G^2 + B^2} = \sqrt{(0.08)^2 + (0.06)^2} = 0.1 \text{ Mho}$$

$$\text{Current, } I_2 = VY_2 = 200 \times 0.1 = 20 \text{ A}$$

$$\text{p.f. } \cos\phi_2 = \frac{R_2}{Z_2} = \frac{8}{10} = 0.8 \textbf{ Ans.}$$

(II) Total conductance of circuit, $G = G_1 + G_2 = 0.06 + 0.08 = 0.14$ Mho

Total susceptance of the circuit, $B = -B_1 + B_2 = -0.08 + 0.06 = -0.02$ Mho

Total circuit admittance, $Y = \sqrt{G^2 + B^2} = \sqrt{(0.14)^2 + (-0.02)^2} = 0.1415$ Mho

$$\text{Total current, } I = VY = 200 \times 0.1415 = 28.30 \text{ A}$$

$$\cos\phi = \frac{G}{Y} = \frac{0.14}{0.1415} = 0.99 \text{ lagging}$$

Example 5.89. An inductive coil, having an inductance of 0.05 H and resistance of 10 Ω is connected parallel with condenser of 100 mfd across 200 V, 50 Hz. Calculate line current and total pf of the circuit. What would be the value of shunting cap to bring the resultant current into phase with applied voltage.

Solution : For Circuit I : $R_1 = 10\ \Omega$

$$L_1 = 0.05 \text{ H}$$

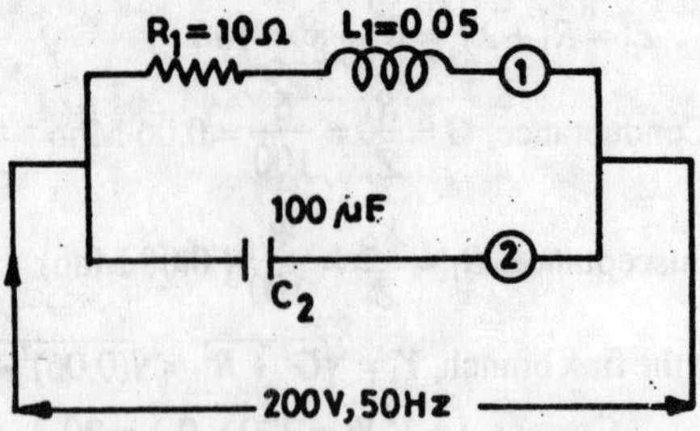

Fig. 5.111.

$$X_{L_1} = 2\pi f L_1 = 2 \times 3.14 \times 50 \times 0.05 = 15.7\ \Omega$$

$$Z_1 = 18.6\ \Omega$$

$$Z_1^2 = 10^2 + 15.7^2 = 346.5\ \Omega$$

$$\therefore \quad G_1 = \frac{R_1}{Z_1^2} = \frac{10}{346.5} = 0.02886 \text{ Mho}$$

$$B_1 = \frac{X_{L_1}}{Z_1^2} = \frac{15.7}{346.5} = -0.045 \quad \text{(Inductive)}$$

$$R_2 = 0$$

$$C_2 = 100\ \mu\text{F}$$

$$X_{C_2} = \frac{1}{2\pi f C_2} = \frac{10^6}{2\pi \times 50 \times 100} = 31.84$$

and $$Z_2^2 = X_C^2$$

$$G_2 = \text{zero}$$

and $$B_2 = \frac{X_C}{Z_2^2} = \frac{X_C}{X_C^2} = \frac{1}{X_C}$$

$$= \frac{1}{\dfrac{10^6}{2\pi \times 50 \times 100}} = 0.0341 \text{ Mho}$$

$$G = G_1 + G_2 = 0.02886 \text{ Mho}$$

$$B = B_1 + B_2 = -0.045 + 0.0341 = 0.0139 \text{ Mho}$$

$$I = VY = 200 \times 0.032 = 6.4 \text{ A}$$

$$\text{p.f. } \cos\phi = \frac{G}{Y} = \frac{0.02886}{0.032} = 0.9$$

$$X_c = \frac{1}{0.0139} = \frac{1}{2\pi f c}$$

$$2\pi f c = 0.0139$$

$$c = \frac{10^6 \times 0.0139}{100\pi} = 44.267 \ \mu F$$

HIGHLIGHTS

1. For an a.c. circuit containing resistance (R) only,

$$v = V_m \sin \omega t$$

$$i = I_m \sin \omega t \quad \text{where } I_m = V_m/R$$

In other words, voltage and current are in phase.

Power consumed, $P = V_m I_m/2 = VI$

where V and I are the rms values of the applied voltage and the circuit current respectively.

2. For an a.c. circuit containing inductance (L) only.

$$v = V_m \sin \omega t$$

$$i = I_m \sin (\omega t - \pi/2) \qquad \text{where } I_m = V_m/X_L$$

That is current lags behind the voltage by $\pi/2$ rad. or 90°.

Inductive reactance, $X_L = \omega L = 2\pi f L$ ohms

Power consumed, $P = 0$

3. For an a.c. circuit containing capacitance (C) only,

$$v = V_m \sin \omega t$$

$$i = I_m \sin (\omega t + \pi/2) \quad \text{where } I_m = V_m/X_C$$

That is current leads the voltage by $\pi/2$ rad or 90°.

Capacitive reactance, $X_C = \dfrac{1}{\omega C} = \dfrac{1}{2\pi f C}$ ohms

Power consumed, $P = 0$

4. **Active Current :** The current component which is in phase with circuit voltage is called active current. It is equal to $I \cos \phi$.

5. **Reactive Current :** The current component which is in quadrature or 90° out of phase to the circuit voltage. It is equal to $I \sin \phi$.

6. **Apparent Power :** The product of rms values of current and rms value of voltage is called apparent power. It is measured in volt-amps.

7. **True Power :** The product voltage, current and cosine angle is called true power.

$$P = VI \cos \phi$$

The unit of true power is watt.

8. **Reactive Power :** The product of apparent power and sine of the angle between voltage and current is called reactive power. Its unit is VAR.

$$\text{Reactive Power, } P = VI \sin \phi$$

9. **Power Factor**

1. It is defined as the cosine of phase angle between voltage and current.

2. Ratio of actual power to apparent power.

3. Ratio of resistance to impedance.

10. Advantages of Improved power factor

(*a*) Smaller conductor size.

(*b*) Better efficiency

(*c*) Less voltage drop, better regulation.

11. In J-method, a vector is represented by complex quantity.

12. Admittance of a circuit is defined as the reciprocal of its impedance. Its unit is siemens (*S*).

$$Y = G^2 + B^2$$

13. Adimittance has two rectangular components :

(*i*) Conductance *G* and

(*ii*) Susceptance (*B*)

14. Conductance *G* is reciprocal of resistance $G = 1/R$ mho.

15. Inductive susceptance is regarded as negative where as capacitive susceptance is regarded as positive.

MULTIPLE CHOICE QUESTIONS

1. In pure resistive a.c. circuit,

(*a*) current is in phase opposition to voltage

(*b*) current is in phase with voltage

(*c*) current is in phase sequence with voltage.

2. In pure resistive circuit

(*a*) Average power = $V_{\max} \times I_{\max}$

(*b*) Average power = $V_{rms} \times I_{rms}$

3. Power in pure inductive a.c. circuit is

(*a*) zero (*b*) maximum

(*c*) equal to pure resistive current.

4. In a pure capacitive a.c. circuit

(*a*) current is in phase with voltage

(*b*) current legs the voltage by $\pi/2$

(*c*) current leads the voltage by $\pi/2$

5. In pure capacitive a.c. circuit

(*a*) $L=\frac{d\phi}{dt}$ (*b*) $L=\frac{dv}{dt}$

(*c*) $L=\frac{dq}{dt}$ (*d*) $L=\frac{di}{dt}$

6. Reciprocal of impedance is known as

(*a*) capacitance (*b*) admittance

(*c*) Inductance (*d*) none of the above

7. The power factor of parallel a.c. circuit is given by

(*a*) G/B (*b*) G/Y

(*c*) G/Y^2 (*d*) Y/G

8. The equivalent series reactance of a parallel a.c. circuit is given by

(*a*) $X=G/Y^2$ (*b*) G/Y

(*c*) B/Y (*d*) B/Y^2

9. Resonance in R-L-C circuit occurs when

(*a*) $R=X_L-X_C$ (*b*) $X_L=X_C$

(*c*) $X<R$ (*d*) $X_L=2X_C$

ANSWERS

1. (*b*) **2.** (*b*) **3.** (*a*) **4.** (*c*) **5.** (*c*)

6. (*b*) **7.** (*a*) **8.** (*a*) **9.** (*b*)

EXERCISE

1. Prove that power consumed in a (*i*) purely inductive circuit (*ii*) purely capacitive circuit is zero when an alternating voltage is applied. Draw the phasor diagrams for *V* and *I*.

2. Explain the following terms as applied to ac circuits :

(*i*) Impedance (*ii*) Power factor (*iii*) Phase angle

3. Define apparent power and reactive power of an a c circuit. Explain how the true power and power factor of the circuit can be determined from them.

4. State, with reasons, whether the following statements are true or false ?

(*a*) In a purely resistive circuit, the current leads the voltage by 90°? [False] (*b*) The power consumed by a purely inductive circuit is very high ? [False]

(*c*) In a purely capacitive circuit, the voltage lags the current by 90°? [True]

(*d*) The max. power factor in a circuit can be unity ? [True]

(*e*) In an ac circuit, the reactive power is $V I \sin \phi$? [True]

5. Distinguish between :

(*i*) Form factor and power factor.

(*ii*) Apparent power and reactive power.

(*iii*) Apparent power and true power.

(*iv*) True power and reactive power.

6. What do you understand by power factor of a circuit ?

7. What do you understand by active and reactive components of current in an a.c. circuit?

8. Answer the following :

(*i*) What is phase relationship between current and voltage of purely resistive circuit?

(*ii*) What is phase relationship between current and voltage of purely inductive circuit ?

(*iii*) What is phase relationship between current and voltage of purely capacitive circuit ?

(*iv*) What is the effect of frequency on capacitive reactance ?

(*v*) What is the effect of frequency on inductive reactance ?

(*vi*) What are the values of power factor for (*i*) purely resistive circuit, (*ii*) purely inductive circuit and (*iii*) purely capacitive circuit ?

(*vii*) What is relationship between phase angle and power factor of an ac circuit ?

(*viii*) What is relationship between apparent power, true power and reactive power of an circuit ?

(*ix*) What is the relationship between apparent power, true power and phase angle of an a c circuit?

9. Discuss the practical importance of power factor.

10. What are the causes of low power factor ? Discuss the disadvantages of poor power factor.

11. What do you understand by Conductance, Impedance and Admittance in A.C. circuit.

12. Define the terms Conductance, Susceptance and Admittande. What are their units?

PROBLEMS

1. Two impedances $Z_1 = (10 + J\,5)\ \Omega$ and $Z_2 = (25 - J\,10)\ \Omega$ are connected in parallel to a 100 V, 50 Hz supply. Find *(i)* circuit admittance, *(ii)* circuit current, *(iii)* the phase angle between circuit current and the applied voltage.
   (Ans. *(i)* 0.1174 Mho, *(ii)* 11.74 A, *(iii)* 12.9° lagging**)**
2. Three impedances $65 \angle 45^\circ\ \Omega$, $42 \angle -20^\circ\ \Omega$ and $30 \angle -35^\circ\ \Omega$ are joined in parallel and connected to 240 V, 60 Hz supply. Determine *(i)* total power, *(ii)* circuit power factor and *(iii)* series equivalent circuit.
   (Ans. *(i)* 3613.5 W, *(ii)* 0.965 lead, *(iii)* $R = 15.38\ \Omega$, $X_C - 4.16\ \Omega$**)**
3. For the circuit shown in Fig. 8.30 *(a)* the instantaneous value of supply voltage is 340 sin 314 *t*. Find *(i)* total rms current flowing from the source *(ii)* currents in L_2 and C_2.

   Draw the phasor diagram of the current showing applied voltage, total current and voltage, across choke. Use Admittance method.
   (Ans. 0.906 A, 0.685 A, 1.593 A**)**
4. Tow parallel circuits comprise respectively *(i)* a coil of resistance 20 Ω and inductance 0.07 H *(ii)* a condenser of capacitance 60 μF in series with a resistance of 50 Ω. Calculate the current in the mains and the power factor of the arrangement when connected across a 200 V, 50 Hz supply.
   (Ans. 7.06 A, 0.907**)**

6

Polyphase System

6.1. CONCEPT OF THE POLYPHASE IN THE POWER SYSTEM

Single phase system so far we have considered involving single phase currents and voltages. The single phase system is quite satisfactory for domestic applications. The motors employed in domestic applications are mostly single phase, for example, motors for coolrs, fans, refrigerators. However, the single phase system has its own limitations and, therefore, has been replaced by three phase system.

For generation, transmission and distribution of electric power 3-phase system has been universally adopted. Moreover, for general supply, three phase system is universally used. Three phase supply is having same frequency and voltages but they have definite phase difference. The amount of phase difference depends upon the number of circuits or phases. Such circuits are called POLY PHASE CIRCUITS (Poly means many and phase means winding).

POLYPHASE SYSTEMS

An a.c. system having a group of equal voltages arranged to have phase difference between adjacent e.m.f's is called a polyphase system.

A polyphase system is essentially a combination of single phase voltage having same magnitude and frequency but displaced from one another by equal electrical angle. The electrical displacement depends upon the number of phases and is given by the following relation :

$$\text{Electrical displacement} = \frac{360^\circ \text{ electrical}}{\text{number of phases}}$$

The above relation does not hold good for two phase system where the voltages are displaced by 90° electrical.

Poly means many and *phase* means winding or circuit so that a polyphase system has "*many phases*" or *circuits*. Each phase has a single alternating voltage of the same magnitude and frequency. However, these voltages are displaced from one another by equal electrical angle.

6.1.1. Types of Polyphase System

There are several polyphase arrangements and some of them are given below:

(i) Two phase system

(ii) Three phase system

(iii) Six phase system

(i) **Two Phase System :** In two phase system, there are two equal voltages of the same magnitude and frequency having a phase difference of 90° electrical.

Fig. 6.1 *(a)* and 6.1 *(b)* show e.m.f. waves and vector diagram respectively of two phase system. It may be seen that the two voltages are of the same magnitude and frequency but have a phase difference of 90° electrical. Their equations are :

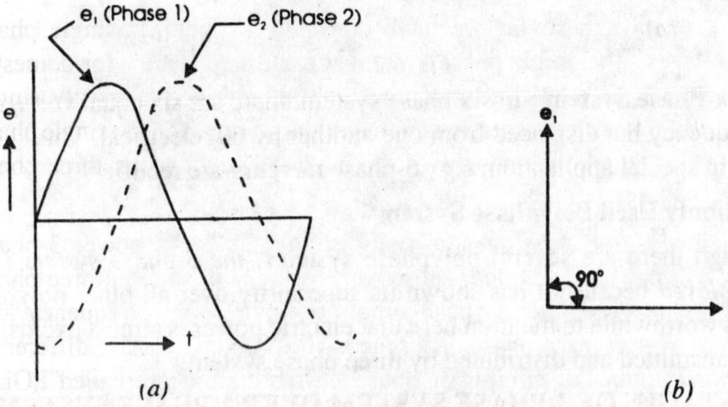

Fig. 6.1.

$$e_1 = E_m \sin \omega t$$

$$e_2 = E_m \sin (\omega t - 90^\circ)$$

(ii) **Three Phase System :** In three phase system, there are three equal voltages of the same magnitude and frequency but displaced from one another by *120° electrical.

Fig. 6.2 *(a)* and 6.2 *(b)* show e.m.f. waves and vector diagram respectively of a 3-phase system. It may be seen that the three voltages are of the same magnitude and frequency but displaced from one another by 120° electrical. Their equations are :

$$e_1 = E_m \sin \omega t$$

$$e_2 = E_m \sin (\omega t - 120^\circ)$$

$$e_3 = E_m \sin (\omega t - 240^\circ)$$

---

* Electrical displacement = $\dfrac{360^\circ \text{ electrical}}{\text{number of phases}} = \dfrac{360^\circ}{3} = 120^\circ$

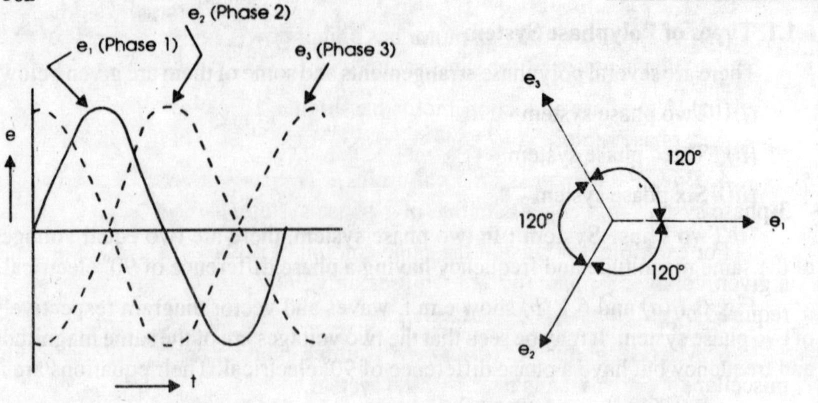

(a) (b)

Fig. 6.2.

(iii) **Six Phase System :** In six phase system, there are six equal voltages of the same frequency but displaced from one another by 60° electrical. This system is used only in special application, *e.g.*, 6-phase mercury are rectifiers.

6.1.2. Commonly Used Polyphase System

Although there are several polyphase systems, *the 3-phase system is invariably employed* because it has shown its superiority over all other polyphase systems. It is worthwhile to mention here that electric power is almost exclusively generated, transmitted and distributed by three phase system.

6.2. ADVANTAGES OF 3-PHASE SYSTEM OVER 1-PHASE SYSTEM

The following are the advantages of 3-phase system over 1-phase system :

1. Constant Power : In a single phase circuit, the power delivered is pulsating. Even when the voltage and current are in phase, the power is zero twice in each cycle. The pulsating nature of power in single phase circuits is objectionable for many applications.

In polyphase system, power delivered is constant if the loads are balanced. This makes polyphase system highly desirable, particularly for power loads.

2. Higher Rating : The rating of a machine increases with the number of phases. Thus, the output of a 3-phase motor is 1.5 times the output of a single phase motor of the same size.

3. Superiority of 3-phase Induction Motor : The most potent reason for adopting 3-phase system is the widespread applications of 3-phase induction motor. A 3-phase induction motor has the following advantages over single phase induction motor :

(i) A 3-phase induction motor is self-starting. On the other hand, a single phase motor has no starting torque and, therefore, an auxillary means is provided for starting it.

(ii) A 3-phase induction motor has higher power factor than that of a single phase motor of the same rating.

(iii) A 3-phase induction motor has higher efficiency than that of single phase motor.

4. Power Transmission Economics : Transmission of electric power by 3-phase system is cheaper than that of 1-phase system.

For example, to transmit the same amount of power over a fixed distance at a given voltage, 3-phase system requires only 3/4th the weight of copper of that required by single phase system.

5. Miscellaneous Advantages : A three phase system has the following miscellaneous advantages over 1-phase system :

(i) Polyphase system can set up rotating magnetic field in stationary winding.

(ii) Parallel operation of 3-phase generators is simpler than that of single phase generators.

6.3. GENERATION OF 3-PHASE E.M.Fs.

In the three phase system, there are three equal voltages of the same frequency but displaced from one another by 120° electrical. These voltages are produced by a three phase generator. A 3-phase generator has three identical windings or phases displaced 120° electrical apart. When these windings are rotated* in a magnetic field, e.m.f. is induced in each winding or phase. These e.m.fs are of same magnitude and frequency but displaced from one another by 120° electrical.

Consider three identical coils, a_1a_2, b_1b_2 and c_1c_2 mounted on the same axis but displaced from each other by 120° electrical. Let these three coils be rotated in anticlockwise direction in a bipolar magnetic field with an angular velocity ω radians/sec as shown in Fig. 6.3 *(a)*. Here a_1, b_1 and c_1 are the start terminals and a_2, b_2 and c_2 are the finish terminals of the three coils. Note corresponding terminals a_1, b_1 and c_1 are 120° electrical apart. Likewise, terminals a_2, b_2 and c_2 are 120° electrical apart.

Magnitude and direction of e.m.f. induced in the various coils is as under :

(a) e.m.f. induced in coil a_1a_2 is zero and is increasing in the positive direction. This is indicated by $e_{a_1a_2}$ wave in Fig. 6.3 *(b)*.

(b) The coil b_1b_2 is 120° electrically behind coil a_1a_2. The e.m.f. induced in this coil is negative and is approaching maximum negative value. This is indicated by $e_{b_1b_2}$ wave in Fig. 6.3 *(b)*.

(c) The coil c_1c_2 is 240° electrically behind a_1a_2 or 120° electrically behind coil b_1b_2. The e.m.f. induced in this coil is positive and is decreasing. This is indicated by $e_{c_1c_2}$ wave in Fig. 6.3 *(b)*.

---

* In practice, the phase windings arre usually placed on the *stator* and field poles are carried by the *rotor*. This arrangement is preferred due to technical reasons.

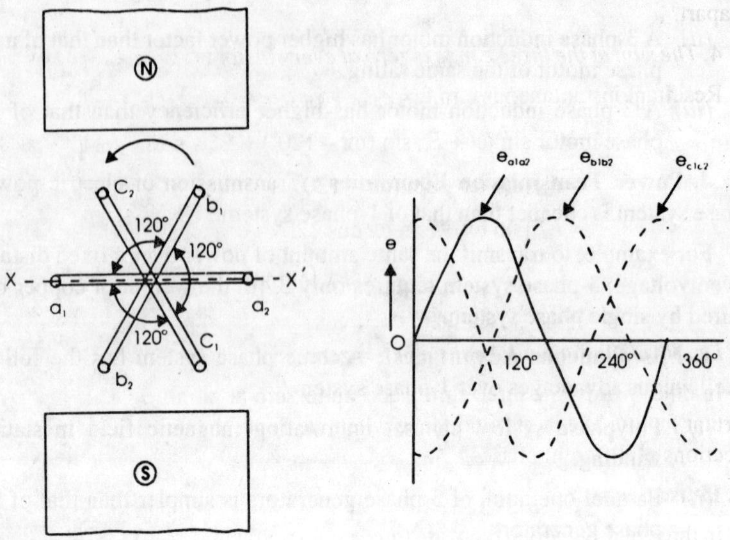

(a) (b)

Fig. 6.3.

It follows, therefore, that e.m.fs induced in the three coils are of the same magnitude and frequency but displaced 120° electrical from one another.

Wave Diagram : The wave diagram of the three voltages is shown in Fig. 6.3 *(b)*. It is clear that voltage $e_{a_1 a_2}$ in coil a_1a_2 is leading the voltage $c_{b_1 b_2}$ in coil b_1b_2 by 120° and that of $e_{c_1 c_2}$ in coil $c_1 c_2$ by 240°.

Vector Diagram : The three phase voltages have been shown vectorially* in Fig. 6.3 *(c)*. It is clear that $e_{a_1 a_2}$ is leading $e_{b_1 b_2}$ by 120° and that $e_{c_1 c_2}$ by 240°.

Fig. 6.3. (c)

Equations : The equations of three voltages are :

$$e_{a_1 a_2} = E_m \sin \omega t$$

$$e_{b_1 b_2} = E_m \sin (\omega t - 120^\circ)$$

$$e_{c_1 c_2} = E_m \sin (\omega t - 240^\circ)$$

Conclusion : The following important conclusions are drawn from the above discussion :

1. There ar' ·'..ee phases or windings displaced 120° electrical apart.

2. The three phases produce three e.m.fs. which are successively 120° out of phase.

* R.M.S. values are shown.

3. The three e.m.fs. are represented by vectors $E_{a_1 a_2}$, $E_{b_1 b_2}$ and $E_{c_1 c_2}$ which are 120° apart.

4. *The sum of the three e.m.fs is zero at every instant.* This is proved as under:

Resultant instantaneous e.m.f. = $e_{a_1 a_2} + e_{b_1 b_2} + e_{c_1 c_2}$

$$= E_m \sin \omega t + E_m \sin (\omega t - 120^\circ) + E_m \sin (\omega t - 240^\circ)$$

$$= E_m [\sin \omega t + \sin (\omega t - 120^\circ) + \sin (\omega t - 240^\circ)]$$

$$= E_m [\sin \omega t + (\sin \omega t \cos 120^\circ - \cos \omega t \sin 120^\circ) + (\sin \omega t \cos 240^\circ - \cos \omega t \sin 240^\circ)]$$

$$= 0$$

i.e., Resultant instantaneous e.m.f. = zero

In other words, the three e.m.fs add up to zero at all times. This is the most important feature and is of importance when we come to consider the methods of connections of different phases.

6.3.1. Phase Sequence

In three phase system, there are three equal voltages displaced 120° electrical from one another. It is very important to know the order in which the voltages in the phases reach their maximum positive values. This is called *phase sequence.*

*Hence the order in which the voltages in the phases reach their maximum positive values is called the phase sequence.**

For example, in Fig. 6.3 *(a)*, the three coils a_1a_2, b_1b_2 and c_1c_2 are rotating in anticlockwise direction in the magnetic field. The coil a_1a_2 is 120° electrical *ahead* of coil b_1b_2 and 240° electrical *ahead* of coil c_1c_2. Therefore, e.m.f. in coil a_1a_2 leads the e.m.f. in coil b_1b_2 by 120° and that in coil c_1c_2 by 240°. It is clear from Fig. 6.3 *(b)*, that $e_{a_1 a_2}$ attains maximum positive value first, then $e_{b_1 b_2}$ and then $e_{c_1 c_2}$. In other words, the order in which the e.m.fs in the three phases a_1a_2, b_1b_2 and c_1c_2 attain their maximum positive values is *a, b, c*. Hence, phase sequence is *a, b, c*.

If the direction of rotation of coils is reversed (*i.e.*, clockwise), then the order in which the three phases attain their maximum positive values would be *a, c, b*. Hence, the phase sequence is now *a, c, b*.

Applications : The knowledge of phase sequence is essential in the following important applications :

1. The direction of rotation of 3-ϕ induction motor depends upon phase sequence of 3-ϕ supply.

2. The parallel operation of 3-ϕ alternators and transformers is only possible if phase sequence is known.

---

* It may be order in which the rotating voltage vector pass a given point.

6.3.2. Naming The Phases

The three phases may be numbered (1, 2, 3) or lettered (*a, b, c*) or given colours (RYB) for identification. It is customary to give the natural colours red (R), yellow (Y) and blue (B) to the three phases and in that case, the phase sequence is RYB.

It is worthwhile to mention here that there are only two possible phase sequences, namely RYB and RBY. By convention, sequence RYB is taken as positive and RBY as negative.

Note : Throughout, the phase sequence taken is RYB unless stated otherwise.

6.3.3. Three Phase Four Wire System

It is usually employed for Alternating Current distribution. The three phases of the A.C. generator or transformer are connected in star and three line wires are taken from the terminals 1, 2 and 3. In addition a fourth wire or neutral wire is taken from the star point. This system enables lighting loads to be supplied at phase voltage E_p, *i.e.*, they are connected between any line conductor and neutral and at the same time it allows motors to be supplied at line voltage ($\sqrt{3}\ E_p$). Motor performance is improved at higher voltages. Three phase motors constitute a balanced load but lighting loads across the phases may not be balanced. In that case the line conductors carry unequal currents and the out of balance current is carried by the neutral. The above system is very extensively used in practice. The *line* voltage being 400 volts and the phase voltage 230 volts.

In specifying the current and voltage of a 3 phase system, line quantities and total 3 phase power is used unless stated otherwise.

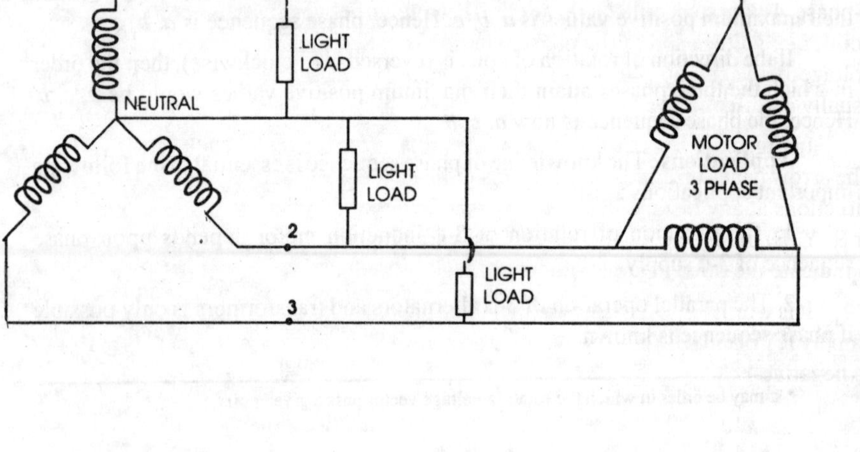

6.3.4. Inter-connection of Three Phases

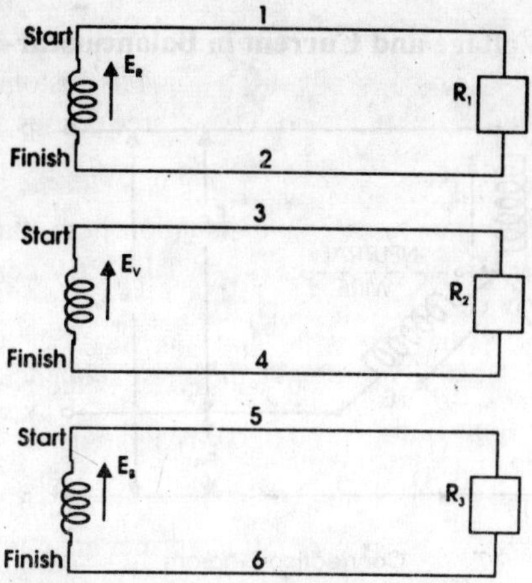

Fig. 6.5.

In the three phase generator, there are three windings or phases. Each phase has two terminals (*start* and *finish*). If a separate load is connected across each phase winding as shown in Fig. 6.5, then each phase supplies an independent load through a pair of leads. Clearly, in such a system (called *non- inter-linked 3-phase system*), we would need a total of six wires to connect the generator and the load. This will make the whole system complicated as well as expensive.

In order to reduce the number of line conductors, the three phase windings are suitably inter-connected. The following are the two universally adopted methods of inter-connecting the three phases :

1. Star or wye (Y) connection
2. Mesh or delta (Δ) connection

6.3.5. Star or WYE (Y) Connected System

This system is obtained by joining together similar ends, either the start or finish, the other ends are joined to the line wires, as shown in Fig. 6.6 *(a)*. The common point N, at which similar (start or finish) ends are connected, is called the *neutral or star point.* Ordinarily only three wires are carried to the external circuit giving 3-phase 3-wire star connected system but sometimes a fourth wire, known as *neutral wire* is carried to the neutral point of the external load circuit giving 3-phase, 4-wire star connected system. The voltage between any line and the neutral point, *i.e.*, voltage across the phase winding is called the *phase voltage*; while the voltage between any two outers is called the *line voltage*. The neutral point is usually earth-connected.

In Fig. 6.6 *(a)* positive directions of e.m.fs. are taken from star point outwards. The arrow heads on e.m.fs and currents indicate the positive direction and not their directions at any instant. The three phases be either numbered as 1, 2, 3 or *a, b, c* or R, Y, B. In Fig. 6.6 *(a)* the three-phases are numbered as usually done; R, Y and B indicate the three natural colours red, yellow and blue respectively.

It is worthwhile to state here that there are only two possible phase sequences, namely RYB and RBY. By convention sequence RYB is taken as positive and RBY as negative.

Voltage and Current in Balance Star-connection

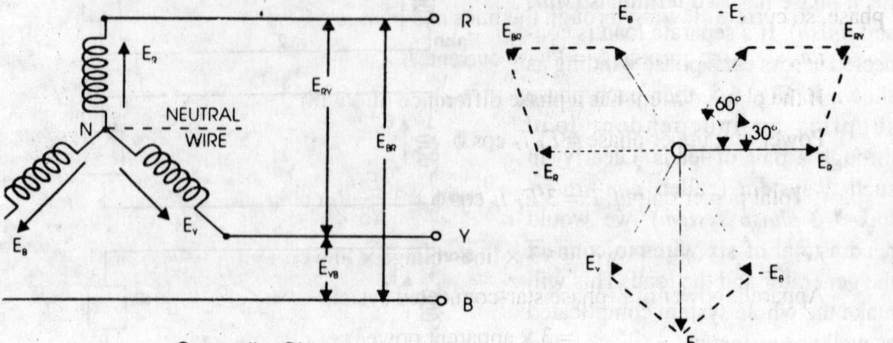

Connection Diagram

Phasor Diagram of Line and Phase Voltages

(a) *(b)*

Fig. 6.6.

In Fig. 6.6 *(b)* the e.m.fs induced in 3-phases are shown by phasor. In star-connection, there are two windings between each pair of outers and due to joining of similar ends together the e.m.fs induced in them are in opposition.

Hence the potential difference between the two outers known as *line voltage* is the phasor difference of phase e.m.fs of the two phases concerned. For example, the potential difference between outers R and Y or line voltage E_{RY} is the difference of phase e.m.fs E_R and E_Y or phasor sum of phase e.m.fs E_R and $(-E_Y)$.

i.e., $E_{RY} = E_R - E_Y$ (phasor difference)

or $E_{RY} = E_R + (-E_Y)$ (phasor sum)

Since phase angle between phasors E_R and $(-E_Y)$ is 60°.

∴ From phasor diagram shown in Fig. 1.6 *(b)*

$$E_{RY} = \sqrt{E_R^2 + E_Y^2 + 2\,E_R\,E_Y \cos 60^\circ}$$

Let $E_R = E_Y = E_B = E_P$ (phase voltage)(Assume balanced system)

Then line voltage $E_{RY} = \sqrt{E_P^2 + E_P^2 + 2\,E_P\,E_P \times (1/2)} = \sqrt{3}\,E_P$

Similarly potential difference between outers Y and B

or line voltage, $E_{YB} = E_Y - E_B = \sqrt{3}\,E_P$ in magnitude

and potential difference between outers B and R

or line voltage, $E_{BR} = E_B - E_R = \sqrt{3}\,E_P$ in magnitude

In balanced star (Y) system E_{RY}, E_{YB} and E_{BR} are equal in magnitude and are known as *line voltages*. Let each of them be equal to E_L.

$\therefore$ $E_L = \sqrt{3}\, E_P$ (in balanced star- connected system)

Since in star connected system each line conductor is connected to separate phase, so current flowing through the lines and phases are same.

i.e., Line current, I_L = Phase current, I_P

If the phase current has a phase difference of ϕ with the phase voltage

Power output per phase = $E_P\, I_P \cos\phi$

$$\text{Total power output, } P = 3\, E_P\, I_P \cos\phi = 3\,\frac{E_L}{\sqrt{3}}\, I_L \cos\phi = \sqrt{3}\, E_L\, I_L \cos\phi$$

i.e., Power = $\sqrt{3}$ × line voltage × line current × power factor

Apparent power of 3-phase star-connected system

$$= 3 \times \text{apparent power per phase}$$

$$= 3\, E_P\, I_P = 3\,\frac{E_L}{\sqrt{3}} \times I_L = \sqrt{3}\, E_L\, I_L$$

It should be noted that in balanced star-connected system

(i) Line voltages are 120° apart.

(ii) Line voltages are 30° ahead of the respective phase voltages.

(iii) Line voltages are $\sqrt{3}$ times of phase voltages.

(iv) Line currents are equal to phase currents.

(v) The angle between line currents and the corresponding line voltages is (30° ϕ); +ve for lagging currents and –ve for leading currents.

(vi) True power output = $\sqrt{3}\, E_L\, I_L \cos\phi$, where ϕ is the angle between the respective phase current and phase voltage (not between line current and line voltage).

(vii) Apparent power = $\sqrt{3}\, E_L\, I_L$

(viii) In balanced system, the potential of neutral or star point is zero because potential at neutral or star point,

$$E_{NR} + E_{NY} + E_{NB} = 0$$

Example 6.1. A star-connected 3-phase load has a resistance of 6 Ω and an inductive reactance of 8 Ω in each branch. A line to line voltage of 220 V is imprssed through a 3- ϕ auto-transformer. Find the voltage across each branch, line voltages and line currents and total active power.

Solution : Line voltage, V_L = Line to line voltage = 220 V **Ans.**

$$\text{Phase voltage, } V_P = \frac{V_L}{\sqrt{3}} = \frac{220}{\sqrt{3}} = 127 \text{ V } \textbf{Ans.}$$

$$\text{Impedance per phase, } Z_P = \sqrt{R_P^2 + X_P^2} = \sqrt{6^2 + 8^2} = 10\ \Omega$$

$$\text{Phase currents, } I_P = \frac{V_P}{Z_P} = \frac{127}{10} = 12.7 \text{ A}$$

$$\text{Total active power, } P = 3 \times \text{power of each phase}$$

$$= 3\, I_P^2\, R_P = 3 \times (12.7)^2 \times 6 = 2{,}904 \text{ watts } \textbf{Ans.}$$

Total active power can also be determined from the formula

$$P = \sqrt{3}\, V_L\, I_L \cos\phi$$

where $\cos\phi = \frac{R}{Z} = \frac{6}{10} = 0.6$

$$\text{Line current, } I_L = I_P = 12.7 \text{ A}$$

and $P = \sqrt{3}\, V_L\, I_L \cos\phi$

$$= \sqrt{3} \times 220 \times 12.7 \times 0.6$$

$$= 2{,}904 \text{ watts, as before } \textbf{Ans.}$$

Example 6.2. The phase voltage of 3 phase 60 MVA star connected alternator is 6000 volts. Calculate the line voltage and line current at full load delivered by the alternator.

Solution : Phase voltage, $E_P = 6000$ volts

$$\text{Line voltage} = \sqrt{3} \text{ phase voltage}$$

$$E_L = 1.732 \times 6000 = 10.392 \text{ KV}$$

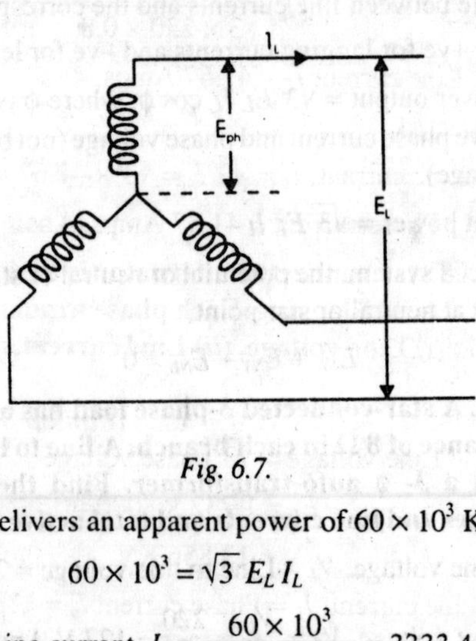

Fig. 6.7.

The alternator delivers an apparent power of 60×10^3 KVA at full load.

So $\quad 60 \times 10^3 = \sqrt{3}\, E_L\, I_L$

or $\quad \text{Line current, } I_L = \frac{60 \times 10^3}{\sqrt{3} \times 10.392} = 3333.4$ amperes. **Ans.**

Example 6.3. A three phase Induction motor having an output of 40 H.P. takes a line current of 60 amp. at a power factor of 0.8. Line voltage is 400 volts. Calculate the efficiency of the motor.

Solution : Output of motor = 40 × 735.5 watts = 29420 watts

Input of motor = $\sqrt{3}\, E_L I_L \cos\phi$ watts

$= 1.732 \times 400 \times 60 \times 0.8 = 33254$ watts

Efficiency of motor = Output/Input

$$= \frac{29420}{33254} = 0.8847 = 88.47\% \text{ Ans.}$$

Example 6.4. A three phase system supplies 23 kw at a power factor of 0.8 lagging. The phase voltage of the system is 230 volts. Calculate the line current and the phase current when the load is *(a)* Star connected, *(b)* Delta connected.

Solution : *(a)* Star connected

Power = $3\, E_p I_p \cos\phi$

Power, P = 23 kw = 23000 watts

Phase voltage, E_p = 230 volts

Power factor $\cos\phi = 0.8$

So $23000 = 3 \,.\, 230 \,.\, I_p \,.\, 0.8$

or $$I_p = \frac{23000}{3 \times 230 \times 0.8} = 41.67 \text{ Amps.}$$

Line current I_L = 41.67 Amps.

(b) Delta connected

Line current, $I_L = \sqrt{3}\, I_p = \sqrt{3} \times 41.67 = 72.167$ Amps.

Phase current = 41.67 Amps. **Ans.**

Example 6.5. A star connected alternator is designed to operate at full load with a power factor 0.8 lagging. Each phase winding is rated at 8000 volts and 418 amp. Find : *(i)* Line voltage, *(ii)* Line current, *(iii)* Full load power in kw.

[Punjab State Board Nov., 1975]

Solution : Line voltage = $\sqrt{3}$ phase voltage

(i) $E_L = 1.732 \times 8000 = 13856$

= 13.856 KV

(ii) Line current, I_L = Phase current I_p = 418 Amps.

(iii) Full load power = $\sqrt{3} \,.\, E_L I_L \cos\phi$

$= \sqrt{3} \times 13.856 \times 418 \times 0.8$ watts = 8025.6 kw **Ans.**

Example 6.6. A balanced three phase load consists of three resistances of 5 ohms each. Determine the total power when the resistance are

(*i*) Star connected, (*ii*) Delta connected.

The supply voltage is 400 volts 50 Hz.

[*Haryana State Board June, 1975*]

Solution : (*i*) Star connected

Line voltage, $E_L = 400$ volts

Phase voltage, $E_p = \frac{400}{\sqrt{3}} = 230$ volts

Phase current, $I_p = \frac{E_p}{R} = \frac{230}{5} = 46$ Amps.

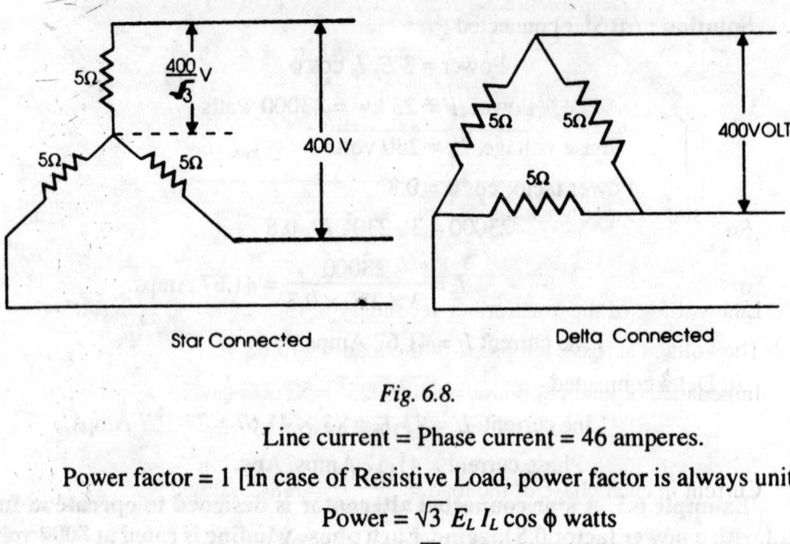

Fig. 6.8.

Line current = Phase current = 46 amperes.

Power factor = 1 [In case of Resistive Load, power factor is always unity]

$$\text{Power} = \sqrt{3}\ E_L\ I_L \cos\phi \text{ watts}$$

$$= \sqrt{3} \times 400 \times 46 \times 1 = 31.869 \text{ kw}$$

(*ii*) Delta Connected

Phase current, $I_p = \frac{400}{5} = 80$ Amps.

Line current, $I_L = \sqrt{3} \times 80 = 138.56$ Amps.

$$\text{Power} = \sqrt{3}\ E_L\ I_L \cos\phi$$

$$= \sqrt{3} \times 400 \times 138.56 \times 1 = 96 \text{ kw Ans.}$$

Example 6.7. A three phase delta connected load, each phase of which has an inductive reactance of 40 ohm and a resistance of 25 ohm is fed from

the secondary of a three phase-star connected transformer which has a phase voltage of 240 volts. Draw the circuit diagram of the system and calculate

(a) The current in each phase of the load.

(b) The voltage across each phase of the load.

(c) The current in the transformer secondary winding.

(d) The total power taken from the supply and its power factor.

[Haryana State Board May, 1975]

Solution :

Fig. 6.9 shows the circuit diagram of the system.

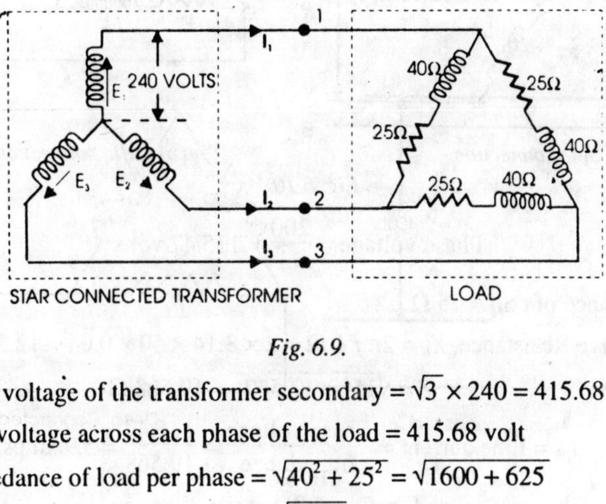

Fig. 6.9.

Line voltage of the transformer secondary = $\sqrt{3} \times 240 = 415.68$ volt

The voltage across each phase of the load = 415.68 volt

Impedance of load per phase = $\sqrt{40^2 + 25^2} = \sqrt{1600 + 625}$

$= \sqrt{2225} = 47.169\ \Omega$

Current in each phase of the load $= \dfrac{415.68}{47.169} = 8.812$ Amps.

Line current $I_1 = 8.8 \times \sqrt{3} = 15.26$ Amps.

Current in the secondary winding of transformer = I_1 = 15.26 Amps.

Line power factor $= \dfrac{R}{Z} = \dfrac{25}{47.2} = 0.53$ lagging

Power supplied to load = $\sqrt{3}\ E_L\ I_L \cos \phi$

$= 1.73 \times 415.68 \times 15.2 \times 0.53 = 5800$ watts **Ans.**

Example 6.8. Three similar coils each having a resistance of 15 Ω and an inductance of 0.04 henery are connected in star to a 3-phase 50 Hz supply, 200 volts between lines. Calculate the line current. If they are now connected in

delta, calculate the phase current, line current and the total power absorbed in each case.

Solution : *(a)* Star Connected

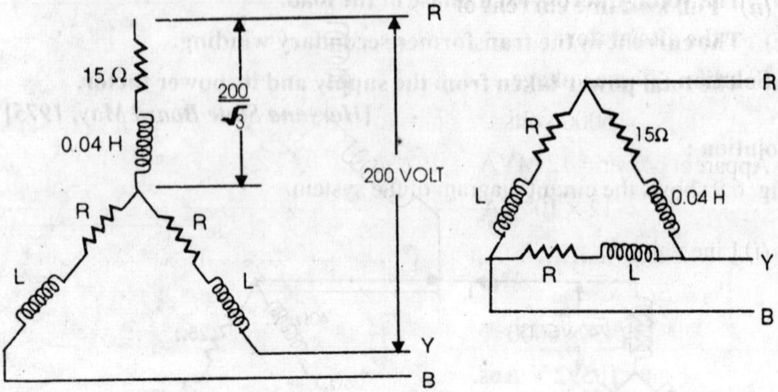

(a) Star connection *(b) Delta connection*

Fig. 6.10.

$$\text{Phase voltage} = \frac{200}{\sqrt{3}} = 115.47 \text{ volts}$$

Resistance of coil = 15 Ω

Inductive Resistance, $X_L = 2\pi f L \,\Omega = 2 \times 3.14 \times 50 \times 0.04 = 12.567\ \Omega$

$$Z = \sqrt{15^2 + 12.567^2} = 19.568\ \Omega$$

$$I_{ph} = \text{Line current} = \frac{\text{Phase voltage}}{\text{Impedance}} = \frac{115.47}{19.568} = 5.9\text{Amps.}$$

$$I_L = I_{ph} = \mathbf{5.9\ A\ Ans.}$$

$$\text{Power factor, } \cos\phi = \frac{R}{Z} = \frac{15}{19.568} = 0.7665 \text{ lagging}$$

$$\text{Power} = \sqrt{3}\, V_L I_L \cos\phi = \sqrt{3} \times 200 \times 5.9 \times 0.767$$

$$= 1567.6 \text{ watts} = 1.567 \text{ kw } \textbf{Ans.}$$

***(b)* Delta Connected**

$$\text{Phase voltage, } V_{ph} = V_L = 200 \text{ V}$$

$$\text{Phase Current, } I_{ph} = \frac{V_{ph}}{Z} = \frac{200}{19.58} = 10.215 \textbf{ A Ans.}$$

$$\text{Line Current, } I_L = \sqrt{3}\, I_{ph} = \sqrt{3} \times 10.215 = 17.69 \textbf{ A Ans.}$$

$$\text{Power} = \sqrt{3}\, V_L I_L \sin\phi$$

$$= \sqrt{3} \times 200 \times 17.69 \times 0.7665$$

$$= 4697.1 \text{ watts or } 4.697 \text{ kw } \textbf{Ans.}$$

Example 6.9. The phase voltage of a 3-phase 12 MVA star connected alternator is 6000 volts. Calculate ;

***(i)* Line voltage**

***(ii)* Full load line current of the alternator.**

Solution : Phase voltage,

$$E_{ph} = 6000 \text{ volts}$$

Apparent power = 12 MVA

$$= 12 \times 10^6 \text{ VA}$$

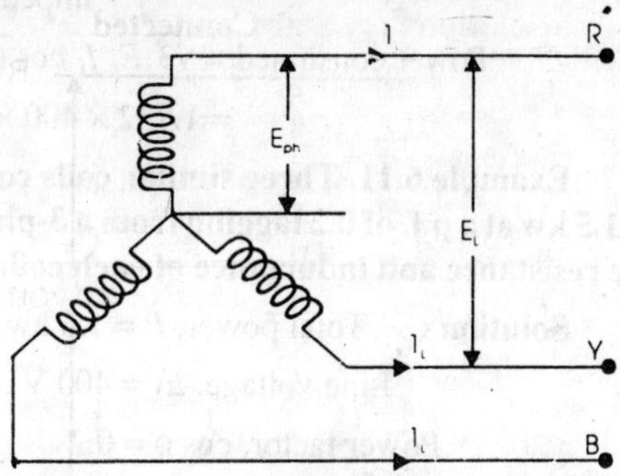

Fig. 6.11.

(i) Line voltage,

$$E_L = \sqrt{3}\, E_{ph}$$

$$= \sqrt{3} \times 6000$$

$$= \mathbf{10392\ V\ Ans.}$$

(ii) Apparent power

$$= \sqrt{3}\, E_L I_L$$

or $$12 \times 10^6 = \sqrt{3} \times 10392 \times I_L$$

$\therefore$ $$I_L = \frac{12 \times 10^6}{\sqrt{3} \times 10392} = \mathbf{666.67\ A\ Ans.}$$

Example 6.10. Three equal impedances are star connected to a 3-phase, 400 V, 50 Hz supply. If the inductive reactance and resistance of each branch are 8 Ω and 6 Ω respectively, find :

***(i)* Line current**

***(ii)* Power consumed.**

Solution : Line voltage,

$$E_L = 400 \text{ V}$$

Phase voltage,

$$E_{ph} = \frac{E_L}{\sqrt{3}} = \frac{400}{\sqrt{3}} = 231 \text{ V}$$

(i) Impedance of each phase,

$$Z_{ph} = \sqrt{R^2 + X_L^2} = \sqrt{8^2 + 6^2}$$

$$= 10\ \Omega$$

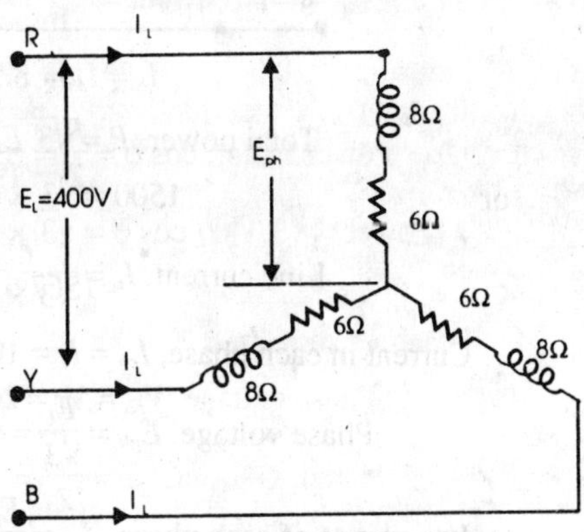

Fig. 6.12.

$\therefore$ Phase current,

$$I_{ph} = \frac{E_{ph}}{Z_{ph}} = \frac{231}{10} = 23.1 \text{ A}$$

$\therefore$ Line current, $I_L = I_{ph} = $ **23.1 A Ans.**

(ii) Power factor, $\cos\phi = \dfrac{\text{resistance per phase}}{\text{Impedance per phase}} = \dfrac{6}{10} = 0.6$

$\therefore$ Power consumed $= \sqrt{3}\, E_L I_L \cos\phi$

$= 1.732 \times 400 \times 23.1 \times 0.6 = $ **9600 watts. Ans.**

Example 6.11. Three similar coils connected in star, take a total power of 1.5 kw at a p.f. of 0.2 lagging from a 3-phase, 400 V, 50 Hz supply. Calculate the resistance and inductance of each coil.

Solution : Total power, $P = 1.5$ kw $= 1500$ watts

Line voltage, $E_L = 400$ V

Power factor, $\cos\phi = 0.2$

Supply frequency, $f = 50$ Hz

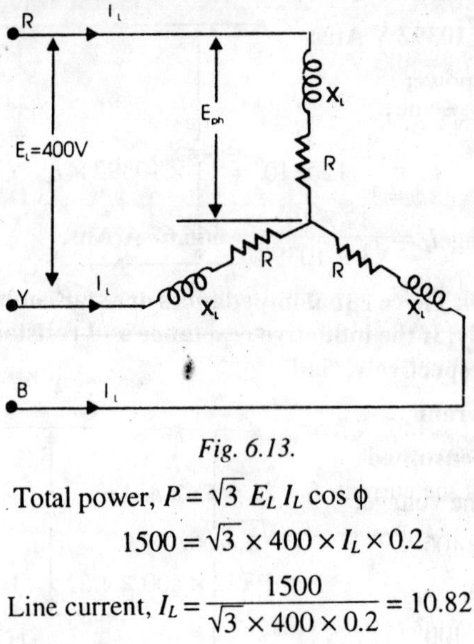

Fig. 6.13.

Total power, $P = \sqrt{3}\, E_L I_L \cos\phi$

or $1500 = \sqrt{3} \times 400 \times I_L \times 0.2$

$\therefore$ Line current, $I_L = \dfrac{1500}{\sqrt{3} \times 400 \times 0.2} = 10.82$ A

$\therefore$ Current in each phase, $I_{ph} = I_L = 10.82$ A

Phase voltage, $E_{ph} = \dfrac{E_L}{\sqrt{3}} = \dfrac{400}{\sqrt{3}} = 231$ V

$\therefore$ Impedance of each phase, $Z_{ph} = \dfrac{E_{ph}}{I_{ph}} = \dfrac{231}{10.82} = 21.33\ \Omega$

Power factor, $\cos\phi = \dfrac{\text{resistance per phase}}{Z_{ph}}$

or $$0.2 = \frac{\text{resistance per phase}}{21.33}$$

$\therefore$ Resistance per phase, $R_{ph} = 21.33 \times 0.2 = \mathbf{4.27\ \Omega}$ **Ans.**

$\therefore$ Inductive reactance per phase, $X_L = \sqrt{Z_{ph}^2 - R_{ph}^2}$

$$= \sqrt{(21.33)^2 - (4.27)^2} = 20.9\ \Omega$$

$\therefore$ Inductance per phase, $L = \frac{X_L}{2\,\pi f} = \frac{20.9}{2\pi \times 50} = \mathbf{0.665\ H}$ **Ans.**

Example 6.12. Three 50 ohm resistors are connected in star across 400 V, 3-phase supply. Find :

***(i)* Phase current, line current and power taken from the mains.**

***(ii)* What would be the above values if one of the resistors were disconnected ?**

Solution :

Line voltage, $E_L = 400$ V

Each phase resistance,

$$R_{ph} = 50\ \Omega$$

Power factor, $\cos \phi = 1$

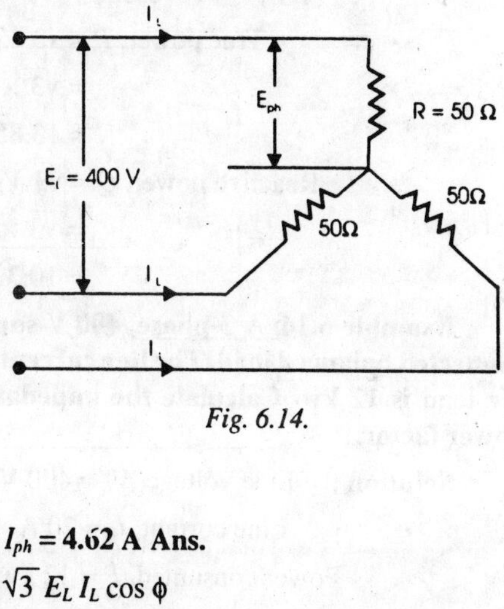

Fig. 6.14.

(i) Phase voltage,

$$E_{ph} = \frac{E_L}{\sqrt{3}} = \frac{400}{\sqrt{3}} = 231\ \text{V}$$

$\therefore$ Phase current,

$$I_{ph} = \frac{E_{ph}}{R_{ph}} = \frac{231}{50} = \mathbf{4.62\ A}\ \textbf{Ans.}$$

Line current, $I_L = I_{ph} = \mathbf{4.62\ A}$ **Ans.**

Power, $P = \sqrt{3}\ E_L\, I_L \cos \phi$

$$= 1.732 \times 400 \times 4.62 \times 1 = \mathbf{3200\ watts}\ \textbf{Ans.}$$

(ii) When one of the resistors disconnected : Refer to Fig. 6.15, where one of the resistors is disconnected. The remaining two resistors behave as if they were connected in series across line voltage.

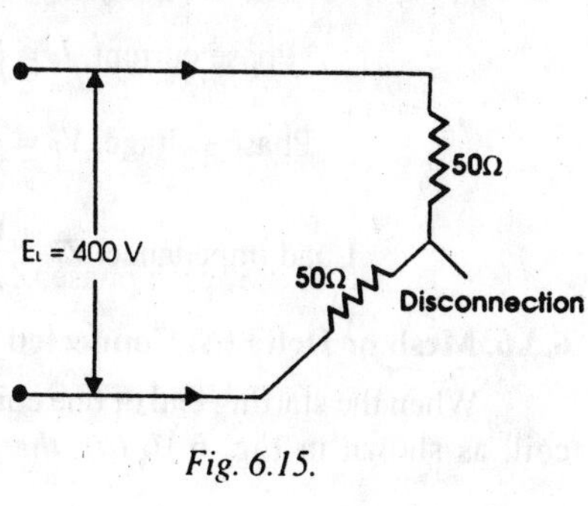

Fig. 6.15.

$$\therefore I_{ph} = I_L = \frac{400}{50 + 50}$$

$$= \frac{400}{100} = \mathbf{4\ A}\ \textbf{Ans.}$$

$$\text{Power} = E_L I_L \cos\phi$$
$$= 400 \times 4 \times 1 = 1600 \text{ watts. } \textbf{Ans.}$$

Example 6.13. A balanced 3-phase star-connected load draws a current of 25 A at a lagging power factor of 0.8 when connected to a 3-phase 400 V (line-to-line) system. Calculate : true power, reactive power and voltage across each phase of load.

Solution : Line voltage, $V_L = 400$ V

$$\text{Voltage across each phase of load} = V_P = \frac{V_L}{\sqrt{3}} = \frac{400}{\sqrt{3}} = \textbf{231 V Ans.}$$

Line current, $I_L = 25$ A

Power factor, $\cos\phi = 0.8$ (lag)

$$\sin\phi = \sqrt{1-\cos^2\phi} = \sqrt{1-0.8^2} = 0.6$$

$$\text{True power, } P = \sqrt{3}\, V_L I_L \cos\phi$$
$$= \sqrt{3} \times 400 \times 25 \times 0.8$$
$$= 13{,}856 \text{ W or } \textbf{13.856 kw Ans.}$$

$$\text{Reactive power, } Q = \sqrt{3}\, V_L I_L \sin\phi$$
$$= \sqrt{3} \times 400 \times 25 \times 0.6$$
$$= 10{,}392 \text{ VAR or } \textbf{10.392 kVAR Ans.}$$

Example 6.14. A 3-phase, 400 V supply is connected to a 3-phase star-connected balanced load. The line current is 20 A and the power consumed by the load is 12 kw. Calculate the impedance of the load, phase current and power factor.

Solution : Line voltage, $V_L = 400$ V

Line current, $I_L = 20$ A

Power consumed, $P = 12$ kw $= 12{,}000$ W

$$\text{Power factor, } \cos\phi = \frac{P}{\sqrt{3}\, V_L I_L} = \frac{12{,}000}{\sqrt{3} \times 400 \times 20} = \textbf{0.866 Ans.}$$

Phase current, $I_P = I_L =$ **20 A Ans.**

$$\text{Phase voltage, } V_P = \frac{V_L}{\sqrt{3}} = \frac{400}{\sqrt{3}} = 231 \text{ V}$$

$$\text{Load impedance, } Z_P = \frac{V_P}{I_P} = \frac{231}{20} = \textbf{11.55 A Ans.}$$

6.3.6. Mesh or Delta (Δ) Connected System

When the starting end of one coil is connected to the finishing end of another coil, as shown in Fig. 6.16 *(a)*, *the delta or mesh connection* is obtained. The direction of e.m.fs in the coils have been taken as positive from start end to finish end in Fig. 6.16 *(a)*.

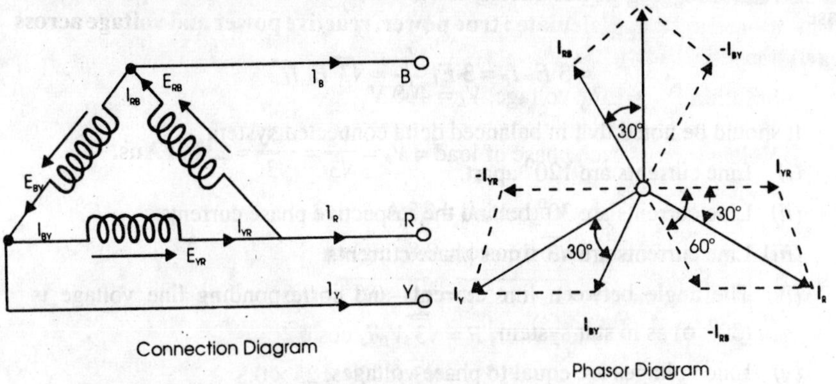

Fig. 6.16.

Voltage and Current in Balanced Δ Connection

From Fig. 6.16 *(b)* it is obvious that line current is phasor difference of phase currents of two phases concerned. For example the line current in red outer I_R will be equal to the phasor difference of phase currents I_{YR} and I_{RB}. The current phasors are shown in Fig. 6.16 *(b)*.

From Figs. 6.16 *(a)* and 6.16 *(b)*, we have

Line current, $I_R = I_{YR} - I_{RB} = I_{YR} + (-I_{RB})$

Since phase angle between phase current phasors I_{YR} and $-I_{RB}$ is 60°.

$$\therefore \quad I_R = \sqrt{I_{YR}^2 + I_{RB}^2 + 2\, I_{YR}\, I_{RB} \cos 60°}$$

Assuming balanced load

The phase current in each winding is equal and let each be equal to I_P.

$$\text{Line current } I_R = \sqrt{I_P^2 + I_P^2 + 2\, I_P\, I_P \times \frac{1}{2}} = \sqrt{3}\, I_P$$

Similarly line current, $I_Y = I_{BY} - I_{YR} = \sqrt{3}\, I_P$ in magnitude

and line current, $I_B = I_{RB} - I_{BY} = \sqrt{3}\, I_P$ in magnitude.

Since in delta connected system, only one phase is included between any pair of line outers, therefore, potential difference between the line outers, called the *line voltage*, is equal to phase voltage.

i.e., Line voltage, E_L = Phase voltage, E_P

Power output per phase = $E_P\, I_P \cos \phi$

where $\cos \phi$ is the power factor of the load.

$$\text{Total power output} = 3\, E_P\, I_P \cos\phi = 3\, E_L \frac{I_L}{\sqrt{3}} \cos\phi = \sqrt{3}\, E_L\, I_L \cos\phi$$

i.e., Total power = $\sqrt{3}$ × line voltage × line current × power factor

Apparent power of 3-phase delta-connected system = 3 × apparent power per phase

$$= 3\, E_P\, I_P = 3\, E_L \frac{I_L}{\sqrt{3}} = \sqrt{3}\, E_L\, I_L$$

It should be noted that in balanced delta connected system

(i) Line currents are 120° apart.

(ii) Line currents are 30° behind the respective phase currents.

(iii) Line currents are $\sqrt{3}$ times phase currents.

(iv) The angle between line currents and corresponding line voltage is (30° ϕ) as in star system.

(v) Line voltages are equal to phase voltages.

(vi) Power current = $\sqrt{3}\, E_L\, I_L \cos\phi$, where ϕ is the phase angle between respective phase current and phase voltage.

(vii) Apparent power = $\sqrt{3}\, E_L\, I_L$

(viii) In balanced system, the resultant e.m.f. in the closed circuit will be zero. *i.e.,* $E_{RY} + E_{YB} + E_{BR} = 0$

Hence there will be no circulating current in the mesh if no load is connected to the lines.

Example 6.15. A 3-phase motor operating on a 400 V balanced system develops 18.65 kw at an efficiency of 0.87 per unit and a power factor of 0.82. Calculate the line current and phase current if the windings are delta connected.

Solution : Motor output = 18.65 kw = 18,650 watts

Motor efficiency, $\eta = 0.87$

$$\text{Motor input, } P = \frac{\text{Motor output}}{\eta} = \frac{18{,}650}{0.87} = 21{,}437 \text{ watts}$$

Line voltage, $V_L = 400$ V

Power factor, $\cos\phi = 0.82$

So, $$\text{line current, } I_L = \frac{P}{\sqrt{3}\, V \cos\phi} = \frac{21437}{\sqrt{3} \times 400 \times 0.82} = 37.73 \text{ A}$$

$$\text{Phase current, } I_P = \frac{I_L}{\sqrt{3}} = \frac{37.73}{\sqrt{3}} = \mathbf{21.8\ A\ Ans.}$$

Example 6.16. A delta-connected balanced 3-phase load is supplied from a 3-phase, 400 V supply. The line current is 20 A and the power taken by the load is 10 kw. Find *(i)* impedance in each branch, *(ii)* the line current, power factor and power consumed if the same load is connected in star.

Solution : *(i)* When load is delta-connected

Phase voltage, V_P = Line voltage, $V_L = 400$ V

Phase current, $I_P = \dfrac{I_L}{\sqrt{3}} = \dfrac{20}{\sqrt{3}} = 11.55$ A

Phase impedance, $Z_P = \dfrac{V_P}{I_P} = \dfrac{400}{11.55} = 34.64\ \Omega$ **Ans.**

Power factor, $\cos\phi = \dfrac{\text{Power taken by the load}}{\sqrt{3}\ V_L I_L}$

$= \dfrac{10{,}000}{\sqrt{3} \times 400 \times 20} = 0.7217$

***(ii)* When load is star-connected**

Phase voltage, $V_P = \dfrac{V_L}{\sqrt{3}} = \dfrac{400}{\sqrt{3}} = 231$ V

Phase current, $I_P = \dfrac{V_P}{Z_P} = \dfrac{231}{34.64} = 6.667$ A

Line current, $I_L = I_P =$ **6.667 A Ans.**

Power factor, $\cos\phi =$ **0.7217, as above Ans.**

Power consumed $= \sqrt{3}\ V_L I_L = \sqrt{3} \times 400 \times 6.667 \times 0.7217$

$=$ **3,333 W Ans.**

Example 6.17. Three similar resistors connected in star draw a line current of 5 A from a 400 V, 3-phase mains. To what value should the line voltage be changed to obtain the same line current with the resistors connected in delta ?

Solution : When star-connected

Phase voltage, $V_P = \dfrac{V_L}{\sqrt{3}} = \dfrac{400}{\sqrt{3}} = 231$ V

Phase current, $I_P = I_L = 5$ A

Phase resistors, $R_P = \dfrac{V_P}{I_P} = \dfrac{231}{5} = 46.2\ \Omega$

When delta-connected

Line current, $I_L = 5$ A, the same as in star connection

Phase current, $I_P = \frac{I_L}{\sqrt{3}} = \frac{5}{\sqrt{3}} = 2.887$ A

Phase voltage, $V_P = I_P R_P = 2.887 \times 46.2 = 133.33$ V

Line voltage, $V_L = V_P =$ **133.33 V, one-third of line voltage for star-connection. Ans.**

Example 6.18. A balanced 3-phase delta connected load consists of three resistance each of 4 ohms connected to a 400 V, 3-phase, 50 Hz supply. Find :

(i) Phase voltage, (ii) Phase current, (iii) Line current, (iv) Power consumed.

Solution : Line voltage, $E_L = 400$ V

(i) Phase voltage,

$$E_{ph} = E_L = \textbf{400 V Ans.}$$

(ii) Phase current,

$$I_{ph} = \frac{E_{ph}}{\text{phase resistance}}$$

$$= \frac{400}{4} = \textbf{100 A Ans.}$$

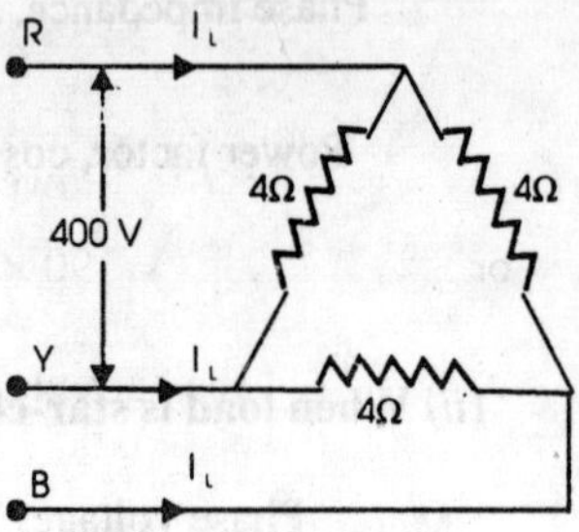

Fig. 6.17.

(iii) Line current,

$$I_L = \sqrt{3}\, I_{ph} = \sqrt{3} \times 100 = \textbf{173.2 A Ans.}$$

(iv) Power, $P = \sqrt{3}\, E_L I_L \cos\phi$

$$= 1.732 \times 400 \times 173.2 \times 1 = \textbf{120000 watts Ans.}$$

Example 6.19. The star connected secondary of a transformer supplies a delta connected motor taking a power of 90 kw at lagging power factor 0.9. If the voltage between lines is 600 V;

(i) Calculate the current in the transformer windings and in the motor windings.

(ii) Draw phasor diagram showing all voltages and currents in the transformer secondary and the motor.

Solution : Fig. 6.18 shows the circuit diagram. From the star connected secondary of the transformer, supply is given to a delta connected motor.

Line voltage, $E_L = 600$ V

Power, $P = 90$ kw $= 90 \times 10^3$ W

Power factor, $\cos\phi = 0.9$

(i) Delta connected motor

Phase voltage, $E_{ph_2} = E_L = 600$ V

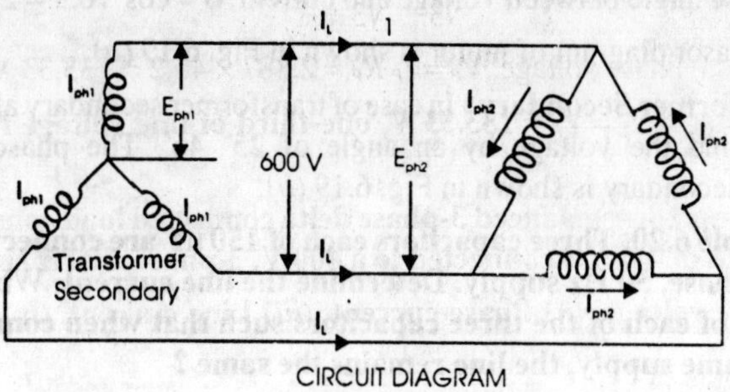

Fig. 6.18.

$$\text{Power, } P = \sqrt{3}\, E_L I_L \cos\phi$$

or $$90 \times 10^3 = 1.732 \times 600 \times I_L \times 0.9$$

$\therefore$ $$\text{Line current, } I_L = \frac{90 \times 10^3}{1.732 \times 600 \times 0.9} = \mathbf{96.2\ A\ Ans.}$$

$$\text{Motor phase current, } I_{ph_2} = \frac{I_L}{\sqrt{3}} = \frac{96.2}{1.732} = \mathbf{55.55\ A\ Ans.}$$

Transformer Secondary

$$\text{Phase current, } I_{ph_1} = I_L = \mathbf{96.2\ A\ Ans.}$$

$$\text{Phase voltage, } E_{ph_1} = \frac{E_L}{\sqrt{3}} = \frac{600}{1.732} = 346.4\ \text{V}$$

(ii) Motor

Power factor of motor, $\cos\phi = 0.9$

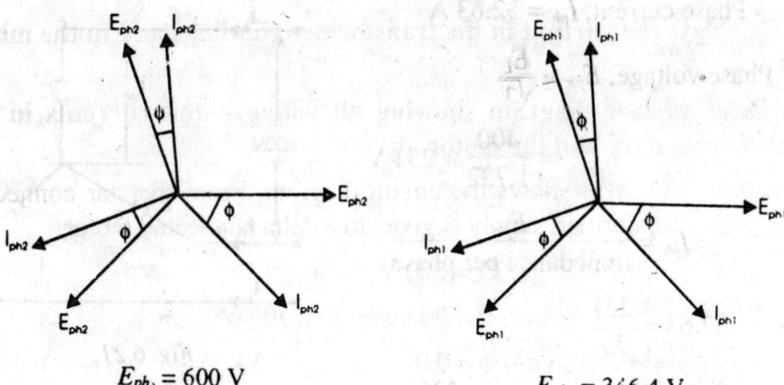

$E_{ph_2} = 600$ V
$I_{ph_2} = 55.55$ A
$\phi = 25^\circ\ 46'$

$E_{ph_1} = 346.4$ V
$I_{ph_1} = 96.2$ A
$\phi = 25^\circ\ 46'$

Fig. 6.19.

$\therefore$ Phase angle between voltage and current, $\phi = \cos^{-1} 0.9 = 25^\circ 46'$

The phasor diagram of motor is shown in Fig. 6.19 *(a)*.

Transformer Secondary : In case of transformer secondary also, the current will lag behind the voltage by an angle of 25° 46′. The phasor diagram of transformer secondary is shown in Fig. 6.19 *(b)*.

Example 6.20. Three capacitors each of 150 μF are connected in delta to a 400 V, 3-phase, 50 Hz supply. Determine the line current. What will be the capacitance of each of the three capacitors such that when connected in star across the same supply, the line remains the same ?

Solution : ***Delta connection***

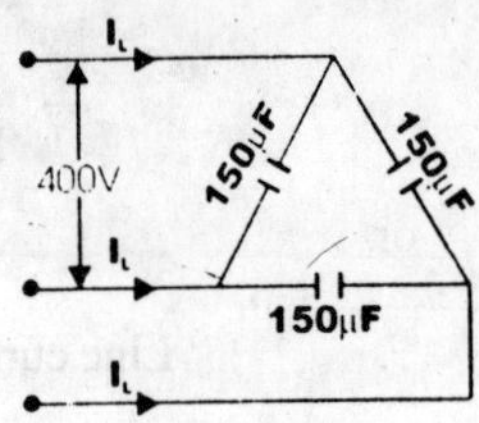

Fig. 6.20.

Line voltage, $E_L = 400$ V

$\therefore$ Phase voltage, $E_{ph} = E_L = 400$ V

Impedance of each phase,

$$Z_{ph} = \frac{1}{2\pi fC} = \frac{10^6}{2\pi \times 50 \times 150} = 21.23\ \Omega$$

$\therefore$ Phase current,

$$I_{ph} = \frac{E_{ph}}{Z_{ph}} = \frac{400}{21.23} = 18.84 \text{ A}$$

$\therefore$ Line current, $I_L = \sqrt{3}\, I_{ph} \equiv 1.732 \times 18.84 =$ **32.63 A Ans.**

Star connection

Let C farads be the required capacitance of each capacitor when connected in star.

Line current, $I_L = 32.63$ A (given)

$\therefore$ Phase current, $I_{ph} = 32.63$ A

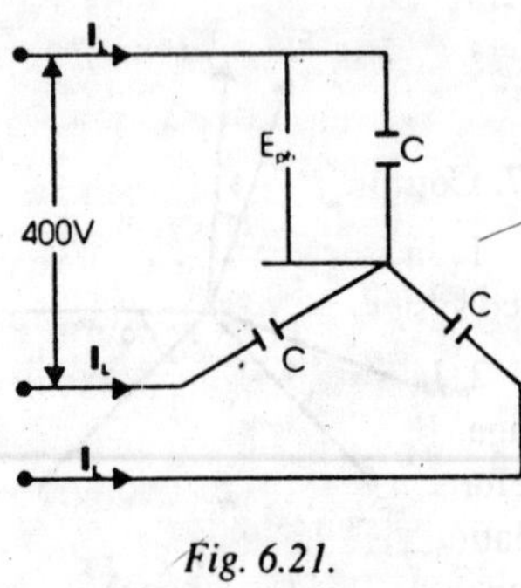

Fig. 6.21.

Phase voltage, $E_{ph} = \dfrac{E_L}{\sqrt{3}}$

$$= \frac{400}{1.732} = 231 \text{ V}$$

$$\therefore \qquad I_{ph} = \frac{E_{ph}}{\text{impedance per phase}}$$

or $$32.63 = \frac{231}{X_C}$$

$$\therefore \qquad X_C = \frac{231}{32.63}$$

or $$\frac{1}{2\pi fC} = \frac{231}{32.63}$$

$$\therefore \quad C = \frac{32.63}{2\pi \times 50 \times 231} = 450 \times 10^{-6}\ \text{F} = \mathbf{450\ \mu F\ Ans.}$$

Example 6.21. Three coils each of resistance 4 Ω and inductive reactance 3 Ω are connected in delta across 400 V, 50 Hz supply. Find the current in each coil, line current, active, reactive and apparent power.

Solution : Resistance in each phase, $R_P = 4\ \Omega$

Inductive reactance in each phase, $X_P = 3\ \Omega$

Impedance in each phase, $Z_P = \sqrt{R_P^2 + X_P^2} = \sqrt{4^2 + 3^2} = 5\ \Omega$

Line voltage, $V_L = 400$ V

Phase voltage, $V_P = V_L = 400$ V

Current in each coil $= I_P = \dfrac{V_P}{Z_P} = \dfrac{400}{5} = \mathbf{80\ A\ Ans.}$

Line current, $I_L = \sqrt{3}\ I_P = \sqrt{3} \times 80 = \mathbf{138.6\ A\ Ans.}$

Power factor, $\cos\phi = \dfrac{R_P}{Z_P} = \dfrac{4}{5} = 0.8$

$\sin\phi = \dfrac{X_P}{Z_P} = \dfrac{3}{5} = 0.6$

Active power, $P = \sqrt{3}\ V_L I_L \cos\phi = \sqrt{3} \times 400 \times 138.6 \times 0.8$

$= 76{,}800$ w or **76.8 kw Ans.**

Reactive power, $Q = \sqrt{3}\ V_L I_L \sin\phi$

$= \sqrt{3} \times 400 \times 138.6 \times 0.6$

$= 57{,}600$ VAR or **57.6 KVAR Ans.**

Apparent power $= \sqrt{3}\ V_L I_L$

$= \sqrt{3} \times 400 \times 138.6$

$= 96{,}000$ VA or **96 KVA Ans.**

6.3.7. Comparison Between Star and Delta Systems

1. In star-connected system similar ends are joined whereas in delta-connected system dissimilar ends are joined together.

2. In a star-connected system phase voltage is equal to $1/\sqrt{3}$ times the line voltage, but in delta-connected system phase voltage is equal to line voltage, therefore, star-connected alternator needs less number of turns per phase and less insulation than a delta-connected alternator for the same line voltage. These are the reasons that 3-phase alternators are usually star-connected.

3. In star-connected system, phase current is equal to line current, *i.e.*, $I_P = I_L$ but in delta-connected system, the phase current is $1/\sqrt{3}$ times of line current, *i.e.*, $I_P = I_L/\sqrt{3}$.

4. It is possible to carry neutral wire to the load in star- connected system only, *i.e.,* only star-connected system can give 3-phase 4-wire arrangement and hence only star-connected system can be used for lighting as well as power loads, *i.e.,* for an unbalanced load. Lighting loads are connected between neutral and on any one of the outers and power circuits can be connected to the lines.

5. Neutral point of a star-connected system can be connected to earth, so relays and protective devices can be provided in star-connected system for protection of the system against ground faults and it is usually done in practice.

6. Delta-connections are mostly used in transformers for operating small low voltage 3-phase motors and are best suited for rotary convertors. Most of the three phase induction motors are delta-connected. 3-phase loads are also usually delta-connected as it provides flexibility for addition or removal of a load on a single phase in case of an unbalanced load.

6.4. CONSEQUENCES OF DISCONNECTING NEUTRAL IN A 3-PHASE, 4-WIRE SYSTEM

In case of a balanced three-phase, four-wire system when the neutral is disconnected, no change is produced. But in case of unbalanced 3-phase, 4-wire, system when the neutral is disconnected, the loads which are connected between any two line conductors and the neutral are connected in series and potential difference across the combined load becomes equal to line voltage. The potential difference across each load is thus changed as per rating of the load.

The effect of disconnecting the neutral in a 3-phase, 4-wire unbalanced system will be clear from the following solved example.

Example 6.22. In a 3-phase, 4-wire, 400/230 V system, a lamp of 100 watts is connected to one phase and neutral and a lamp of 200 watts is connected to the second phase and the neutral. Calculate the voltage across each lamp when the neutral wire is disconnected.

Solution : Rating of lamp $L_1 = 100$ watts

Rating of lamp $L_2 = 200$ watts

$$\text{Resistance of lamp, } L_1, R_1 = \frac{(V_P)^2}{\text{wattage of lamp } L_1} = \frac{(230)^2}{100} = 529\ \Omega$$

$$\text{Resistance of lamp, } L_2, R_2 = \frac{(230)^2}{200} = 264.5\ \Omega$$

When the neutral wire is disconnected, as shown in Fig. 1.23 the lamp L_1 (a resistance of 529 Ω) and lamp L_2 (a resistance of 264.5 Ω) are connected in series and the potential difference across them would be equal to line voltage, *i.e.,* 400 V.

$$\text{Current flowing through lamps, } I = \frac{V_L}{R_1 + R_2} = \frac{400}{529 + 264.5} = 0.505\ \text{A}$$

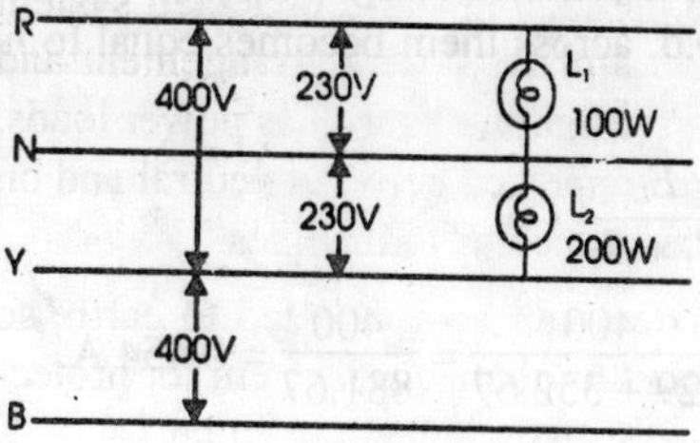

Fig. 6.22.

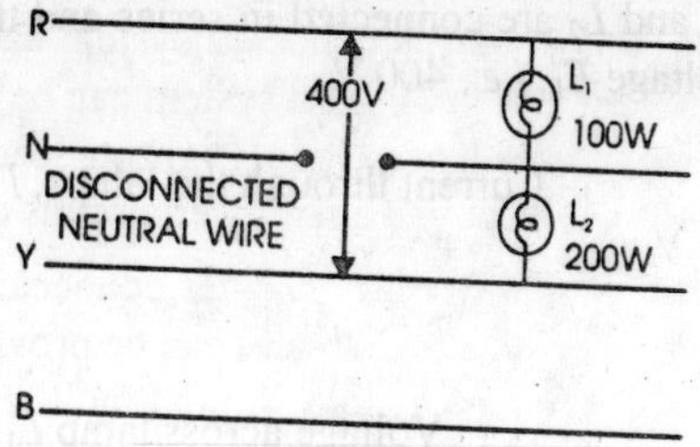

Fig. 6.23.

Voltage across lamp, $L_1 = IR_1 = 0.505 \times 529 =$ **266.67 V Ans.**

Voltage across lamp, $L_2 = IR_2 = 0.505 \times 264.5 =$ **133.33 V Ans.**

Conclusion : So when the neutral wire is disconnected, voltage across lamp L_1 is increased to 266.67 volts; whereas voltage across lamp L_2 is decreased to 133.33 volts. The lamp L_1 will become more bright and lamp L_2 becomes dim. If 100 watt lamp L_1 happens to be rated at 230 or 250 volt, it may be fused, as now 264.5 volt will be applied across this lamp.

Example 6.23. In a 3-phase, 4-wire 400/230 V system, a lamp of 100 watts is connected to one phase and neutral and a lamp of 100 watts is connected to the second phase and the neutral. Calculate the voltage across each lamp when the neutral wire is disconnected. Comment on the result.

Solution : The lamp L_1 of 100 watts is connected between R phase and neutral and lamp L_2 of 150 watt is connected between Y phase and neutral as shown in Fig. 6.24 *(a)*.

$$\text{Resistance of lamp } L_1,\ R_1 = \frac{E_{ph}^2}{\text{wattage}} = \frac{(230)^2}{100} = 529\ \Omega$$

$$\text{Resistance of lamp } L_2,\ R_2 = \frac{E_{ph}^2}{\text{wattage}} = \frac{(230)^2}{150} = 352.67\ \Omega$$

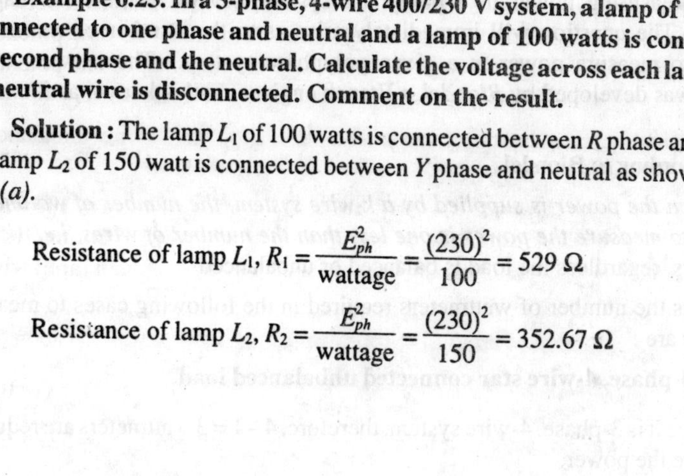

(a) *(b)*

Fig. 6.24.

When the neutral wire is disconnected as shown in Fig. 6.24 *(b)*, the lamps L_1 and L_2 are connected in series and the p.d. across them becomes equal to line voltage E_L, *i.e.*, 400 V.

$$\text{Current through the lamps, } I = \frac{E_L}{R_1 + R_2}$$

$$= \frac{400}{529 + 352.67} = \frac{400}{881.67} = 0.454 \text{ A}$$

Voltage across lamp $L_1 = IR_1 = 0.454 \times 529 =$ **240 V Ans.**

Voltage across lamp $L_2 = IR_2 = 0.454 \times 352.67 =$ **160 V Ans.**

Comments on the Result : The above example shows that when the neutral wire is disconnected, voltage across 100 watt lamp is increased to 240 V, whereas that across 150 watt is decreased to 160 V. Therefore, 100 watt bulb becomes more bright and 150 watt bulb becomes dim. It may be mentioned here that if 100 watt bulb happens to be rated at 230 V, it may be fused due to 240 V coming across this bulb.

6.5. POWER MEASUREMENT IN 3-PHASE CIRCUITS

In measuring power in a polyphase system, several wattmeters may be required. The question will immediately arise as to the number of wattmeters required to measure power in a given polyphase system. The answer to this question was developed by *Blondel,* a French engineer and is known as *Blondel's theorem.*

According to Blondel,

When the power is supplied by a k-wire system, the number of wattmeters required to measure the power is one less than the number of wires, i.e., $(k - 1)$ wattmeters, regardless the load is balanced or unbalanced.

Thus the number of wattmeters required in the following cases to measure the power are :

***(i)* 3-phase, 4-wire star connected unbalanced load.**

Since it is 3-phase, 4-wire system, therefore, $4 - 1 = 3$ wattmeters are required to measure the power.

∴ Required number of wattmeters = **3 Ans.**

(ii) 3-phase, 4-wire star connected balanced load.

It is a 3-phase, 4-wire system. Therefore, according to Blondel's theorem 4 $- 1 = 3$ wattmeters are required to measure power. Since the load is balanced therefore, current carried by the neutral wire is zero i.e. it can be supposed that neutral is not present so that the system is virtually reduced to 3-phase, 3-wire.

∴ Required number of wattmeters $= 3 - 1 = 2$ **Ans.**

***(iii)* 3-phase, 3-wire star connected balanced or unbalanced load.**

It is a 3-phase, 3-wire system.

$\therefore$ Number of wattmeters required = 3 – 1 = 2 **Ans.**

***(iv)* 3-phase, 3-wire delta connected balanced or unbalanced load.**

It is a 3-phase, 3-wire system.

$\therefore$ Number of wattmeters required = 3 – 1 = 2 **Ans.**

6.5.1. Three-Wattmeter Method

This method is employed for measurement of power in a three phase system is to use three wattmeters so that each wattmeter measures the power absorbed by one phase. Fig. 6.25 *(a)* and Fig. 6.25 *(b)* shows the use of three wattmeters for star connected and delta connected loads.

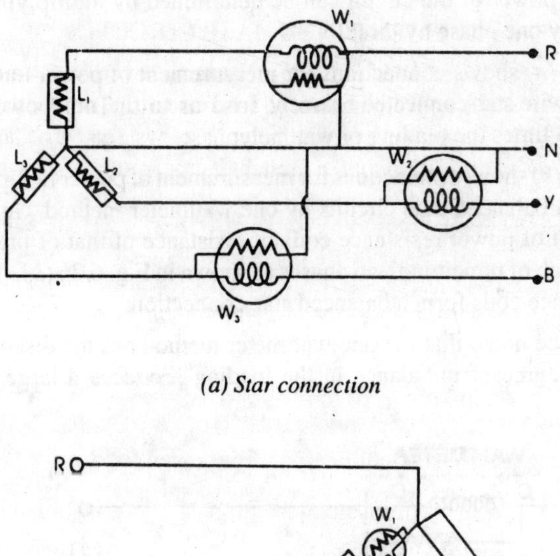

(a) Star connection

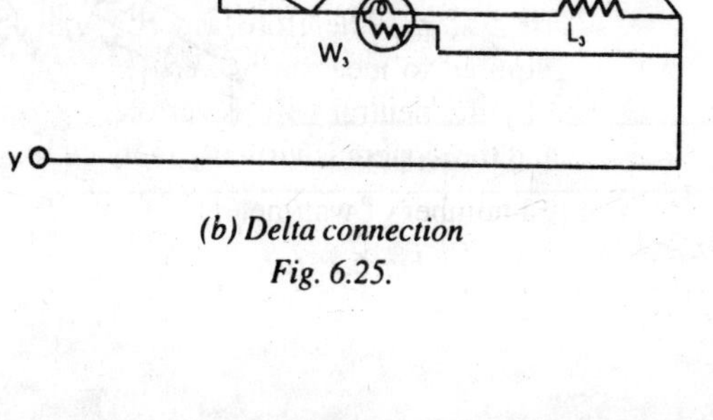

(b) Delta connection

Fig. 6.25.

In case of star connected loads it appears as if it is necessary to bring out a fourth wire from neutral or star point. However, in the case of 3-phase 3-wire system when load is star connected, the three ends of potential coils of three wattmeters will form a star point themselves. Each wattmeter measures the power in the phase in which it is connected. The total power is given by the algebraic sum of the readings of three wattmeters viz.

$$\text{Total 3-phase power } W = W_1 + W_2 + W_3$$

The method of power measurement has one particular advantage, *i.e.*, it indicates power contributed to each phase. However, two wattmeter method, which has been discussed, requires only two wattmeters instead of three.

But in the case of 3-phase 4-wire unbalanced load, three wattmeter method will have to be used.

6.5.2. One Wattmeter Method

In balanced 3-wire, 3-phase load circuit the power in each phase is equal and, therefore, total power of the circuit can be determined by multiplying the power measured in any one phase by three.

Fig. 6.26 *(a)* shows connections for measurement of power in one phase of the 3-phase, 3-wire star-connected balanced load circuits. Total power of the load circuit will be 3 times the reading of wattmeter.

Fig. 6.26 *(b)* shows connections for measurement of power in 3-phase, 3-wire delta connected balanced load circuits by one wattmeter method. In this method of measurement of power resistance coils of resistance of that of pressure coil is connected in each of remaining two phases as shown in Fig. 6.26 *(b)*. The pressure coil and resistance coils form a balanced star-connection.

It should be noted that the one-wattmeter method has the disadvantage that even a slight degree of unbalance in the loading produces a large error in the measurement.

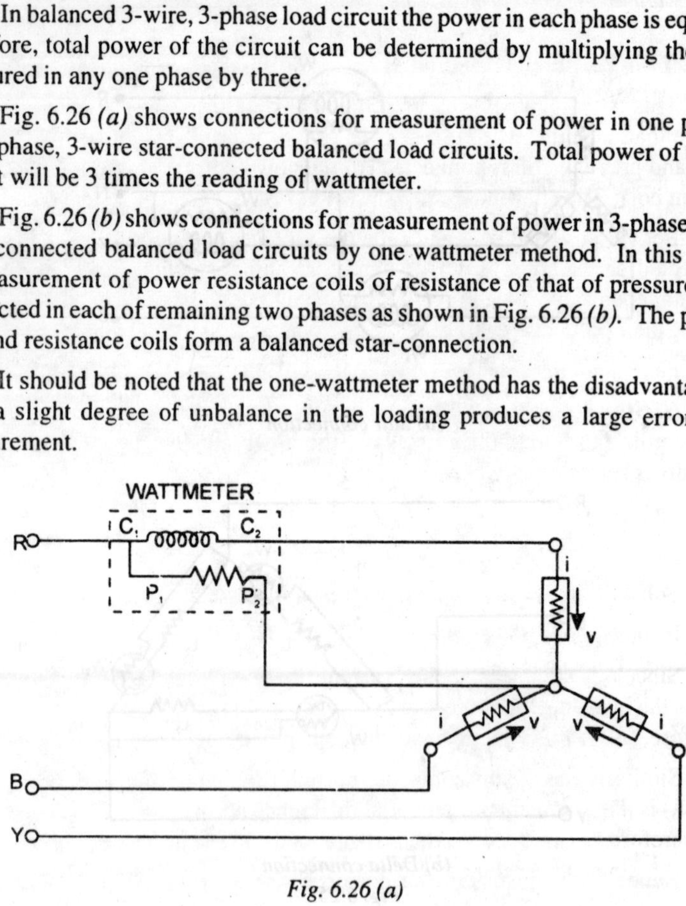

Fig. 6.26 (a)

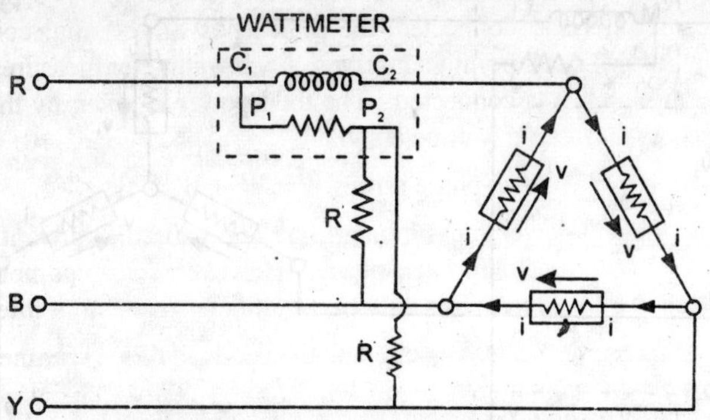

One wattmeter method of measuring power in 3-phase, 3-wire Balanced Load circuits.
Fig. 6.26 (b)

6.5.3. Two-Wattmeter Method

This is the generally used method for measurement of power in 3- phase, 3-wire load circuits. The current coils of two wattmeters are inserted in any two lines and pressure coil is connected from its own current coil to the line without a current coil. The connections are shown in Fig. 6.27.

Let v_1, v_2, v_3 and i_1, i_2 and i_3 be the voltages and currents of the three loads connected across three different phases at any particular instant. These being instantaneous values, the power at the instant under consideration is equal to the sum of their products regardless of power factor.

i.e., instantaneous power, $p = v_1 i_1 + v_2 i_2 + v_3 i_3$ watts ...*(i)*

(i) **Star-connected System :** Since all the three phases meet at a star point so according to Kirchhoff's first law, the algebraic sum of three instantaneous currents is zero.

i.e., $$i_1 + i_2 + i_3 = 0$$

or $$i_3 = -(i_1 + i_2)$$

Substituting $i_3 = -(i_1 + i_2)$ in expression *(i)* we get

Instantaneous power, $p = v_1 i_1 + v_2 i_2 - v_3 (i_1 + i_2) = i_1 (v_1 - v_3) + i_2 (v_2 - v_3)$

Since i_1 is the instantaneous current flowing through the current coil and $(v_1 - v_3)$ is the instantaneous potential difference across pressure coil of wattmeter W_1, therefore, $(v_1 - v_3)\, i_1 = w_1$ the instantaneous power measured by wattmeter W_1.

Similarly i_2 is the instantaneous current flowing through the current coil and $(v_2 - v_3)$ is the instantaneous potential difference across pressure coil of wattmeter W_2, therefore, $(v_2 - v_3)\, i_2 = w_2$, the instantaneous power measured by wattmeter W_2.

Hence total instantaneous power, $p = w_1 + w_2$

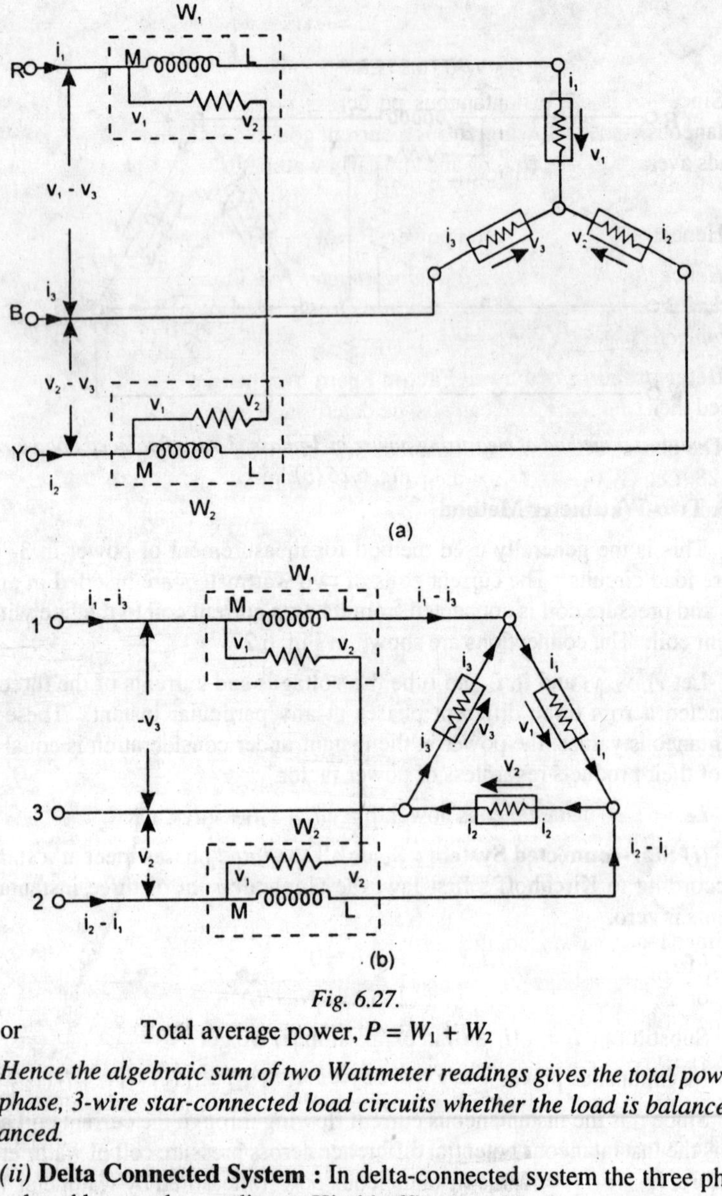

Fig. 6.27.

or Total average power, $P = W_1 + W_2$

Hence the algebraic sum of two Wattmeter readings gives the total power in the 3-phase, 3-wire star-connected load circuits whether the load is balanced or unbalanced.

(ii) **Delta Connected System :** In delta-connected system the three phases form a closed loop and according to Kirchhoff's second law,

$$v_1 + v_2 + v_3 = 0$$

or

$$v_1 = -(v_2 + v_3)$$

Instantaneous power, $p = v_1 i_1 + v_2 i_2 + v_3 i_3$

$$= -(v_2 + v_3)\,i_1 + v_2\,i_2 + v_3\,i_3$$
$$= -v_3\,(i_1 - i_3) + v_2\,(i_2 - i_1)$$

Since $-v_3$ is the instantaneous pd across pressure coil and $(i_1 - i_3)$ is the instantaneous current flowing through current coil of wattmeter W_1, so wattmeter W_1 reads average of $-v_3\,(i_1 - i_3)$ and similarly wattmeter W_2 reads average of $v_2\,(i_2 - i_1)$

Hence total power, $P = W_1 + W_2$

Hence the algebraic sum of two wattmeter readings gives the total power of the circuit irrespective of the fact that the circuit is balanced or unbalanced and star-connected or delta-connected.

Determinaiton of Power Factor From Wattmeter Readings : If load is balanced then of p.f. the load can also be determined from the wattmeter readings.

The phasor diagram for a balanced star-connected inductive load shown in Fig. 6.28. Let V_1, V_2 and V_3 be the rms values of phase voltages and I_1, I_2 and I_3 be the rms values of phase currents.

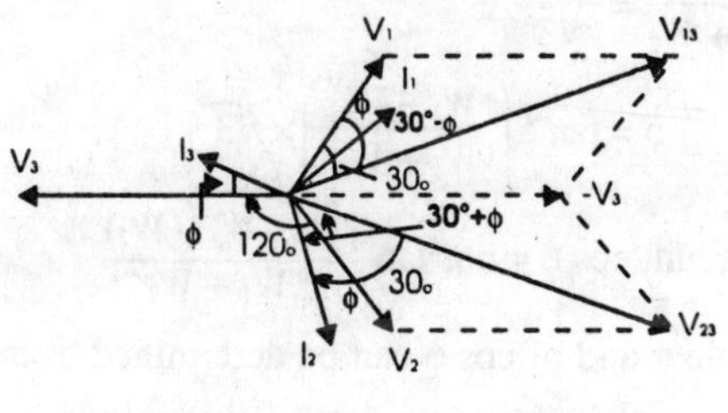

Phasor diagram for balanced Star-connected Inductive Load

Fig. 6.28.

Since load is balanced, therefore,

(i) Phase voltage V_1, V_2 and V_3 will be equal (say, equal to V_p)

(ii) Phase currents I_1, I_2 and I_3 will be equal (say, equal to I_P or I_L)

(iii) Phase angles between respective phase voltages and phase currents will be equal, say ϕ.

The current in current coil of wattmeter $W_1 = I_1 = L$ lagging behind V_1 by ϕ.

The pd across pressure coil of wattmeter $W_1 = V_{13} = \sqrt{3}\,V_P = V_L$ lagging behind V_1 by 30°.

Therefore, phase angle between voltage across potential coil and current through current coil of wattmeter W_1 is $(30° - \phi)$ and reading of wattmeter

$$W_1 = V_L I_L \cos (30^\circ - \phi)$$

The current in current coil of wattmeter $W_2 = I_2 = I_L$ lagging behind V_2 by ϕ.

The pd across pressure coil of wattmeter $W_2 = V_{23} = \sqrt{3}\, V_P = V_L$ leading V_2 by 30°.

Therefore, phase angle between pd across potential coil and current through current coil of wattmeter W_2 is $(30^\circ + \phi)$.

Hence reading of wattmeter $W_2 = V_L I_L \cos (30^\circ + \phi)$.

The sum of two wattmeter readings

$$W_1 + W_2 = V_L I_L \cos (30^\circ - \phi) + V_L I_L \cos (30^\circ + \phi)$$

$$= V_L I_L \times 2 \cos 30^\circ \cos \phi$$

$$= \sqrt{3}\, V_L I_L \cos \phi = \text{True power of load} \quad ...(i)$$

and $$W_1 - W_2 = V_L I_L \cos (30^\circ - \phi) - V_L I_L \cos (30^\circ + \phi)$$

$$= V_L I_L \times 2 \sin 30^\circ \sin \phi = V_L I_L \sin \phi \quad ...(ii)$$

Dividing expression *(ii)* by expression *(i)*, we get

$$\frac{W_1 - W_2}{W_1 + W_2} = \frac{\tan \phi}{\sqrt{3}}$$

or $$\phi = \tan^{-1}\left(\frac{W_1 - W_2}{W_1 + W_2}\right) \times \sqrt{3}$$

and pf of load circuit, $\cos \phi = \cos \tan^{-1} \dfrac{\sqrt{3}\,(W_1 - W_2)}{W_1 + W_2}$

Hence phase angle ϕ and pf $\cos \phi$ can be determined from readings of two wattmeters.

Wattmeter Readings For Capacitive Loads : In the above discussion inductive load has been considered. If the circuit is capacitive the expression for readings of two wattmeters are obtained by substituting ϕ by $(-\phi)$ in the above expressions for wattmeter readings.

$$W_1 = V_L I_L \cos [30^\circ - (-\phi)] = V_L I_L \cos (30^\circ + \phi)$$

$$W_2 = V_L I_L \cos [30^\circ + (-\phi)] = V_L I_L \cos (30^\circ - \phi)$$

i.e., readings of wattmeters are interchanged.

Example 6.24. In a two wattmeter method of measurement of 3-phase power, the readings of the wattmeters are 3 kw and 1 kw respectively. Calculate the power and power factor for a balanced circuit.

Solution : Reading of wattmeter $W_1 = 3$ kw

Reading of wattmeter $W_2 = 1$ kw

Total power of the circuit $P = W_1 + W_2 = 3 + 1 =$ **4 kw Ans.**

$$\text{Power factor of the circuit, } \cos\phi = \cos\tan^{-1}\left(\frac{W_1 - W_2}{W_1 + W_2}\right)\sqrt{3}$$

$$= \cos\tan^{-1}\left(\frac{3-1}{3+1}\right)\sqrt{3}$$

$$= \cos\tan^{-1}\frac{\sqrt{3}}{2} = \mathbf{0.756\ lag\ Ans.}$$

Example 6.25. Two wattmeters connected to measure the power input to a three phase circuit indicate 30 kw and 10 kw. Find *(a)* the power, *(b)* the power factor assuming the load to be balanced.

Solution : Power of 3 phase circuit $= W_1 + W_2 = 30 + 10 = 40$kw

Power factor

We know that $\tan\phi = \sqrt{3}\,\frac{W_2 - W_1}{W_1 + W_2} = \sqrt{3}\,\frac{(30-10)}{(30+10)}$

$$= \frac{\sqrt{3}\times 20}{40} = \frac{\sqrt{3}}{2} = 0.866$$

or $\phi = 40^\circ\ 54'$

The power factor = cos 40° 54′ = **0.7559 lagging Ans.**

Example 6.26. Two wattmeters connected to measure the power input to a three phase circuit indicate 8 kw and 0.8 kw, the latter reading being obtained after reversing the current coil connections. Find the power and power factor of the load. Why it was essential to reverse the connections of current coil of one wattmeter. Draw also the phasor diagram.

Solution : Reading of one wattmeter $W_2 = 8$ kw

Reading of second wattmeter $W_1 = 0.8$ kw (–ve)

The reading has been taken as (–ve) as it was obtained by reversing the current coil connections.

(a) Total Power $W_2 - W_1 = 8 + (-0.8) = 7.2$ kw

(b) Power factor $\tan\phi = \sqrt{3}\,\frac{W_2 - W_1}{W_2 + W_1}$

$$= \sqrt{3}\left[\frac{8-(-0.8)}{8+(-0.8)}\right] = 1.73 \times \frac{8.8}{7.2} = 2.144$$

$$\phi = 65^0$$

Power factor = cos φ = cos 65° = 0.4226 lagging.

(c) As the power factor of the load is less than 0.5 lagging so one wattmeter will given –ve reading. In order to read the scale either the connections of current coil or potential coil should be reversed.

(*d*) Fig. 6.29 shows the phasor diagram. In this case one wattmeter which reads the power $I_R\, V_{RNy} \cos(30 + \phi)$ will read –ve as $\phi = 65°$. So the current coil connections will have to be reversed.

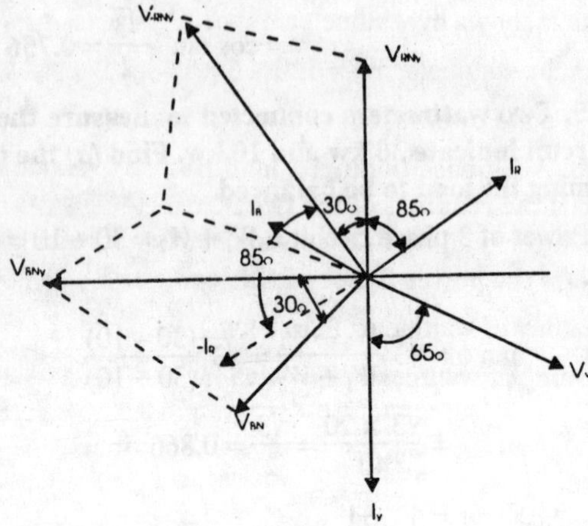

Fig. 6.29.

Wattmeter W_1 will give the reading as per above

$$V_{RNy}\, I_R \cos(30° + \phi)$$

or
$$W_1 = V_{RNy}\, I_R \cos(30° + 65°)$$
$$= V_{RNy}\, I_R \cos 95° = - V_{RNy}\, I_R \cos 85°$$

So to get reading on the scale of the wattmeter, the connections of either the current coil or potential coil will have to be reversed.

Example 6.27. A three phase 4 wire star connected system supplies only non inductive load. The current in line R is 8 Amps. The current in line *y* is 10 Amp. and the current in line B is 6 Amps. The voltage from each line to neutral wire is 120 volts.

(*i*) Find wattage as shown by each of the three wattmeters.

(*ii*) Total power taken by the lighting load.

Solution :

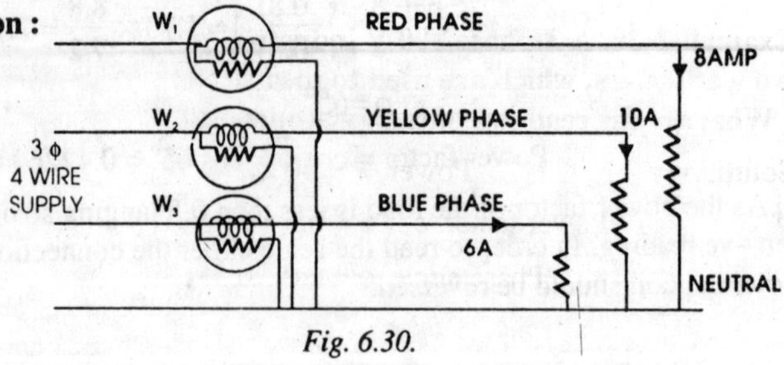

Fig. 6.30.

Phase voltage = 120 volts

$\cos\phi = 1$ (Resistive Load)

Current in Red Phase = 8 Amp.

Wattage shown by wattmeter $W_1 = 120 \times 8 \times 1 =$ **960 watts Ans.**

Similarly wattage shown by wattmeter $W_2 = 120 \times 10 \times 1 =$ **1200 watts Ans.**

Wattage shown by wattmeter $W_3 = 120 \times 6 \times 1 =$ **720 watts Ans.**

Total Power taken = 960 + 1200 + 720 = **2880 watts Ans.**

Example 6.28. A 3-phase balanced load power was measured by two wattmeter method. If the readings of the two wattmeters so connected are 5.0 kw and 0.5 kw, the latter reading being obtained after reversal of current coil connections, calculate the power factor of the load.

Solution : Reading of wattmeter $W_1 = 5$ kw

Reading of wattmeter $W_2 = -0.5$ kw

$$\text{Power factor of the load, } \cos\phi = \cos\tan^{-1}\frac{W_1 - W_2}{W_1 + W_2}\sqrt{3}$$

$$= \cos\tan^{-1}\frac{5-(-0.5)}{5+(-0.5)}\sqrt{3} = \cos\tan^{-1}\frac{5.5}{4.5}\sqrt{3} = \mathbf{0.427\ (lag)\ Ans.}$$

Example 6.29. Phase voltage and current of a star-connected inductive load is 150 V and 25 A. Power factor of the load is 0.707 (lag). Assuming that the system is 3-wire and power is measured using two wattmeters, find the readings of wattmeters.

Solution : Line voltage, $V_L = \sqrt{3}\,V_P = \sqrt{3} \times 150 = 259.8$ V

Line current, $I_L = I_P = 25$ A

Power factor, $\cos\phi = 0.707$ lag

Phase angle, $\phi = \cos^{-1} 0.707$ (lag) $= 45^\circ$

Reading of wattmeter $W_1 = V_L I_L \cos(30^\circ - \phi)$

$= 259.8 \times 25 \cos(30^\circ - 45^\circ) =$ **6,274 watts Ans.**

Reading of wattmeter $W_2 = V_L I_L \cos(30 + \phi)$

$= 259.8 \times 25 \cos(30^\circ + 45^\circ) =$ **1681 watts Ans.**

Example 6.30. A 3-phase 500 V industrial motor has a power factor of 0.5. Two wattmeters, which are used to measure input, show the input to be 50 kw. What are the readings of each wattmeter ?

Solution : Power, $P = 50$ kw

Power factor, $\cos\phi = 0.5$

Phase angle, $\phi = \cos^{-1} 0.5 = 60^\circ$

$$\therefore \quad V_L I_L = \frac{P \times 1000}{\sqrt{3} \cos \phi} = \frac{50 \times 1000}{\sqrt{3} \times \cos 60^\circ} = \frac{100000}{\sqrt{3}} \text{ VA}$$

$$\text{Reading of wattmeter } W_1 = V_L I_L \cos (30^\circ - \phi)$$

$$= \frac{100000}{\sqrt{3}} \times \cos (30^\circ - 60^\circ)$$

$$= \mathbf{50000\ W\ or\ 50\ kw\ Ans.}$$

$$\text{Reading of wattmeter } W_2 = V_L I_L \cos (30^\circ + \phi)$$

$$= \frac{100000}{\sqrt{3}} \cos (30^\circ + 60^\circ) = \mathbf{0\ Ans.}$$

Example 6.31. The power input to a 2000 V, 50 Hz, 3-phase motor running on full load at an efficiency of 90% is measured by two wattmeters which indicate 300 kw and 100 kw respectively. Calculate the *(a)* input; *(b)* the p.f. and *(c)* the line current.

Solution : Reading of wattmeter $W_1 = 300$ kw

Reading of wattmeter $W_2 = 100$ kw

(a) Input power, $P = W_1 + W_2 = 300 + 100 =$ **400 kw Ans.**

$$\text{Phase angle, } \phi = \tan^{-1}\left(\sqrt{3}\, \frac{W_1 - W_2}{W_1 + W_2}\right)$$

$$= \tan^{-1}\left(\sqrt{3}\, \frac{300 - 100}{300 + 100}\right) = 40.89^\circ \text{ (lag)}$$

(b) Power factor, $\cos \phi = \cos 40.89^\circ =$ **0.756 (lag) Ans.**

(c)

$$\text{Line current, } I_L = \frac{P}{\sqrt{3}\, V_L \cos \phi}$$

$$= \frac{400 \times 1000}{\sqrt{3} \times 2000 \times 0.756} = \mathbf{152.74\ A\ Ans.}$$

6.5.4. Balanced and Symmetrical System

Balanced System : A polyphase system is said to be balanced when e.m.f.'s, currents and phase angles are equal for each phase.

Symmetrical System : A polyphase system is said to be symmetrical when several e.m.f.s are equal and are displaced from one another by equal angles. Three phase is the most important symmetrical system.

Balanced Load : A load is said to be balanced load in which phase and line voltages; phase and line currents drawn by load are equal in magnitude and the phase and line voltages and line currents are displaced from one another by 120° respectively.

6.5.5. Concept of Unbalanced Loads

So far we have considered balanced loads, *i.e.*, the loads that have the same impedance and power factor in each phase. The problems on balanced loads can be solved by considering one phase only; the conditions in the other two phases being similar. However, we may come across a situation where loads are unbalanced, *i.e.*, each load phase has different impedance and/or power factor. In that case, current and power in each phase will be different. In practice, we may come across the following unbalanced loads :

(i) Four-wire star-connected unbalanced load

(ii) Unbalanced Δ-connected load

(iii) Unbalanced 3-wire, *Y*-connected load.

The 3-phase, 4-wire system is widely used for distribution of electric power in commerical and industrial buildings. The single phase load is connected between any line and neutral wirer while a 3-phase load is connected across the three lines. The 3-phase, 4-wire system invariably carries *unbalanced loads. In this chapter, we shall only discuss this type of unbalanced load; the other two types of unbalanced loads being beyond the scope of this book.

Four-Wire Star-Connected Unbalanced Load

We can obtain this type of load in two ways. First, we may connect a 3-phase, 4-wire unbalanced load to a 3-phase, 4-wire supply as shown in Fig. 6.31. Note that star point *N* of the supply is connected to the load star point *N*. Secondly, we may connect single phase loads between any line and the neutral wire as shown in Fig. 6.32. This will also result in a 3-phase, 4-wire, **Unbalanced load because it

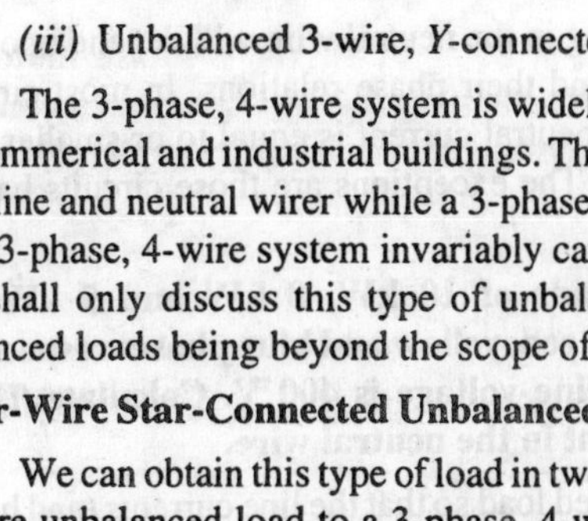

Fig. 6.31.

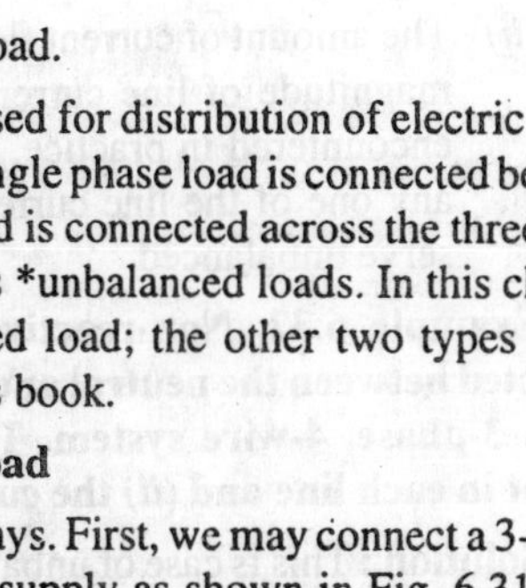

Fig. 6.32.

* No doubt 3-phase loads (e.g., 3-phase motors) connected to this supply are balanced but when we add single phase loads (e.g., lights, fans etc.), the balance is lost. It is because it is rarely possible that single phase loads on all the three phases have the same magnitude and power factor.

** In actual practice, we never have an unbalanced 3-phase, 4- wire load. Most of the 3-phase loads (e.g., 3-phase motors) are 3-phase, 3-wire and are balanced loads. In fact, these are the single phase loads on the 3-phase, 4-wire supply which constitute unbalanced, 4-wire Y load.

is rarely possible that single phase loads on all the three phases have the same magnitude and power factor. Since the load is unbalánced, the line currents will be different in magnitude and displaced from one another by unequal angle. The current in the neutral wire will be the phase sum of the three line currents, *i.e.*,

Current in neutral wire, $I_N = I_R + I_Y + I_B$ ...phasor sum

The following points may be noted carefully :

(i) Since the neutral wire has neglitible resistance, supply neutral *N* and neutral *N* will be at the same potential. It means that voltage across each impedance is equal to the phase voltage to the supply. However, current in each phase (or line) will be different due to unequal impedances.

(ii) The amount of current flowing in the neutral wire will depend upon the magnitude of line currents and their phase relations. In most circuits encountered in practice, the neutral current is equal to or smaller than any one of the line currents. The exceptions are those circuits having serve unbalanced.

Example 6.32. Non-reactive loads of 10 kW, 8 kW and 5 kW are connected between the neutral and the red, yellow and blue phases respectively of a 3-phase, 4-wire system. The line voltage is 400 V. Calculate *(i)* the current in each line and *(ii)* the current in the neutral wire.

Solution : This is case of unbalanced load so that the line currents (and hence the phase currents) in the three lines will be different. The current in the *neutral wire will be equal to the phase sum of the three line currents as shown in Fig. 6.33.

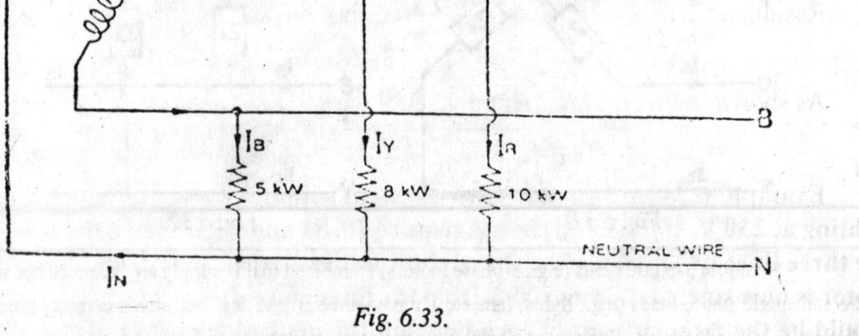

Fig. 6.33.

* Had the load been balanced (i.e., each phase having identical load), the current in the neutral wire would have been zero.

(i) Phase voltage $= \dfrac{400}{\sqrt{3}} = 231$ V

$$I_R = 10 \times \frac{10^3}{231} = 43.3 \text{ A } \textbf{Ans.}$$

$$I_Y = 8 \times \frac{10^3}{231} = 34.5 \text{ A } \textbf{Ans.}$$

$$I_B = 5 \times \frac{10^3}{231} = 21.65 \text{ A } \textbf{Ans.}$$

(ii) The three line currents are represented by the respective phasors in Fig. 6.34. Note that the three line currents are of different magnitude but displaced 120° from one another. The current in the neutral wire will be the phasor sum of the three line currents.

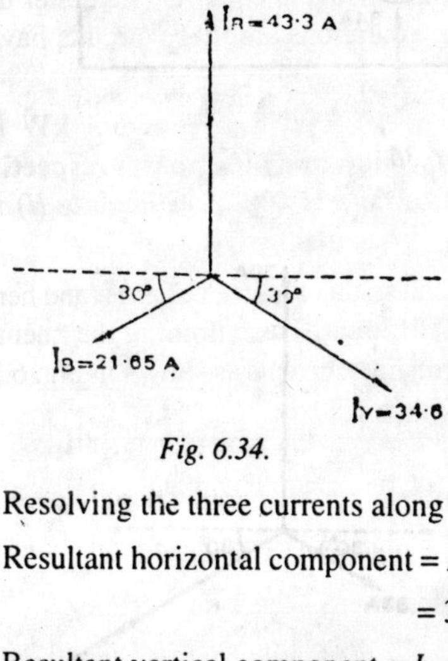

Fig. 6.34.

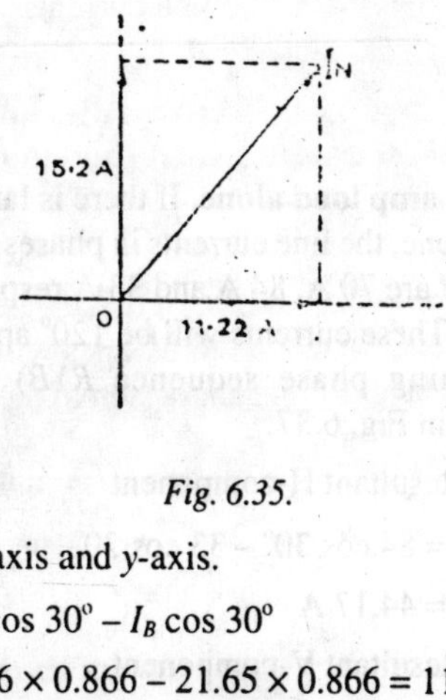

Fig. 6.35.

Resolving the three currents along *x*-axis and *y*-axis.

Resultant horizontal component $= I_y \cos 30° - I_B \cos 30°$

$$= 34.6 \times 0.866 - 21.65 \times 0.866 = 11.22 \text{ A}$$

Resultant vertical component $= I_R - I_y \cos 60° - I_B \cos 60°$

$$= 43.3 - 34.6 \times 0.5 - 21.65 \times 0.5 = 15.2 \text{ A}$$

As shown in Fig. 6.35, current in neutral wire,

$$I_N = \sqrt{(11.22)^2 + (15.2)^2} = 18.9 \text{ A } \textbf{Ans.}$$

Example 6.33. A 3-phase, 4-wire system supplies power at 400 V and ighting at 230 V. If the lamps in use require 70, 84 and 33 amperes in each of he three lines, what should be the current in the neutral wire ? If a 3-phase motor is now started, taking 200 A from the lines at a p.f. of 0.2 lagging, what hould be the total current in each line and the neutral wire ? Find also the otal power supplied to the lamps and the motor.

Solution : Fig. 6.36 shows the lamp load and motor load on 400 V/240 V 3-phase, 4-wire supply.

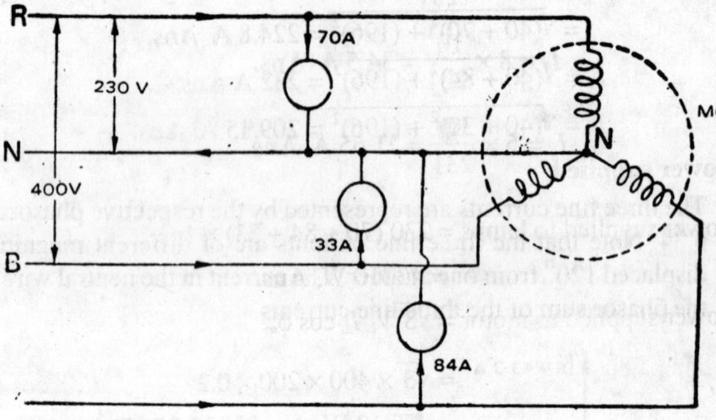

Fig. 6.36.

Lamp load alone. If there is lamp load alone, the line currents in phases, *R*, *Y* and *B* are 70 A, 84 A and 33 A respectively. These currents will be 120° apart (assuming phase sequence *RYB*) as shown in Fig. 6.37.

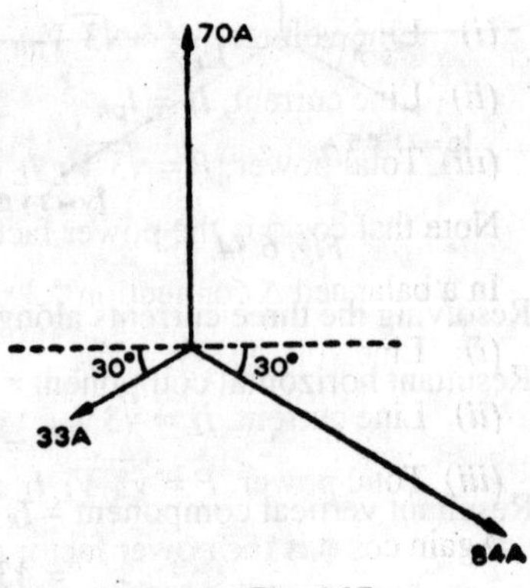

Fig. 6.37.

Resultant H-component

$= 84 \cos 30^\circ - 33 \cos 30^\circ$

$= 44.17$ A

Resultant V-component

$= 70 - 33 \cos 60^\circ - 84 \cos 60^\circ$

$= 111.5$ A

∴ Neurtal current,

$$I_N = \sqrt{(44.17)^2 + (11.5)^2} = 45.64 \text{ A} \quad \textbf{Ans.}$$

Both lamp load and motor load. When motor load is also connected along with lighting load, there will be not change in current in the neutral wire. It is because the motor load is balanced and hence no current will flow in the neutral wire due to this load.

∴ Neutral current, $I_N = 45.64$ A **Ans.** ...same for before

The current in each line is the phase sum of the line currents due to lamp load and motor load.

Active component of motor current = $200 \times \cos\phi = 200 \times 0.2 = 40$ A

Reactive component of motor current = $200 \times \sin\phi = 200 \times 0.98 = 196$ A

$$\therefore \quad I_r = \sqrt{(\text{sum of active comp.})^2 + (\text{reactive comp.})^2}$$

$$= \sqrt{(40+70)^2 + (196)^2} = 224.8 \text{ A} \quad \textbf{Ans.}$$

$$I_Y = \sqrt{(40+84)^2 + (196)^2} = 232 \text{ A } \textbf{Ans.}$$

$$I_B = \sqrt{(40+30)^2 + (196)^2} = 209.15 \text{ A} \quad \textbf{Ans.}$$

Power supplied

Power supplied to lamps = 230 (70 + 84 + 33) × 1 $\qquad (\because \cos\phi_L = 1)$

= 43010 W **Ans.**

Power supplied to motor = $\sqrt{3}\, V_L\, I_L \cos\phi_m$

$= \sqrt{3} \times 400 \times 200 \times 0.2 \qquad (\because \cos\phi_m = 0.2)$

= 27712 W **Ans.**

HIGHLIGHTS

1. In a balanced star connection,

   (i) Line voltage, $V_L = \sqrt{3}\, V_{ph}$

   (ii) Line current, $I_L = I_{ph}$

   (iii) Total power, $P = \sqrt{3}\, V_L\, I_L \cos\phi = 3V_{ph}\, I_{ph} \cos\phi$

   Note that cos ϕ is the power factor of each load phase.

2. In a balanced Δ connection,

   (i) Line voltage, $V_L = V_{ph}$

   (ii) Line current, $I_L = \sqrt{3}\, I_{ph}$

   (iii) Total power, $P = \sqrt{3}\, V_L\, I_L \cos\phi = 3V_{ph}\, I_{ph} \cos\phi$

   Again cos ϕ is the power factor of each load phase.

3. A single phase wattmeter connected in an a.c. circuit to measure power gives the reading as :

   Reading of wattmeter = $VI \cos\phi$

   where V = r.m.s. value of voltage across the potential coil

   I = r.m.s. value of current in current coil

   ϕ = phase angle between V and I.

4. In two-wattmeter method of measuring power of a 3-phase load (balanced of unbalanced), the total power is the algebraic sum of the wattmeter readings, *i.e.*, Total power = $W_1 + W_2$

5. If two wattmeter method is used to measure power in a *balanced* 3-phase load, then the wattmeter readings will be :

$$W_2 = V_L I_L \cos (30^\circ - \phi)$$

$$W_1 = V_L I_L \cos (30^\circ + \phi)$$

where ϕ is load p.f. angle.

(i) The wattmeter whose deflection is proportional to $(30^\circ - \phi)$ *i.e.*, W_2 in this case is always gives positive reading.

(ii) The wattmeter whose deflection is proportional to $(30^\circ + \phi)$ *i.e.*, W_1 in this case is always gives negative reading. If the p.f. is less then 0.5 the wattmeter reads zero.

(iii) If the lower-reading wattmeter gives downscale reading, it means that (connections being correct) load p.f. is less then 0.5. reverse either potential or current coil connection to obtain upscale reading. The reading obtained after reversal of coil connection should be taken as negative.

(iv) If the p.f. is unity with wattmeters indicate equal readings.

6. For leading power factor of a balanced 3-phase load, the two- wattmeter readings are interchanged, *i.e.*,

$$W_1 = V_L I_L \cos (30^\circ - \phi)$$

$$W_2 = V_L I_L \cos (30^\circ + \phi)$$

Now W_1 is the higher reading wattmeter.

7. The power factor of a balanced 3-phase load can be determined frorm the readings of the two wattmeters as under :

$$\tan \phi = \sqrt{3}\, \frac{W_2 - W_1}{W_2 + W_1} \quad \text{...lagging p.f.}$$

$$\tan \phi = \sqrt{3}\, \frac{W_1 - W_2}{W_1 + W_2} \quad \text{...leading p.f.}$$

The value of ϕ and hence $\cos \phi$ can be determined.

OBJECTIVE TYPE QUESTIONS

1. Polyphase system is adopted in

(a) a.c. supply *(b)* d.c. supply

2. Electric power is almost exclusively generated, transmitted, and distributed by 3-ϕ system because it

(a) is more capable and reliable .

(b) is more efficient and costs less.

(c) delivers constant power (not pulsating).

(d) all of the above.

3. The commonly used phase sequence is

(a) BRY (b) RYB

(c) YBR

4. The interconnection of three phases reduces the number of line conductors

(a) True (b) false

5. The voltages induced in the three windings of a 3-phase alternator are .... apart in the phase.

(a) 60° (b) 90°

(c) 120° (d) 150°

6. A 3-phase star-connected alternator is feeding the given load at 400 V, 50 Hz. The phase voltage is

(a) 400 V (b) 200 V

(c) $400/\sqrt{3}$ V (d) 200 V

7. In star-connected system

(a) only line current is equal to phase current.

(b) only line voltage is equal to phase voltage.

(c) both line currents and line voltages are equal to phase currents and phase voltages respectively.

8. Which of the following statements associated with 3-phase delta connected circuits is TRUE ?

(a) line voltage is equal to phase voltage.

(b) line current is equal to phase current.

(c) line voltage is $\sqrt{3}$ time of phase voltage.

(d) line currents are 60° apart.

9. If the line current in delta connected system is I_L, then phase current will be equal to

(a) I_L (b) $I_L/\sqrt{3}$

(c) $\sqrt{3}/I_L$ (d) $I_L/\sqrt{2}$

10. A 3-phase delta connected motor is supplied at a line voltage of 400 V, the voltage across each winding of the motor will be

(a) 400 V (b) 400/V

(c) 200 V (d) 200/V

11. The power consumed in a balanced star or delta connected load is given by the expression

(*a*) $P = 3\ V_L\ I_L$ (*b*) $P = \sqrt{3}\ V_L\ I_L \cos\phi$

(*c*) $P = \sqrt{3}\ V_L\ I_L \sin\phi$

12. The power in a 3-phase system is given by $\sqrt{3}\ V_L\ I_L \cos\phi$ where ϕ is the phase angle between

(*a*) line voltage and line current. (*b*) phase voltage and phase current.

(*c*) line voltage and phase current.

13. The minimum number of wattmeters to measure power in a 3-phase unbalanced star-connected load is

(*a*) one (*b*) two

(*c*) three (*d*) four

14. Two wattmeter method of measuring power in 3-phase system can be used for

(*a*) only star-connected balanced system.

(*b*) only delta-connected balanced system.

(*c*) both star and delta-connected balanced system or unbalanced system.

15. In three phase power measurement by two wattmeter method, the power factor of the load is given by

(*a*) $\cos \tan^{-1} \sqrt{3}\ \dfrac{W_1 - W_2}{W_1 + W_2}$ (*b*) $\cos \tan^{-1} \sqrt{3}\ \dfrac{W_1 + W_2}{W_1 - W_2}$

(*c*) $\cos \tan^{-1} \sqrt{3} \left(\dfrac{W_1 + W_2}{W_1 - W_2} \right)$

16. During the measurement of power in a 3-phase load by two wattmeter method one wattmeter reads negative, the power factor of the circuit is

(*a*) 0.5 (*b*) greater than 0.5

(*c*) less than 0.5 (*d*) unity.

17. If the power factor of the 3-ϕ load circuit is unity, the readings of two wattmeters connected properly to measure the power

(*a*) W_1 will be +ve and W_2 –ve (*b*) both W_1 and W_2 +ve and unequal.

(*c*) both W_1 and W_2 +ve and equal.

(*d*) none of the above.

18. If the power factor of the 3-ϕ load circuit is 0.5, then one of the wattmeter will read

(a) zero (b) $\frac{W}{2}$

(c) $\frac{W}{\sqrt{3}}$ (d) $\frac{\sqrt{3}\,W}{2}$

ANSWERS

1.	(a)	2.	(d)	3.	(b)	4.	(b)	5.	(c)
6.	(c)	7.	(a)	8.	(a)	9.	(b)	10.	(a)
11.	(b)	12.	(b)	13.	(b)	14.	(c)	15.	(a)
16.	(c)	17.	(c)	18.	(a)				

FILL IN THE BLANKS

1. The most commonly used polyphase system is .....
2. The correct phase sequence is .....
3. In star connection, phase voltage is equal to .....
4. In a three phase system the phase voltage differ by .....
5. The number of wattmeters required to measure power in a 3-phase 3-wire system is .....

ANSWERS

1. 3-ϕ system 2. RYB 3. $\sqrt{3}$ line voltage 4. 120°

5. Two

EXERCISE

1. What are the advantages of 3-phase system over 1-ϕ system.
2. What do you understand by phase sequence of 3 phase voltages ? What would be the effect of phase sequence reversal on *(i)* direction of rotation of an Induction Motor, *(ii)* voltage across 3 phase balanced load ?
3. Discuss the advantages of inter connecting the 3 phases.
4. Deduce the relationship between the phase and line voltages of a three phase star connected alternator.
5. Derive both for star and delta connected systems, an expression for the total power input for a balanced three phase load in terms of line voltage, line current and power factor.
6. What will happen in the following cases ?
   (i) 3 phase windings of a transformer designed for star connections are put in delta across same voltage supply.
   (ii) In a three phase 4 wire system the neutral of the star connected balanced load gets disconnected from the neutral of the supply.
7. What do you understand by a balanced load and a balanced supply system ?

8. Show the connections for the three wattmeter method used with 3 phase 4 wire star connected system. What is the total power as measured by the wattmeter ?

9. Explain the most commonly used method of power measurement in three phase circuit.

10. Explain in brief the concept of unbalanced loads.

NUMERICALS

1. A three phase mesh connected induction motor has a full load output of 100 H.P., the power factor being 0.85 lagging and efficiency 91%. If the line voltage is 1100 volts, calculate the line current. (**Ans.** 50.63 Amp.)

2. A three phase balanced load connected in star takes 1.5 kw at a power factor of 0.2 from three phase 400 volt 50 Hz supply. Calculate the resistance and inductance of the load in each phase. (**Ans.** 4.27 Ω and 0.0666 Henry)

3. Three similar inductors each of resistance 10 Ω and inductance 0.019 henry are delta connected to a three phase 415 volt, 50 Hz supply. Calculate *(i)* The value of line current, *(ii)* Power factor, *(iii)* Power input to the circuit.
(**Ans.** 61.7 Amp., 0.858 lagging, 38 kw)

4. Three capacitors, each of capacitance 100 μF are connected in mesh to a 500 volt, 3 phase 50 Hz supply. Find the line current. What will be the capacitance of each of three capacitors such that when connected in star across the same supply the line current will be the same ? (**Ans.** 27.2 A; 300 μF)

5. A factory has the following loads with power factor 0.9 lagging in each phase Red phase 40 A, Yellow phase 50 A and Blue phase 60 A. If the supply is 400 V, three phase, four wire, calculate the current in the neutral and the total power. Draw a phasor diagram for phase and line quantities. Assume that relative to the current in the red phase, the current in the yellow phase lags by 120° and that in the blue phase leads by 120°.
(**Ans.** 17.3 amperes, 31.2 kw)

6. Two wattmeters connected to read the total power in a 3 phase system supplying a balanced load read 10.5 kw and –2.5 kw respectively. Calculate the total power.

   Draw suitable phasor diagrams and explain the significance of *(i)* equal wattmeter readings, *(ii)* zero reading on one wattmeter. (**Ans.** 8 kw)

7. Two wattmeters connected to measure input to a balanced three phase circuit indicate 2500 watt and 500 watt respectively. Find the power factor of the circuit *(a)* when both readings are positive, *(b)* when the latter reading is obtained after reversing the connections to the current coil of one instrument. Draw the phasor and connection diagrams. (**Ans.** 0.65 and 0.359)

8. Three coils are connected in delta to a three phase three wire 415 volt, 50 Hz supply and takes a line current of 5 amp. at 0.8 lagging power factor. Calculate the resistance and inductance of the coil.

If the coils are star connected to the same supply, calculate the line current and total power.

Calculate the line currents if one coil becomes open circuited when the coils are connected in star.

(**Ans.** 115 Ω, 0.274 Henry; 1.667, 956 W, 1.445 A, 1.445 A.)

9. Two wattmeters connected to a three phase, 500 volt motor indicate the total input to be 10 kw. The p.f. is 0.3 leading. Find the reading of each instrument.

(**Ans.** 14.175 kw, –4.175 kw)

10. In a 3-phase, 4-wire 400/230 V system, a lamp of 200 watt is connected to one phase and neutral and a lamp of 500 watt is connected to the second phase and neutral. Both the lamps are rated at 230 volt. Calculate the voltage across each lamp when neutral wire gets disconnected. Comment on the result. (**Ans.** 285.7 volt across 200 watt lamp, 114.3 volt across 500 watt lamp, 200 watt lamp can fuse due to over voltage)

11. *(a)* Show that total power in a 3-phase star or delta connected circuit is $\sqrt{3}\ VI \cos \phi$. What do V, I and ϕ represent ?

(b) A 440 volts, 3-phase circuit has an input of 86.5 kw at 0.87 power factor lagging. Find the line current and the readings on each of the two wattmeters connected to measure the input.

12. Three identical impedances are connected in Delta to a 3-phase 400 V supply. The line current is 34.65 amperes and total power taken from supply is 14.14 kw. Calculate the values of resistance and reactance of each phase.

13. Three 100 μF capacitors are connected in Delta to a 500 volt three phase 50 Hz supply. Find the line current. What will be the value of capacitors such that when connected in star across the same supply, the line current will be the same ?

14. A 3-phase supply system with 230 volts between each phase and neutral supplies a balanced 3-phase star-connected load having resistances of 6 ohms in each phase. Calculate the current flowing in each phase and the neutral current. Find the total power absorbed.

Laboratory Practicals

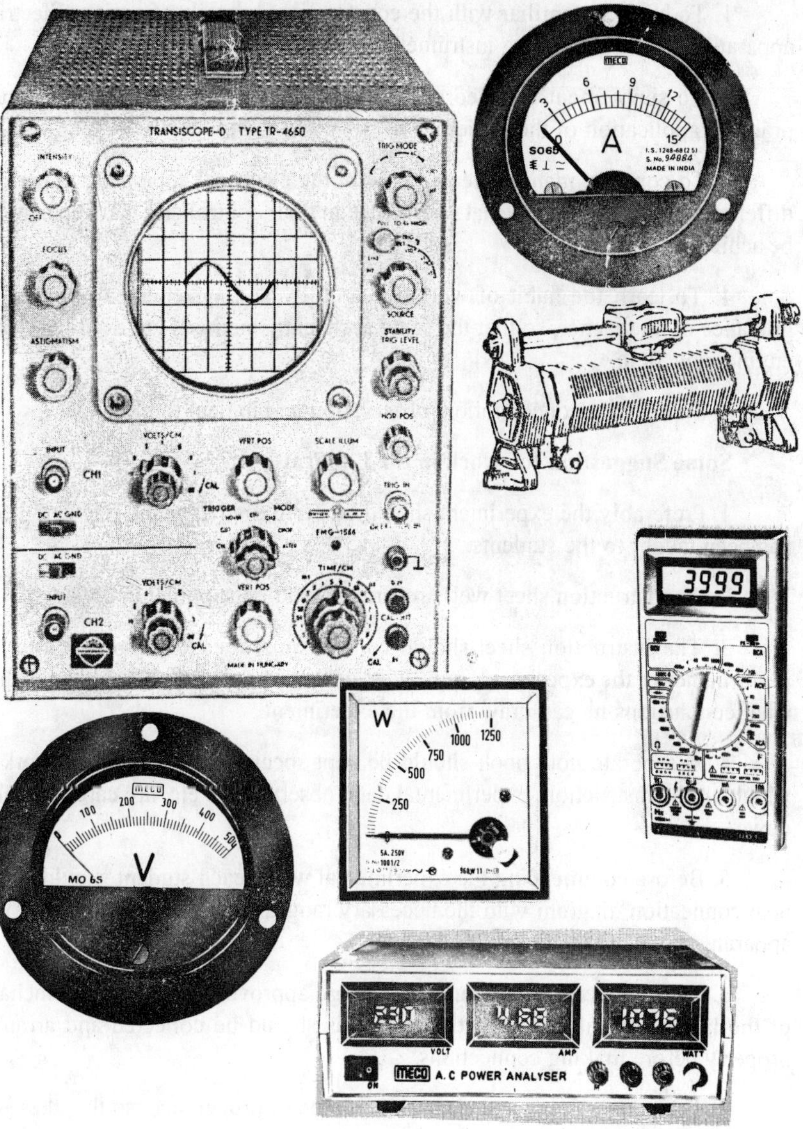

General Instructions

The main purpose of the laboratory work should be :-

1. To become familiar with the constructional details of various electrical apparatus such as measuring instruments, machines etc.

2. To supplement the theoretical study in the class room with the actual practical application of the subject.

3. To connect, manipulate and control the electrical apparatus, to perform different tests upon them so that confidence in their control and management can be achieved.

4. To learn the habit of working carefully and systematically under the guidance of an teacher, so that the same systematic methods are used later on in engineering field.

5. To develop concentration, observation and logical thinking.

Some Suggestion for work in the Laboratory :

1. Preferably the experiment should be performed after the relevant theory has been taught to the students.

2. An instruction sheet well prepared should be supplied in advance.

3. The instruction sheet should contain quiz questions; so that after the performance of the experiment' quiz questions may encourage the student to think what conclusions he can draw from the experiment.

4. A seperate note book should be kept specially for laboratory work, in which all the instruction, experimental data. observations etc. are carefully taken down.

5. Before commencing the experimental work, each student should draw a neat connection diagram with the necessary ranges of the instruments and other apparatus marked clearly on the diagram,

6. The connection diagram should be go approved by the teacher incharge of the laboratory, all the essential apparatus should be collected and arranged properly before making connections.

7. The connecting wires chosen should be of proper size, so that these can carry the expected current.

8. The supply to the circuit should never be switched on till the circuit has been checked and approved by the teacher incharge.

9. Every care must be taken to prevent damage to apparatus, but should such damage occur it should be invariably reported to the teacher incharge of the laboratory.

10. The students should work out all the result of the experiment. They should plot the curves if any and should get the result and curves approved by the teacher.

A Laboratory Journal may be prepared under the following main heading.

1. Name of the experiment.

2. Object of the experiment.

3. A complete list of apparatus required, with particulars about their rating, number and make.

4. A complete circuit diagram should be drawn on left hand of the Journal.

5. Theory in short on which the experiment is based.

6. The procedure of the experiment in short should be written in first person.

7. Precautions if any to be observed in the performance of the experiment.

8. Observation table should be arranged on the left hand side of the laboratory journal.

9. Calculations should also be written down on the left hand side of the laboratory journal.

It is suggested that left hand side of the laboratory journal should be in pencil.

10. Plotting of Curves : All curves should be plotted on graph paper; the graph paper should be inserted laboratory journal.

11. Conclusion and deductions on the results, the nature of curves obtained and the general knowledge gained from the performance of the experiment should be written in detail.

The journal should be kept upto date and submitted regularly for correction and marking to the laboratory incharge.

EXPERIMENT NO. 1

Name of the Experiment : Determination of voltage relationship in a d.c. circuit under specific physical conditions and to draw conclusions.

Objective : Able to establish the voltage-current relationship in electric circuit.

Apparatus :

S.No.	*Name*	*Specifications*	*Quantity*
1.	D.C. Source of supply	0-12 V	1
2.	Ammeter	0-2 A, DC	1
3.	Voltmeter	0-15 V, DC	1
4.	Rheostat or variable Resistance	100 Ω	1
5.	Fixed value Resistance	10 Ω, or 5 Ω	3 or 4

Theory : According to Ohm's Law, the current flowing through a conductor is directly proportional to potential difference applied across its ends, provided the physical conditions, *i.e.*, temperature etc. do not change.

Thus
$$I \propto V$$

or
$$\frac{V}{I} = \text{constant}$$

This constant is resistance R of the conductor.

$$\therefore \quad \frac{V}{I} = R$$

Hence Ohm's law can also be defined as under.

The ratio of potential difference applied across a conductor and current flowing through it remains constant provided physical conditions, *i.e.*, temperature etc. of the conductor do not changes.

$$\frac{V_1}{I_1} = \frac{V_2}{I_2} = \ldots = \frac{V_n}{I_n} = \text{constant} = R$$

Procedure :

1. Connect the apparatus as per circuit diagram shown in Fig. 1.
2. Get your connection checked by your class teacher.
3. Add whole resistance of the rheostat in the circuit before the key is closed.
4. Take the reading of voltmeter and ammeter.
5. **Take five set of reading of ammeter and voltmeter by decreasing the resistance of rheostat.**

6. Record the observations on the observation table.
7. Draw graph between voltage and current.
8. Discuss the nature of graph.

Circuit Diagram :

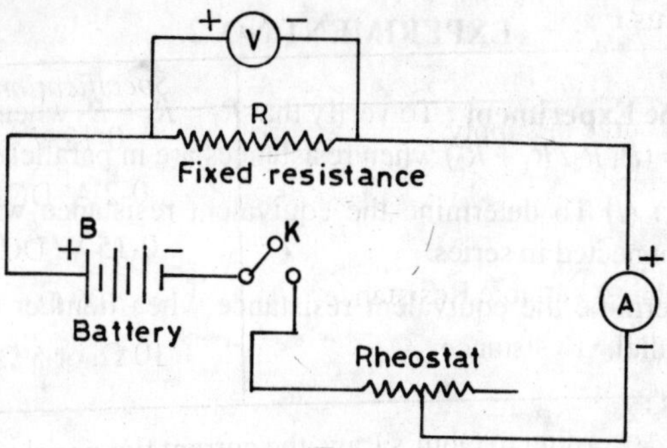

Fig. 1

Observations :

S.No.	*Voltmeter Reading (volts) (V)*	*Ammeter Reading (I)*	*Resistance* $R = V/I\ \Omega$

Conclusion

The ratio of *V* and *I* for any set of readings is constant and is equal to *R*. This relationship $\frac{V}{I} = R$ or $V = IR$ is the verification of Ohm's law.

The nature of the graph is a straight lines passing through the origin and having different slopes. For higher values of *R* the slopes are higher for lower values of *R* slopes are lower.

Test Exercise :

1. Define Ohm's Law.
2. Express Ohm's law in terms of *V*, *I* and *R*.

3. Is there any variations in ratio $\frac{V}{I} = R$, if it is so explain the reasons.

4. State the application of Ohm's law.

5. Is the Ohm's law applicable in A.C.

EXPERIMENT NO. 2.

Name of the Experiment : To verify that $R_T = R_1 + R_2$ when resistances are in series and $R_T = (R_1 R_2/R_1 + R_2)$ when resistances are in parallel.

Objective : *(i)* To determine the equivalent resistance when number of resistances are connected in series.

(ii) To determine the equivalent resistance when number of resistors are connected in parallel.

Apparatus :

S.No.	Name	Specification/Range	Quantity
1.	D.C. supply source	0-12 V	1
2.	Voltmeter	0-15 V, DC	3
3.	Ammeter	0-2 A, DC	2
4.	Fixed resistance	10 Ω	2
5.	Rheostat	100 Ω, 2 A	1

Theory : *(i)* **Series Connections :** When number of resistors are connected in series, each one of them carries the same current, however the supply voltage is divided and different voltages appear across them :

$$\therefore \quad V = V_1 + V_2$$

$$IR_{eq} = IR_1 + IR_2$$

from which the equivalent series circuit resistance is

$$R_{eq} = R_1 + R_2$$

(ii) **Parallel Connections :** When number of resistors are connected in parallel, the voltage across each one of them is same, however the current is divided into number of such resistors :

$$\therefore \quad I = I_1 + I_2$$

If the equivalent value of the circuit resistance is R, then

$$\frac{V}{R} = \frac{V}{R_1} + \frac{V}{R_2}$$

or $$\frac{1}{R} = \frac{1}{R_1} + \frac{1}{R_2}$$

i.e., resiprocal of the equivalent resistance is equal to the sum of resiprocal of individual resistances connected in parallel

Circuit Diagram :

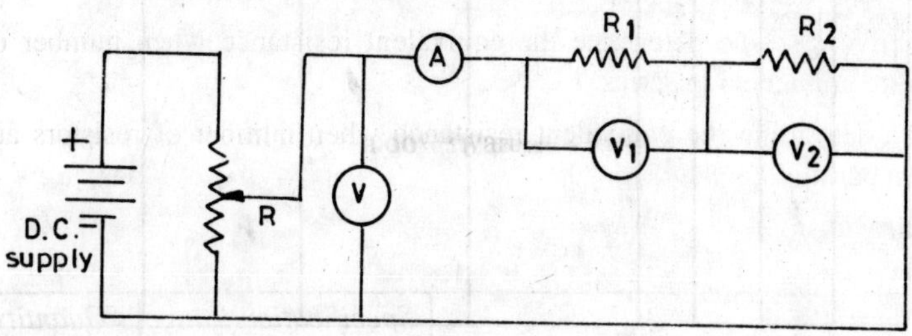

Fig. 2. (a)

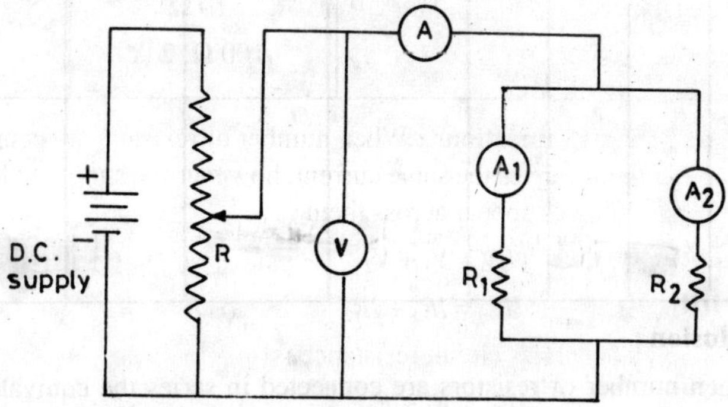

Fig. 2. (b)

Procedure : To perform the experiment proceed as follows :

1. Connect the apparatus as per the circuit shown in Fig. 2 (a) and then as shown in Fig. 2 (b).

2. Get the connections checked by the class teacher.

3. Insert a small value of rheostat resistance in the circuit and switch on the supply.

4. Take the reading from voltmeters and ammeters.

5. Change the resistance of rheostat in steps and take the readings of voltmeter and ammeter for each setting.

6. Record the observations in the observation table given below.

Observations : *(i)* For series connections :

S.No.	V	V_1	V_2	I	$R_1 = V_1/I$	$R_2 = V_2/I$	$R = V/I$	$R = R_1 + R_2$

***(ii)* For parallel connections :**

S.No.	I	I_1	I_2	V	$R_1 = \frac{V}{I_1}$	$R_2 = \frac{V}{I_2}$	$R = \frac{V}{I}$	$R = \frac{R_1 R_2}{R_1 + R_2}$

Conclusion :

1. When number of resistors are connected in series the equivalent value increases.

2. When number of resistors are connected in parallel the equivalent value decreased.

3. When more resistors are connected in parallel, the equivalent value of resistance decreases but the current drawn increases.

Exercises :

1. Four resistances 2, 4, 8 and 16 ohms respectively are connected in parallel, find the equivalent resistance of the circuit.

2. Voltmeter reads 12 V and ammeter reads 3 amps. Find the load resistance of the circuit.

3. Solve the given circuit and find equivalent resistance.

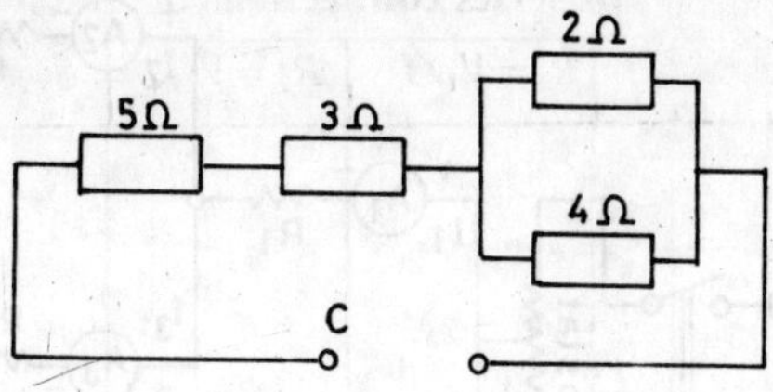

Fig. 3.

EXPERIMENT NO. 3.

Name of the Experiment : Verification of Kirchhoff's Current Law and Kirchhoff's Voltage Law in a d.c. circuit.

Objective :

1. To be able to find out relation between currents coming to a junction and flowing into branches.

2. To be able to establish a relation between different voltage drops in a closed circuit.

Apparatus :

S.No.	*Name*	*Specification*	*Quantity*
1.	D.C. supply adjustable	0-12 V, DC	1
2.	Voltmeter	0-15 V, DC	1
3.	Ammeters	0-2 A, DC	3
4.	Adjustable resistances	100 Ω, 1.5 A	3
5.	Connecting leads	—	—

Theory : In an electric circuit 'Node' or 'Junction' is a point at which a number of branches meet. Kirchhoff's first law states that in d.c. circuit the algebraic sum of currents at a junction is zero.

or Total incoming current at the junction = Total out going current at the junction.

The above statement is true as the current can not accumulate at a junction. Therefore, whatever current flows into a junction that must flow out of the junction.

Kirchhoff's second law which is also known as Kirchhoff's voltage law states that in a closed electrical circuit the algebraic sum of the emf's and the product of current flowing through the conductors and their resistances is equal to zero.

Circuit Diagram :

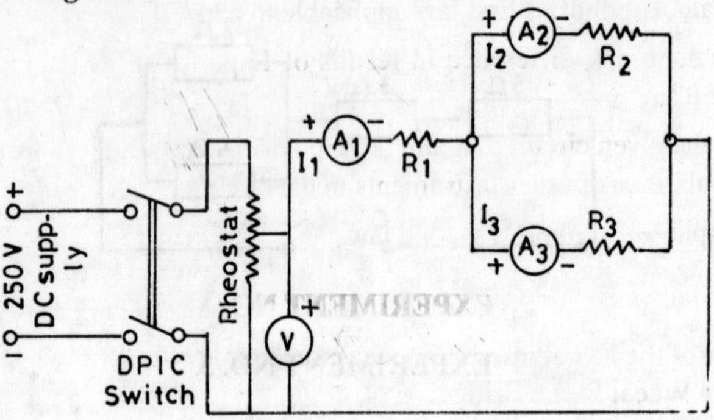

Fig. 4.

Procedure : (Verification of Kirchhoff's Laws)

1. Select the apparatus according to availability in your laboratory, the above given are the only suggested values.

2. Make the circuit connections as shown in the circuit diagram.

3. Keep the polarity of the meters in order.

4. Close the switch after adjusting the potential divider.

5. Note down the readings of various ammeters voltmeter in the observation table.

6. Repeat the process so that a number of readings can be obtained.

Observations :

S.No.	I_1	I_2	I_3	$I_1 = I_2 + I_3$	$I_1 R_1$	$I_2 R_2$	$I_3 R_3$	$V = I_1 R_1 + I_2 R_2 + I_3 R_3$
1.								
2.								
3.								
4.								

Conclusion :

From the above calculation $I_1 = I_2 + I_3$ it is concluded that at any junction Total incoming currents = Total outgoing currents and currents cannot accumulate at a junction.

In a closed circuit the algebraic sum of emf voltage drops is zero.

Test Exercises :

1. State Kirchhoff's first law applicable to d.c. circuits.
2. State Kirchhoff's first law applicable to a.c. circuits.
3. Is there any difference in results of Experiment, if it is so explain the reasons for it?
4. In the given circuit diagram if the polarities of A_2 and A_3 are reversed then show the polarities of other instruments in the circuit.
5. Explain Kirchhoff's second law.

EXPERIMENT NO. 4

Name of the Experiment : To measure the value of unknown resistance with the help of a Wheat Stone Bridge.

Apparatus required :

1. Wheat Stone bridge	1 No.
2. Galvanometer 0-500 μA	1 No.
3. Storage Battery, 6V	1 No.
4. Resistance Box	1 No.
5. Key	1 No.
6. Connecting wires.	

Theory : Wheat Stone bridge consists of a wire of uniform cross- section one metre in length and is stretched between the points *A* and *C*. Contact is made at point *B* on get the balanced condition. This point *B* also divides the wire into the lengths l_1 and l_2 which form the two arms of the bridge. R_C is the resistance box. The unknown resistance of the instrument is R_X. The position of the sliding contact is so adjusted that the galvanometer shows no deflections. When this occurs then

$$R_X = R \times \frac{l_2}{l_1}\ \Omega$$

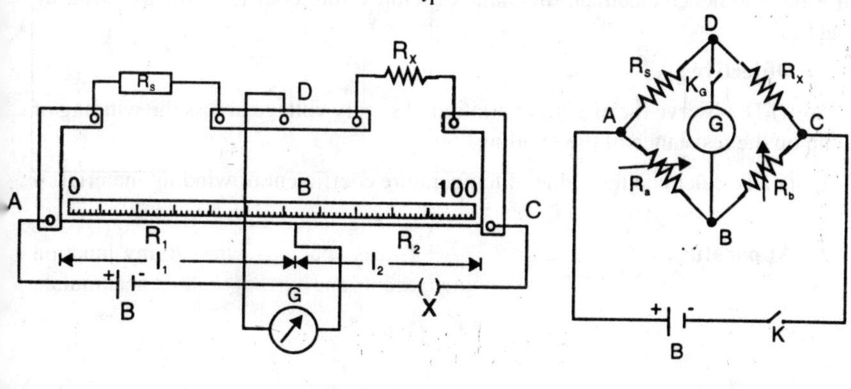

Fig. 5. Circuit diagram

Procedure :

1. Connect the various components as shown in the connection diagram (circuit diagram).

2. Connect an ammeter (unknown resistance) between the terminals C and D.

3. Take out 1 ohm resistance plug from the resistance box and close the key.

4. Adjust the sliding contact B on the wire in such a way that galvanometer shows zero deflection on the scale.

5. Now take out ammeter from the circuit, and connect voltmeter between the same terminals.

6. Put 1 ohm resistance plug in the resistance box and take out 5 K and 10 K ohm resistance plugs from the resistance box.

7. Again adjust the sliding contact B, so that galvanometer shows zero deflection. Note down distance l_1 and l_2 (between ABandBC).

Observation Table :

S. No.	Instrument	$R\,\Omega$	l_1 cm	l_2 cm	$R_X = R\frac{l_2}{l_1}\,\Omega$
1.	Ammeter				
2.	Voltmeter				

Results : The resistance of an ammeter as per calculations is found very small. Due to this reason it is always connected in series. It has very small voltage drop and also consumes small power.

The resistance of the voltmeter has been found to be very high. Due to this reason it is connected in parallel. The voltmeter takes a little current from the circuit and consumes the least power (V^2/R).

EXPERIMENT NO. 5.

Name of the Experiment : To observe the effect of temperature on resistance of a fan and hence calculate the value of temperature coefficient of the winding material.

Objective :

(i) To observe the effect of variation of supply voltage across the winding of a fan on the resistance of the winding.

(ii) To calculate the value of temperature coefficient of winding material.

Apparatus :

S.No.	Name	Specifications	Quantity
1.	Auto transformer	0-270 V, 4 A	1
2.	Voltmeter	0-300 V, AC	1
3.	Ammeter	0-1 A, AC	1
4.	Wattmeter	0-500 W	1
5.	Electric fan	-	1
6.	Connecting Leads	-	-

Theory : The winding of an electric fan is mostly made of copper wire. At normal working voltage the winding gets heated to some temperature. Resistance of the most conducting materials increases with the rise of temperature and accordingly the resistance of the fan winding also increases when it is connected across the supply which results into change the temperature coefficient of the winding material. When the fan winding is connected across the supply, the power will be consumed by the winding which depends upon the size of the fan winding. The power consumed by the winding can be measured with the help of wattmeter.

In this experiment will study the effect of variation of voltage on resistance of the fan winding and power consumed by the winding and the value of temperature coefficient of winding material.

Circuit Diagram :

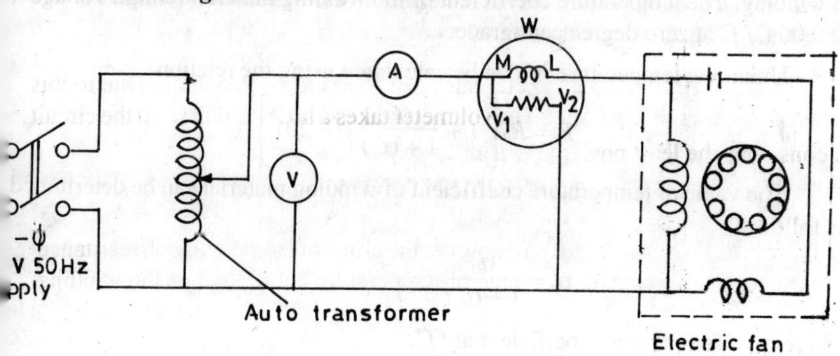

Fig. 6.

Procedure :

1. Connect the auto transformer.

2. Connect other circuit elements as shown in Fig.

3. Switch on the supply and set auto transformer at zero.

4. Vary the input voltage in steps starting from 100 V to the rated fan voltage.

5. Take reading of ammeter, voltmeter and wattmeter for each voltage setting.

6. Take five such sets of readings.

7. Calculate the value of resistance, power consumed and temperature coefficient of resistance.

Observations and Results :

S.No.	*V*	*I*	*W*	$R = \frac{W}{I^2}$	*Temp* (°C)	$\alpha_t = \frac{\alpha_0}{1 + \alpha_0 t}$ /°C

The value of temperature of the fan winding at different voltages can be calculated as follows.

Assume the room temperature as 20°C. Suppose at 20°C resistance of the winding is R_{20} and at any other temperature t° C is say R_t. The resistance of the fan winding at 20°C may be measured with the multimeter before connecting the supply to winding. The temperature coefficient of the winding material (copper) is taken as 0.0043/°C at zero degree centigrade.

Unknown temperature '*t*' can be calculated using the relation

$$R_t = R_0 \left[1 + \frac{\alpha_0}{1 + \alpha_0 t} (t - 20) \right]$$

The value of temperature coefficient of winding material can be determined as follows;

$$\alpha_t = \frac{\alpha_0}{1 + \alpha_0 t}$$

where α_t = Temperature coefficient at t°C.

Draw on a graph paper variation of resistance with voltage, power input with voltage and temperature with resistance.

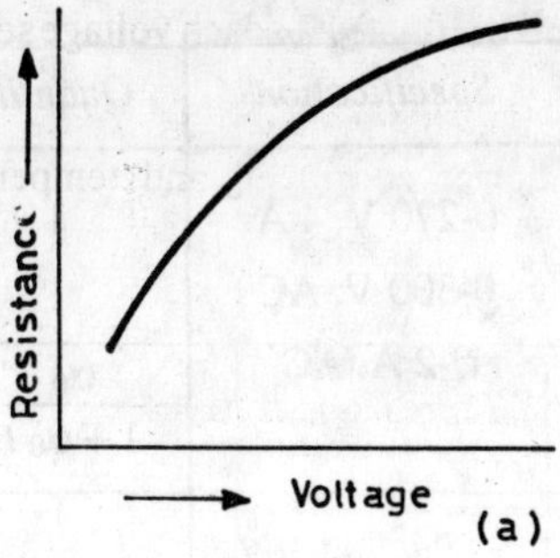

Fig. 7.(a)

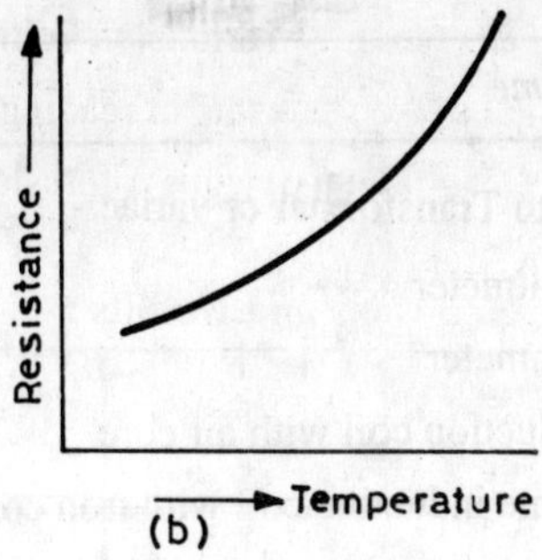

Fig. 7. (b)

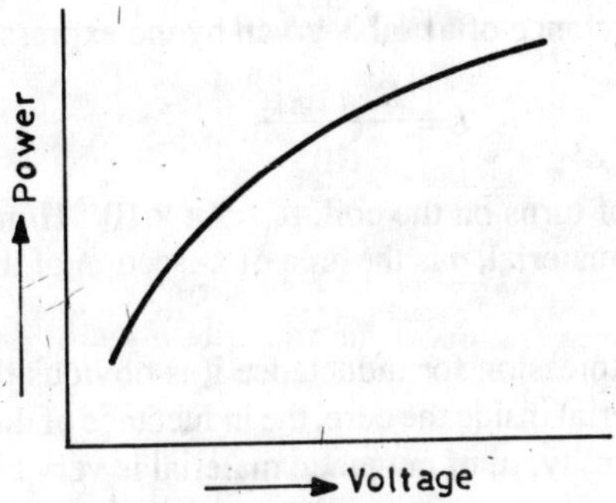

Fig. 8.

Conclusion :

1. From the result it is concluded that resistance of winding material varies with the variation of supply voltage. It increases with the increase in voltage as the temperature coefficient of resistance of the winding material increase.

2. Power is proportional to the square of voltage.

Test Exercise :

1. Explain, why the resistance of the winding material increases with the rise in voltage.

2. State the various sources of error in this experiment.

EXPERIMENT NO. 6.

Name of the Experiment : To find the ratio of inductance values of a coil having air-core and iron-core respectively and to observe the effect of introduction of a magnetic core on coil inductance.

Objective : 1. To measure the inductance of a coil having air core and iron core respectively.

2. Introduction of magnetic material inside the coil increases the inductance value.

Apparatus :

S.No.	Name	Specification	Quantity
1.	Auto Transformer or variac	0-270 V, 4 A	1
2.	Voltmeter	0-300 V, AC	1
3.	Ammeter	0-2 A, AC	1
4.	Induction coil with air core		1
5.	Same inductive coil with iron core		1

Theory : The inductance of a coil is given by the expression

$$L = \frac{N^2 a \mu_0 \mu_r}{l}$$

where N is the number of turns on the coil, $\mu_0 = 4\pi \times 10^{-7}$ H/m, μ_r is the relative permeability of the core material, a is the area of x- section of the core and l is the length of the core.

From the above expression for inductance it is obvious that with the introduction of magnetic material inside the core, the inductance of the coil will increase because relative permeability, μ_r of magnetic material is very high.

Circuit Diagram :

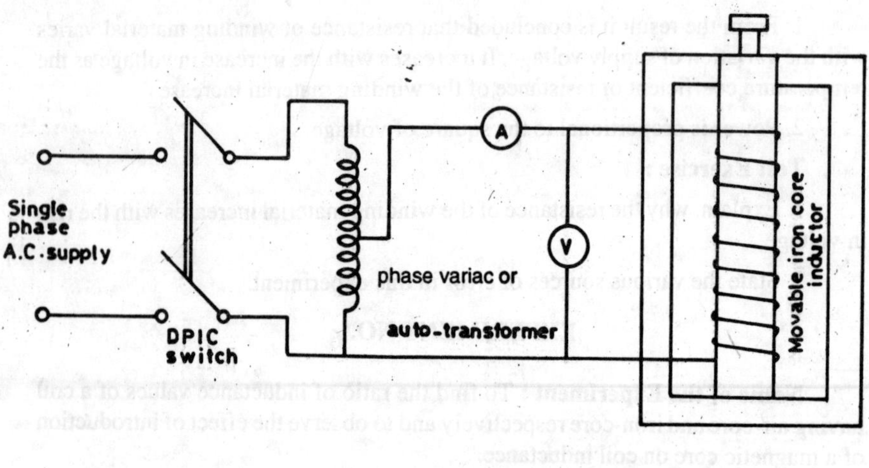

Fig. 9. Determination of Inductance of coil.

Procedure :

1. Movable iron core inductor, voltmeter, ammeter are connected through single phase variac or auto transformer and DPIC switch to the a.c. supply mains as shown in the circuit diagram.

2. First of all the core is entirely removed from the coil with the help of the knob provided for this purpose.

3. Knob of the single phase variac is set at suitable position.

4. DPIC switch is closed and the readings of the voltmeter and ammeter are noted.

5. This process is repeated for another two positions of the variac.

6. The iron core is moved inside the coil to some extent and again the above whole process is repeated.

7. Now the iron core is inserted in the coil to the full extent and again the process is repeated as mentioned in the preceding paragraph.

Observations :

Supply frequency, $f = 50$ Hz.

S. No.		1.	2.	3.
Core Entirely Removed From Coil	Voltmeter Reading in Volts, V			
	Ammeter Reading in Amperes, I			
	Reactance of coil X_L = V/I Ohms			
	Coil Inductance. $L=X_L/2\pi f$ Heneries			
	Mean Value of L in Henries			
Core Partly Inside Coil (Say Half)	Voltmeter Reading in Volts, V			
	Ammeter Reading in Amperes, I			
	Coil Reactance X_L =V/I Ohms			
	Coil Inductance $L=X_L/2\pi f$ Henries			
	Mean value of L in Henries			
Core Entirely Inside The Coil	Voltmeter Reading in Volts, V			
	Ammeter Reading in Amperes, I			
	Coil Reactance X_L =V/I Ohms			
	Coil Inductance $L=X_L/2\pi f$ Henries			
	Mean value of L in Henries			

Conclusion :

It will be observed that the inductance of the coil increases as the core inserted inside the coil.

Test Exercise :

1. Define impedance.

2. Is it possible to have a pure inductance coil.

3. Is there any difference in the theoretical and practical values of inductance, if so explain the reasons for it ?

EXPERIMENT NO. 7. *(a)*

Name of the Experiment : To find the voltage current relationship in a single phase R-L series circuit. Draw the impedance triangle and determine the power factor.

Objective :

1. To find the voltage current relationship in an R-L series circuit.

2. To draw the impedance triangle of R-L series circuit.

3. To determine the power factor of the circuit.

Apparatus :

S.No.	*Name*	*Specifications*	*Quantity*
1.	A.C. Supply	230 V, 1 ϕ, 50 Hz	
2.	A.C. voltmeters	0-250 V	3
3.	A.C. Ammeter	0-2 A	1
4.	Inductor	40 W	1
5.	Variable Resistance	100 Ω	1
6.	Variac	0-270 V, 4 A	1
7.	Watt meter	200 W	1

Theory : In an a.c. circuit same current flows through the resistance and inductance. The voltage drops across the resistance and inductor are V_R and V_L respectively. The phasor sum of these voltages will be the applied voltage V and relation $\frac{V}{I}$ will give the impedance of the series circuit. The cosine of angle between V and I will give the power factor and calculation the power factor, $\cos\phi = \frac{W}{VI}$ because $VI \cos\phi = W$.

Circuit Diagram :

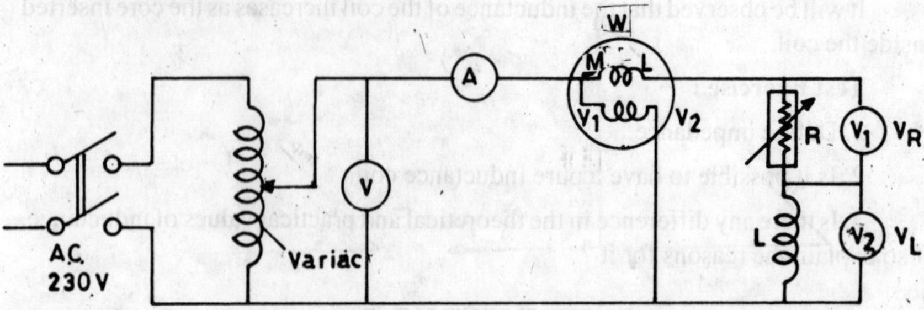

Fig. 10.

Procedure :

1. Choose the apparatus according to availability in the laboratory.
2. Make connections as per circuit diagram shown in Fig. 8.
3. Take readings of voltmeters, wattmeter and ammeter.
4. Vary the value of *R* and take another set of readings.
5. Draw vector diagrams by choosing a voltage scale and measure the angle ϕ in both cases and find the power factor cos ϕ.
6. Calculate power factor from readings.
7. Calculate Impedance of circuit from readings.
8. Compare the values obtained from readings and vector diagram.
9. Draw the impedance triangle from the voltage drops across *R*, V_R and across *L* is V_L.

Observations :

S.No.	V	I	V_R	V_L	W	$R = \frac{W}{I^2}$	$p.f. = \frac{W}{VI}$	$X_L = \sqrt{Z^2 - R^2}$	$Z = \frac{V}{I}$

Expected Shape of Vector Diagram :

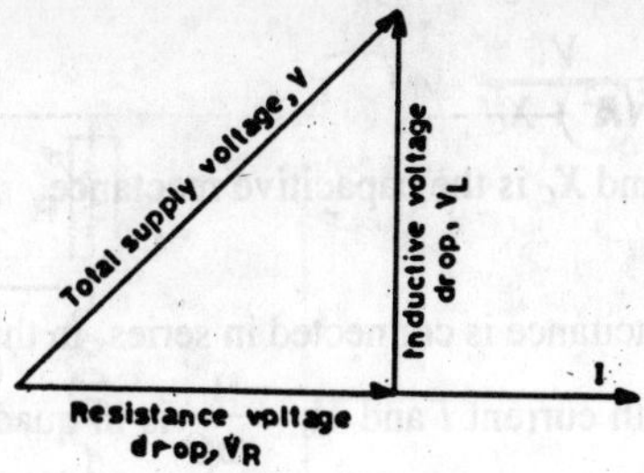

Fig. 11.

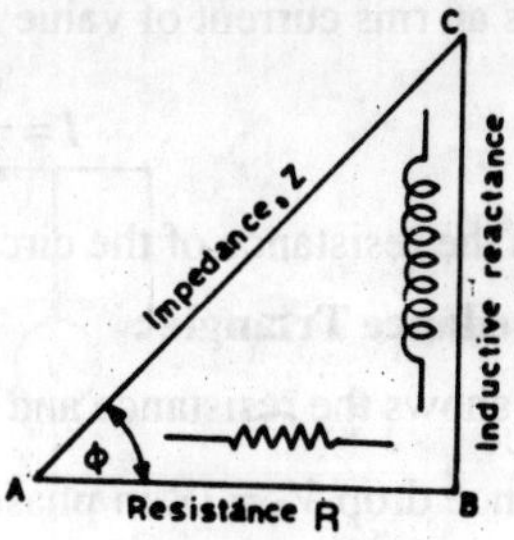

Fig. 12.

Conclusion :

It is concluded that by varying the resistance of the circuit, impedance changes and it is the ratio of voltage and current of an a.c. circuit.

Angle between current and applied voltage is determined by the vector sum of V_R and V_L and cos ϕ is the power factor of the circuit.

Value of power factor by vector method and by calculating from the readings is almost the same, smallest variation is due to the inference of the instruments.

Test Exercise :

1. What is the angle of lead or lag of current from voltage for a pure resistive circuit ?

2. At what angle the current will lag from voltage in a pure inductive circuit?

3. What will be the power factor of an R-L circuit when the value of *R* is made zero.

EXPERIMENT NO. 7. *(b)*

Name of the Experiment : To find the current and voltage relationship in R-C series circuit. Draw the impedance triangle and determine the power factor.

Apparatus :

S.No.	*Name*	*Specifications*	*Quaitity*
1.	Ammeter	0-5 A	1
2.	Voltmeter	0-250 V	1
3.	Wattmeter	0-200 W	1
4.	Rheostat	100 Ω, 5 A	1
5.	Auto transformer	270 V, 4 A	1
6.	Capacitor	2.5 μ F, 250 V	1
7.	Connecting Wires	——	——

Theory : When a.c. supply of voltage V is applied to an R-C series, it establishes an rms current of value I, which is given by

$$I = \frac{V}{Z} = \frac{V}{\sqrt{R^2 + X_C^2}}$$

where R is the resistance of the circuit and X_C is the capacitive reactance.

Impedance Triangle :

Fig. shows the resistance and capacitance is connected in series. In this case the resistance drop $V_R = IR$ in phase with current I and $V_C = \frac{1}{\omega C}$ is in quadrature with the current I. The capacitive reactance X_C being regarded as negative V_C as a vector is shown downwards.

$$V^2 = V_R^2 + V_C^2$$

$$V = \sqrt{(IR)^2 + (I_{X_C})^2}$$

$$V = I\sqrt{R^2 + X_C^2}$$

$$I = \frac{V}{Z}$$

and

$$Z = \sqrt{R^2 + X_C^2}$$

Here Z is called impedance of the circuit. By taking drops across R and C, impedance triangle can be constructed.

Circuit Diagram :

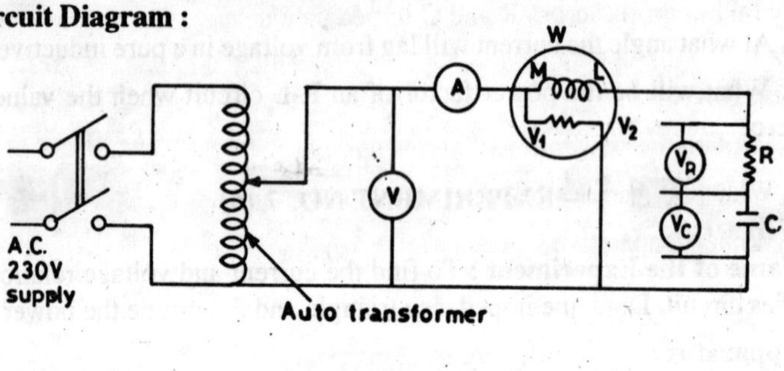

Fig. 13.

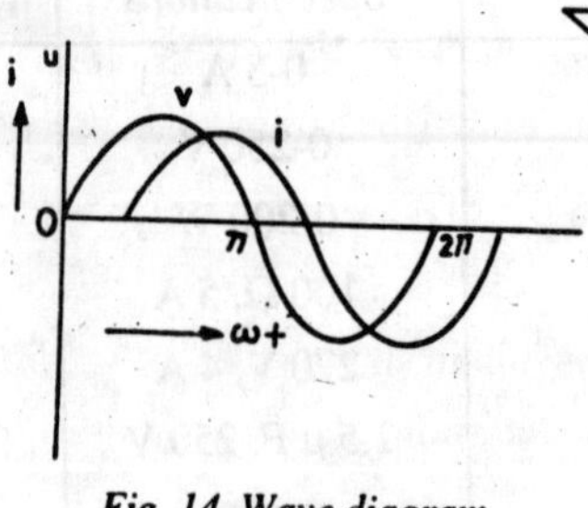

Fig. 14. Wave diagram

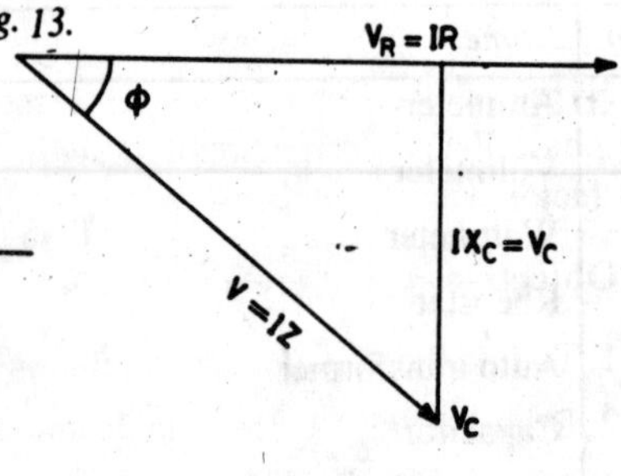

Fig. 15. Vector diagram

Procedure :

1. Make the connection as shown in Fig.
2. The resistance and capacitance is connected in series.
3. Adjust the current value to one ampere with the help of auto- transformer.
4. After this note down the reading of wattmeter, ammeter and voltmeter.
5. Take another set of reading by varying the voltage by auto transformer.
6. Record the readings in the observation table.

Observations :

S.No.	V	I	$Z=\frac{V}{I}$	W	$R=\frac{W}{I^2}$	$X_L=\sqrt{Z^2-R^2}$	$\cos\phi=\frac{R}{Z}$

Conclusion :

By taking drops across R and C impedance triangle and the power factor is less than 1.

Test Exercise :

1. Explain the power in R-C series circuit.
2. What is difference between active power and apparent power?
3. What is capacitive reactance?

EXPERIMENT NO. 8.

Name of the Experiment : To find the voltage-current relationship in a single phase R-L-C series circuit, draw its impedance triangle and determine the power factor.

Objectives :

1. To find the voltage current relationship in single phase RLC series circuit.
2. Draw the impedance triangle and determine the power factor.

Apparatus :

S.No.	*Name*	*Specifications*	*Quaitity*
1.	Capacitor	5 μ F	1
2.	Inductor (Choke)	40 W	1
3.	Resistance	100 Ω	1
4.	Ammeter	0-2 A, AC	1
5.	Voltmeter	0-300 V, AC	1
6.	Wattmeter	0-750 W	1
7.	Connecting leads	—	—

Theory : When an AC, voltage V (r.m.s. value) is applied to an R-L-C series circuit, it establishes an r.m.s. current of value I, which is given by the relation

$$I = \frac{V}{z} \text{ where } Z = \sqrt{R^2 + (X_L - X_C)^2}$$

Here $X_L > X_C$. The effect of $X_L > X_C$ is a lagging power factor because current I lags the voltage V by an angle ϕ.

In an inductor, the copper losses take place due to the resistance of the coil and core losses take place in magnetic core (in case of *'iron cored'* inductors). In the capacitor, the losses take place in the *di-electric medium* used for making it. The losses taking place in the magnetic core of the inductor or di-electric medium of the capacitor are usually ignored. The only losses considered are only when the resistance of the coil is shown connected in series with the inductance.

Circuit Diagram :

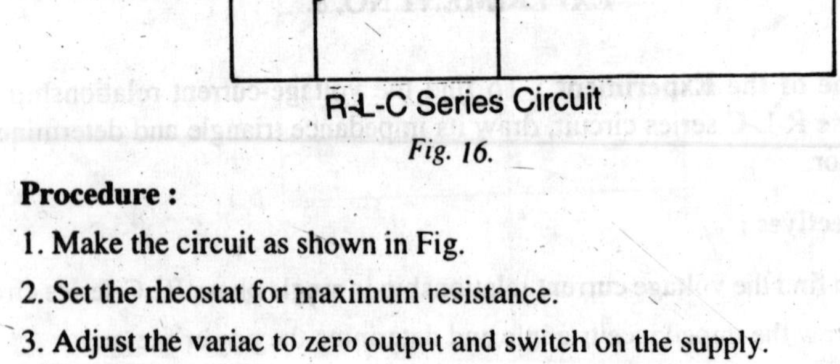

Fig. 16.

Procedure :

1. Make the circuit as shown in Fig.

2. Set the rheostat for maximum resistance.

3. Adjust the variac to zero output and switch on the supply.

4. Apply the variable voltage by adjusting the variac (dimmerstat), so that a uitable current flows. Measure and note down the voltage across *R* , *L* and *C* and lso the output voltage of the variac. Note down the readings of ammeter and attmeter.

5. Take different readings by applying different voltages. It is possible to take ifferent readings by varying resistance *R*. If a variable inductor is available, dditional readings can be obtained.

6. Enter the observations in the table and calculate the results as shown in the ble.

Observations :

	Observations						Calculations					
	1	2	3	4	5	6	7	8	9	10	11	12
	V	I	V_R	V_L	V_C	P	$R = \frac{V_R}{I}$	$X_L = \frac{V_L}{I}$	$X_C = \frac{V_C}{I}$	$\cos\phi = \frac{P}{VI}$	$Z = \frac{V}{I}$	$\phi = \cos^{-1}\frac{P}{VI}$
S. No.	Volts	Amp.	Volts	Volts	Volts	Watts						
1												
2												
3												
4												
5												

Conclusion :

From the voltage drops obtained from *R, L* and *C* series circuit impedance triangle can be drawn. The power factor calculated from the readings is less than 1.

Test Exercise :

1. What will be power factor of R-L-C series circuit.
2. Can the value of power factor of a circuit be negative?
3. In R-L-C series, the voltage applied is 100 volts a.c. The voltage measured across resistance is 30 volts and across *C* is 20 volts. What will be voltage across the inductor?

EXPERIMENT NO.9.

Name of the Experiment : To test a lead acid storage battery for charged and discharged condition and to make connections for its recharging.

Objective : To be able to :

1. Determine the conditions of a battery.
2. Prepare and connect a battery for charging?

Apparatus :

S.No.	*Name*	*Specification*	*Quantity*
1.	Battery	12 V	1
2.	Battery charger	12 V, 5 A	1
3.	Hydrometer	—	1
4.	Cell tester	15-0-15 V	1
5.	Voltmeter	0-15 V, DC	1
6.	Ammeter	0-5 A, DC	1
7.	Distilled water battery solution	—	2 bottles

Theory : In a battery chemical energy is stored to convert it into electrical energy. It is a secondary cell, known as a storage cell and is chemically reversible. Its active materials are not consumed. Such cells may be recharged by passing current through the battery in the reverse direction. These cells are rated in voltage and ampere-hours capacity. Battery is discharged at 10 hour rate to the end voltage of 1.80 volts per cell.

A lead acid battery has positive plates of lead peroxide of choclate brown colour and negative plates of sponge lead, grey in colour. The plates are separated by separators of prorous insulated and nonreactive with H_2SO_4 materials. During discharge both the plates become lead peroxide and whitish in colour. The charging processes store them to their original make up of lead peroxide and sponge lead.

The open circuit voltage of a fully charged cell is about 2.2 volts at normal temperature and specific gravity.

Circuit Diagram :

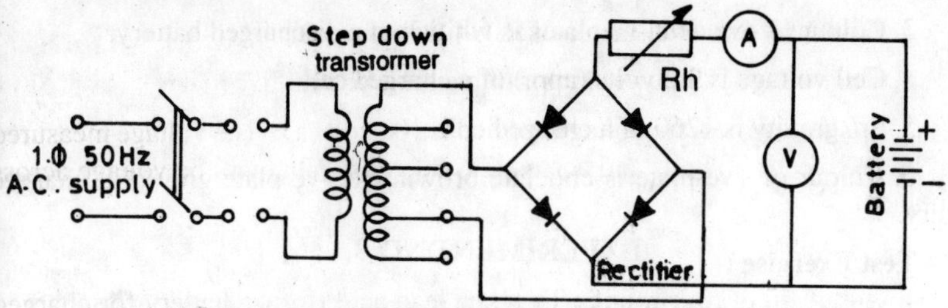

Fig. 17. Battery charger connect4ed with battery.

Procedure :

1. Check the voltage of each cell of a battery with the help of cell tester.
2. Check the specific gravity of the electrolyte in the cell.
3. Observe the colour of the +ve and –ve plates.
4. Clean the terminals and top of the battery.
5. Grease the terminals of the battery with petroleum jelly.
6. Connect the battery with battery charger +ve terminal of battery with +ve of the charger and –ve terminal of battery with –ve of the charger.
7. Set the charging current by varying the variable resistance.
8. Put the battery for charging and on trickle charge if the battery condition s above half charger.
9. Check the state of re-charge for cell voltage with a cell tester.
10. Check for the specific gravity of the electrolyte with a hydrometer.
11. Observe the colour of plates of the battery.

Observations :

Before Charging		*After Charging*	
Terminal voltage of each cell	*Sp. Gravity of electrolyte of each cell*	*Terminal Voltage of each cell*	*Sp. Gravity of electrolyte of each cell*

Conclusion :

It is concluded in a lead acid battery.

1. Cell voltage of a discharged battery is about **1-1.8 volts.**

2. Sp. gravity of electrolyte is about 1140 when the cell is discharged.

3. Colour of +ve and –ve plates is whitish of a discharged battery.

4. Cell voltage is 2.2 volts (app.) of a charged cell.

5. Sp. gravity is 1260 of a charged cell.

6. Colour of +ve plate is choclate brown and –ve plate grey of a charged battery.

Test Exercise :

1. What will happen if distilled water is not used in the electrolyte?

2. Write the precautions taken to maintain the battery in good condition.

3. Why battery is only charged with d.c. supply?

4. Why battery is not allowed to remain in discharged condition for a long time?

EXPERIMENT NO.10.

Name of the Experiment : To find out resonance frequency in a devices R-L-C circuit.

Objectives :

1. To determine the resonance frequency.

2. To measure the bandwidth and selectivity of the R-L-C circuit.

Apparatus :

S.No.	*Name*	*Specifications*	*Quantity*
1.	A.C. source of varying frequency	100 KHz	1
2.	Inductive coil	100 mH	1
3.	Resistance	100 Ω	1
4.	Capacitor	10 μ F	1
5.	Wattmeter	500 W	1
6.	Ammeter	0-2 A	1
7.	Voltmeter	0-300 V	1
8.	Frequency meter	—	1
9.	Connecting Wires	—	—

Theory : In R-L-C series circuit as shown in Fig. 17, the impedance of this circuit is given by

$$Z = \sqrt{R^2 + (X_L - X_C)^2}$$

where X_L = inductive reactance and

X_C = capacitive reactance.

Then $(X_L - X_C) = X$

$\therefore$
$$Z = \sqrt{R^2 + X^2}$$

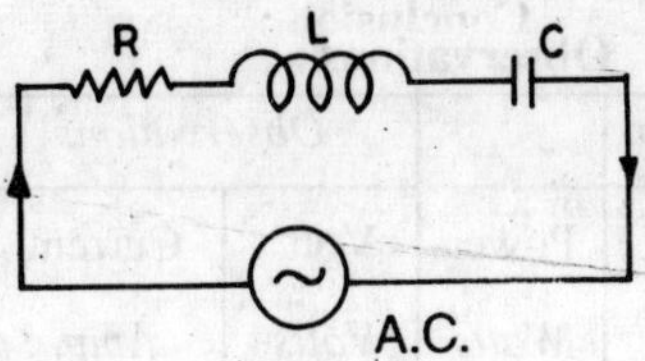

Fig. 18.

Circuit Diagram :

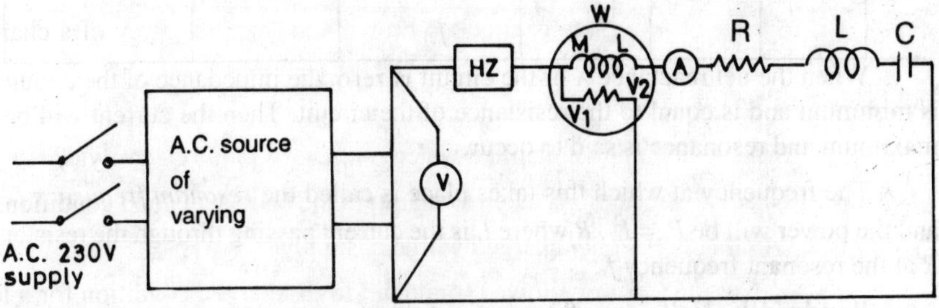

Fig. 19.

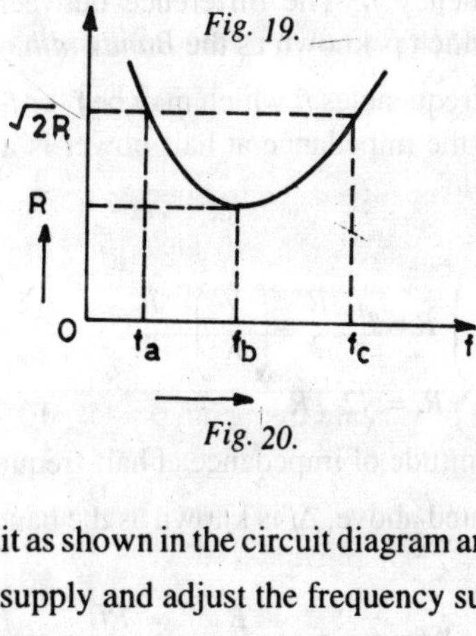

Fig. 20.

Procedure :

1. Make the circuit as shown in the circuit diagram and make the connections.

2. Switch on the supply and adjust the frequency such that $X_L = X_C$.

3. Note down the readings of the various instruments. The frequency meter at this moment gives the resonant frequency f_r and current $I_r = V/R$.

4. Vary the frequency, keeping the applied voltage constant to the upper half frequency f_2 and lower half frequency f_1.

5. Measure these frequencies on the frequency meter and find the bandwidth $(\Delta f = f_2 - f_1)$.

6. Plot the graph between resistance and frequency as shown.

7. Find the degree of selectivity of a resonant circuit.

Observations :

S.No.	Observations				Calculations	
	Power *Watts*	Volt. *Volts*	Current *Amp.*	Frequency *Hz*	$R=\frac{V}{I_r}$	$\Delta f=f_2-f_1$
				f_r		
				f_1		
				f_2		

When the net reactance X of the circuit is zero, the impedance of the circuit is minimum and is equal to the resistance of the circuit. Then the current will be maximum and resonance is said to occur.

The frequency at which this takes place is called the *resonant frequency* f_r, and the power will be $P_r = I_r^2 \,.\, R$ where I_r is the current passing through the resistor R at the resonant frequency f_r.

Bandwidth : Half power frequencies f_1 and f_2 are those frequencies at which the power across the series R-L-C circuit is exactly half of the power across the circuit at resonant frequency f_r. The difference between the two- half power frequencies $f_2 - f_1 = \Delta f$ which is known as the *Bandwidth*.

The power at half frequencies f_x which may be f_1 or f_2 is exactly the half, *i.e.*, $P_r/2$. The magnitude of the impedance at half power is also $R/2$ and the power consumed will be $I_r^2 \,.\, \frac{R}{2}$.

$$\therefore \quad \left(\frac{V}{R_x}\right)^2 R = I_r^2 \,.\, \frac{R}{2} = \left(\frac{V}{R}\right)^2 \,.\, \frac{R}{2}$$

$$\therefore \quad R_x = \sqrt{2} \,.\, R$$

Here R_x is the magnitude of impedance at half frequencies (f_1 and f_2).

Selectivity : As stated above, Δf is known as the bandwidth in cycles/sec. or hertz.

$$\therefore \quad \Delta f = f_2 - f_1 = \frac{R}{2\pi L} = \frac{Rf_r}{2\pi f_r \,.\, L} = \frac{f_r}{Q}$$

where $Q = \frac{2\pi f_r L}{R} = \frac{\omega_r \,.\, L}{R}$ is known as the quality factor. The largeness of Q is an indicator of the quality of the circuit as far as frequency selectivity is concerned. The degree of selectivity of a resonant circuit is represented by the narrowness of bandwidth given by $\frac{f_r}{\Delta f} = Q$ of the inductor of the circuit at resonant frequency.

EXPERIMENT NO. 11.

Name of the Experiment : Measurement of power and power factor in a single phase R-L-C derives circuit, and to calculate active power and reactive power.

Objectives :

To be able to establish the relation that the power factor is the ratio of active power and reactive power and power in an a.c. circuit to the product of apparent power and power factor.

Apparatus :

S.No.	*Name*	*Specification*	*Quantity*
1.	A.C. supply source	230 V, 1-ϕ, 50 Hz	
2.	Voltmeter	0-300 V	1
3.	Ammeter	0-2 A	1
4.	Wattmeter	0-200 W	1
5.	A lamp bank	0-500 W	1
6.	Inductive load (choke)	40 W	1
7.	Variac	270 V, 4 A	1

Theory : Power in d.c. circuit is the product of voltage and current whereas in a.c. circuit power is the product of voltage, current and power factor. In an a.c. circuit product of voltage, current and sine angle between the two is called the reactive power which is different from active power consumed by the circuit is measured by the wattmeter. Power factor in an a.c. circuit is the ratio of Actual power to apparent power. Also it is the product of voltage and current and cosine angle between the two.

Circuit Diagram :

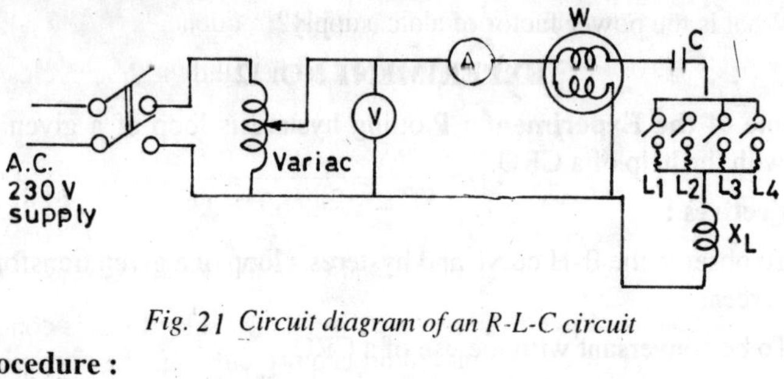

Fig. 21 Circuit diagram of an R-L-C circuit

Procedure :

1. Select the apparatus and material for the experiment according to availability in the laboratory.

2. Make the connections according to circuit diagram as shown in Fig. 20.

3. Switch on lamp L and take the readings of V, I and W.

4. Switch on all the lamps one by one and take the set of readings.

5. Calculate the p.f. and power.

6. Compare the calculated power with recorded power.

Observations :

S.No.	V	I	W	$\cos\phi = \frac{W}{VI}$	Active Power $= VI\cos\phi$	Reactive Power $P = VI\sin\phi$

Conclusion :

It is concluded that power factor is the ratio of actual power the apparent power. Calculated power is equal to the measured power. Power increases as the current increases in the circuit.

Test Exercise :

1. Explain why power increases with the increase of current in the circuit?

2. If the d.c. supply is given to the same a.c. circuit will the power be same. Support your answer with reasons.

3. What will be the power factor of a pure resistive circuit?

4. What is the power factor of a d.c. supply?

EXPERIMENT NO. 12

Name of the Experiment : Plotting hysteresis loop of a given magnetic material with the help of a CRO.

Objectives :

1. To observe the B-H curve and hysteresis loop of a given transformer core on CRO screen.

2. To be conversant with the use of a CRO.

Brief Theory : The B-H (or magnetisation) curve of a feromagnetic material is not a straight line due to variation in the permerability of iron at different flux densities. The iron also undergoes magnetic saturation at higher values of flux densities. The core of a transformer is carrying alternating flux and therefore, undergoes magnetic reversal during each cycle. Complete cycle of magnetic reversal is called hysteresis loop. Hysteresis loop of a transformer's core can be observed on an oscilloscope by the use of following circuit of Fig. 22.

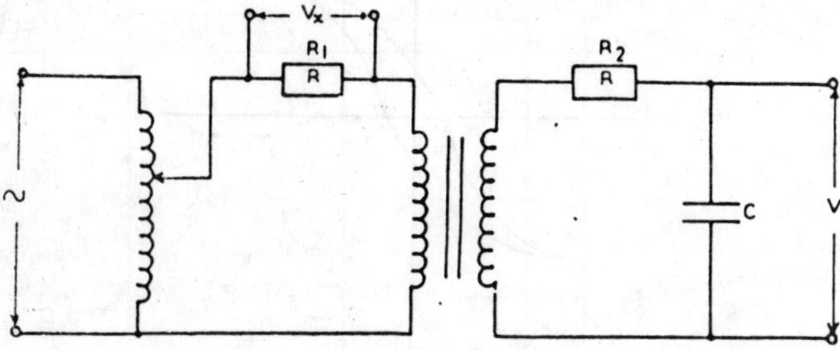

Fig. 22. Connection diagram for observing hysteresis loop on CRO

A capacitor C and a high resistance R_2 is connected across the secondary load because of negligible loading effect of $R_2 - C$ network. Current in the transformer primary will, therefore, comprise of only magnetising current and voltage droop across R_1 will be proportional to the magnetising intensity H. This voltage, if connected to the oscilloscope's horizontal input, will deflect the beam which will be proportional to H.

The magnitude of i_2 (*i.e.*, secondary current), however, small it may be, will be in phase with secondary induced emf e_2 because R_2 is of the order of megaohms and X_C of the order of kilo ohms ($R_2 >> X_C$) and $Z_2 \approx R_2$

Therefore, $i_2 \propto e_2$

or $$i_2 = \frac{e_2}{R_2}$$

also $$V_y = \frac{1}{C}\int i_2\,dt = \frac{1}{CR_2}\int e_2\,dt \qquad \left(\text{put } e_2 = \frac{N_2\,d\phi}{dt}\right)$$

$$= \frac{N_2}{CR_2}\int \frac{d\phi}{dt}\,dt = N_2\,CR_2 \int d\phi = k\phi$$

Thus, voltage across the capacitor at any instant is proportional to the flux in the core and is in phase with it.

Since $\phi = BA_i$ V_y represent B in the core as the cross-sectional area A_i is constant. Thus V_y if connected to vertical input of the oscilloscope will deflect the beam proportional to the flux density B.

Thus, the trace on CRO screen is the B-H loop of the transformer core material. The slope of the BH loop or the hysteresis loop will be as shown in Fig. 23.

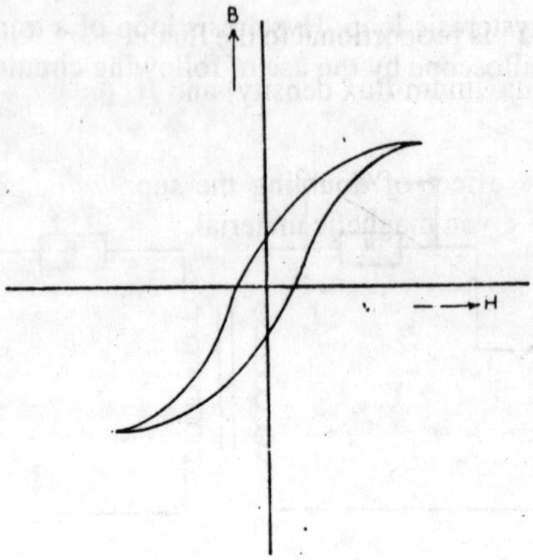

Fig. 23. Hysteresis loop of a transformer core

Circuit Diagram : Connect the instruments as per circuit diagram of Fig. 22. Choose suitable values of R_1 , R_2 and C

Suggested values are $R_1 = 100$ ohm, $R_2 = 1$ M ohm, $C = 5 - 10\ \mu$ F

Equipment/Instruments Required

S.No	*Name of the equipment/instrument*	*Range/rating*	*Make*
1.	CRO – 1 No.		
2.	Transformer – 1 No.		
3.	Variac – 1 No.		
4.	Resistance – 2 Nos.		
5.	Capacitor – 1 No.		

Procedure :

1. Connect the CRO with the transformer, etc. as shown in the Fig. 22.

2. Adjust the horizontal and vertical gain of the CRO to observe the BH loop on the screen.

Observations and Results : Trace the B-H loop observed on the CRO on tracing paper and comment on its shape.

Comments/Discussions on the Results : Write your comments on the results obtained and discuss the discrepancies, if any.

Questions : Answer the following questions in your report :

1. Estimate the shape of B-H curve if dc instead of ac supply of appropriate value is applied across the primary of the above transformer.

2. Justify that V_x is proportional to the magnetising force H in the transformer core.

3. Justify that V_y is proportional to the flux density B in the transformer core.

4. Mark B_m (maximum flux density) and H_c (coercive force) on the tracing of hysteresis loop.

5. Discuss the effect of doubling the supply frequency on the shape of hysteresis loop for a given magnetic material.

JULY 1998
ELECTRICAL SCIENCE-III

Time Allowed : 3 hours *Max. Marks : 75*

Solve any FIVE out of Eight

1. (a) Distinguish between power of energy. 3

   (b) Define temperature coefficient of resistance. 2

   (c) Apply Thevemin's theorem to find Thevemin's equivalent circuit for the following : 10

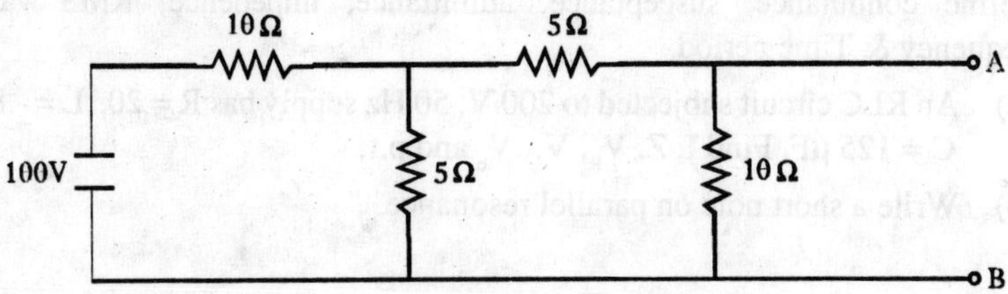

Fig. 1.

2. (a) Explain a proper charging method for batteries. 5

   (b) Find the currents I_1, I_2 & I_3 in Fig. 2, in magnitude. 10

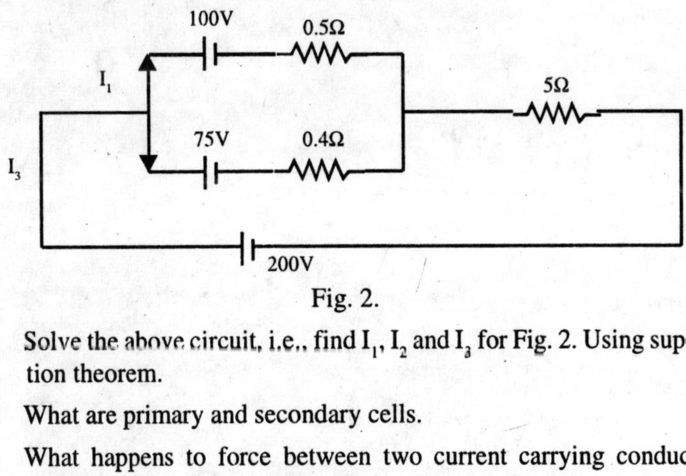

Fig. 2.

3. (a) Solve the above circuit, i.e., find I_1, I_2 and I_3 for Fig. 2. Using superposition theorem. 12

   (b) What are primary and secondary cells. 3

4. (a) What happens to force between two current carrying conductors if distance between them is doubled and current in them is halved. 4

   (b) Find the mutual inductance if a current of 100 A is suddenly (within a millisecond) dropped to 50A in a coil and as a consequence an emf of 10V is induced in an adjoining coil. 8

   (c) State Faraday's Law. 3

5. (a) State the utility of Lenz's Law and Fleming's Right & Left hand rules. 6

(b) Find the percentage change in stored energy in a coil if current through it is doubled and its inductance halved ? 9

6. (a) An iron ring has permeability of 550, is 10 cm in dia and 10 cm^2 in X-section, is would with 200 turns of wire. For a flux density of 1 Wb/m^2 in the ring, find the exciting current and inductance. Also find these if 2 mm air-gap is made in the ring. 10

(b) How can phasors represents sinusoidally varying quantities. 5.

7. Define condutance, susceptance, admittance, impedence, RMS value, frequency & Time period. 15

8. (a) An RLC circuit subjected to 200 V, 50 Hz supply has R = 20, L = -.8 H, C = 125 μF. Find I, Z, V_R, V_L, V_c and p.f. 10

(b) Write a short note on parallel resonance. 5

INDEX

Win Prizes!

ATTENTION READERS

We request you for your frank assessment regarding some of the aspects of the book given as under :

ELECTRICAL SCIENCE

By TARLOK SINGH

Please fill up the given space in neat capital letters.

(i) What topics of your syllabus that are important from you examination point of view are not covered in the book ?

....

....

....

....

....

(ii) What are the chapetrs wherein the treatment of the subject matter is not systematic or organised or updated ?

....

....

....

....

....

(iii) Have you come across misprints/mistakes/factual inaccuracies in the book ? Please specify the chapters and the page numbers.

....

....

....

....

....

(iv) What other books on the same subject have you found/heard better that the present book ? Please specify in terms of price and quality.

....

....

....

....

....

(v) Further suggestions and comments for the improvement of the book:

....

....

....

....

....

(PS: If need be please attach a separate sheet to write your views.

Other Details :

(i) Who recommended you the book ? (Please tick in the box near the option relevant to you.)

☐ Teacher ☐ Friends ☐ Bookseller

....

....

....

....

....

(ii) Name of the recommending teacher, his designation and address :

....

....

....

....

....

(iii) Name and address of the bookseller you purchased the book form :

....

....

....

....

....

(iv) Name and address of your institution (Please mention the University or Board, as the case may be)

....

....

....

....

....

(v) Your name and complete postal address :

....

....

....

....

....

(vi) Course you are enrolled in :

....

....

....

The best assessment will be awarded each month. The award will be in the form of our publications as decided by the Editorial Board.

Please mail the filled up coupon at your earliest to :

S.K. KATARIA & SONS®
4885/109, Prakash Mahal, Dr. Subhash Bhargav Lane,
Opposite Delhi Medical Association, Daryaganj, New Delhi-110002 (INDIA)
Phone: +91-11-23243489, 23269324, 43551243; Telefax: +91-11-23243489
e-mail: katariabooks@yahoo.com; katariabook@gmail.com
Website : www.skkatariaandsons.com; For ebooks: www.katsonbooks.com